The Micro-World Observed by Ultra High-Speed Cameras

Kinko Tsuji
Editor

The Micro-World Observed by Ultra High-Speed Cameras

We See What You Don't See

Springer

Editor
Kinko Tsuji
Shimadzu Europa GmbH
Duisburg
Germany

ISBN 978-3-319-61490-8 ISBN 978-3-319-61491-5 (eBook)
DOI 10.1007/978-3-319-61491-5

Library of Congress Control Number: 2017945220

Printed on acid-free paper

This Springer imprint is published by Springer Nature
The registered company is Springer International Publishing AG
The registered company address is: Gewerbestrasse 11, 6330 Cham, Switzerland

Preface

My first encounters with an ultra high-speed camera were two video films, which showed how a balloon is broken, when it is pricked. One movie was taken with 30 fps (frames per second) and the other with 100,000 fps. In the movie of 30 fps we see: a girl has a balloon in her left hand and a needle in her right hand. She pricks the balloon, which immediately disappears from the view. A few moments later her face is distorted and she starts to cry. It was funny to notice (sorry for the girl) the time lag between the rupture of the balloon and her frightened reaction: she was frightened, after everything had finished. So I knew that our reaction time can be measured at about 30 fps, but that the rupture of the balloon cannot be seen with this frame rate. In the second movie with 100,000 fps I saw a series of images that were completely different from my expectation. The balloon was not simply blown away, but its surface was cut into two halves. Each of them moved away from the point of needle impact, thus opening a glimpse into the inside of the balloon.

Soon afterwards I obtained more movies showing many other fast phenomena, like a destruction test of concrete, spark ignition, development of shock waves after an explosion and others. The maximum camera speed was 1 million fps. Full of enthusiasm I showed these movies to my colleagues and friends. The most popular movie was (and is) not among the movies that were taken with the maximum speed but "only" with 4600 fps: a movie of a fast train passing through a station at a speed of 180 km/h. A high-speed camera and, for comparison, a conventional video camera were installed on the platform. With the camera of 4600 fps the faces of passengers sitting in the train can be clearly recognized. In the movie taken with the conventional video camera you do not see any passengers, but the opposite platform through wide windows. After seeing this movie, many people said "Wow!" and suggested me to take images of, for example, a flying insect, movements of an eye lid, muscles of gymnasts jumping and landing, and similar scenes. In contrast to this film of the moving train they were not so excited about spark ignition or shock waves.

Why is that so? To see movements of a train, insects or our body in slow motion is, of course, interesting. However, they are just an extension of movements which we see. It is difficult to imagine something which we have never seen, not because it does not exist, but because it is too fast to be seen with our eyes. To my offer to use an ultra high-speed camera with up to 1 million fps for testing, some of them said indeed that they have a camera which is good enough to take images of 1000–5000 fps, and that they do not need any faster cameras, because they have no processes faster than those to be seen at a few thousand fps. But how do they know that there are no faster phenomena? Isn't it worth seeking phenomena of which we do not know their existence?

This question reminds me of the electron microscope, which offers much higher spatial resolution than the optical microscope. Optical microscopes can detect only objects larger than the wavelength of visible light. Therefore, it was not possible to visualize objects smaller than the visible wave lengths, typical for most of viruses, even though one was sure that something smaller than bacteria causing diseases had to exist. Therefore, they tried to find the objects which were not detectable by optical microscopes. And finally electron microscopes were invented, and one was able to observe viruses.

Analogous to the spatial resolution, we can expect something to see, which we do not see yet, by using equipment of higher temporal resolution. If something (for example, an explosion) happens, we see the results (many broken pieces). To see how it happens, we need equipment which has higher temporal resolution. There would be many fast movement/phenomena/steps which we have not observed yet. With an ultra high-speed camera it is possible to observe them, something that we have never seen before.

I would like to show to readers of this volume, not only to the high-speed specialists but also to people who have no experiences with high-speed cameras, what we can see and what kind of inspirations one can get for things to be done in the future. Similar to many other excellent technical devices, high-speed cameras could be a tool like a double-edged sword. I hope that this tool will be used in science and technology for human welfare and environmental protection.

Acknowledgements

The editor thanks all contributors of this book for accepting my wish to make a volume on high-speed cameras. Each of them, excellent in their specific field, spent a lot of time to prepare valuable manuscripts. Thomas Ditzinger of the Springer Publishers is acknowledged for his support to realize my wish to create a book on high-speed cameras. Finally I thank my husband, Stefan C. Müller, and my daughter, Sho Tsuji, for encouraging and supporting me.

Duisburg, Germany Kinko Tsuji
May 2017

Contents

Contributors

Marcus Aldén Division of Combustion Physics, Lund University, Lund, Sweden

Thomas Arnold Institut für Strahlwerkzeuge (IFSW), Universität Stuttgart, Stuttgart, Germany

Meiko Boley Institut für Strahlwerkzeuge (IFSW), Universität Stuttgart, Stuttgart, Germany

Amine Bouterf Laboratoire de Mécanique et Technologie (LMT), ENS Paris-Saclay/CNRS/Université Paris-Saclay, Cachan cedex, France

Richard Browning Institute of Biomedical Engineering, University of Oxford, Oxford, UK

Dario Carugo Institute of Biomedical Engineering, University of Oxford, Oxford, UK

Robert Eckersley Biomedical Engineering Department, Kings College London, London, UK

Takeharu G. Etoh Graduate School of Science and Engineering, Ritsumeikan University, Kusatsu, Japan

Sebastian Faas Institut für Strahlwerkzeuge (IFSW), Universität Stuttgart, Stuttgart, Germany

Florian Fetzer Institut für Strahlwerkzeuge (IFSW), Universität Stuttgart, Stuttgart, Germany

Pascal Forquin Laboratoire 3SR, Université Grenoble Alpes, Domaine Universitaire, Saint Martin d'Hères, France

Christian Freitag Institut für Strahlwerkzeuge (IFSW), Universität Stuttgart, Stuttgart, Germany

Valeria Garbin Department of Chemical Engineering, Imperial College London, London, UK

Thomas Graf Institut für Strahlwerkzeuge (IFSW), Universität Stuttgart, Stuttgart, Germany

Christian Hagenlocher Institut für Strahlwerkzeuge (IFSW), Universität Stuttgart, Stuttgart, Germany

Andreas Heider Institut für Strahlwerkzeuge (IFSW), Universität Stuttgart, Stuttgart, Germany; CR/APJ1, Robert Bosch GmbH, Stuttgart, Germany

Hartmut Hieronymus Bundesanstalt für Materialforschung und -prüfung (BAM), Berlin, Germany

François Hild Laboratoire de Mécanique et Technologie (LMT), ENS Paris-Saclay/CNRS/Université Paris-Saclay, Cachan cedex, France

Michael Jarwitz Institut für Strahlwerkzeuge (IFSW), Universität Stuttgart, Stuttgart, Germany

Benjamin Jodar UMR 6251—CNRS/Université de Rennes 1, Rennes, France

Nico de Jong Department of Biomedical Engineering, Erasmus MC, Rotterdam, The Netherlands

Nobuyuki Kawahara Department of Energy Systems Engineering, Okayama University, Okayama, Japan

Harald Kleine School of Engineering and IT, University of New South Wales, Canberra, Australia

Nobuki Kudo Graduate School of Information Science and Technology, Hokkaido University, Sapporo, Japan

Rihito Kuroda Graduate School of Engineering, Tohoku University, Sendai, Japan

Thomas Kurz Drittes Physikalisches Institut, Georg-August-Universität Göttingen, Göttingen, Germany

Guillaume Lajoinie Physics of Fluids Group, University of Twente, Enschede, The Netherlands

Kenneth R. Langley King Abdullah University of Science and Technology (KAUST), Thuwal, Saudi Arabia

Werner Lauterborn Drittes Physikalisches Institut, Georg-August-Universität Göttingen, Göttingen, Germany

Er Q. Li King Abdullah University of Science and Technology (KAUST), Thuwal, Saudi Arabia

John Morse European Synchrotron Radiation Facility (ESRF), Grenoble, France

Helen Mulvana School of Engineering, University of Glasgow, Glasgow, UK

Stefan C. Müller Institut für Experimentelle Physik, Otto-von-Guericke-Universität Magdeburg, Magdeburg, Germany

Quang A. Nguyen Graduate School of Science and Engineering, Ritsumeikan University, Kusatsu, Japan

Margie Olbinado European Synchrotron Radiation Facility (ESRF), Grenoble, France

Joshua Owen Institute of Biomedical Engineering, University of Oxford, Oxford, UK

Alexander Rack European Synchrotron Radiation Facility (ESRF), Grenoble, France

Paul Rademeyer Institute of Biomedical Engineering, University of Oxford, Oxford, UK

Malte von Ramin Fraunhofer Institute for High-Speed Dynamics, Ernst Mach Institute (EMI), Efringen-Kirchen, Germany

Mattias Richter Division of Combustion Physics, Lund University, Lund, Sweden

Stéphane Roux Laboratoire de Mécanique et Technologie (LMT), ENS Paris-Saclay/CNRS/Université Paris-Saclay, Cachan cedex, France

Mario Scheel Synchrotron Soleil, Gif-Sur-Yvette, France

Daniel Schmitt Fraunhofer Institute for High-Speed Dynamics, Ernst Mach Institute (EMI), Efringen-Kirchen, Germany

Alexander Stolz Fraunhofer Institute for High-Speed Dynamics, Ernst Mach Institute (EMI), Efringen-Kirchen, Germany

Eleanor Stride Institute of Biomedical Engineering, University of Oxford, Oxford, UK

Shigetoshi Sugawa Graduate School of Engineering, Tohoku University, Sendai, Japan

Kazuyoshi Takayama Institute of Fluid Science, Tohoku University, Sendai, Japan

Mengxing Tang Department of Bioengineering, Imperial College London, London, UK

Sigurdur T. Thoroddsen King Abdullah University of Science and Technology (KAUST), Thuwal, Saudi Arabia

Kinko Tsuji Shimadzu Europa GmbH, Duisburg, Germany

Michel Versluis Physics of Fluids Group, University of Twente, Enschede, The Netherlands

Rudolf Weber Institut für Strahlwerkzeuge (IFSW), Universität Stuttgart, Stuttgart, Germany

Abbreviations

ADC	Analog-to-digital converter
ADV	Acoustic droplet vaporization
AOM	Acousto-optic modulator
ATDC	After top dead center
BBB	Blood brain barrier
BDC	Bottom dead center
BEOL	Back-end-of-line
BSI	Backside illuminated
BTB	Bromothymol blue
BZ	Belousov–Zhabotinsky
CAD	Crank angle degree
CCD	Charge coupled device
CDS	Correlated double sampling
CFD	Computational fluid dynamics
CFRP	Carbon fiber reinforced plastic
CMOS	Complementary metal oxide semiconductor
CUSA	Cavitation ultrasonic surgical aspirator
CW	Continuous wave
DIC	Digital image correlation
DISI	Direct-injection spark ignition
DMOS	Depletion-type MOS
DPSS	Diode pumped solid state
DSPC	1, 2-Distearoyl-sn-glycero-3-phosphocholine
ESWL	Extracorporeal shock wave lithotripsy
FD	Floating diffusion
FEOL	Front-end-of-line
FOPH	Fiber optics hydrophone
fps	Frames per second

FSI	Frontside illuminated
FWC	Full well capacity
GPS	Global positioning system
HCCI	Homogeneous compression charge ignition
HIFU	High intensity focused ultrasound
IC	Internal combustion
iCCD	intensified CCD
IED	Improvised explosive device
ISAS	Image signal accumulation sensor
ISIS	In-situ storage image sensor
LED	Light emitting diode
LIF	Laser-induced fluorescence
LII	Laser-induced incandescence
LSI	Large scale integrated circuit
MCG	Multi-collection-gate
MFC	Mass flow controller
MOS	Metal oxide semiconductor
NA	Numerical aperture
OPO	Optical parametric oscillator
PA	Photoacoustic
PD	Photo diode
PDA	Phase Doppler anemometer
P-I-	Pressure-impulse-
PIV	Particle image velocimetry
PLIF	Planar laser-induced fluorescence
PMT	Photomultiplier tube
PRF	Pulse repetition frequency
PVB	Polyvinylbutyral
Q4	4-noded quadrilaterals
RC	Register-capacitor
RO	Ring oscillator
ROI	Region of interest
SDOFS	Single degree of freedom system
SHB	Split-Hopkinson bar
SI	Spark ignition
SNR	Signal-to-noise ratio
SPS	Serial-parallel-serial
STAMP	Sequentially timed all-optical mapping photography
STAR	Scattering to attenuation ratio
TCAD	Technical computer aided design
TDC	Top dead center
THG	Third-harmonic-generating
TMR	Transitional Mach reflection

UHS	Ultra high-speed
VBIED	Vehicle borne improvised explosive device
VEGF	Vascular endothelial growth factor
vNMR	von Neumann–Mach reflection
ZOI	Zone of interest

Part I
Introduction

History of Curiosity

Kinko Tsuji

Abstract A brief introduction to high-speed cameras is presented. The most frequently asked question is “Why do we need high-speed cameras?” To this we answer with three typical situations of an object: (1) it remains still, (2) it is moving, and (3) it is observed under a microscope. For the next question “How can we acquire high-speed images?” two important factors (1. illumination and 2. trigger) are discussed. One example is shown, where it is very difficult to find an appropriate trigger as well as illumination device. As the last part of this chapter there follows a summary of this book.

1 How Does a Horse Gallop?

Curiosity is one of the most essential elements of the human mind. When we encounter something extraordinary, we like to know “what” at first, “how” as the next, and then “why”. This roughly corresponds to a famous scenario of “Copernicus—Kepler—Newton”: Copernicus said that the earth revolves around the sun, Kepler showed how the earth revolves around the sun, and Newton said why it does so. Often we have to know the “what” more precisely in order to get an answer to the “how”, and the “how” more precisely in order to understand the “why”. One of the close-to-life examples may be a sudden sound of an explosion which you hear, when you are walking in the street. What? You look around. Nothing is broken, but you see an airplane in the sky. So, you can guess the “what”: the airplane flies faster than sound. To know “how” and “why” is, however, not so simple, and scientists have worked and are working on that.

Electronic supplementary material The online version of this chapter (doi:10.1007/978-3-319-61491-5_1) contains supplementary material, which is available to authorized users.

K. Tsuji (✉)
Shimadzu Europa GmbH, Albert-Hahn-Straße 6-10, 47269 Duisburg, Germany
e-mail: kts@shimadzu.eu

K. Tsuji (ed.), *The Micro-World Observed by Ultra High-Speed Cameras*,
DOI 10.1007/978-3-319-61491-5_1

Without technical help our knowledge on "what" and "how" is very limited. We obtain information from our five senses: sight, hearing, taste, smell and touch. For the sight we can see light of wavelength from about 380 to 750 nm. The ultraviolet or infrared light we cannot see. As detectors the human eyes are limited not only due to their detectable wavelength range, but also their detectable spatial resolution, temporal resolution and intensity. However, this kind of limitation is not always disadvantageous for our daily life. For example, human eyes cannot follow an intensity change faster than 15 Hz. Therefore, light from lamps which work with a frequency of 50/60 Hz looks continuous for our eyes. Classical silent movies are made with 16–24 Hz to show us continuous movement [1]. On the contrary, flickering of 10 Hz may cause seizure symptoms. An example is the Pocket Monster incident that happened in Japan in 1997 [2]: Many children who watched a TV program of the popular animation film "Pocket Monster" got indescribable feelings, and about 700 of them were treated in hospitals for seizure syndromes.

In order to satisfy our curiosity plenty of equipment and devices have been developed. For the light of wavelength shorter than 380 nm or longer than 750 nm we need ultraviolet or infrared detectors. Development of microscopes and electron microscopes offer a better spatial resolution. Photomultipliers and photodiodes can amplify weak signals to become detectable. For improving temporal resolution we need additional equipment, namely high-speed cameras. In this book we like to show with high-speed cameras, what we can see, that we do not see just with our own eyes.

The first concrete high-speed images were taken by Eadweard Muybridge in 1887 [3–5]. He placed 12 cameras along a galloping way, and the shutter of each camera was connected with taut wire, which released the shutter electrically when

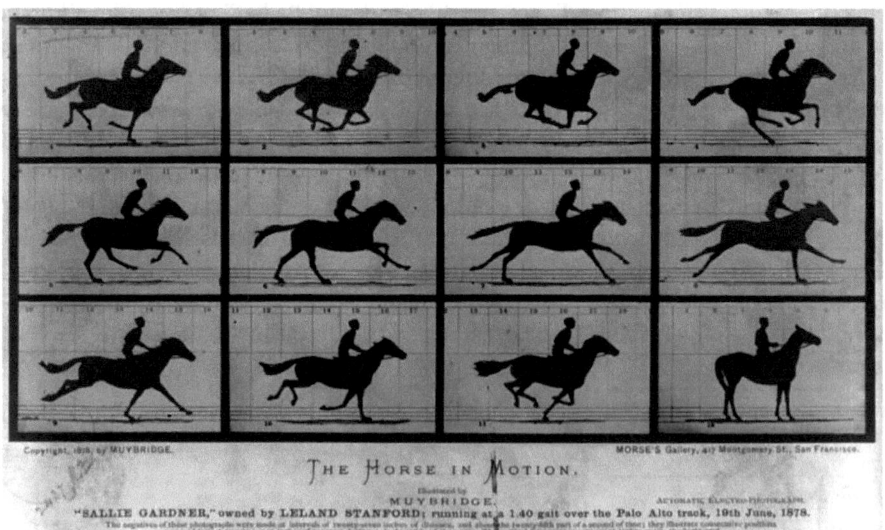

Fig. 1 A series of images of a galloping horse taken by Eadweard Muybridge in 1887

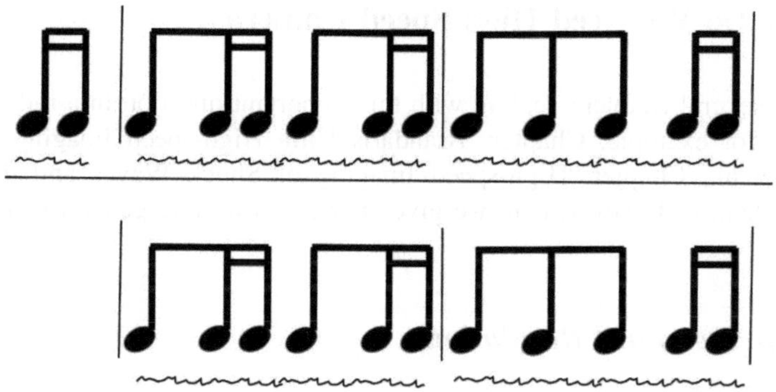

Fig. 2 Basic rhythms of the second half of the William Tell Overture. Irregular lines under the notes express noises

the horse passed. The 12 cameras were successively triggered so that they produced a series of images of a galloping horse. At that time, there was an exciting discussion on whether at least one of the four feet touches the ground during galloping. Muybridge's photographs (Fig. 1) showed, in fact, that there is a moment when all four legs are in the air. Moreover, they showed that a very complicated coordination of the legs is involved: there is no symmetry between front legs and hind legs, and no strongly synchronized movement either of the two front legs or of the two hind legs.

Perhaps Gioacchino Rossini knew this before Muybridge. He composed the William Tell Overture in 1829. When you hear the second half of the overture, you have a vivid recollection of galloping horses. He managed to express the gallop acoustically from rather simple rhythms by using various musical instruments together (see Fig. 2). If you beat out these rhythms, probably you are disappointed. It does not sound like gallop. But if you listen to an orchestra playing it, it is gallop! We hear many additional "noises" between the main rhythms. Interestingly, the temporal resolution of our ear is better than that of our eyes (human ears: 20–100 Hz depending on the band width, main frequency, as well as the amount of involved noises) [6]. Although such psycho-acoustics is a very interesting field, we will not discuss this matter here.

To come back to the high-speed imaging, since Muybridge's work the methods to take high-speed images have been continuously developed, using strobes, laser pulses on the one hand, rotating drums, rotating mirrors and streak tubes on the other hand. The development of light sensitive semiconductors has drastically changed the world of high-speed imaging. Note that the terms "high-speed cameras", as well as "high-speed imaging" used in this volume do not relate to any specific definition of the SMPTE (Society of Motion Pictures and Television Engineers).

2 Why Do We Need High-Speed Cameras?

There are several chapters dealing with this important question in a rather specific way (see, for example, Chapter "Brandaris Ultra High-speed Imaging Facility": Lajoinie et al., Chapter "High-speed Imaging of Shock Waves and their Flow Fields": Kleine). However, here we give answers of a more general nature.

2.1 For Temporal Resolution

The Pocket Monster incident [2] indicated that the human eye certainly can detect a frequency of 10 Hz. This corresponds to a camera which takes an image every 0.1 s. If we like to see the light intensity changes of a fluorescent lamp lighted by a power source with alternating current of 50 Hz, we need a camera which works with a time interval shorter than 0.02 s, as long as the lamp itself does not move. If the change of the light intensity is faster, then we need a correspondingly faster camera: for instance, for observing light intensity changes of 10^6 Hz, we need a camera which records images faster than every 1 μs.

2.2 For Spatio-Temporal Resolution

When an object is moving (relative to the camera), then we have to take a spatial factor into consideration. In order to illustrate such a case, we often show a movie of a fast train passing with 180 km/h in front of a camera fixed on the platform. In a movie taken with a conventional video camera with 25–30 fps you see a kind of moving band with interruptions, whereby you may detect segments of wagons. Here the question is: With which frequency do you have to take images in order to recognize faces of passengers in this train? In our example the train moves in one direction with 50 m/s (=180 km/h). Human eyes with a temporal resolution limit of 10 Hz can observe objects in the train larger than 5 m length along the train. If you have to detect an object of 1 cm length (which is necessary for the recognition of human faces), a frequency 500 times larger (5000 Hz, corresponding to 0.2 ms) is required.

2.3 For Microscopic Observation

When a particle is moving with a velocity of 0.01 m/s, you can observe this movement with your own eyes, as far as the particle is large enough to be seen. However, if the particle is so small that you have to magnify it with a microscope, then the observation of even such a slow movement (0.01 m/s) is not any more so obvious. Under a microscope an apparent velocity is magnified, depending on the magnification factor. If the magnification factor along the particle movement is 100, the apparent velocity of the object moving with 0.01 m/s is 1 m/s. Moreover, the field of view gets small, also depending on the magnification. Therefore, an appropriate high-speed camera is required to observe a moving object under the microscope.

3 How to Take High-Speed Images

Since automatic cameras have been developed, taking pictures in daily life has become very easy. Just to push a button produces more or less acceptable photographs. You do not need to think about exposure, diaphragm, or distance. Even a flash goes off automatically, when it is dark. If you, as a user of automatic cameras, have an opportunity to use a professional non-automatic camera, you may be frustrated: you have to select the correct diaphragm and exposure time, and adjust the focus precisely, in order to take pictures which are worth keeping. Even more complicated is the situation, if you, as a user of normal video cameras, suddenly get a high-speed camera. In the following I describe my own experiences, when I had an opportunity to use a high-speed camera for the first time: I was a normal user of automatic cameras and had no experience with high-speed cameras.

The camera offered to me for use was an ultra high-speed camera, which can take images with a frame rate of up to 1 million frames/s (1 Mfps) and 100 successive images [7]. The camera was a kind of a prototype without any instruction manual. The person who brought the camera to me demonstrated how to take images of the moments during hand clapping. When he clapped his hands, the acoustic signal triggered the camera and a series of images of clapping hands was taken. I could see how the hands were deformed at the clapping moment and right after. I tried to follow what he had shown me and it worked perfectly. The acoustic trigger was very convenient for this purpose. So I told him that I could manage to use the camera now, and he left the laboratory.

Being on my own, I was curious and ambitious to see clapping hands with the maximum frame rate. I set the frame rate to 1 Mfps and clapped my hands. The camera was triggered, captured signals were processed… and I saw 100 successive black pictures. Disappointed I set the frame rate back to 5000 fps and increased it stepwise. I noticed that at the frame rate of 10,000 fps the light which I used was not strong enough even with the maximum diaphragm. So I brought a stronger lamp and increased the frame rate. Fine, I could take 100 successive images with 10,000, 20,000 fps, and more. However, since the number of the images was limited to 100, I could see only the very beginning of hand clapping with such a higher frame rate. How would I see the later phase of hand clapping with the higher frame rate? Can I change the timing of the trigger?

Thus, there are two crucial things that you have to consider: how to illuminate the object and how to take events at the right moment. If these two factors are not well set, then images obtained may be completely dark, or completely still throughout the whole series of frames.

3.1 Illumination

Not only for high-speed photography, but also for normal photography, it is important to choose the right illumination device and corresponding exposure time, depending on the sensitivity of the equipment. For the high-speed photography, the exposure time should be shorter (or, at least, equal) than the time interval of one frame, and subsequently a correspondingly strong light source is required. If you use a flash lamp or pulsed lamp as a strong light source, the timing of the flash or pulse should be synchronized with the course of events.

Note that it is difficult to take high-speed images of light-sensitive objects. Also care should be taken for protecting operators' eyes, when strong lamps are used.

3.2 Trigger

An appropriate trigger device is essential for taking images at the right moment, especially when the number of frames is limited. For example, if the number of frames is limited to 100 and the frame rate is 1 Mfps, the total time of observation is 100 μs. In such a case the manual trigger does not function at all. Even for much slower motion like horse galloping, Muybridge had to use an electrical signal from the taut wire to open the shutters successively. The trigger signal can be electrical, optical, acoustic, pneumatic, or other. You have to select the most convenient trigger for your experiments and build the corresponding experimental set-up. For example, an acoustic trigger with a microphone is most convenient, if you like to observe the moment of a rupturing balloon: when a balloon is ruptured, usually a big bang is caused.

There are two ways to trigger the camera: a trigger for starting to capture images, or one for stopping to capture images. The latter is possible, when the captured signals are saved on chip electronically [7]. While the shutter of the camera is open, images are captured and stored continuously and at the same time overwritten one after the other. When a trigger signal arrives, image capturing and overwriting processes are stopped. Afterward the image signals of all storage elements are read out. Recently, this trigger method is used preferentially, because the timing can be set in a more flexible way, as explained in the following.

Usually a TTL (Transistor-Transistor-Logic) signal from electronic, optical, acoustic or other sensors is used as a trigger signal. It is practical to set a variable delay to stop the image capturing process after an input of the TTL signal. It is also possible to set a number of frames (m), during which images are still captured after the trigger signal arrives. The range of m is from 0 to n, where n is the total number of frames. For example, if m is set to 0, the image capturing is stopped immediately, and all signals kept in the memory are those of the phenomena happening before the trigger (pre-trigger mode). If m is set to n, the image capturing continues for n frames (m = n) after the trigger, and images of n frames after the trigger are recorded (post-trigger mode). The image capturing can be stopped at any frame within n frames (middle-trigger mode). Thus, the on-chip memory has an additional advantage regarding such flexible trigger modes.

3.3 A Difficult Case for Both Illumination and Trigger: Collapsing Beer Bubbles

Figure 3 presents a series of images, how a beer bubble starts collapsing. One bubble marked by a yellow circle has a heptagonal boundary with 7 adjacent bubbles. Images from (a) to (f) show that this heptagonal border is "melting" in time: the borderlines become wider and unclear. The "melting" takes place within 3 ms. It is nice to see such a process. (This series of images is one of the best among

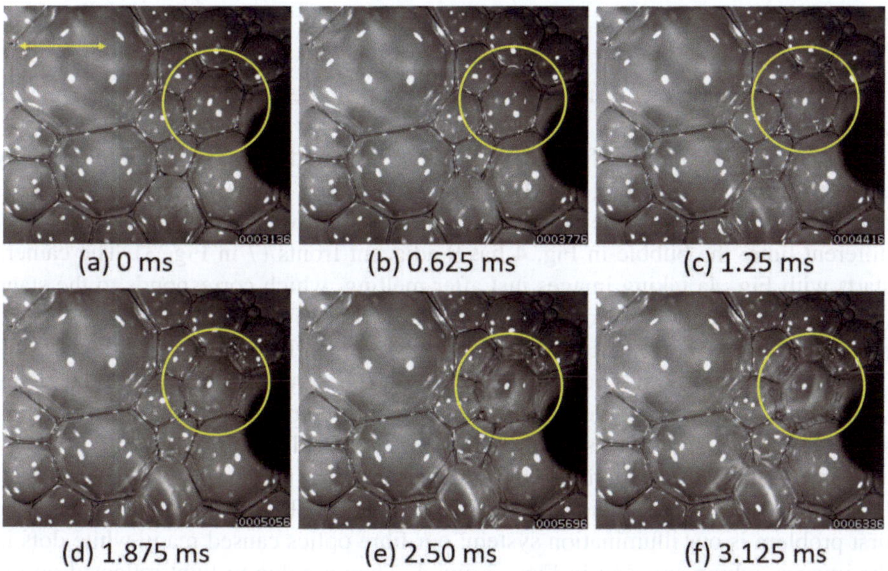

Fig. 3 Beer bubble collapse observed at 16,000 fps. Collapse occurs inside of the *yellow circle*. *White dots* are random reflections of illumination light. A *black semicircle* at the *right* side is the tip of an optical fiber used for illumination. Scale bar: 10 mm. The corresponding video is available online

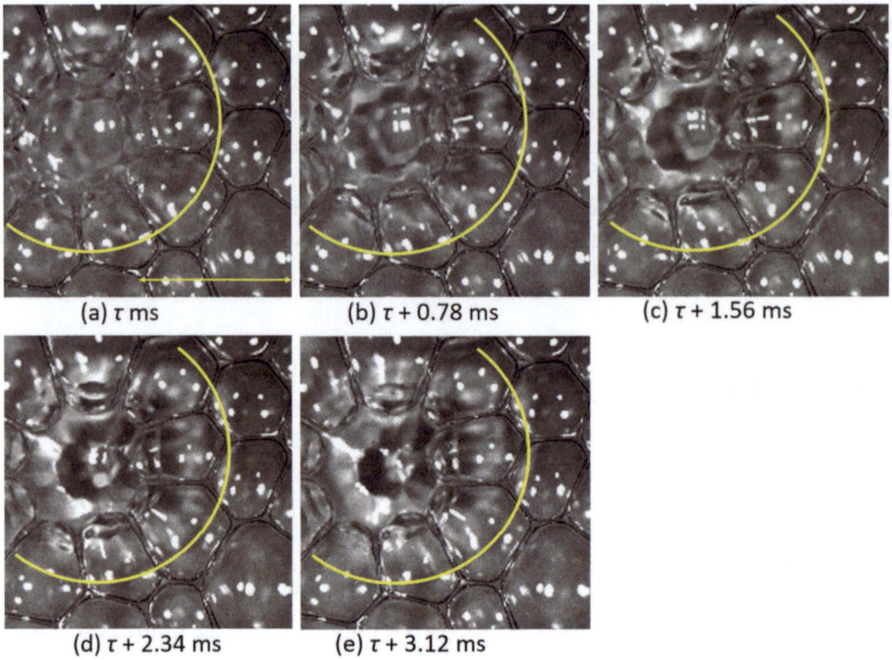

Fig. 4 Beer bubble collapse observed at 32,000 fps. Collapse occurs inside of the *yellow* circular arc. *White dots* are random reflections of illumination light. Scale bar: 10 mm. The corresponding video is available online

about 100 experiments.) But then, what will happen afterward? I wish to see images beyond 100 frames.

In search of examples for the later phase, I found one movie which just by chance started after melting (see Fig. 4). Note that the images shown in Fig. 4 are not directly succeeding the ones of Fig. 3, but they were taken at a completely different time: the bubble in Fig. 4 has 9 adjacent fronts (7 in Fig. 3). The camera starts with Fig. 4a taking images just after melting, which corresponds to the stage (f) of Fig. 3. Since we cannot measure the exact time after the event in Fig. 4 starts, just "t = τ" is given as the time for Fig. 4a: $\tau \approx 3$ ms. It looks as if the fences of 9 houses around a lake sink into an abyss. The surrounding bubbles are torn inside by the central collapsing bubble. This process takes further 3 ms. (Videos of these two series of Figs. 3 and 4 are shown online—see Appendix.)

Carrying out the experiments for observing the process of collapsing beer bubbles, we could not avoid the two big problems: illumination and trigger. The first problem is our illumination system: our fibre optics caused many white dots in the images, which are seen in Figs. 3 and 4. They are due to light reflected on the surface of the foam. These random dots disturb the view of events. If more sophisticated illumination devices were used, the intensity of the reflected light could be reduced.

On the other hand, these dots can be useful. They indicate clearly where an event occurs: no event—no movement. Only the dots inside of the yellow circle change their location and intensity, while the dots outside of this circle stay during a time interval of 3 ms as they are (see Fig. 3a–f). Thus, an appropriate illumination device should be selected not only for the intensity but also for what you like to observe.

The second, and even more difficult problem is how to trigger the camera at the right moment. Honestly speaking, to obtain such images as shown in Figs. 3 and 4 is a matter of luck. First of all we do not know which bubble will start to collapse. If you take a large field of view, then some of the bubbles in this field may collapse at some time during acquiring images. But then, each bubble is too small to allow for observation of what happens at the border to adjacent bubbles. Therefore, it is necessary to magnify the field of view. Then, of course, the probability of a collapse occurring in the smaller area becomes lower.

If ever a collapse luckily occurs in this field of view, there is no way to trigger the camera on time. Collapse happens spontaneously and provides no detectable signal. Can we wait for further luck? Let us see which kind of luck we need, when we do not have any appropriate trigger.

Figure 5a illustrates, how long image acquisition lasts, depending on the frame rate of the camera, when the number of consecutive frames is limited to 100. When the frame rate is 100 fps, images are taken at every 10 ms. Therefore, the total acquisition time is 1000 ms (10 ms × 100 frames). When the frame rate is

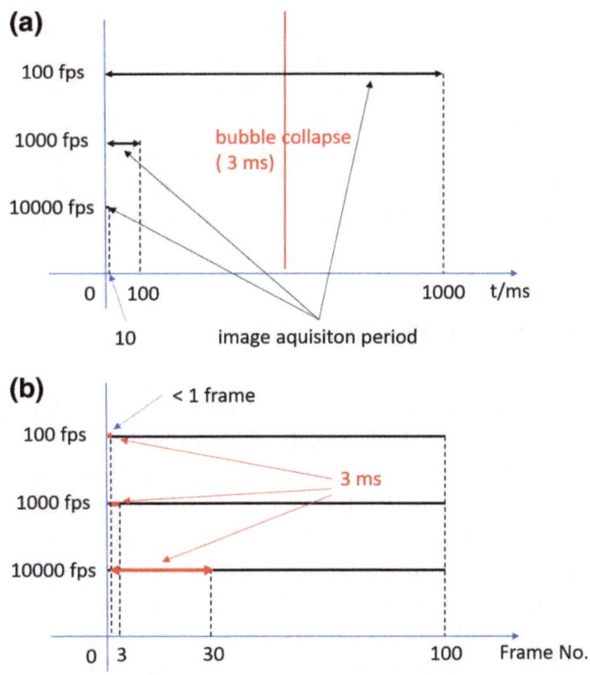

Fig. 5 Probability to catch an event and its temporal resolution

1000 fps, the total acquisition time is 100 ms, and with a frame rate of 10,000 fps, it is reduced to 10 ms. Now, let us assume that a bubble collapse occurs every 1 s within our field of view. Then, theoretically this event should be recorded with the probability of 100%, when the frame rate is 100 fps. At 1000 fps, the probability of a bubble collapse during acquisition period is 10% and at 10,000 fps it is even 1% only.

The next assumption is that this event to be observed lasts 3 ms. Figure 5b shows how many frames correspond to 3 ms, depending on the frame rate. Since images are taken at every 10 ms for the frame rate of 100 fps, an event of 3 ms occurs within one frame. It means that among 100 images you obtain two consecutive images right before and after the event occurs. At 1000 fps, 3 ms corresponds to 3 frames, and at 10,000 fps it corresponds to 30 frames. If you need 30 frames for analyzing the event of 3 ms length in detail, the frame rate of the camera should be 10,000 fps. So one has to carry out experiments under the probability of 1%.

Images of Fig. 4 were taken with 32,000 fps. Was I so lucky to catch a collapse in the field of view within an acquisition period of 3 ms?

I have to confess that I used some tricks in order to satisfy my curiosity. Bubbles should be larger and easier for collapse, so that not so much magnification is required and that the probability to take images at the right time gets higher. After several attempts I found that the best way is to add one drop of kitchen detergent to about 10 ml beer and to bubble up with breath. These bubbles often collapse, when the container (a petri dish in this case) is slightly shaken. Knocking at the table on which the petri dish was placed can trigger a collapse, and also can trigger the camera acoustically. In this way I could take some series of images with a probability of about 10%. And my initial curiosity was satisfied.

On the other hand, as a scientist I am not yet satisfied with these results and procedures. To analyze such bubble collapse I have to carry out more systematic measurements. However, I cannot rely totally on luck in this endeavor. This is the reason why I stopped to work on this subject for a while. An image trigger could solve the problem: a trigger device which gives a signal, when parts of the image start changing. And then, I could resume my work on this subject….

4 About This Book

Following this introductory Chapter, PART II "Pioneering Works on High-Speed Imaging" is presented by Lauterborn and Kurz and Lajoinie et al. Lauterborn and Kurz describe their studies on bubbles and microbubbles along the development of high-speed imaging. Principles of various devices for capturing high-speed images, as developed since the 1960s, are explained, especially in the research field of microbubbles. Important is that their challenges have not yet finished, but they are further developing 3D holography. Lajoinie et al. write about the "Brandaris" camera and its applications. Here the rotating mirror technique and light-sensitive semiconductors are combined, and this combination realizes a high frame rate of up to 25 Mfps together with a high resolution of 500 × 292 pixels and large consecutive frame numbers up to 128. There is one disadvantage of this camera: it is quite large (150 cm length) and heavy (140 kg). This camera and cameras with the same principle are often used now for medical applications.

Two articles on light-sensitive semiconductors follow in PART III "Cameras with CCD/CMOS Sensors". The light-sensitive semiconductors are at the heart of modern high-speed cameras. Two types of light-sensitive semiconductors are involved: the CCD (charge coupled device) and the CMOS (complementary metal oxide semiconductor). Etoh and Nguyen report on the development of high-speed image sensors, especially about a CCD with on-chip memory. With such newly developed CCD sensors, it becomes possible to make ultra high-speed cameras of a handy size, which achieve a high frame rate of 1 Mfps without losing two other crucial specifications: a high resolution of 312 × 256 pixels and a hundred consecutive images. Kuroda and Sugawa describe their improvement of CMOS sensors with on chip-memory. A camera with the most recently developed CMOS sensor is extremely light sensitive and can take 256 consecutive images with up to 10 Mfps even under day light. The resolution is 400 × 250 pixels.

The theme of PART IV is "Shock Waves". Kleine gives an overview of visualization techniques for shock waves. He explains principles of various methods to visualize transparent media: shadowgraph, schlieren technique and interferometry. He compares images taken with different methods and discusses their advantages and disadvantages. While Kleine mainly describes the air flow, Takayama shows the behavior of underwater shock waves: underwater shock waves generated in ultrasound oscillations or in a shock tube, generated by underwater explosions, shock/bubble interaction, reflection of shock waves, and many other phenomena. For the visualization he uses holograms and interferometry.

In the field of material research high-speed cameras have become inevitable. There are three contributions in PART V "Materials Research": Hild et al. for DIC (digital image correlation), Freitag et al. for laser processing, and Rack et al. for X-ray imaging. Hild et al. present analyses of the mechanical behavior of materials under stress (for example, deformation, damage or fracture) by using DIC. Freitag et al. investigated laser processing like laser welding and laser drilling of CFRP (carbon fiber reinforced plastics). Laser processing could overcome difficulties of mechanical processing of CFRP, which is used wherever high strength-to-weight ratio and rigidity are required, such as in aerospace and automotive fields. Rack et al. show how useful it is for visualizing the interior of opaque materials, if the hard X-rays are combined with high-speed imaging technique. As examples of application, observations of the dynamics of aqueous foam and laser processing of polystyrene foam are shown.

One of the most important industrial applications of high-speed cameras are combustion studies, which can contribute to a sustainable and environmentally friendly energy use. The first article of PART VI "Combustion" is written by Aldén and Richter on high-speed laser visualization techniques in combustion research. They developed such visualization techniques with a combination of a Multi YAG laser and a framing camera, or a combination of a high repetition rate laser and a high power burst laser together with CMOS cameras. With these techniques they report on the behavior of turbulent combustion. Kawahara presents in the second article of PART VI combustion processes in car engines. He visualizes fuel injection, spray impingement on a wall, laser ignition processes, as well as an engine knocking cycle in order to understand what is going on inside a car engine.

PART VII "Explosions and Safety" is concerned with two objects which are contrasting in size between each other. One is explosions in a free field and the other is explosions of bubbles in a liquid. In order to protect buildings and infrastructure against explosions, Stolz et al. observe the behavior of flying fragments after explosions: their trajectories were calculated by tracing their launch conditions after explosion. They also studied the resistance of materials under highly dynamic conditions, which is certainly different from that under gentle stresses. Hieronymus observed shock-induced explosions of bubbles and further interaction of exploding bubbles. The influence of different parameters, such as the composition of the bubbles and the initial pressure was analyzed. The results are important for the safety assessment of explosion risks in the systems of bubble containing liquid.

The next subject after explosive bubbles is presented in PART VIII "Droplets". There are two contributions: one is an inactive droplet impacting on a solid surface and the other a chemically active droplet impacting on a liquid surface. Langley et al. present, by using time-resolved interferometry, what happens in detail at the moment when a droplet contacts a solid surface. They discuss the formation and evolution of an air layer between the droplet and the solid surface, compression of the gas, and other newly found phenomena. Contrary to this work, for Müller the droplet is not a direct object to be observed but a tool to initiate chemical reactions (observed as color changes of a pH indicator, as well as the Belousov-Zhabotinsky reaction). He finds spatial inhomogeneity and instability at the early stage of color change of the pH indicator bromothymol blue. Furthermore, he suggests that such color reactions can be useful for studying drop behavior.

PART IX "Microbubbles and Medical Applications" is concerned with microbubbles and their medical applications: three contributions from physics (Garbin), technical characterization (Stride et al.), and therapeutic applications (Kudo) of coated/not coated microbubbles. Garbin describes the dynamics of coated microbubbles in ultrasound, including their stability, behavior in ultrasound, optical trapping, buckling and expulsion and shape oscillations. Stride et al. present a characterization of functionalized microbubbles, which could be applied to ultrasound imaging and targeted drug delivery. They use three complementary methods for characterization: ultra high-speed video microscopy, laser scattering and acoustic scattering/attenuation, and compare the characteristics of coated and non-coated microbubbles. Kudo observes cavitation bubbles generated by an artificial heart valve, which might cause valve failure and valve thrombosis. On the other side, cavitation bubbles are used as surgical equipment, non-invasive therapy and drug delivery. By using high-speed imaging it is possible to visualize cavitation activities in medical fields so that its effects can be minimized or maximized.

Certainly, some readers may wonder why "shock waves" appear not only in PART IV "Shock Waves", or "bubbles" appear not only in PART IX "Microbubbles and Medical Applications", but almost everywhere throughout this book. In fact, many subjects are related to each other: for example, when a shock wave interacts with some bubble, the bubble may explode and create another shock wave.

Since each article of this volume has its own theme, all contributions are allocated at the most suitable place according to the theme. We think that many readers will learn about new ideas from the chapters of this book and start to look for future challenges in wider fields. They may see what we have not seen before, and thus enlarging the scope of modern science and technology.

References

1. J. Brown, Audio-visual palimpsests: Resynchronizing silent films with 'special' music, in *The Oxford Handbook of Film Music Studies*, ed. by D. Neymeyer (Oxford University Press, Oxford, 2014), p. 588
2. T. Takahashi, Y. Tsukahara, Pocket Monster incident and low luminance visual stimuli: Special reference to deep red flicker stimulation. Pediatr. Int. **40**, 631–637 (1997)
3. E. Muybridge, A.V. Mozley, in *Human and Animal Locomotion* (Dover, New York, 1887)
4. M. Leslie, in *The man who stopped time* (Stanford Magazine) (May–June 2001)
5. P. Oridger, *Time Stands Still: Muybridge and the Instantaneous Photography Moment* (Oxford University Press and Stanford University, Stanford, 2003)
6. M. Osaki, T. Ohkawa, T. Tsumura, Temporal resolution of hearing detection of temporal gaps as a function of frequency rasion and auditory temporal processing channels. Japan Soc. Physiol. Anthropol. **3**, 25–30 (1998)
7. T.G. Etoh, D. Poggeman, G. Kreider, H. Mutoh, A.J.P. Theuwissen, A. Ruckelshausen, Y. Kondo, H. Maruno, K. Takubo, H. Soya, K. Takehara, T. Okinaka, Y. Takano, An image sensor which captures 100 consecutive frames at 1,000,000 frames/s. IEEE T. Electron. Dev. **50**, 144–151 (2003)

Part II
Pioneering Work on High-speed Cameras

The Bubble Challenge for High-Speed Photography

Werner Lauterborn and Thomas Kurz

Abstract Bubbles in liquids show a rich set of phenomena ranging from harmless gaseous bubbles rising in a liquid to almost empty cavitation bubbles responsible for the destruction of ship propellers. A special property of bubbles are the extremely high-speed liquid flows they enable by providing almost empty space for acceleration. Thus bubble dynamics presents a challenge for proper investigation, in particular high-speed liquid jet formation and shock wave radiation. One tool developed to study fast dynamics is high-speed photography with suitable cameras. The gradual approach to resolve bubble dynamics, in particular bubble collapse, via ever too slow cameras up to the state of the art of some hundred million frames per second is reviewed. Some ideas on the numerical extension of camera speed limits are put forward. Moreover, the first historic steps into getting three-dimensional images recorded via high-speed holography with up to some hundred thousand holograms per second are reported.

1 Historic Reflections

Already the old Greeks knew that to stop a motion one should have a short glimpse only at an object, e.g. an arrow. In the limit of shorter and shorter glimpses the object comes to a standstill, that is, no motion is to be perceived at all. This led

Electronic supplementary material The online version of this chapter (doi:10.1007/978-3-319-61491-5_2) contains supplementary material, which is available to authorized users.

W. Lauterborn (✉) · T. Kurz
Drittes Physikalisches Institut, Georg-August-Universität Göttingen,
Friedrich-Hund-Platz 1, 37077 Göttingen, Germany
e-mail: Werner.Lauterborn@phys.uni-goettingen.de

T. Kurz
e-mail: Thomas.Kurz@phys.uni-goettingen.de

K. Tsuji (ed.), *The Micro-World Observed by Ultra High-Speed Cameras*,
DOI 10.1007/978-3-319-61491-5_2

Zeno of Elea to argue that there can be no motion in the world, a paradoxon. Of course, there is motion in the world, but by this and other paradoxa Zeno made motion a problem to think about. Zeno could not solve the motion problem himself —in essence the problem of how to define velocity—but in the course of history mankind learned that one must look *two* times at an object to detect a motion and determine its velocity. Then, from the respective locations of the object at the two times and the time difference a velocity can be defined.

The question, however, remains what the object has done in between the two glimpses that today are realized by suitable cameras with short exposure time, the equivalent of a short glimpse for a camera. Sometimes, an object changes appreciably even when the time interval between the two exposures is very short. Then either the time interval has to be shortened or, when technical constraints do not allow this option, an interpolation has to be done—either mathematically, e.g. by linear interpolation, or according to the applicable laws of nature when known. To stay with the case of an arrow launched from a bow, instead of a mathematical linear interpolation a parabola may be fitted to the measurements as gravity is expected to act in this case. A modern example is a tennis or soccer ball that is followed by high-speed photography and subsequent curve fitting to decide, whether the ball has touched or crossed a line. In the case of a bubble in a liquid, as considered here, the Navier-Stokes equation or simpler equations derived from them would be chosen for interpolation.

It was not before the invention of photography and its cameras that whole scenes with fast moving objects could be stopped in their motion by short exposure times. Already a single high-speed photograph is able to reveal details of a process or phenomenon not to be detected by normal human perception. One prominent example is a bullet smashing a bulb. Here, the case of a bubble in a liquid is taken to exemplify the achievements of high-speed photography. Bubbles in a liquid are particularly suited for this purpose because there exists a variety with very peculiar and extremely fast dynamics that presents a challenge till today. In particular, cavitation bubbles—almost empty bubbles that appear in a liquid upon rupture by too strong tension or too high local energy deposit—may collapse so violently that shock waves are radiated, light is emitted and the hardest materials are destroyed. Material pitting was first noticed in the destruction of ship propellers and became known as cavitation erosion. There is ongoing research on this topic and bubble dynamics near surfaces in general. Some examples of bubble dynamics near a flat, solid wall can be found in this article.

Bubbles seldom come alone and thousands of them may occupy a three-dimensional volume. To stop and follow their motion high-speed holography has been invented. Some early devices constructed in connection with bubble dynamics will be discussed.

2 To Stop a Motion

To stop a motion exposure times must be chosen according to the speed of the object in question, the higher the speed the shorter the exposure time to arrive at a sharp image. Object speeds range from slow motions (0 m/s in the lower limit) to the speed of light (about 300,000 km/s in the higher limit). At very low speeds, for instance of a snail, exposure times of seconds may be appropriate. These were the exposure times possible at the beginning of photography because of the then very low sensitivity of the photographic material and the low intensity of the available light sources. At moderate speeds, for instance in athletics sports, below a millisecond (10^{-3} s) is often needed to catch specific instants of human motion. Exposure times in this range can be realized by mechanical shutters.

At high speeds, for instance of a bullet, lower exposure times in the order of microseconds (10^{-6} s) or less are necessary to stop the motion for inspection. These motions and faster ones are beyond human perception speed. Exposure times in this range and beyond are usually realized by short flashes of light as can be obtained by sparks or nowadays by short pulses of laser light.

At very high speeds, for instance in explosions or motions with the speed of sound (about 1500 m/s in water) or with supersonic speeds (shock waves), even lower exposure times of down to nanoseconds (10^{-9} s) may be needed to stop the motion. An example of how such a motion can be caught is given in Fig. 1. A bar (the black rectangle in Fig. 1) oscillating at 2.4 kHz up and down is generating a cavitation bubble cloud below it that periodically or almost periodically collapses (dark spot in the middle of the picture) with emission of shock waves from the individual bubbles. The exposure time in this case is determined by an illuminating

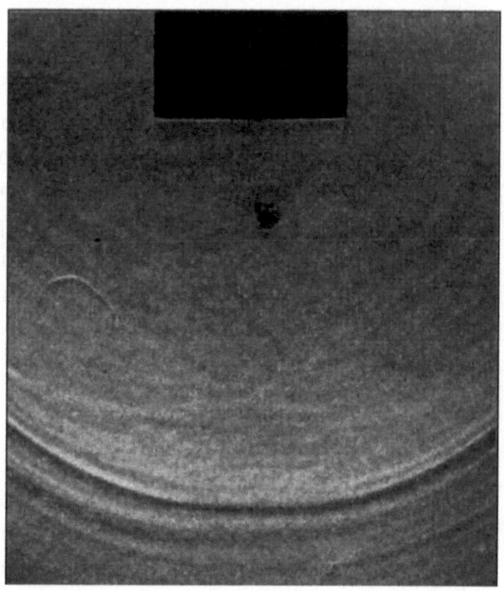

Fig. 1 Shock waves in water from a collapsing bubble cloud in an acoustic field of frequency 2.4 kHz. Photography with 500 ns exposure time from a light flash. See also Kuttruff 1962 [1]

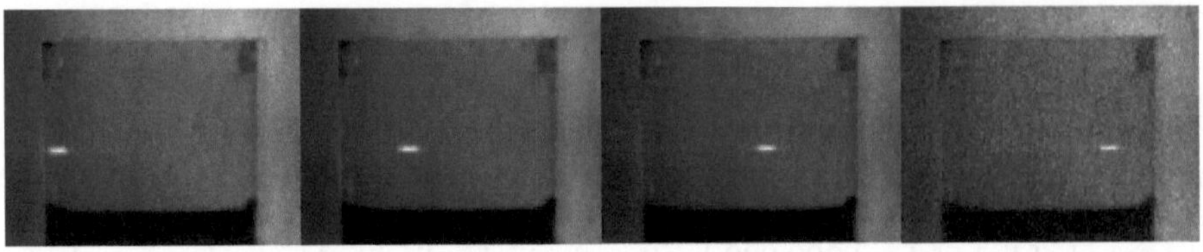

Fig. 2 A bullet of light propagating in water. The motion has been stopped with exposure times in the femtosecond range

flash of 500 ns duration from a spark source in a schlieren arrangement to make the shock waves visible (the dark and bright arcs below the collapsed bubble cloud). This way, the event has been caught and can be inspected on a larger, humanly accessible time scale. Here, the shock waves and the light emitted by the bubbles upon collapse (sonoluminescence) were of interest [1]. This topic surfaced about thirty years later because of the incredible violence and speed of the process of bubble collapse and the phenomena related with it [2].

At the ultimate speed, the speed of light, exposure times also approach their limit. Picoseconds (10^{-12} s) to femtoseconds (10^{-15} s) must be employed to stop motions near or at the speed of light. *Light in flight* is an extreme example. In Fig. 2 a bullet of light (a very short light pulse in the femtosecond range or shorter) is almost stopped in its motion through a cuvette filled with water and scattering particles by a light activated shutter. A succession of single pictures with increasing delay between repeated propagating ultrashort light pulses and an ultrashort shutter light pulse (of actually the same duration) demonstrates the propagation of light by stopping its motion by the very short *glimpses* thus provided. For details of the arrangement see Fig. 11.22 in the book by Lauterborn and Kurz [3].

This example leads us to the topic of how a motion may be followed. When a motion can be repeated identically, then a proper succession of pictures with time delay between repeated motions gives the illusion of a real motion, actually it can be considered the real motion itself. This principle is used in sampling oscilloscopes in the case of identically repeating electronic signals. However, it may be a fake motion, as the backwards rotating wheels of vehicles in movies reveal. This means that the time difference between one exposure and the next one must be short enough with respect to the real motion or properly chosen in the case of repeating motions. When this demand cannot be fulfilled as in a movie with a fixed time difference between exposures, false motion impressions may be the result.

3 To Follow a Motion

As already stated, the exposure time to get a sharp image from a fast motion must be sufficiently short. But also the time difference between one picture (frame) and the next must be sufficiently short (usually larger than the exposure time). Otherwise the object (a light bullet, for instance) may have escaped the field of view. Or a whole scene with different objects may have altered so strongly that the different motions in the scene cannot be discerned. For slow object motions, the film material can be transported from frame to frame for the next exposure as in a normal camera or in movie cameras. This principle soon reaches its speed limit, as the film has to be moved and stopped at a faster and faster pace for faster and faster motions.

3.1 Rotating Drum Camera

Higher speeds of the film material can be reached without stopping the film for each exposure. Then the illumination of the scene must be short enough so that the additional motion of the film does not matter. This principle is realized in the rotating drum camera (Fig. 3). There, the film is placed around the outer periphery of a drum in a light-tight housing and the drum is set into rotation. A slit in the housing behind an objective lens can be opened for the time of one rotation by a mechanical shutter and a series of flashes is illuminating the object in question. The repetition rate of the flashes has to be set according to the speed of the film and the width of the slit for placing one image aside the neighbouring one. The flashes themselves have to be short enough for not to blur the image on the moving film.

An example of a relatively slow motion captured with the help of a rotating drum camera is given in Fig. 4, left four columns [4]. A drop is falling down onto a water surface, produces a splash and pushes water aside to leave about a half sphere of air below the water surface. When this half sphere closes, a water jet is shooting upwards with a drop on top. The picture series had been taken at 300 frames/s, but already at this low framing rate the succession of the events caught is beyond human perception speed. In this case, the frames have individually been taken from the film material and arranged for most compact presentation. For the latest developments in drop impact see Visser et al. [5].

An example of a faster process that could be resolved with the help of a rotating drum camera is the visualization of the gross succession of events following the focusing of a short light pulse of a ruby laser (about 30–50 ns duration) into water (Fig. 4, right two columns). The process has been slowed down by lowering the static pressure in the flask filled with water down to about half the ambient pressure. Then 6000 frames/s are sufficient to provide an overview of the motion. The intense light leads to a breakdown of the liquid and the formation of a bubble,

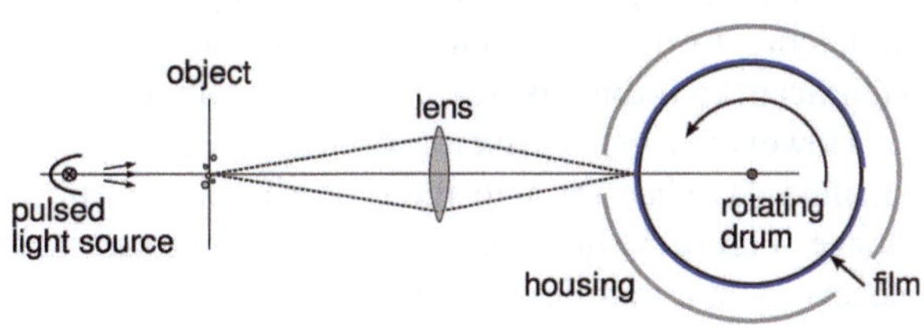

Fig. 3 Principle of a rotating drum camera

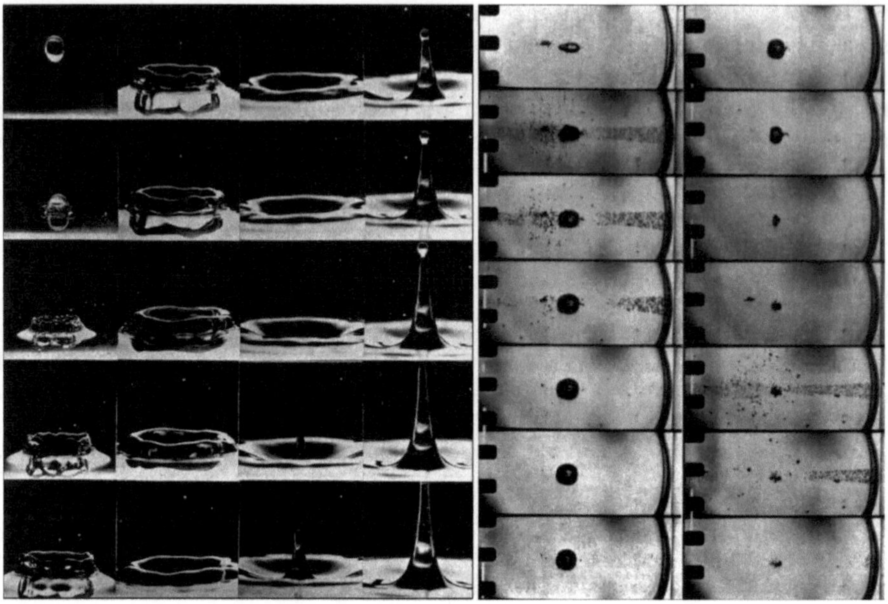

Fig. 4 *Left* Water drop impinging on a water surface. Rotating drum camera, 300 frames/s, front illumination. *Right* The successions of events when a high intensity light pulse of a ruby laser is focused into water (laser beam from the *right*). Rotating drum camera, 6000 frames/s, backlight. Time is proceeding from *top* to *bottom* and then to the next column *right*. See also Lauterborn [4]

a phenomenon called optic cavitation. The bubble expands up to a maximum size and collapses later to a smaller size. The focusing light cone becomes visible by a dense cloud of small bubbles. The bubble cloud reappears later after bubble collapse. The appearance of the bubble cloud is connected with the strong shock wave that is generated at breakdown of the liquid together with the bubble and again at bubble collapse. The shock waves themselves could not be captured optically here, because of the speadiness of the shock wave propagation and therefore blurring of its motion and/or escape from the frames.

Also the rotating drum camera has its framing rate limit. The film material is moving at high speed on the circumference of a drum and is subjected to enormous centrifugal forces from its circular motion. Indeed, it happens at extreme speed that the film material ruptures and only flakes remain in the housing. Up to about 100,000 frames/s can be reached at maximum, when the slit is brought down to about 1 mm height. That means, the height of the image has to be diminished more and more to achieve higher framing rates until it becomes too small to obtain significant two-dimensional (2D) information.

However, a new principle is then emerging. When the slit is becoming very small, a 1D slice through the object is imaged in rapid succession called a streak image. Figure 5 shows an approach to a streak image with a small image height, necessary at a framing rate of 50,000 frames/s. When following the outer boundary

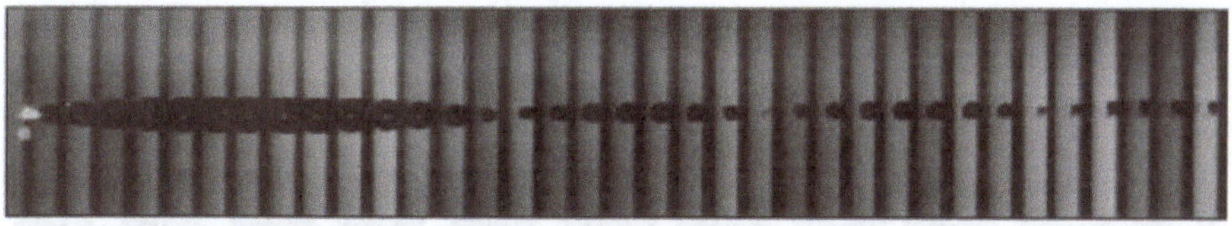

Fig. 5 Bubble formation and oscillations after a high intensity light pulse of a ruby laser is focused into water. Rotating drum camera, 50,000 frames/s

Fig. 6 Principle of a rotating mirror camera. Redrawn after Güth [7]

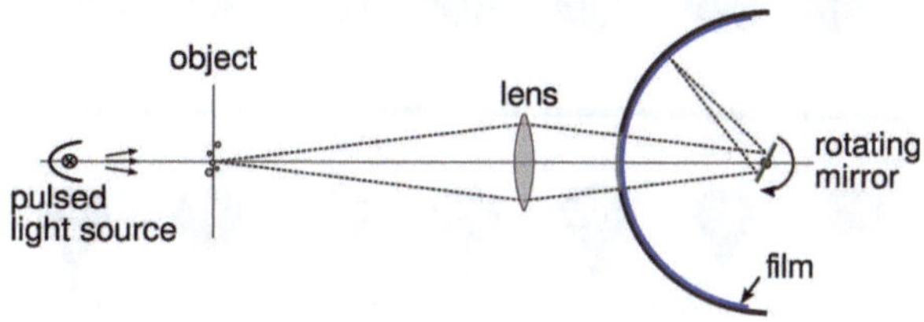

of the bubble, its growth with subsequent collapse, rebound (regrowth of the bubble to a second, smaller maximum) with second collapse, etc., can intuitively be followed. When the slit is very small, there is no longer a need for short pulses, as the slit together with the moving film acts as a fast shutter. This principle is used advantageously to visualize the propagation of spherical shock waves, where a 1D slice is sufficient for determining the (may be varying) speed of the shock wave. A modern example can be found, for instance, in the work of Noack and Vogel [6]. Very fast 1D motions can be measured by this type of high-speed photography. Up to 6 km/s have been found for the propagation speed of the shock wave emitted during laser-induced breakdown in water [6].

3.2 *Rotating Mirror Camera*

Higher framing rates than with a rotating drum camera can be obtained by a new principle, when not the film material is moved, but only a mirror rotated that projects the image of a scene onto a stationary film. The principle of a rotating mirror camera is given in Fig. 6. An objective lens images an object via a mirror onto a stationary stripe of film material that is located on a circle around the axis of rotation of the mirror. When the illumination consists of short enough flashes, one image is laid aside the previous one and a whole sequence of images can be obtained as in the case of the rotating drum camera. But now up to about one million frames per second can be reached. Moreover, also at slower framing rates, but high ones for a rotating drum camera, there is the advantage of larger individual images of a series and thus greater resolution of details.

Figure 7 gives an example of a comparably *slow* framing rate for a rotating mirror camera: 75,000 frames/s. Shown is a sequence of the motion of a

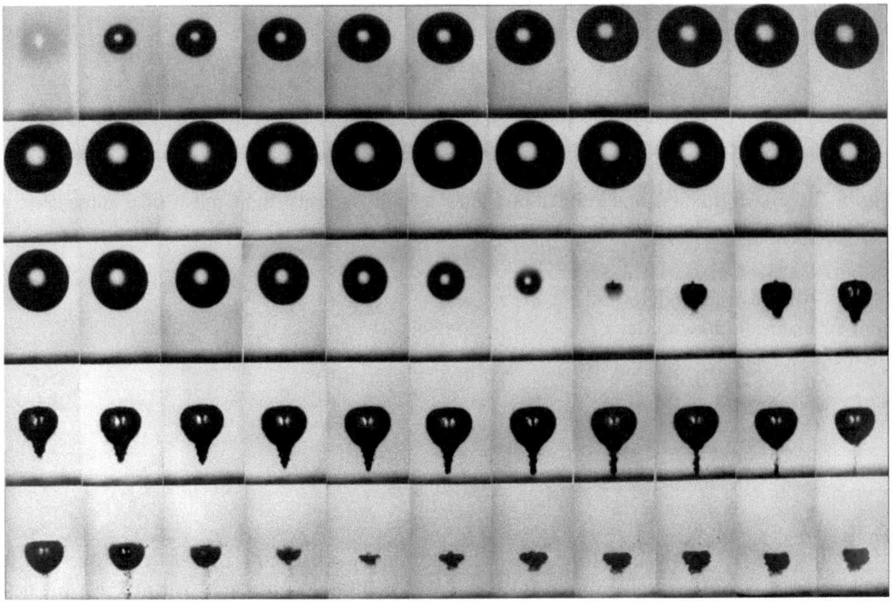

Fig. 7 Bubble dynamics near a flat, solid boundary (*lower dark line* in each frame). Rotating mirror camera, 75,000 frames/s, backlight. The maximum bubble radius is 2.0 mm and the distance of the bubble centre at maximum from the boundary is 4.9 mm. Frame size is 7.2 mm × 4.6 mm. Time runs from *left* to *right* and *top* to *bottom*. See also Lauterborn and Bolle [8]

laser-induced bubble near a flat, solid wall as taken by Lauterborn and Bolle [8]. The bubble appears black on a bright background, because the light from the back is reflected at the bubble wall, but with a bright spot at the centre, where the light passes undeflected. Bubbles near walls are very common and encountered with ship propellers and cleaning baths, for instance. For long time, their motion could not be followed by lack of suitable high-speed cameras and thus also the process of cavitation erosion could not reasonably be investigated. With laser-induced bubbles and the progress in framing speed high resolution bubble dynamics studies became possible in the nineteen seventies [4, 8] and flourished in the next decades [9], Philipp and Lauterborn [10]. In Fig. 7 the bubble first expands as in Fig. 4 and after reaching its maximum radius collapses in a characteristic way. The collapse proceeds aspherically. One part of the liquid shoots through the bubble (to be seen as the dark line in the bright spot in the centre of the bubble) and sticks out at the opposite side, a phenomenon called jet formation. Near a flat, solid wall in a stationary liquid the jet is directed through the bubble towards the wall. This has been found—in the course of time—to be the most common type of bubble collapse, when absolutely spherical conditions cannot be guaranteed. The initiation of the jet is not really resolved in this sequence. Rather, the jet is only to be seen after its formation and best upon expanding and shrinking of the bubble in the last but one row.

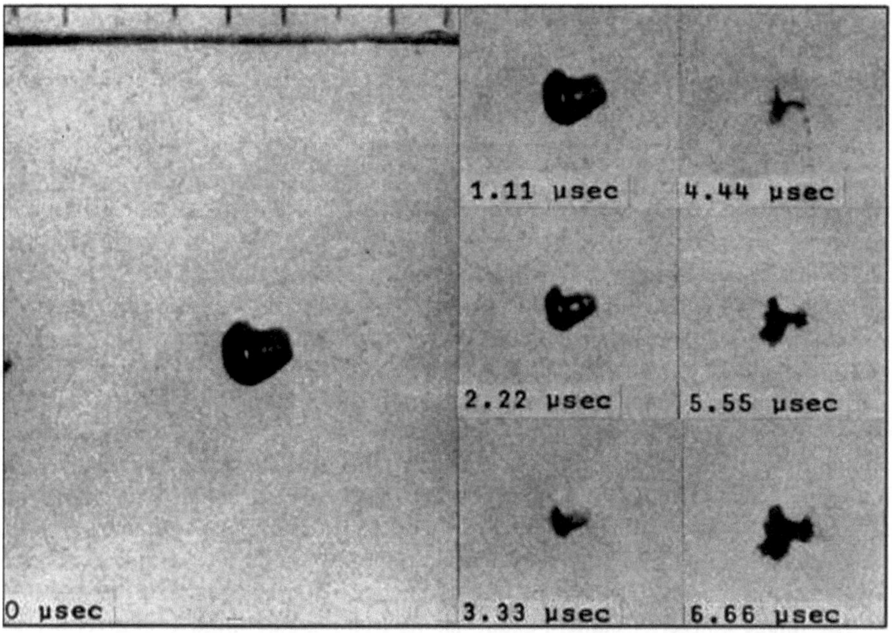

Fig. 8 Collapse of a laser-induced bubble in water. Rotating mirror camera, 900,000 frames/s. The *left* frame shows the bubble shortly before collapse together with a scale bar (1 mm distance between adjacent bars). Time runs from *top* to *bottom* and *left* to *right*. See also Lauterborn [4]

Not only jet formation, also the very collapse phase is not resolved in Fig. 7. Higher framing rates are necessary and possible with the rotating mirror camera. An example is given in Fig. 8. The sequence is taken at 900,000 frames/s. Now the collapse phase is resolved better in time and the jet formation process can be followed more closely, for instance to determine jet velocities. As discussed in the introduction, from two locations (here of the jet tip) and the two corresponding instants of time (giving the time difference, here 1.1 µs) a velocity can be determined. That way, maximum jet velocities of 100–200 m/s were obtained, depending on the bubble size and the distance of the bubble to the wall.

A review on the application of high-speed cameras to bubble dynamics up to 1985 can be found in Lauterborn and Hentschel [11].

3.3 *Image Converter Camera*

From Fig. 8 it seems that the collapse of a bubble can be fairly well resolved with the help of a rotating mirror camera at its limit of about one million frames per second. However, there are collapse configurations, when this framing rate is not sufficient to follow the motion (dynamics) of a bubble. Fortunately, soon higher framing rates became available, only limited by electronics, i.e. dispensing with mechanically movable parts and later with film material to be processed.

The development of electronic cameras with their fast electronic shutters also dispensed with the need of short illumination pulses. One of the first electronic high-speed cameras was the image converter camera. In this type of camera the light is "converted" to electrons that are multiplied in a succession of photomultiplier stages (the equivalent of the development process of a film) and delivered to a phosphor screen for converting the electrons back to light. By shifting the electrons with a high voltage onto different places on the output screen one image can be placed aside the other in fast succession. The screen is then photographed and the succession of images thus is permanently recorded. In the first version of the image converter camera the readout was made with an at that time normal camera on film. Later, with the advent of CCD cameras, also this step became fully electronic.

Shutter speeds in the range of nanoseconds can be reached, short enough for stopping the motion of shock waves. Figure 9 shows the shock waves radiated upon bubble collapse obtained by Lauterborn and Timm [12] at one million frames per second. One drawback of the image converter camera is the limited number of frames, typically eight to ten only. Therefore triggering on the collapse event by the bubble itself is necessary, see [12] and also Vogel et al. [13] and the examples therein.

The image converter camera again has its framing rate limit. It is given by the high voltage necessary to shift the electrons from one image to the next. This limit of about 20 million frames/s has been exploited by Ohl et al. [14], as a million frames per second proved not sufficient for resolving interesting details of bubble

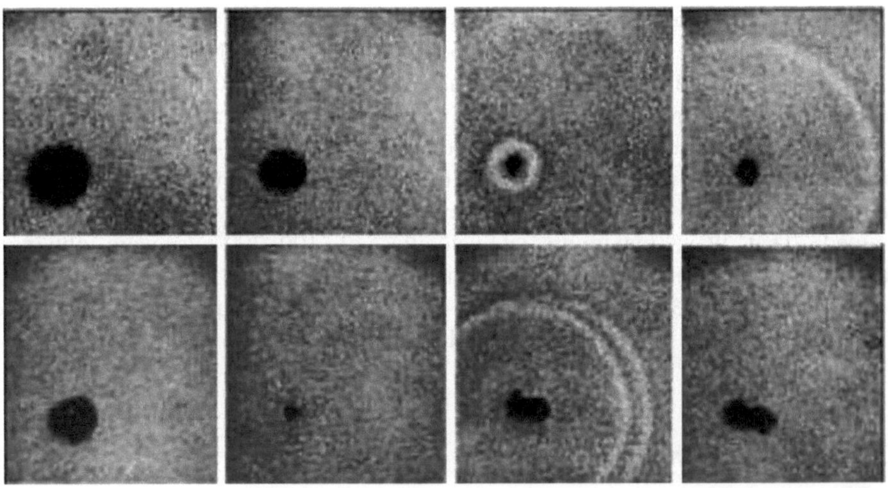

Fig. 9 Shock wave from a spherically collapsing laser-induced bubble (*upper row*) and two shock waves from an aspherically collapsing bubble with bubble splitting at collapse and merging during rebound (*lower row*). Image converter camera, one million frames per second. Width of the frames is 2.7 mm. See also Lauterborn and Timm [12]

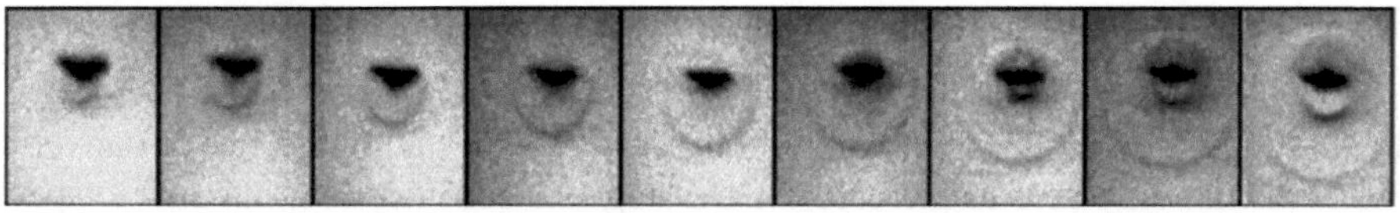

Fig. 10 Shock waves from a laser-induced bubble collapsing near a flat, *solid wall*. Image converter camera, 20.8 million frames/s. The maximum bubble radius has been 1.4 mm, the distance from the wall is 2.8 mm, the size of the frames is 1.5 × 1.9 mm. See also Ohl et al. [14]

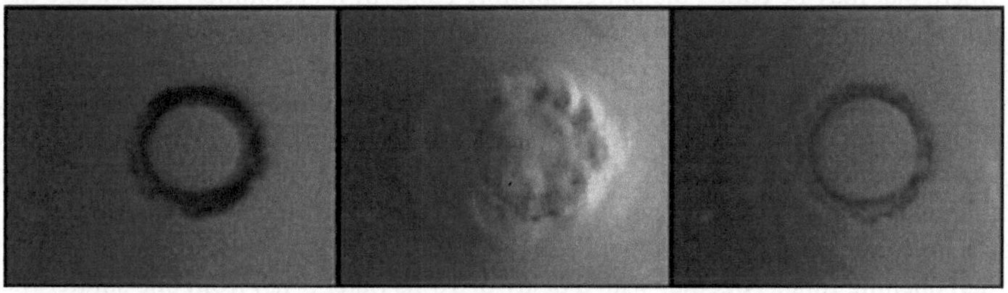

Fig. 11 Shock waves from a laser-induced bubble collapsing near a flat, *solid wall* in *bottom* view. Image converter camera, one million frames per second. Frame width 4.2 mm. See also Philipp and Lauterborn [10]

collapse. Figure 10—taken at 20.8 million frames/s—reveals that the collapse phase is quite occupied with different phenomena. At least two shock waves are emitted that are separated by about 300 ns. The first one is related to the impact of the jet onto the opposite bubble wall from the interior. This event changes the topology of the bubble from simply connected (spherelike) to doubly connected (toruslike). At this instant of time the bubble still collapses further, i.e. diminishes its volume. At collapse of the torus, the second shock wave is emitted from the sudden stop of the inward motion. Only the speed-up from one million frames per second to about 20 million frames/s allowed this distinction to be made. It should be noted that it is not easy to exactly catch the collapse phenomena at this high speed in one sequence of eight frames. Therefore the sequence of Fig. 10 is a composition from different sequences with (largely) identical laser-induced bubbles.

In rare instances it happens that singular events of the collapse can be caught by chance. Figure 11 gives a sequence of three frames taken at one million frames per second with an image converter camera and CCD readout of the very moment of collapse of the torus bubble. In this case the bubble has not been photographed in side view, but from the bottom through a transparent solid boundary. The many shock waves on a ring show that the torus collapse is unstable and gives rise to many separate collapse events.

3.4 Image Splitting Camera

The—still ongoing—digital revolution that started with the hybrid image converter camera is not only able to increase the framing rate considerably, but also to speed up experimental investigations by dispensing with the time consuming processing of the film material and production of photoprints. Instead, the image sequences are available instantly and can be used to immediately direct the next measurements.

Thereby the combination of old ideas and new technology led to a new type of camera that may be called the *image splitting camera*. It is common knowledge that an image can be multiplied by partly reflecting mirrors or a scene be looked at via several mirrors and that these images can be photographed altogether *at the same time*. The new technology admits photographing each image of a mirror at slightly different times, whereby the electronic shutter and delay times between images may be as short as 5 ns. In one commercial device this is done with the help of a pyramidal image splitter to get eight (later sixteen) individual image paths. Each path is connected to a CCD camera equipped with a fast shutter down to 10 ns (later 5 ns) opening time. Together with a delay of 10 ns (later 5 ns) from one shutter (i.e. CCD camera) to the next, frame rates up to 100 (later 200) million frames per second are possible. This type of camera needs strong illumination, as only 1/(number of images) of the light is available per image. But this demand normally is achievable by a strong xenon or LED flash. An advantage of the image splitting camera versus the image converter camera is that each image has the full frame resolution of the respective CCD camera and that each image can have its separate delay to the previous image. An example of the image quality that can be attained at a million frames per second is given in Fig. 12. It consists of two combined sequences, as only eight frames were available per measurement. Seen is a bubble at an angle of 45° from above a solid wall. The bubble is contracting with jet formation (involution from the top) passing through its collapse phase and rebounding. When looking at the collapse phase (frame 12–14), it is seen that within one microsecond the bubble considerably alters its shape, i.e. the collapse is not resolved.

A better resolution in time has already been explored with the image converter camera. However, the image splitting camera gives us the possibility to look at bubble collapse with unprecedented time and spatial resolution in the quest to reveal the secrets of bubble collapse. Figure 13 gives a composite image sequence taken at 100 million frames/s of the very collapse phase of a laser-induced bubble near a flat, solid wall in side view. The frames are labelled with the instant of time they were taken with respect to the first frame shown. The sequence covers 260 ns plus one additional frame at 300 ns. Now the jet impact phase is better time resolved and reveals much of its complex nature. As can be seen from frame 12 of Fig. 12, shortly before the jet hits the lower bubble wall, the bubble attains the shape of a (tiny) bowl. This intermediate shape lasts for a small fraction of a microsecond only. The soon following jet impingement itself proves to be an extended process in time as well as in space. Instead of touching the lower bubble

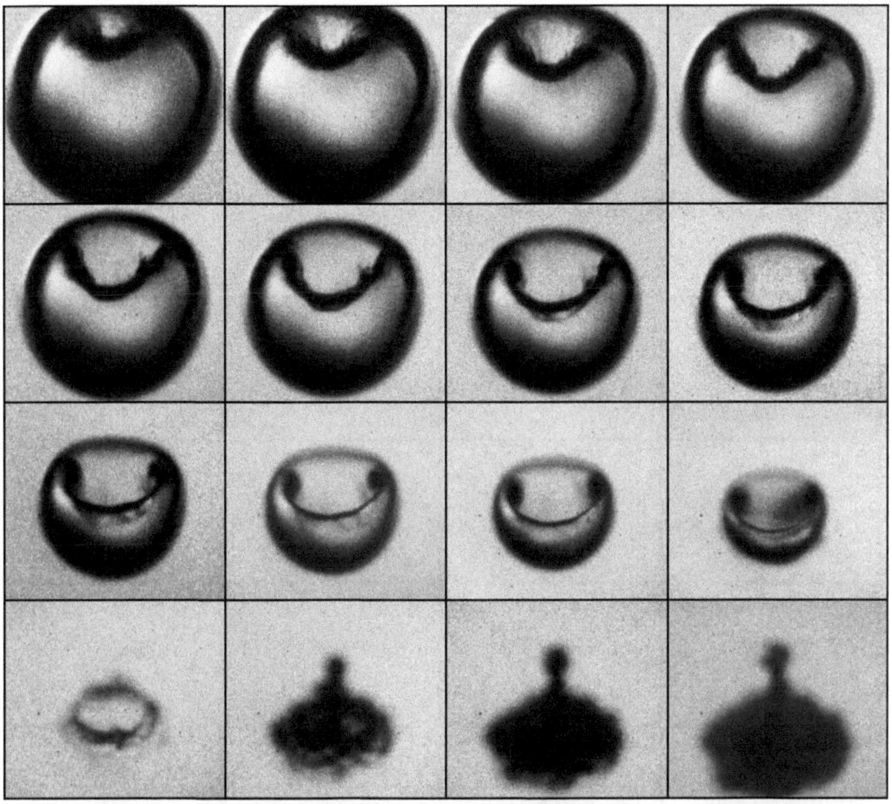

Fig. 12 Jet and counterjet formation from a laser-induced bubble collapsing near a flat, solid wall seen at an angle of 45° from above the wall. Image splitting camera, one million frames per second. Frame size 1.2 × 1.1 mm. See also Lindau and Lauterborn [15]

wall at a point, the jet hits it on a ring. This can be concluded from the two shock wave rings observed to the left and the right of the bubble in Fig. 13. The rings belong to a torus shock wave generated at the impingement ring. This is a highly sensitive situation, as only in a perfectly axially symmetric geometry a simultaneous touching of the jet at a ring will occur. The subsequent dynamics seems predetermined. The ring closing has punched out a separate bubble besides forming the torus bubble. Both bubbles will be compressed further by the fluid flow and therefore more shock waves can be expected. Moreover, a proliferation of shock waves by further liquid impacts can be taken for granted from the instability of the torus bubble (see Fig. 11).

The situation, however, is not that clear, as the torus shock wave interacts with itself and the bubble. It is not known, what effects the torus shock wave will produce inside the jet itself (not observable here), upon self-penetration along the torus axis and by diffraction around the torus bubble. At about 140 ns the torus shock wave impacts with itself on the bubble axis below the bubble at the side to

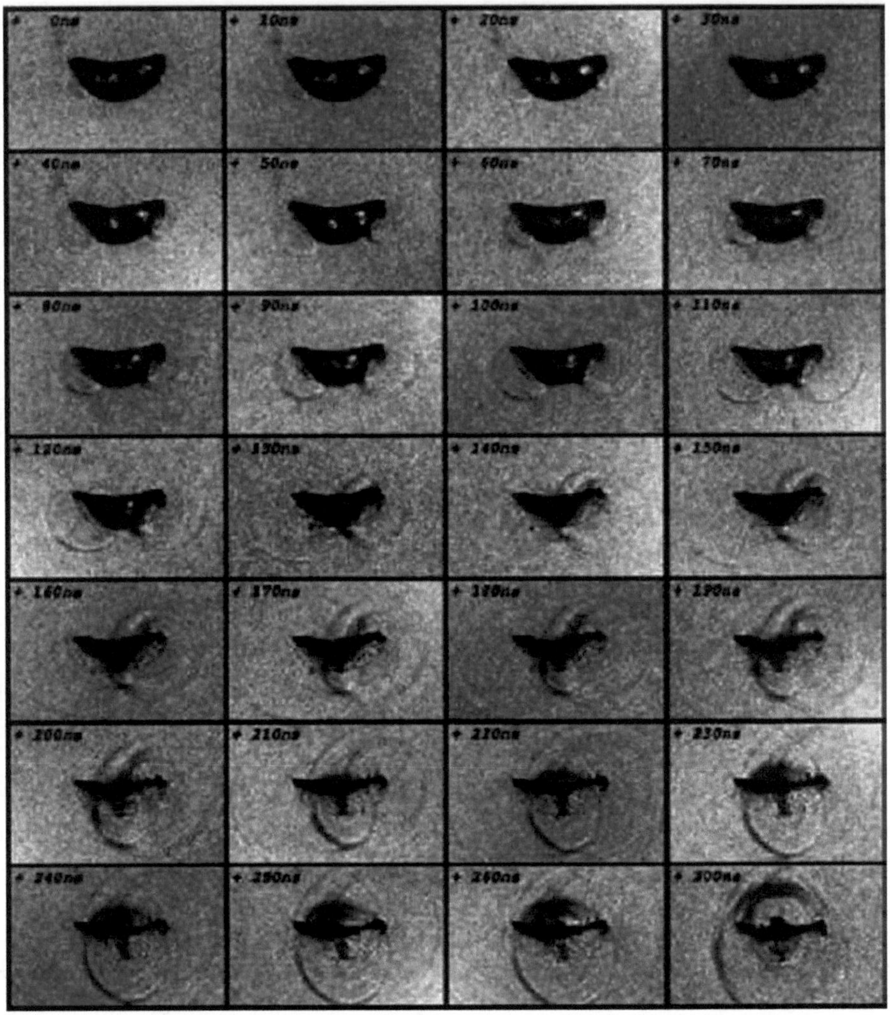

Fig. 13 Bubble dynamics and shock wave emission during bubble collapse near a flat, *solid wall* located below the bubble. Image splitting camera, 100 million frames/s. Side view and shadowgraph. Bubble radius at maximum expansion is 1.5 mm, distance of the bubble centre to the wall is 3.9 mm. Frame size is 1.3 × 0.8 mm. See also Lindau and Lauterborn [15]

the solid wall. It seems to be the origin of a further shock wave, presumably either from compression of the punched bubble or impact of the inflowing liquid from the main bubble collapse with the liquid of the jet. The jet is surrounded with gas and vapour from the bubble and this may be the reason for the dark granular stem to be seen from frame 170 ns onwards. It may be the remnants of the gaseous/vapourous hull of the jet after inflow of liquid from the outside towards the axis. At frame 260 ns a *nose* appears above the now flat bubble. It is of unknown origin and obviously connected with the emission of a further shock wave to be seen expanding above the bubble. It may be connected with the *splash* observed by Tong et al. [16], a tiny ring water wave running up the inner bubble wall and impinging onto itself similar to the torus shock wave. Alternatively, it may be considered as if the inflow of liquid towards the axis is squeezing the gas hull of the main jet upwards to form the nose.

All these phenomena happen within a few hundred nanoseconds. Thereafter, on a much slower time scale, the main jet towards the solid wall forms, when the bubble is rebounding owing to the compressed gas/vapour that acts as a spring.

The main drawback of the image splitting camera is the limited number of frames of eight to sixteen that it shares with the image converter camera. Triggering on a mechanical event down to the nanosecond range for an image is exceedingly difficult. The underlying jitter from tiny deviations then may prevent to catch a fast event. New camera types are evolving that may finally reach the framing rates needed here. An order or better two orders of magnitude larger number of frames without sacrificing resolution by diminishing image height would be welcome for the time being. An unlimited stream of ultra high-speed images with post-event trigger capabilities, of course, would present the final solution for many problems.

3.5 Numerical Extension of High-Speed Photography

With laser-induced bubbles and high-speed photography the first calculations of jet formation of a bubble near a wall reported by Plesset and Chapman [17] could be compared with experiments by Lauterborn and Bolle [8]. The first calculations, however, could only be done up to the moment, when the jet touches the opposite bubble wall. The experiments naturally extend further in time and thus can follow the jet also further in space beyond touching the opposite bubble wall (see Figs. 7 and 8).

To jump from history to the present: it is only nowadays that numerical codes are able to proceed beyond the impact of the jet onto the opposite bubble wall (for the latest examples see Han et al. [18] and Koch et al. [19]). A comparison with the possibilities of today is given in Fig. 14 for the case of a bubble collapsing in front of a flat, solid wall with touching it upon first rebound. The right and lower part of each image shows the experimentally obtained outline of the bubble (one half) taken with a modern high-speed video camera, the left, lightly coloured rectangle the numerically obtained one by solving the Navier-Stokes equation. The lower boundary of the rectangle in each image represents the surface of the solid wall, the darker grey areas below come from the reflection of the bubble in the polished surface. To follow the jet into and in the interior of the bubble is difficult, as the outer bubble wall deflects the light and gives the view free only in the middle of the bubble, where the light passes undeflected. This is different in the numerical calculations, where a cut through the bubble can easily be done and presented as shown here. Then the jet and the formation of a torus are directly visible (each image is to be rotated along the vertical axis of symmetry, here also given by the

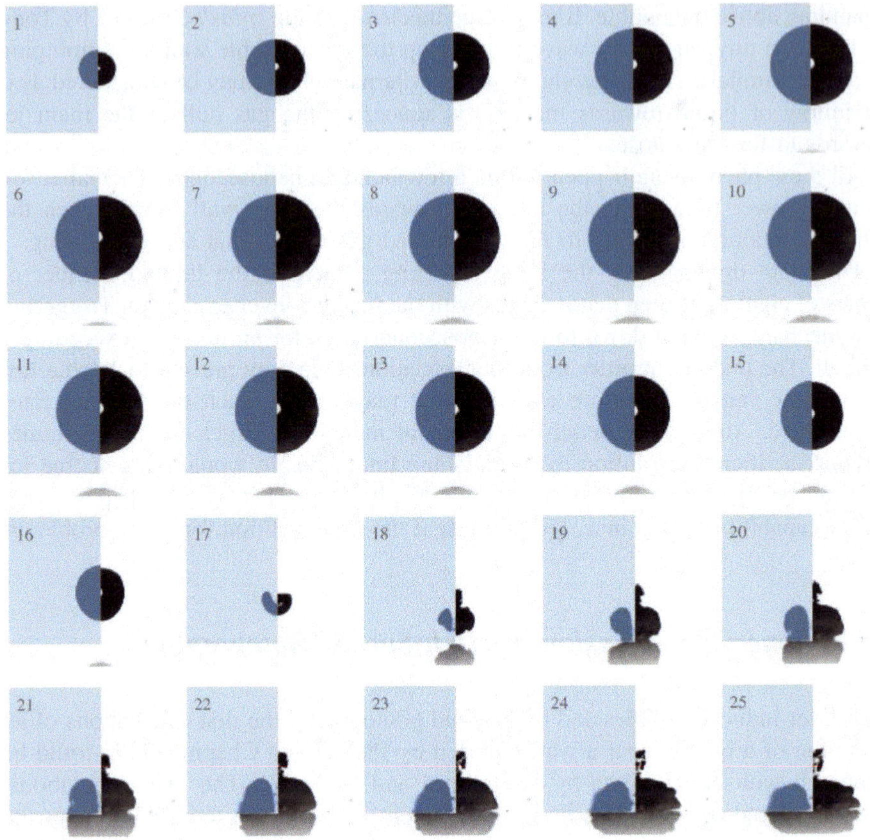

Fig. 14 Comparison of experimentally determined bubble shapes (*right* and *lower* part of each image) upon collapse of a laser-induced bubble in water near a flat, solid wall with numerically calculated shapes (*left* in each image). Modern high-speed video camera, 180,000 frames/s; Navier-Stokes equation solved with OpenFOAM. The maximum bubble radius is 450 µm, the distance to the boundary amounts to 642 µm. See also Koch et al. [19]

vertical numerical–experimental boundary in each composite image). The so-called *counterjet* sticking out above the experimental bubble probably is due to secondary cavitation that is not included in the bubble model. The reflection of the bubble in the boundary (best seen from frame 18 onwards) very nicely fits with the extension of the bubble in the numerical calculation. Thus, the obvious agreement over the complete sequence gives confidence into the numerical model. Then, as already put forward in the introduction, the physical model can be used to interpolate between adjacent images to get an even better time resolution than obtainable with the highest-speed cameras. Calculations of the bubble collapse with a time resolution below 0.1 ns have already been done, corresponding to more than 10,000 million frames/s (M. Koch, private communication 2016). This digital extension allows an even deeper look into what we cannot see.

4 The 3D Challenge

A conventional image is a two-dimensional projection of a three-dimensional scene. As such it contains only part of the reality. The first approach to 3D consisted in simultaneously taking two projections at an angle between them, called stereoscopy. A depth perception is resulting from the parallax between the two projections. Therefore it might be applicable to measure 3D bubble distributions. This approach has been reported by Appel et al. [20]. Figure 15 gives an arrangement for studying bubble dynamics in an ultrasonic field of 22.8 kHz (acoustic cavitation). Three cameras are looking at different angles onto a bubble distribution. The three cameras, called HiSIS1 and HiSIS2 and Pulnix, are synchronized to take images at the same instants of time. Typically they are operated at 2250 frames per second by us. Figure 16 shows a result of the achievement with this arrangement. A 3D bubble distribution taken from many frames is plotted with three orthogonal planes of projection for better perceiving the 3D structure. It is seen that by no means the bubble distribution is homogeneous. Instead, the bubbles are arranged in a characteristic way with many branches (called streamers in the community). This is very unfortunate for ultrasonic cleaning, as cleaning is mediated by bubbles: no bubbles, no cleaning. Moreover, velocities of the bubbles can be deduced from the frames and velocity distributions be given [20].

Stereoscopy is like looking with two eyes, but without being able to turn the head or approaching the scene. Also the cameras have a limited depth of field and can view only a small volume at high resolution. Therefore other solutions were sought for and found in holography to be discussed next.

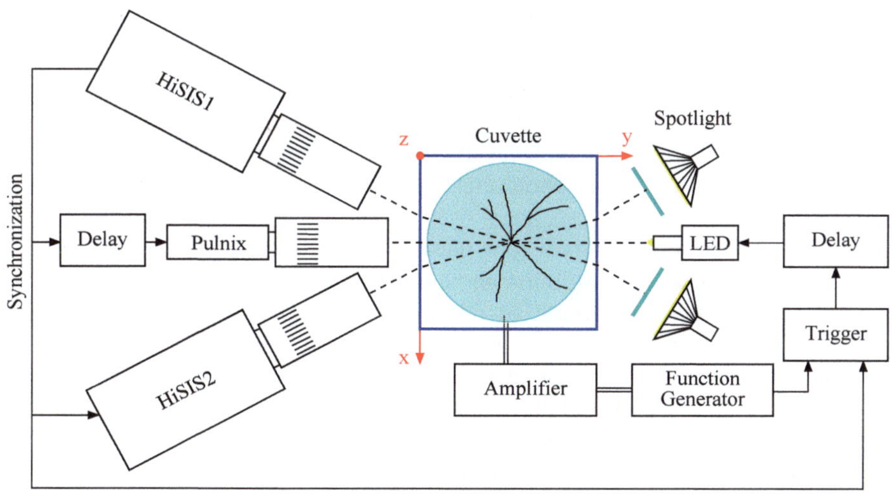

Fig. 15 Arrangement for high-speed stereoscopy of bubble distribution and bubble motion. Courtesy of R. Mettin

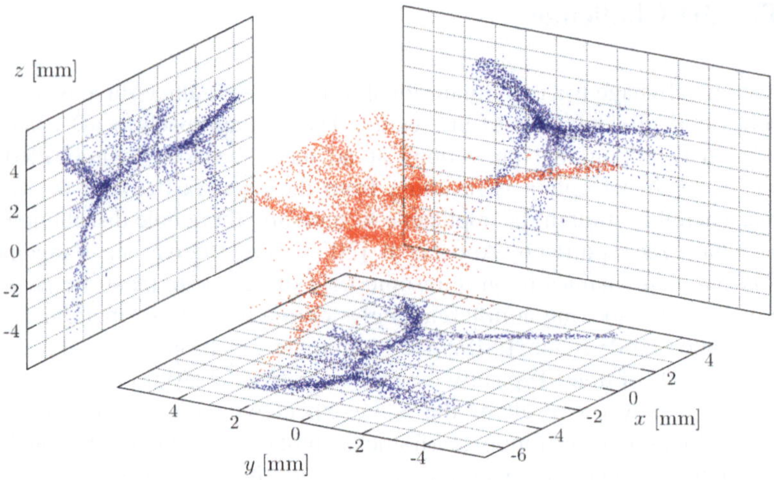

Fig. 16 Bubble distribution obtained by high-speed stereoscopy. High-speed CCD cameras, 2250 frames/s. Bubbles driven acoustically at 22.8 kHz. Courtesy of J. Appel. The corresponding videos are available online

4.1 Holographic Cameras

When a scene can be illuminated with coherent light, one can do much better in principle with a holographic camera. In holography a hologram stores the scene in an interference pattern between the light from the scene and a reference beam. Upon reconstruction from a hologram by illuminating it with the reference beam, the reconstructed 3D image can be photographed at different angles and from different distances at different magnifications. A hologram thus preserves a three-dimensional image, albeit with the restriction of looking through the hologram like through a window. With the help of the conjugate reference beam (backwards propagating reference beam passing the hologram) the 3D image can be projected into real space. Then planes through the 3D image can be intercepted and stored on film or any other photosensitive surface without any lenses or other imaging system.

Figure 17, left column, gives four planes of a reconstructed 3D image from a hologram taken on a high resolution photographic plate with a single exposure from a He–Ne laser at a wavelength of 632.8 nm [21]. The 3D scene holographed contained many bubbles on wires in water. It is seen that a very good quality of photographs can be obtained from the real image projected into space. The diameter of the wires is 40 μm only and the smallest bubbles are about 100 μm in size. Thus holography may be used for bubble studies.

With short enough coherent pulses also high-speed motions can be stopped as in conventional photography, but now in three dimensions. Any short, sufficiently coherent pulse from any laser can be used for a holographic camera. Therefore quite

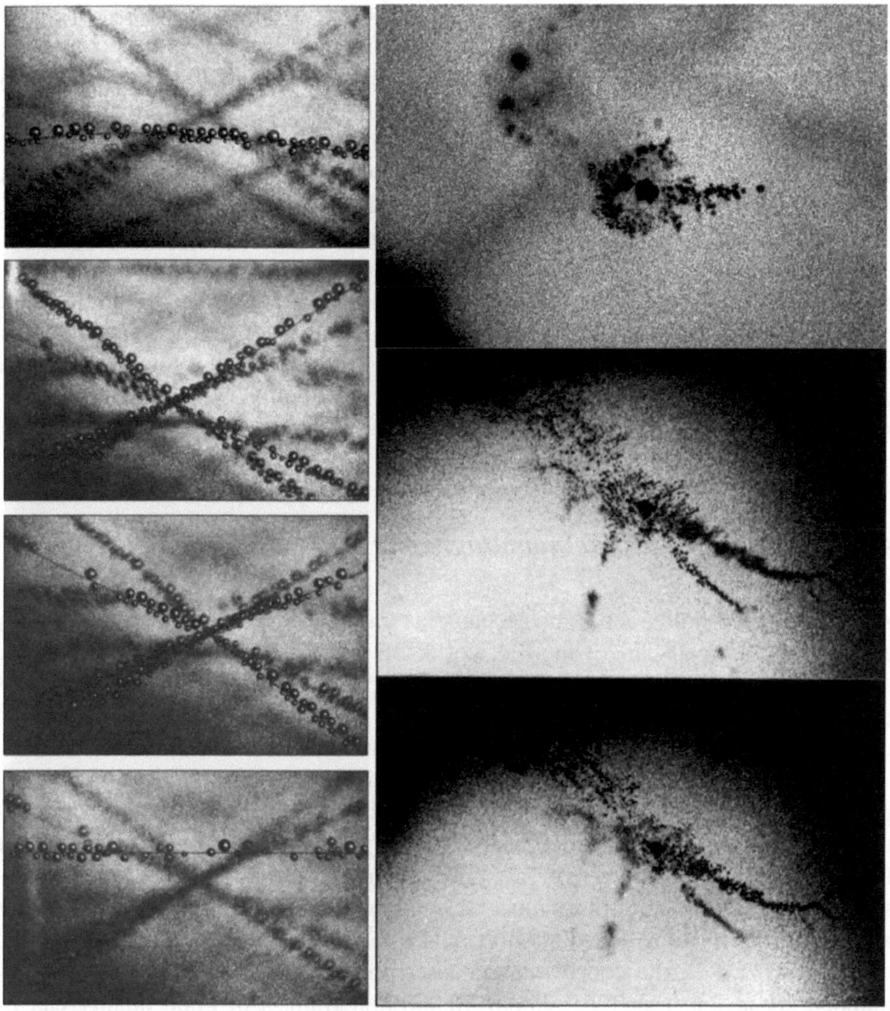

Fig. 17 *Left column* bubbles on wires in water. He–Ne laser holographic camera. Four planes covering 1.5 cm in depth from the reconstructed real image of a hologram. Diameter of a wire is 40 μm. See also Lauterborn et al. [21]. *Right column* cavitation bubble distribution in water inside a cylindrical acoustic transducer driven at 19 kHz. Photographic images from the reconstructed 3D real image of a hologram. Three different planes in depth and from different locations in the 3D image. The two lower images are about 5 mm apart. Ruby laser high-speed holographic camera. Exposure time is about 20 ns. Hologram taken in collaboration with K. Hinsch and F. Bader in 1972

different types exist depending on the wavelength. The first laser used was the ruby laser at a wavelength of 694.3 nm. By Q-switching a short coherent pulse (or also a series of pulses) can be obtained. An example is given in Fig. 17, right column. A single pulse from a ruby laser was used to holograph the cavitation bubble distribution in water inside a cylindrical piezoelectric acoustic transducer driven at 19 kHz. The high acoustic intensity applied ruptures the water, called acoustic cavitation, and gives rise to the characteristic distribution of bubbles with many branches, the streamers, that are occupied with bubbles. They extend in three dimensions and are not easily photographed in their entirety in high resolution. Three planes from the reconstructed 3D real image are shown at different depth. Two streamers are in focus in their respective planes about 5 mm apart (lower two images). A special, but quite common bubble distribution is given in the upper 2D image from the same hologram: a comet like bubble aggregation with two larger bubbles, a halo of smaller bubbles and a tail, also made up of small bubbles. One large bubble in an acoustic cavitation bubble cloud is often accompanied by a school of small bubbles following the larger one. Presumably the small bubbles are offsprings from the larger ones in the bubble distribution, produced upon non-spherical collapse of a larger bubble.

4.2 High-Speed Holocinematography

Also high-speed holographic movie cameras have been developed. Actually, they are catching a small part of the 4D world: 3D in space and 1D in time. Bubble physics has a demand for 3D images in time. In particular in acoustic cavitation—as used in cleaning baths—thousands of bubbles (that do the cleaning) are distributed in a volume of liquid. The dynamics of these bubbles is of interest.

A straightforward idea for high-speed holocinematography is to use a rotating plate or rotating drum camera, to take a series of short coherent pulses for illumination and to add a reference beam to each of the illumination pulses. As the interference fringes may move fast and should not be blurred for good reconstructions, short enough pulses must be employed. Several different laser systems are able to deliver long series of short pulses for this purpose, e.g. the Q-switched ruby laser [22, 23], the copper vapour laser pumped dye laser [24] or the cavity dumped argon ion laser [25]. About 50,000 holograms/s of good quality can be obtained similarly to high-speed photography. Higher hologram rates have been obtained, when diminishing the hologram size (as with picture size in conventional high-speed cameras) and introducing additional deflection methods.

For short, fast hologram series (about ten holograms) the image splitting principle can be used in the form of partially reflecting mirrors. The time delay between holograms is realized by the spatial path for the image from mirror to mirror and a respective reference light delay, when the coherence length is not large enough. As a long time delay for light requires a long path in space (owing to the high speed of light) only short time delays are realizable. This corresponds to extremely high hologram framing rates. Novaro [26] could take ten holograms at 500 million holograms/s with his image splitting high-speed holographic camera. However, only near light speed dynamics can be studied with this holographic camera.

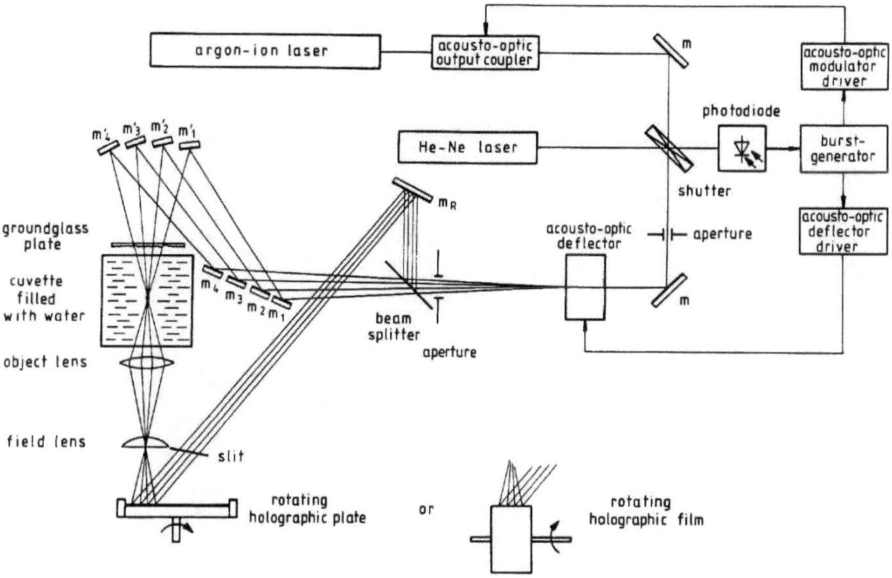

Fig. 18 Argon ion laser high-speed holographic camera with either a rotating plate or a rotating drum and additional acousto-optic beam deflection for spatial hologram separation. Limit about 300,000 holograms/s. See also Hentschel and Lauterborn [27]

Figure 18 gives a holographic arrangement for off-axis high-speed holography with an argon ion laser at a wavelength of 514 nm. It makes use of a rotating holographic plate or a rotating drum and fourfold hologram framing rate extension with a beam deflection unit as constructed by Hentschel and Lauterborn [27]. With this device up to about 300,000 holograms/s are possible [28]. An example of a holographic sequence taken at 200,000 holograms/s is given in Fig. 19. The laser-induced breakdown from focusing a Q-switched pulse of a ruby laser into water and the subsequent events are holographed. Reconstructed images from consecutive holograms are shown and numbered. The image from hologram 4 shows the shock waves radiated upon breakdown and the subsequent reconstructed images (with shifted views in columns one and four) show the growing bubble cluster and further tiny bubbles in the path of the laser beam and around the growing bubble cluster.

Image quality suffers from small hologram sizes, more than the photographic counterpart. Firstly, there is speckle in the image from the coherent light. Speckles appear at the limit of resolution, give the image a granular appearance and disturb the perception. Secondly, the hologram contains depth information and a single plane in the reconstructed 3D image therefore cannot contain the full resolution as the same plane would have in direct photography. Thus a hologram generally suffers from a resolution loss for 2D information per same area compared to conventional photography.

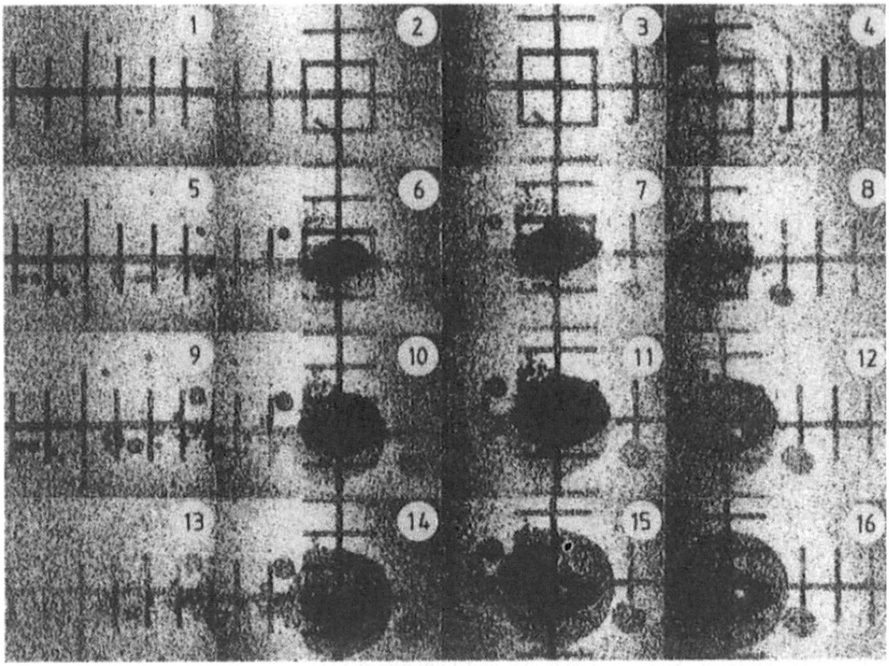

Fig. 19 Dynamics of a laser-induced bubble cloud in water. Argon ion laser high-speed rotating drum holographic camera, 200,000 holograms/s. Time runs from *left* to *right* and from *top* to *bottom*. The numbers show consecutive hologram numbers, i.e. adjacent reconstructed images are 5 μs apart. The two *middle columns* show the growing bubble cluster, the images in the first and last column are shifted in space to the *left* and to the *right* of the main bubble cluster. Millimetre scale placed behind the bubbles. See also Hentschel and Lauterborn [28]

A similar arrangement as given in Fig. 18 has been used to holograph bubbles inside a piezoelectric cylindrical transducer (76 mm in length and diameter, wall thickness 5 mm) driven at 23.1 kHz. Inside the transducer a cloud of cavitation bubbles is generated with thousands of individual bubbles oscillating and moving around in the three-dimensional volume. The interesting fact is that with increasing driving pressure amplitude the sound emitted from the bubbles undergoes a period-doubling sequence to chaos [29]. The question was, whether the complete bubble cloud does the same. This is indeed the case as could be proven by Lauterborn and Koch [30] with the help of high-speed holocinematography (for an example of a chaotically oscillating bubble cloud taken holographically see Lauterborn and Holzfuss [31]). Holograms were taken at a rate of 69,300 holograms/s (three holograms per period of the sound field), and 3D images were reconstructed for inspection. Figure 20 shows one plane of the reconstructed images per period of the driving at the same phase of the sound field. The images are full of speckles from the coherent light, but the bubbles stand out by their clustering. When going from one image in a row to the next one, it is seen that the

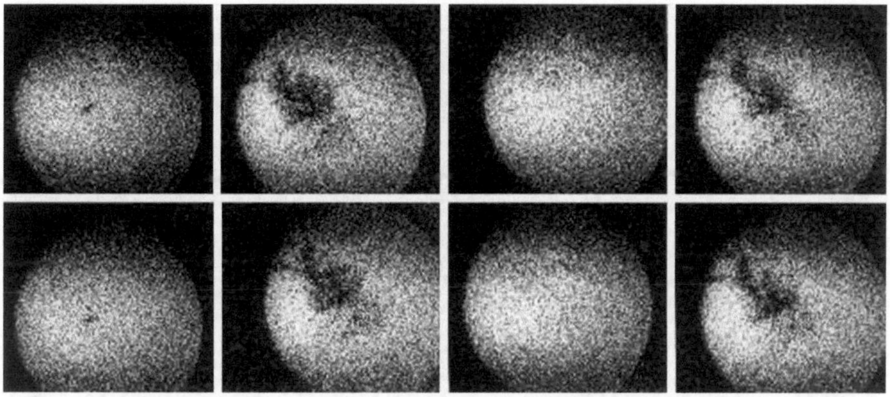

Fig. 20 Reconstructed images from a high-speed holographic sequence taken at 69,300 holograms/s of a bubble cloud driven acoustically at 23.1 kHz. Argon ion laser high-speed holographic camera. Every third image selected at a fixed phase to give one image per period of the driving. The sequence is running from *left* to *right* and from *top* to *bottom*. The bubble cloud has period doubled to at least period four. See also Lauterborn et al. [34]

bubble cloud needs at least four periods of the sound field to repeat. By careful inspection and comparison with the simultaneously taken sound emitted it is found that it needs even eight periods of the sound field for the bubble cloud to repeat.

A review on the application of high-speed holographic cameras to bubble dynamics up to 1986 can be found in Lauterborn and Hentschel [32], later in the overview by Lauterborn and Hentschel [33] and the reviews by Lauterborn et al. [34] and Lauterborn and Kurz [35].

4.3 Digital Holography

The advent of the digital computer, digital printers and digital cameras has made possible a complete digitization of the holographic process, from calculation of a hologram and plotting it, direct digital recording of a light produced hologram (the interference fringes) to numerical reconstruction from a digitally recorded hologram or from a digitized hologram on film. Subsequent digital object recognition is necessary in all cases to make (scientific) use of it.

A hologram can be computed by calculating the interference pattern from a given scene, realized by plotting it on a digital printer and reconstructed either digitally again (then no plot needed) or by illuminating it with coherent light. The last option will be a direct proof of correct calculation.

An example of a calculated binary hologram (interference fringes in black and white) is given in Fig. 21. Twenty eight points in space are coded in it, arranged in four letters, each in a different plane in depth. This can be tested by a real reconstruction with coherent light waves. To this end, the holographic pattern has to be

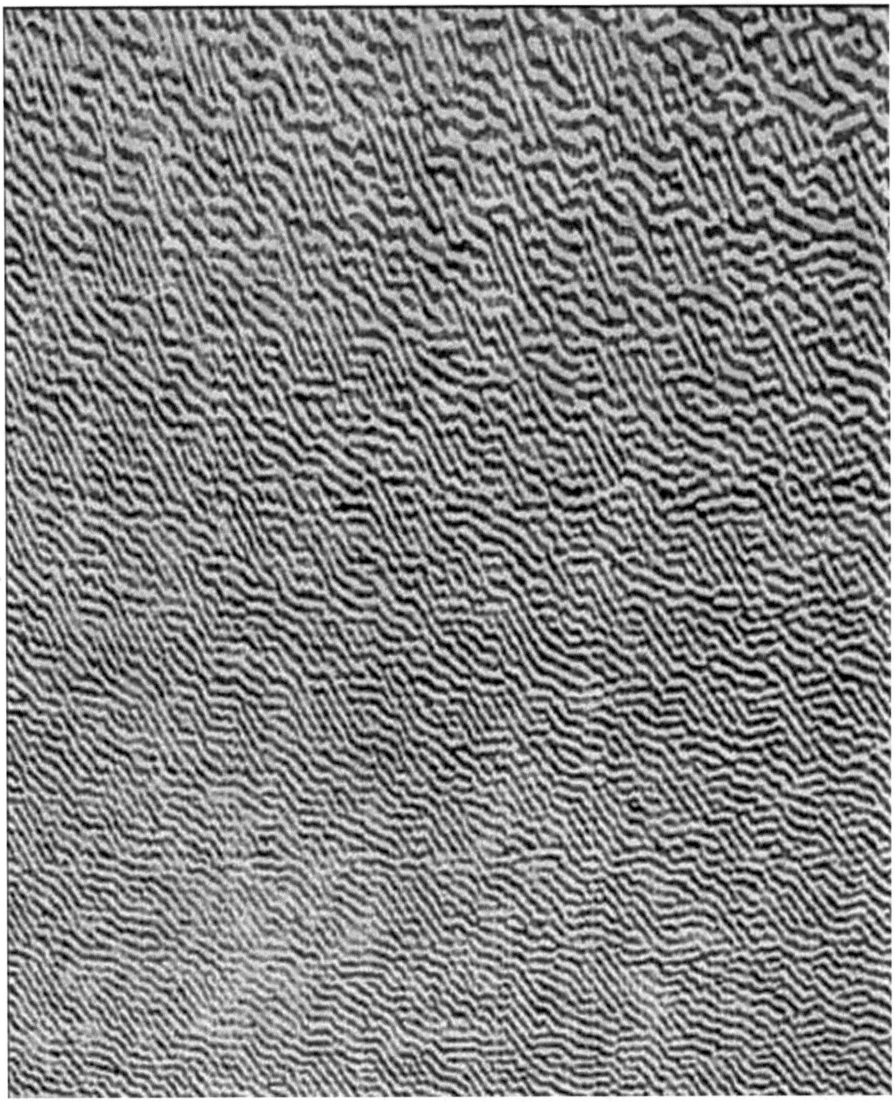

Fig. 21 Part of a computed and printed off-axis hologram in *black* and *white* (a binary hologram). It contains the word HOLO with the four letters in different planes in depth. See also Lauterborn [36]

reduced for stronger diffraction at a wavelength of visible light. In the present case, the hologram of Fig. 21 has been reconstructed from a reduction to about 5 mm edge length and illuminated with the continuous beam of a He–Ne laser (wavelength = 632.8 nm) to get the 3D image. Four photographs from the reconstructed letters are shown in Fig. 22. The 3D property of the image is clearly demonstrated.

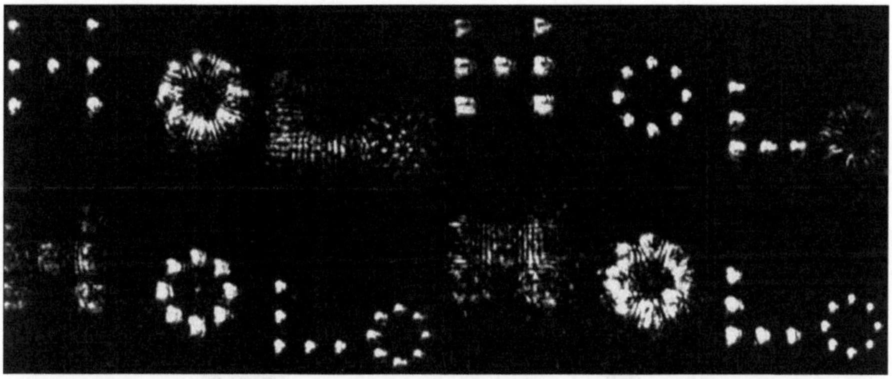

Fig. 22 Reconstructed 3D image HOLO from the binary hologram in Fig. 21 in four different views. The four different letters are located in different depths. See also Lauterborn [36]

Fig. 23 Arrangement for multiple laser-induced breakdown sites in a liquid with phase holograms. See also Lauterborn [37]

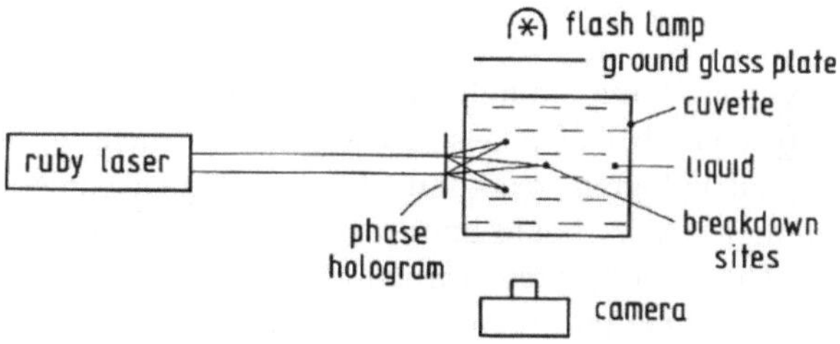

This type of digital holograms has been studied for use as lenses with multiple focal points. Then many-bubble systems can be produced in a liquid, when high intensity laser pulses are used for reconstruction. First attempts are described by Lauterborn [37] and Hentschel and Lauterborn [38]. The principle of focusing through a holographic lens is given in Fig. 23. It is a very simple arrangement, once the phase hologram is fabricated so that it can withstand the high intensities. With a similar arrangement, but with a Fourier transform hologram and an additional focusing lens, five-bubble systems have been generated in high-viscosity silicone oil (Fig. 24). Then the bubbles stick for some time to the same place and can even be seen with the naked eye.

Nowadays, digital holograms can be formed with a digital spatial light modulator (SLM) and, when properly illuminated, can be used for bubble production. Examples are given by Lim et al. [39].

The interference pattern of a hologram from a real scene can also be recorded digitally instead of on film. A proof of principle together with numerical reconstruction has been given by Schnars and Jüptner [40]. They used a CCD camera with 1024 × 1024 pixels with a pixel area of 6.8 μm × 6.8 μm. The small hologram size of about 7 mm × 7 mm leads to an image with pronounced speckles. And the coarseness of the pixels with respect to the grain size of film material only allows for holographing quite far away objects at small off-axis angles.

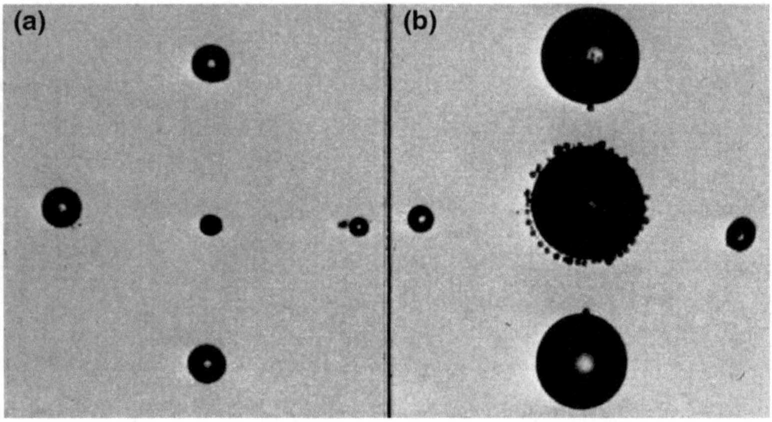

Fig. 24 Two five-bubble systems in silicone oil. Q-switched ruby laser pulse focused through a binary phase Fourier transform hologram. See also Hentschel and Lauterborn [38]

By magnifying the hologram to obtain coarser interference fringes and by using a larger number of pixels (2048 × 2048) Sheng et al. [41] considerably improved digital hologram recording. Modern digital cameras with their fourty and more million pixels should improve the resolution further.

The numerical reconstruction of the 3D image from a digital (or other) hologram needs further processing as to detect objects or structures in a 3D volume [42]. Also this detection process has been advanced over the years [43]. The progress over time in digital holography is documented by Schnars and Jüptner [44] and Katz and Sheng [45].

5 Summary and Conclusion

The use of different types of now historic high-speed cameras with increasing framing rates is reviewed in their application to resolve the violent motion of bubbles in liquids. Collapsing bubbles belong to the fastest mechanical objects. It was not before 100 million frames/s became available that the sequence of events during bubble collapse could approximately be resolved. High-speed holography was developed up to several hundred thousand holograms per second for observing the dynamics of spatially separated objects (bubbles in liquid) at the same instants of time. The digital revolution has brought more comfortable high-speed cameras than the historic ones and will serve for a proliferation of high-speed studies. However, the resolution of film material is still unsurmounted by a factor of ten, at least in high-speed applications.

When the time resolution cannot be increased further experimentally, a numerical extension may be possible by sophisticated interpolation. A prerequisite is that the physical laws underlying the dynamical process are known. This interpolation is presently done routinely with following tennis and soccer balls for more precise determination of impact on or aside a line. It is under way with bubble collapse impacts by using Navier-Stokes-equation interpolation and elsewhere with other physical equations. Movies from bubble collapse with the equivalent of 10,000 million frames/s are already done. Then we can see even more what we cannot see directly.

References

1. H. Kuttruff, Über den Zusammenhang zwischen der Sonolumineszenz und der Schwingungskavitation in Flüssigkeiten (On the connection between sonoluminescence and acoustic cavitation). Acustica **12**, 230–254 (1962)
2. D.F. Gaitan, L.A. Crum, C.C. Church, R.A. Roy, Sonoluminescnece and bubble dynamics for a single, stable cavitation bubble. J. Acoust. Soc. Am. **91**, 3166–3183 (1992)
3. W. Lauterborn, T. Kurz, *Coherent Optics*, 2nd edn. (Springer, Berlin, 2003), pp. 234–236
4. W. Lauterborn, Kavitation durch Laserlicht (Laser-induced cavitation). Acustica **31**, 51–78 (1974)
5. C.W. Visser, P.E. Frommhold, S. Wildeman, R. Mettin, D. Lohse, C. Sun, Dynamics of high-speed micro-drop impact: Numerical simulations and experiments at frame-to-frame times below 100 ns. Soft Matter **11**, 1708–1722 (2015)
6. J. Noack, A. Vogel, Single-shot spatially resolved characterization of laser-induced shock waves in water. Appl. Opt. **37**, 4092–4099 (1998)
7. W. Güth, Kinematographische Aufnahmen von Wasserdampfblasen (Cinematographic recording of water vapour bubbles). Acustica **4**, 445–455 (1954)
8. W. Lauterborn, H. Bolle, Experimental investigations of cavitation-bubble collapse in the neighbourhood of a solid boundary. J. Fluid Mech. **72**, 391–399 (1975)
9. Y. Tomita, A. Shima, Mechanisms of impulsive pressure generation and damage pit formation by bubble collapse. J. Fluid Mech. **169**, 535–564 (1986)
10. A. Philipp, W. Lauterborn, Cavitation erosion by single laser-produced bubbles. J. Fluid Mech. **361**, 75–116 (1998)
11. W. Lauterborn, W. Hentschel, Cavitation bubble dynamics studied by high-speed photography and holography: Part one. Ultrasonics **23**, 260–268 (1985)
12. W. Lauterborn, R. Timm, Bubble collapse studies at a million frames per second, in *Cavitation and Inhomogeneities in Underwater Acoustics*, ed. by W. Lauterborn (Springer, Berlin, 1980), pp. 42–46
13. A. Vogel, W. Lauterborn, R. Timm, Optical and acoustic investigations of the dynamics of laser-produced cavitation bubbles near a solid boundary. J. Fluid Mech. **206**, 299–338 (1989)
14. C.-D. Ohl, A. Philipp, W. Lauterborn, Cavitation bubble collapse studied at 20 million frames/s. Ann. Physik. **4**, 26–34 (1995)
15. O. Lindau, W. Lauterborn, Cinematographic observation of the collapse and rebound of a laser-produced cavitation bubble near a wall. J. Fluid Mech. **479**, 327–348 (2003)
16. R.P. Tong, W.P. Schiffers, S.J. Shaw, J.R. Blake, D.C. Emmony, The role of 'splashing' in the collapse of a laser-generated cavity near a rigid boundary. J. Fluid Mech. **380**, 339–361 (1999)
17. M.S. Plesset, R.B. Chapman, Collapse of an initially spherical vapour cavity in the neighbourhood of a solid boundary. J. Fluid Mech. **47**, 283–290 (1971)

18. B. Han, K. Köhler, K. Jungnickel, R. Mettin, W. Lauterborn, A. Vogel, Dynamics of laser-induced bubble pairs. J. Fluid Mech. **771**, 706–742 (2015)
19. M. Koch, C. Lechner, F. Reuter, K. Köhler, R. Mettin, W. Lauterborn, Numerical modeling of laser generated cavitation bubbles with the finite volume and volume of fluid method, using OpenFOAM. Comp. Fluids **126**, 71–90 (2016)
20. J. Appel, P. Koch, R. Mettin, D. Krefting, W. Lauterborn, Stereoscopic high-speed recording of bubble filaments. Ultrason. Sonochem. **11**, 39–42 (2004)
21. W. Lauterborn, K. Hinsch, F. Bader, Holography of bubbles in water as a method to study cavitation bubble dynamics. Acustica **26**, 170–171 (1972)
22. K.J. Ebeling, Hochfrequenzholografie mit dem Rubinlaser (High-speed holography with the ruby laser), Optik **48**, 383–397, 481–490 (1977)
23. W. Lauterborn, K.J. Ebeling, High-speed holography of laser-induced breakdown in liquids. Appl. Phys. Lett. **31**, 663–664 (1977)
24. W. Lauterborn, A. Judt, E. Schmitz, High-speed off-axis holographic cinematography with a copper-vapor-pumped dye laser. Opt. Lett. **18**, 4–6 (1993)
25. W. Hentschel, Hochfrequenzholografie mit dem Argon-Ionen Laser (High-speed holocinematography with the argon ion laser). Optik **68**, 283–310 (1984)
26. M. Novaro, Camera holographique ultra-rapide (Ultra high-speed holographic camera), C. R. Acad. Sci. Paris, B. **273**, 941–943 (1971)
27. W. Hentschel, W. Lauterborn, New speed record in long series holographic cinematography. Appl. Opt. **23**, 3263–3265 (1984)
28. W. Hentschel, W. Lauterborn, High speed holographic movie camera. Opt. Eng. **24**, 687–691 (1985)
29. W. Lauterborn, E. Cramer, Subharmonic route to chaos observed in acoustics. Phys. Rev. Lett. **47**, 1445–1448 (1981)
30. W. Lauterborn, A. Koch, Holographic observation of period-doubled and chaotic bubble oscillations in acoustic cavitation. Phys. Rev. A. **35**, 1974–1976 (1987)
31. W. Lauterborn, J. Holzfuss, Acoustic chaos. Int. J. Bif. Chaos **1**, 13–26 (1991)
32. W. Lauterborn, W. Hentschel, Cavitation bubble dynamics studied by high-speed photography and holography: Part two. Ultrasonics **24**, 59–65 (1986)
33. W. Lauterborn and W. Hentschel, Holografische Hochgeschwindigkeitskinematografie (Holographic high-speed cinematography), in *Praxis der Holografie (Practical holography)*, ed. by H. Marwitz et al. (Expert Verlag, Ehningen, 1989), pp. 354–370, 474–475
34. W. Lauterborn, T. Kurz, R. Mettin, C.D. Ohl, Experimental and theoretical bubble dynamics. Adv. Chem. Phys. **110**, 295–380 (1999)
35. W. Lauterborn and T. Kurz, Physics of bubble oscillations, Rep. Prog. Phys. 106501 (88 pp) (2010)
36. W. Lauterborn, Optische Kavitation (Optic cavitation). Physikalische Blätter **32**, 553–563 (1976)
37. W. Lauterborn, Cavitation and coherent optics, in *Cavitation and Inhomogeneities in Underwater Acoustics*, ed. by W. Lauterborn (Springer, Berlin, 1980), pp. 3–12
38. W. Hentschel, W. Lauterborn, Holographic generation of multi-bubble systems, in *Cavitation and Inhomogeneities in Underwater Acoustics*, ed. by W. Lauterborn (Springer, Berlin, 1980), pp. 47–53
39. K.Y. Lim, P.A. Quinto-Su, E. Klaseboer, B.C. Khoo, V. Venugopalan, C.-D. Ohl, Nonspherical laser-induced cavitation bubbles, Phys. Rev. E. **81**, 016308 (9 pp) (2010)
40. U. Schnars, W. Jüptner, Direct recording of holograms by a CCD target and numerical reconstruction. Appl. Opt. **33**, 179–181 (1994)
41. J. Sheng, E. Malkiel, J. Katz, Digital holographic microscope for measuring three-dimensional particle distributions and motions. Appl. Opt. **45**, 3893–3901 (2006)
42. G. Haussmann, W. Lauterborn, Determination of size and position of fast moving gas bubbles in liquids by digital 3-D image processing of hologram reconstructions. Appl. Opt. **19**, 3529–3535 (1980)

43. E. Malkiel, J.A. Abras, J. Katz, Automated scanning and measurement of particle distributions within a holographic reconstructed volume. Meas. Sci. Technol. **15**, 601–612 (2004)
44. U. Schnars, W. Jüptner, Digital recording and numerical reconstruction of holograms. Meas. Sci. Technol. **13**, R85–R101 (2002)
45. J. Katz, J. Sheng, Applications of holography in fluid mechanics and particle dynamics. Annu. Rev. Fluid Mech. **42**, 531–555 (2010)

Brandaris Ultra High-Speed Imaging Facility

Guillaume Lajoinie, Nico de Jong and Michel Versluis

Abstract High-speed imaging is in popular demand for a broad range of scientific applications, including fluid physics, and bubble and droplet dynamics. It allows for a detailed visualization of the event under study by acquiring a series of images captured at high temporal and spatial resolution. The challenge here is the combination of microscopic length scales and ultrashort time scales associated with the mechanisms governing fluid flows. In this chapter, ultra high-speed imaging at frame rates exceeding 10 million frames per second (fps) is briefly reviewed, including the emerging ultrafast sensor technologies and ultrashort nanoseconds flash illumination techniques. We discuss in detail the design and applications of the Brandaris 128 ultra high-speed imaging facility. The high-speed camera combines the optical frame of a rotating mirror camera with 128 CCD sensors and can record at a maximum frame rate of 25 Mfps. Six acquisitions can be stored in the on-board memory buffer, while in a segmented mode images are acquired in subsets, e.g. 24 × 32 frames, allowing parametric studies to be performed. We also discuss how the Brandaris camera is operated to capture details of bubble dynamics, droplet vaporization, and inkjet printing.

Electronic supplementary material The online version of this chapter (doi:10.1007/978-3-319-61491-5_3) contains supplementary material, which is available to authorized users.

G. Lajoinie (✉) · M. Versluis
Physics of Fluids Group, University of Twente, 217 7500 AE, Enschede, The Netherlands
e-mail: g.p.r.lajoinie@utwente.nl

M. Versluis
e-mail: m.versluis@utwente.nl

N. de Jong
Department of Biomedical Engineering, Erasmus MC, Rotterdam, The Netherlands
e-mail: n.dejong@erasmusmc.nl

K. Tsuji (ed.), *The Micro-World Observed by Ultra High-Speed Cameras*,
DOI 10.1007/978-3-319-61491-5_3

1 Introduction

In the last two decades high-speed imaging has emerged as an important tool for a variety of applications. These include car crash testing, air bag deployment, machine vision technology for packing and sorting, high-speed impact and material testing, sport science, ballistics and (nuclear) detonation and explosions. In fluid dynamics [1], high-speed imaging has been used in propulsion and cavitation, combustion, turbines and supersonic flows, sprays and jets, and shock waves. Emerging applications in microfluidics, biomedicine, and biomechanics require top-of-the-line high-speed imaging systems, i.e. high frame rates at high spatial resolution at a high number of frames to resolve the microscopic details at microseconds time resolution.

Digital high-speed imaging devices include specially devised charge-coupled device (CCD) and complementary metal-oxide semiconductor (CMOS) sensors. The CCD sensor is popular for its low noise characteristics and has a corresponding sensitivity coupled to the noise floor within the individual pixels. Pixels can be as small as 2.5 μm, making up a very compact chip. While CCD chips have a better image quality because of their reduced noise component, chips fabricated using CMOS technology are by far less expensive and faster and require less power to operate. Each CMOS pixel can be read out individually, and no shifting of the rows is necessary (as with CCDs), which dramatically reduces the readout time of a subframe of the chip. As also the data size extracted from the region of interest is reduced, CMOS chips allow for a much faster readout, up to 1 Mfps, before the upper limit of the throughput is reached. CMOS chips are, literally *a bit* more noisy and the pixels in CMOS high-speed imaging systems are often larger, typically of a size of 20 μm, to increase their sensitivity. For a typical microscopic applications this can be problematic as a higher magnification will be needed to obtain a similar field of view.

The interesting questions that need to be answered before we perform the actual high-speed imaging experiment are as follows. First, what is the required frame rate that we need to resolve our event. Second, what exposure time are required to reduce motion blur for example, and finally, how do frame rate and exposure time relate to the desired magnification, in other words how are the spatial sampling resolution and the temporal sampling resolution connected. For the first question, the key feature is the relevant timescale of the event. In a cyclic event with a known (driving) frequency or repetition rate, following the Nyquist sampling theorem, the required sampling rate should be more than twice, or more conveniently ten times, the cycle frequency. However, most applications are non-cyclic and the relevant timescale may not be obvious. One may suggest to use the typical velocity to estimate the relevant timescale, but the timescale also scales with the inverse length scale. Thus, the optimum frame rate f can be determined [1] from the ratio of the typical velocity V and the typical length scale ℓ, i.e. the relevant frequency or rate:

$$f = \frac{N \cdot V}{\ell} \tag{1}$$

with N sufficiently high, but at least larger than 2, not to undersample the event under study. It is evident from Eq. 1 why high-speed imaging in microfluidic applications is important, as at microscopic level ℓ is very small and as a consequence the frame rate f needs to be large.

One example depicted in Fig. 1(left) shows the head of an ink-jet droplet shortly after ejection from a drop-on-demand micronozzle. The typical volume contained within the single microdroplets is 15 pL, which corresponds to a typical size of approximately 10 μm. The speed with which the droplets are ejected is 10 m/s. While such a velocity may seem rather non-extreme, Eq. 1 dictates that a frame rate of 10 Mfps is required to image the flight of the microdroplet because of the microscopic length scale of the problem. Another way to look at this problem is to note that the droplet flies with a speed of 10 m/s, which corresponds to 10 μm/μs. Thus, the typical timescale of a 10-μm droplet flying at 10 m/s is 1 μs, and correspondingly it needs to be captured at a fraction of a microsecond.

A second example is that of an oscillating microbubble, see Fig. 1(right). These stabilized microbubbles are used clinically as ultrasound contrast agents for ultrasound imaging for their superior echogenicity due to gas compressibility, which results in a nine orders higher acoustic scattering cross section as compared to particles of the same size. The bubbles resonate when driven near their eigenfrequency [2, 3], which leads, first, to the generation of an even stronger echo, as the bubble oscillates with higher amplitude, and, second, to the generation of harmonics, emitted at multiples of (second, third harmonic) [4, 5] and half (subharmonic) the driving frequency [6, 7]. These nonlinear features of bubble dynamics provide high contrast and allow to discriminate bubble echoes from linear tissue echoes. Blood, being a very poor ultrasound scatterer, can be seeded with an

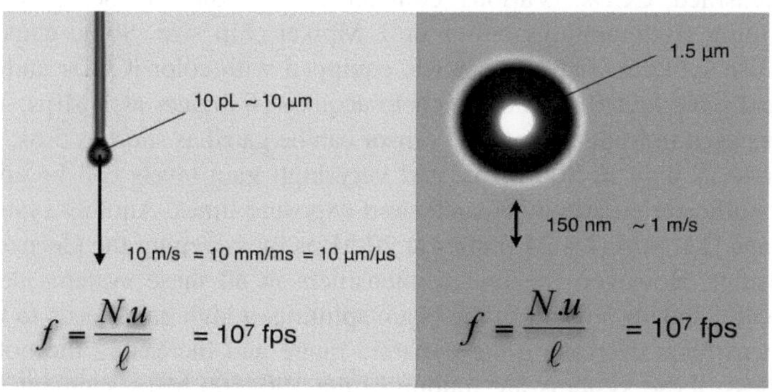

Fig. 1 Cartoon used to estimate the required frame rate for high speed imaging based on typical length scales and typical velocities. *Left* example for inkjet printing. *Right* example for an oscillating microbubble. Both examples are at microscopic scale and a simple calculation shows that ultra high-speed imaging is required at 10 Mfps

intravenous injection of ultrasound contrast microbubbles. The bubbles then act as a blood pool agent and the perfusion of organs can be visualized in real-time, as well as the hypervascularization of malignant lesions. A typical radius R of these microbubbles is 3 μm, and at frequencies near their eigenfrequency ω_0 of 1–10 MHz their oscillation amplitude is around 10% of the ambient radius. This 300-nm amplitude can be temporally resolved, based on simple arguments on a multiple of the frequency, resulting in 10 Mfps. Based on sinusoidal driving frequency $P_A \cdot \sin(\omega_0 t)$, a response of the bubble $R = R_0 + \Delta R \cdot \sin(\omega_0 t)$ a velocity of the oscillation of $\dot{R} = \omega_0 \Delta R \cdot \cos(\omega_0 t)$ with magnitude $\omega_0 \Delta R = 1$ m/s. Equation 1 then also gives 10 Mfps as the optimum frame rate.

2 Ultra High-Speed Imaging

2.1 Ultra High-Speed Imaging Camera Technology

To adequately image these event requires pushing the limits of current CCD and CMOS camera technology and perform high-speed imaging at frame rates higher than 1 Mfps with a sufficient number of frames (typically one hundred or more). We therefore need to resort to unconventional designs. The simplest solution is to follow the classic examples of using a set of multiple cameras. This principle in fact predates cinematography and was used in 1878 by Muybridge [8] in the famous first framing recording of a galloping horse, employing 12 separate cameras. Later, Brixner [9] used a similar technique with 37 cameras to record the first nuclear explosion at a 1 ms interframe time. An 8-channel image-intensified ultra high-speed framing camera for high-speed shock wave imaging was developed [10], which was later commercialized (Imacon 468). It utilizes a pyramidal beam splitter with an octagonal base that redirects the incoming image to eight individual image-intensified CCDs. Various companies now offer cameras with similar beam-splitting configurations and 1 or 2 Mpixel chip size. Some camera manufacturers use splitters for four channels equipped with color iCCDs and then split electronically the four RGGB channels to acquire 16 images at 1 Mfps. The image intensifiers used in front of the CCD sensor can be gated as short as 5 ns, leading to a frame rate as high as 200 Mfps, and very high gain levels can be applied that allow for sufficient sensitivity at such short exposure times. Another system, called the Ultranac [11] provides 24 frames at 20 Mfps by sweeping the electron beam in the intensifier. However, the image intensifiers in all these systems degrade the image quality and, as a result of the beam splitting, a high gain needs to be applied to the intensifiers, decreasing the dynamic range and increasing the noise levels. Nevertheless, these systems are unique in the 100–200 Mfps segment. A second class of imaging systems utilizes in situ storage (ISIS) of image data on the CCD chip itself. This type of sensor can record 103 frames in a single experiment at a maximum frame rate of 1 Mfps with a fixed resolution of 312 × 260 pixels. Recent developments in CMOS technology include backside-illuminated (BSI) image sensors with increased quantum efficiency, a fill factor of nearly 100% and frame rates reaching 10–20 Mfps. For details, see Chap. 2.

In the presented cases one would like to image directly at the timescale of the event. When the timescales become as small as nanoseconds the current camera technologies present an insurmountable problem, with the exception of reproducible and repetitive events. In this special case, stroboscopic imaging can be used even without the use of (expensive) ultra-high-speed camera machinery. In stroboscopic imaging the high-speed flash photography technique is applied, and all consecutive recordings are delayed by a fixed time spacing that is tuned such that it spans the interframe time of the camera. Then, playing the captured frames in a video sequence provides us with a stroboscopic movie of the event, which in many cases cannot be distinguished from a real-time high-speed movie, and moreover with a dramatically improved spatial image resolution.

The application of flash photography in experiments is primarily aimed at obtaining precise information about the position and dimensions of the studied object at a certain instant in time. It is therefore of prime importance to capture the smallest details in both space and time with the highest contrast. To define this more quantitatively, three criteria can be defined [12]. First, based on the spatial resolution and pixel size, the spatial Nyquist criterion defines the magnification, avoiding oversampling or a too small field-of-view. Secondly, to ensure maximum image contrast, the intensity of the illumination should be adjusted to cover the full dynamic range of the camera sensor. The extent to which the intensity can be varied is, however, limited due to the reciprocal relationship between illumination intensity and exposure time. Finally, minimizing motion blur is accomplished by minimizing the temporal resolution of the imaging system. The temporal resolution is determined by the duration of the illumination pulse or by the camera exposure time; whichever is shorter defines the degree of motion blur.

Here we shorten the temporal resolution with a laser pulse to obtain nanoseconds illumination precision. Exciting a laser dye solution with a pulsed laser resulted in a short flash. The resulting high-intensity fluorescence removes the coherence of the laser light, which would lead to interference and speckle. The illumination time of the light source is only 7 ns. To record the images, a sensitive, but otherwise, standard CCD camera is used. The short illumination time of the fluorescence light source then gives the possibility to choose a time step resolution corresponding to the smallest timescale of the experiment.

As an example of the iLIF technique, Fig. 2 shows the formation of inkjet droplets captured in flight. As before, these picoliter droplets are ejected from the inkjet printhead at a speed of 10 m/s. Imaging these droplets at microscopic resolution demands a spatiotemporal resolution that cannot be provided by real-time ultra high-speed imaging, however owing to the near-perfect reproducibility of the inkjet system a the stroboscopic principle can be applied to capture these micro-droplets with superior clarity. Using this method a perfect delineation of the droplet contour can be obtained, from which the one-dimensional velocity profile inside a single droplet is extracted during drop formation. To that purpose a novel

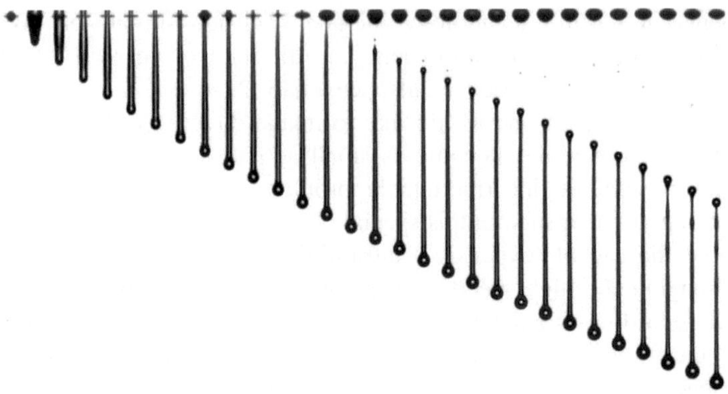

Fig. 2 Stroboscopic imaging of drop-on-demand inkjet droplets using illumination by laser-induced fluorescence (iLIF) [12]. Each image corresponds to a different droplet imaged in a subsequent cycle with a small running delay. The excellent reproducibility of inkjet printing allows to capture the high-speed droplet dynamics in flight at high resolution

experimental approach was used to capture two detailed images of the very same droplet with a small time delay. The obtained velocity profiles were compared to a numerical simulation based on the slender jet approximation of the Navier-Stokes equation and a very good agreement with the model was found [13]. Despite the high optical quality of these images, there are transients or sudden events to which stroboscopic imaging cannot be applied, e.g. to study the details of the breakup of a secondary tail, where we need ultra high-speed imaging in an operating regime exceeding 10 Mfps.

2.2 Rotating Mirror Cameras

Rotating mirror cameras do not suffer from the drawbacks of beam splitting or image intensifying and offer a higher resolution, dynamic range, and frame count. A short exposure time is obtained by sweeping the image during the rapid rotation of a turbine-driven mirror across a photographic film mounted on an arc [14]. The original concepts were devised in the late 1930s and the first camera dates back to the first hydrogen bomb explosion in 1952, see Fig. 3. The high rotation speed of the mirror allows up to 130 frames to be recorded at a maximum frame rate of 25 Mfps. Photographic films offer excellent image quality, however, they have some disadvantages. They have limited light sensitivity, reduced flexibility due to film handling and long turnover time due to film development and processing time. In addition, highly specialized shutters, which are both expensive and inconvenient to use, are required at high speeds to prevent multiple exposure of the film.

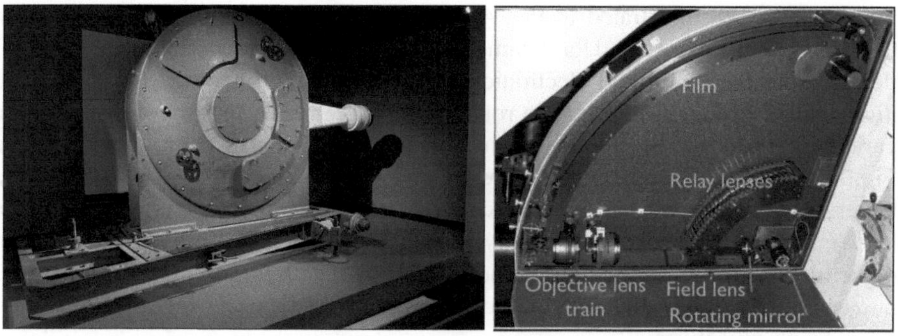

Fig. 3 Rotating mirror cameras have been around since the 1950s. *Left* The C4 rotating mirror high speed camera was developed at the end of World War II to study explosive reactions at the British Atomic Weapons Research Establishment (AWRE)—now exhibited at London's Science Museum. (Reprint permitted by the Science Museum.) *Right* Inside of rotating mirror framing camera with rotating mirror turbine in the lower right corner and the film track spanning the inside of the quarter arc

3 Brandaris 128

In a novel concept of the rotating mirror camera introduced in 2003 by Chin et al. [15], a series of CCD sensors replace the traditional negative film thereby greatly improving the sensitivity. Moreover, CCD sensors are highly flexible, can be accurately timed, and allow for multiple and repetitive exposures in short succession. The Brandaris 128 camera [15, 16] is equipped with 128 CCD sensors, each of which can store six full-frame images in its onboard RAM. The helium-driven turbine can rotate at 20,000 rps, completing a sweep across the image arc in just 5 μs, thereby providing an interframe time of 40 ns. The Brandaris 128 camera can therefore acquire a total of 768 frames at a frame rate of 25 Mfps in a set of 6×128 frames, and in its segmented mode it acquires 12×64 frames, or 24×32 frames, or any permutation of the above, provided that a 20-ms readout time is allowed once all 128 channels are filled.

The Brandaris 128 camera has been dedicated predominantly to the study of medical ultrasound contrast agents. The understanding of the interaction of the bubbles and ultrasound has allowed improvement of their performance for which a detailed insight of the microbubble dynamics proved essential. Despite the numerous novel scientific results obtained with the Brandaris 128 camera, until recently the camera still had a few limitations. A few modifications to the camera design in 2012 improved the imaging capabilities, notably a region of interest mode (ROI), a higher number of collected frames (now 16,000) and ultra high-speed fluorescence capabilities [16]. Another ultra high-speed imaging system, named UPMC Cam, based on identical rotating mirror technology was recently introduced with a special emphasis on in vitro and in vivo fluorescence imaging where photon availability is limited and high sensitivity is required [17].

Brandaris 128, illustrated in Fig. 4, is developed from the Cordin 119 (Cordin Company, Salt Lake City, Utah) camera frame. The system combines the superior flexibility and sensitivity of electronic CCD detectors with the high frame rate and high number of frames available in rotating mirror cameras. In front of the camera a variety of lenses can be mounted to suit the need of the experiment, as an example Fig. 4 shows the Brandaris camera equipped with a high-resolution microscope. Other objectives, such as long working distance microscopes and video camera lenses, have also been used. Generally, a suitable field lens is needed to couple such objective lenses to the Brandaris camera and to reduce light loss due to unmatched numerical apertures, which is a particular problem when using high-magnification microscope objectives.

Once an image is formed at the entrance plane of the camera, or primary image plane (Fig. 5), a set of relay lenses projects the target image onto a three-faced mirror prism. The high-speed turbine (Cordin Company Model 1220) is driven by high-pressure bottled helium. The maximum turbine rotation speed is 20,000 rps, or 1.2 million rpm, resulting in a maximum frame rate of 25 Mfps. The gas flow is controlled by a mass flow controller (MFC), which in turn is controlled by a PC. The frame rate and timing parameters are set by internal software, the MFC actively

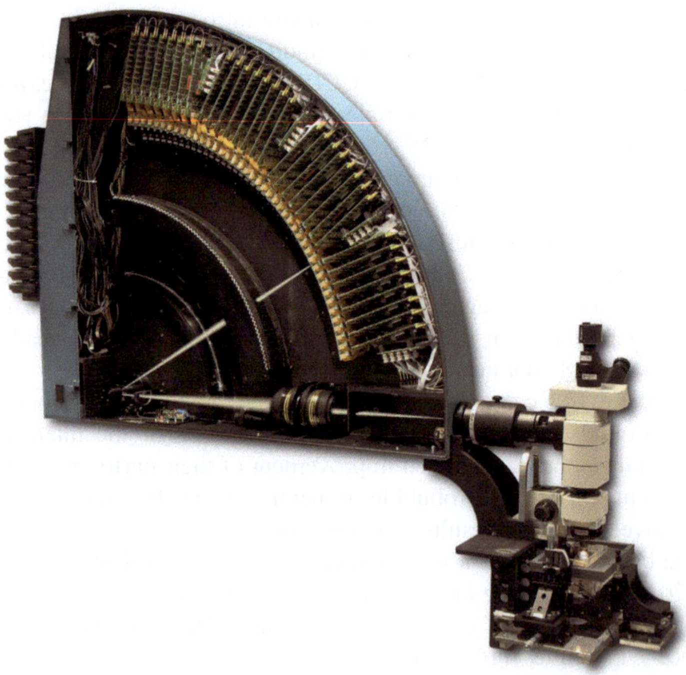

Fig. 4 The Brandaris 128 ultra high-speed camera equipped with a high-resolution microscope. Inside the camera the film track is replaced by 128 high-sensitivity CCDs. The height of the camera is 1.5 m [1], with permission of Springer

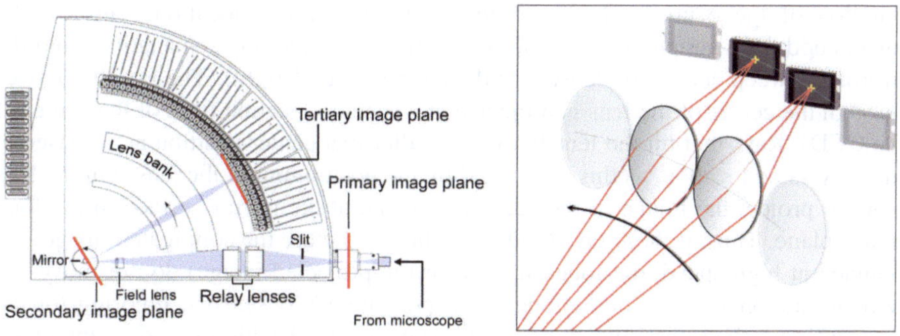

Fig. 5 Optical configuration of the Brandaris 128 camera. *Left* locations of primary, secondary and tertiary image planes. *Right* ray tracing helps to understand why images are stagnant at the sensor plane while the image that is reflected from the rotating mirror is swept across the lens bank

regulates the flow rate between 1 and 100 m^3/h. The active control is needed as the rotation speed and helium flow rate are coupled by a nonlinear relation due to fluid dynamical drag of the mirror assembly in the helium environment, which is the main reason for using helium in the first place; the drag in air would limit the turbine speed to 5000 rps. The response of the turbine also changes over time due to wear and tear on the turbine ball bearings system. The speed of the turbine is monitored by an infrared laser–photodiode pair mounted at 30° below the optical axis. It generates three mirror pulses per prism rotation and these provide accurate timing of the rotation rate. The internal software also utilizes the mirror pulses as a master to control the experiment including the timing of the CCDs and the illumination sources. The field lens in Fig. 5(left) is shifted to either of two preset positions in order to compensate for the change in reflective index while operating in different gases (air or helium).

3.1 Optical Configuration and Miller Principle

Figure 5 shows the optical path upon reflection from the secondary image plane at the rotating mirror. The image passes through an optical lens bank system containing spherical lens pairs (to control focus and magnification) that project the image onto the tertiary and final image plane where the sensors are located. To comply with the small pitch in the arc of the lens bank, the lenses are cut to a narrow slice with a width of 9 mm. The total magnification factor from the primary image plane to the final image arc is 0.44.

The unique optical feature and working principle of the camera is contained in this fine piece of optics and is based on the Miller principle for high-speed cinematography [18, 19]. While the reflected image is swept along the lenses of the lens bank, the image formed in the sensor plane is static and does not move for the duration of the exposure before it moves to next frame location. This can be understood from simple geometrical optics, if we consider rays emanating from the mirror to sweep across the lenses, as depicted in Fig. 5(right). The ray that passes through the center of the lens is projected at a specific location (that of the center of the CCD). Rays that hit the lens before and after that center position are focused at the very same location, as this is dictated by the positioning of the lenses in the lens bank to project the image in the secondary plane at the mirror surface to the final image plane. Thus, if these rays (and in the bigger picture the conjugated image) are scanned at high-speed, the focused rays (or the projected image) do not move but remains stationary, and long enough to expose the CCD sensors. Ignoring for the moment any change of the dynamic picture due to the high-speed event under study, the image experiences a small residual change as the mirror rotates. This leads to a small deterioration of the image, termed residual mirror drag [14], which is of the order of a few micrometer. The typical drag is calculated to be smaller than the pixel size of the CCD detectors and as a matter of fact smaller than the overall optical resolution of the system using a fixed mirror.

Figure 6(left) illustrates an alternative view of the Miller principle based on the conjugated image in the Fourier plane of the optical system. When the conjugated image is scanned across the aperture of the focusing lens pairs in the lens bank, the sweeping motion is filtered out by the inverse Fourier transform, and the image slowly builds up in amplitude (light intensity) and in resolution, as more complex phase information is gathered within the aperture of the channel. This system then also dictates the triangular illumination profile at the final image plane, see Fig. 6 (right). Here two fast photodiodes mounted in the sensor plane measure the time evolution of the image brightness. It is seen that the triangular profiles [20] overlap between frames, however when the 10–90% integrated illumination intensity is evaluated [15], it was found that the exposure time ($t_{90} - t_{10}$) perfectly fits the

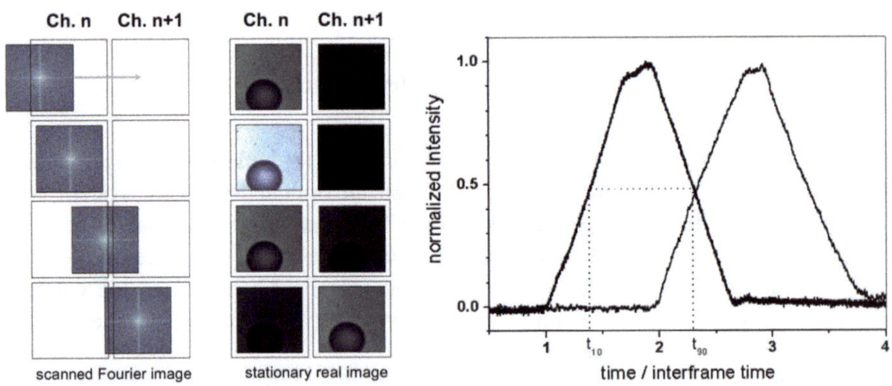

Fig. 6 The Miller principle, which is the key optical feature of rotating mirror cameras, can be understood from scanning the conjugated Fourier transformed image across the apertures at the lens bank position. By inverse Fourier transform at the sensor plane the build up of the real image and it intensity profile is recovered, reprinted from [15], with the permission of AIP Publishing

interframe time (1/frame rate), which in turn is directly defined by the channel pitch. Figure 6(right) also shows that each individual frame is indeed recorded discretely in time, as the intensity level at which t_{10} and t_{90} occur coincides with the crossing of the intensity profiles of the two channels.

3.2 CCD Camera Technology

The images are recorded on 128 un-intensified high-sensitivity CCDs (ICX055AL, Sony Corp., Tokyo, Japan) mounted on an image arc in the final image plane of the camera. Thirty-two CCD Controller Cards, named C^3, built around a microcontroller chip control the CCDs. Each C^3 performs the following tasks: flush and charge transfer of 4 CCDs, readout of the CCDs; storage of the data into a RAM buffer, and command and data input/output, in particular transfer the images to a PC for storage through eight USB 2.0 hubs. Three hardware signal inputs, start, flush, and transfer, provide synchronization to all the C^3 with 40 ns precision. The C^3 s host a memory buffer to accommodate six full recordings. The CCD camera modules measure 27 mm × 40 mm and since the channel pitch at the final image plane is only 9 mm, a surface mirror block assembly was designed to accommodate 3 CCD sensors, one mounted on the optical axis image arc (no mirror required), one mounted on a top plane perpendicular to the original image plane and one mounted on a bottom plane also perpendicular to the original image plane.

The CCD sensor contains 500 × 584 photosensitive cells, each measuring 9 μm × 6 μm. The CCD are read out by a standard flush and charge transfer capture protocol. There is one transport channel for two photosensitive cells, resulting in 500 × 292 effective pixels. The readout of the transport channel is performed by analogue-to-digital conversion. Beginning and end of exposure are therefore in principle defined by flush and charge transfer, respectively. Flush and charge transfer are performed on all pixels in parallel and requires a few microseconds. Readout is performed sequentially on a pixel by pixel basis and requires 20 ms to complete.

Mirror pulses are extracted from the rotating turbine assembly and since the turbine operates at a stable rotation rate for a set value of the mass flow, pretriggers can be derived. From the mirror pulse, flush and transfer pulses are derived for the C^3 s. Moreover, the mirror pulse is used to calculate pre and post-triggers for illumination, wave form generators, switches and multiplexers.

Figure 7(top) shows the standard timing diagram for a Brandaris experiment. Here a single event is captured in a full sequence of 128 frames. Synchronized exposure of the CCDs is ensured by letting the C^3 s perform a flush and charge transfer (xfer) when triggered by the master trigger, typically coupled to the mirror trigger. All CCDs are flushed before channel 1 is illuminated, and a charge transfer is performed right after channel 128 is illuminated. Lead and trail times are ten and a few microseconds, respectively. Charge transfer prevents multiple exposures of the CCD. A readout is performed after charge transfer.

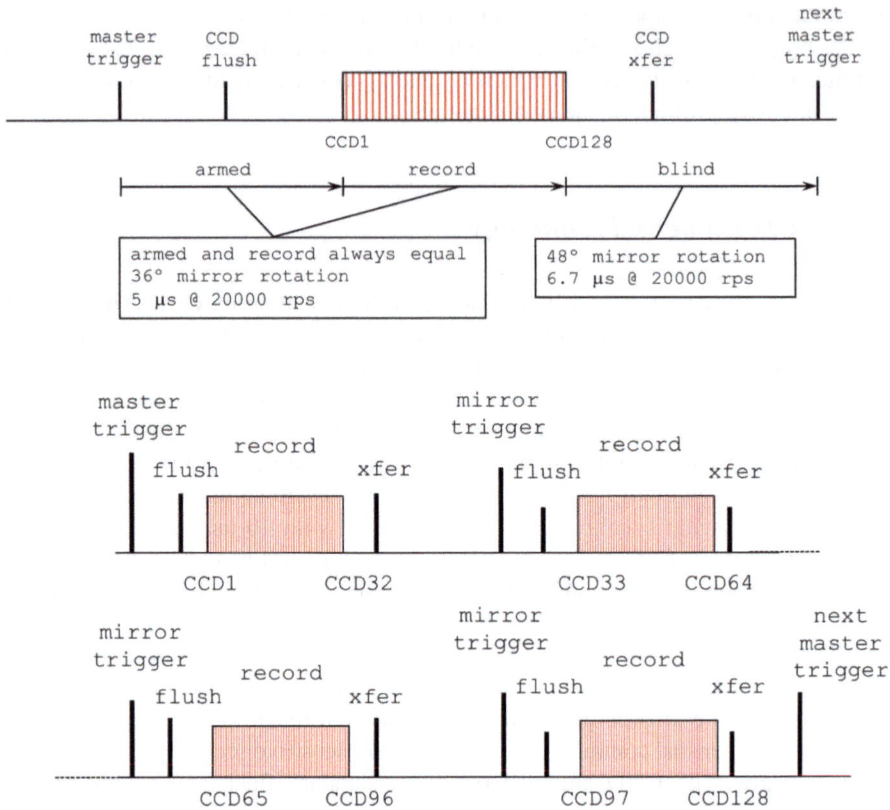

Fig. 7 Brandaris 128 timing diagram. *Top* standard experiment with a single acquisition of 128 frames. *Bottom* segmented mode acquiring 4 segments of 32 frames

The readout time of 20 ms severely limits the speed at which experiments can be repeated. Higher repetition rates are achieved through the use of a segmented mode, see Fig. 7(bottom). In this example, the experiment is designed to capture 4 series of 32 consecutive frames in four mirror periods. During the first mirror period, the C^3 s flush all the CCDs before illumination and charge transfer channels 1–32 after illumination. During a subsequent mirror period, e.g. at the next mirror face, the C^3 s again flush all the CCDs before exposure and then charge transfer channels 33–64 after illumination. As before, the images stored in the CCD transport channels of channels 1–32 are not affected by subsequent illuminations, nor by repetitive flushes after charge transfer.

After acquisition, image frames are realigned using affine transformations to correct for the misalignment between the detector channels, which arise from imperfect positioning of the relay lenses and CCDs (order 50 μm). The transformation matrix is calculated by analyzing the recording of a calibration grid target, to correct for $x - y$ translation, rotation, magnification and perspective.

3.3 Fluorescence Imaging

The Brandaris 128 camera system, as well as the UPMC Cam, host the capability to perform fluorescence microscopy on nanoseconds timescale. To assess the feasibility for high-speed fluorescence imaging a thorough analysis was made, taking into account all parts that contribute to obtain an acceptable signal-to-noise ratio at 25 Mfps: the image sensor, the imaging optics, the fluorescent dye and the excitation light source. The original CCD sensors were found to outperform any new technologies, including back-illuminated CCD's with a factor 2 higher quantum efficiency, and EMCCDs that use electron multiplication which may introduce higher signal levels at the expense of higher noise levels. Optimization of the optical configuration, e.g. by improving the limited numerical aperture within the camera frame, or redesign of the channel pitch left little room for improvement, which means that an increase in sensitivity must come from the fluorescence emission alone. Brighter fluorescence emission means a stronger excitation, and the use of more efficient fluorophores, with increased saturation levels where the photon flux becomes independent of the excitation intensity. It was shown [16, 17] that the fluorescence of microparticles and oil-filled polymeric microcapsules can be imaged successfully at frame rates exceeding 20 Mfps, while microbubbles with a less efficient loading of fluorescent dye molecules could be imaged up to 5 Mfps in fluorescence.

Excitation of the fluorophores is best accomplished with pulsed lasers, however, the repetition rate of pulsed lasers is limited (typically 10 Hz, or 40 kHz for a Cu vapor laser), or the laser fluence drops down dramatically at high repetition rate (200 kHz for Nd:YLF lasers). Thus, for ultra high-speed imaging running at 10 Mfps a separate laser cavity would be required for each individual frame to be able to perform fluorescence imaging frame rates exceeding 1 Mfps, see e.g. the 8-frame approach of Ding et al. [21], with major cost per additional channel. CW (continuous-wave) lasers, e.g. diode-pumped solid-state lasers, can be utilized to excite the fluorescent molecules for a prolonged period of time, using accurate exposure control with an acousto-optic modulator (AOM). Care should be taken, however, for proper dye selection, as less efficient fluorophores can induce non-negligible local heating through non-radiative absorption of laser light. The small form factor of these laser allows the mounting of the laser underneath the camera, as shown in Fig. 8. Coupling of the laser light into the optical path of the microscope was accomplished via a dichroic mirror. The beam is passed through a spatial filtering pinhole and expanded to control the spot size at the sample level. A notch filter located above the dichroic mirror filters out any reflected excitation light. The fluorescence emission is then passed on to the Brandaris camera. The dichroic mirror has a 50% transmission of the Xe-flash light for bright-field illumination, which means that bright-field imaging and fluorescence imaging can be performed in a single experiment, or run, of the camera system. Digital delay control give full control over triggers to laser, AOM, and flash illumination, which gives access to alternating bright field and fluorescence imaging or combined

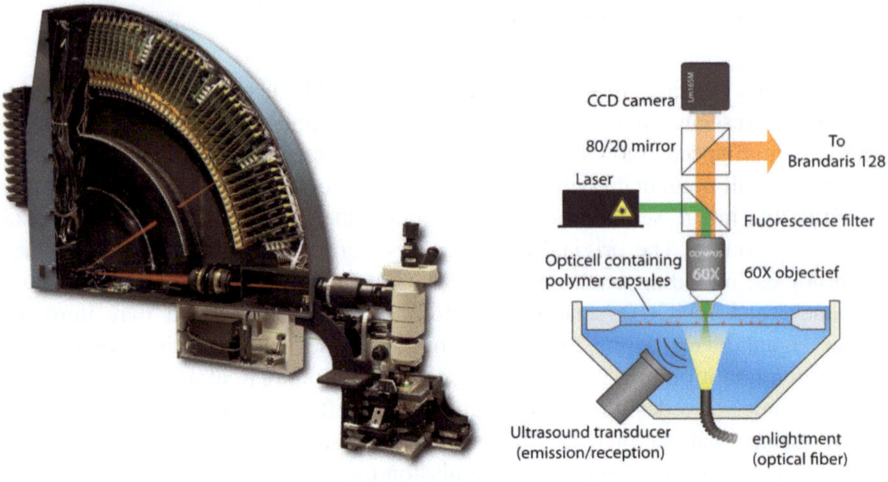

Fig. 8 *Left* Brandaris 128 camera in fluorescence mode, reprinted from[16], with the permission of AIP Publishing. The laser used for illumination is mounted underneath the camera and coupled into the microscope using a dichroic mirror (*right*). Fluorescence is transmitted in the epifluorescent setup and transmitted to the Brandaris camera

recordings where the first 64 frames are recorded in bright field and the second 64 frames are recorded in fluorescence.

3.4 Performance

Optical resolution, image quality and sensitivity of the Brandaris system is governed by the intrinsic f-number of the camera, and the objective lenses used before the primary imaging plane. The combined system was tested using a glass resolution test target with photo-etched lines down to 200 nm separation. At the plane of the image sensor the optical resolution of the camera was 36 line pairs per millimeter horizontally and 16 lp/mm vertically on the image sensor. With a typical magnification for a water-immersion high-resolution microscope objective of 60 × (NA of 0.9) and a 2 × magnifier, and 2.2 × demagnification in the Brandaris, objects as small as 400 nm wide can be clearly defined, see Fig. 9. Optical quality is limited at ultra high speed framing rates by the increased noise level set by the gain of CCD sensor or the image intensifier voltage, in the case of image intensified-CCDs, which, from the start, already present a more grainy image, that only deteriorates for higher gain settings. The un-intensified CCDs in Brandaris 128, that can be used due to the gating function provided by the rotating mirror turbine, help in achieving higher grade optical quality images. For Brandaris the optical sensitivity was expressed in ISO numbers. While quantifying the exact

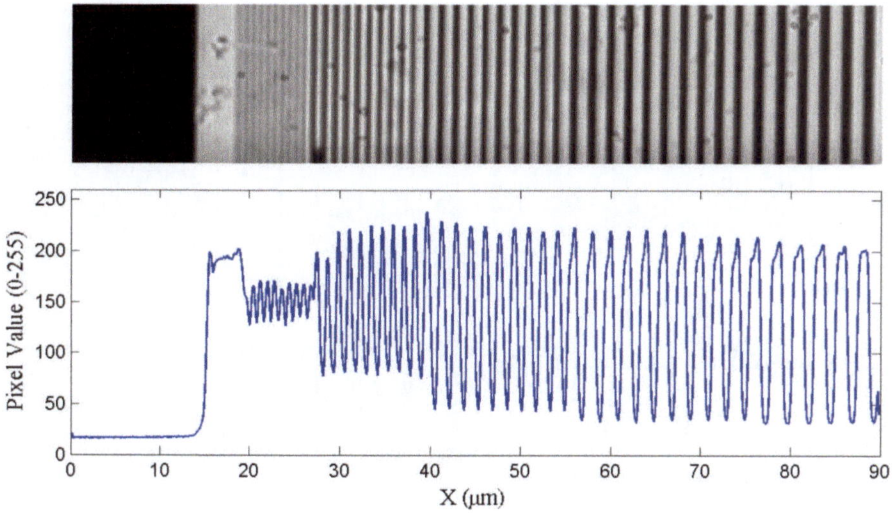

Fig. 9 Resolution measurement of the Brandaris 128 camera using a target grid with a resolution down to 200 nm, reprinted from [15], with the permission of AIP Publishing

ISO number is complex and needs painstaking calibration and a range of spectral illumination boundary conditions, the measured ISO number ranges from 400 to 100,000 depending on the gain setting. Finally, the dynamic range of a camera system is defined as the ratio between the highest light intensity resulting in unsaturated images and the lowest light intensity resulting in detectable images above the noise level. Brandaris 128 gives a near 8-bit dynamic range of 250, but it should be kept in mind that the dynamic range within an image frame is lower than the full dynamic range, since the gain within a single frame is fixed and a typical Brandaris 128 image frame has a dynamic range of 30.

4 Applications

4.1 Microbubble Dynamics

A suspension of ultrasound contrast agents contains a polydisperse size distribution of microbubbles with a size ranging from 1 to 10 μm in diameter. The bubbles resonate when driven near their resonance frequency boosting their radial response. The eigenfrequency of the bubbles follows directly from their size following the Minnaert frequency $f_0R_0 = 3.3$ μMHz [2, 3], neglecting in first approximation the effect of surface tension and shell elasticity. Thus bubbles of a size of a few micrometer resonate at a driving frequency of a few MHz, right in the heart of the

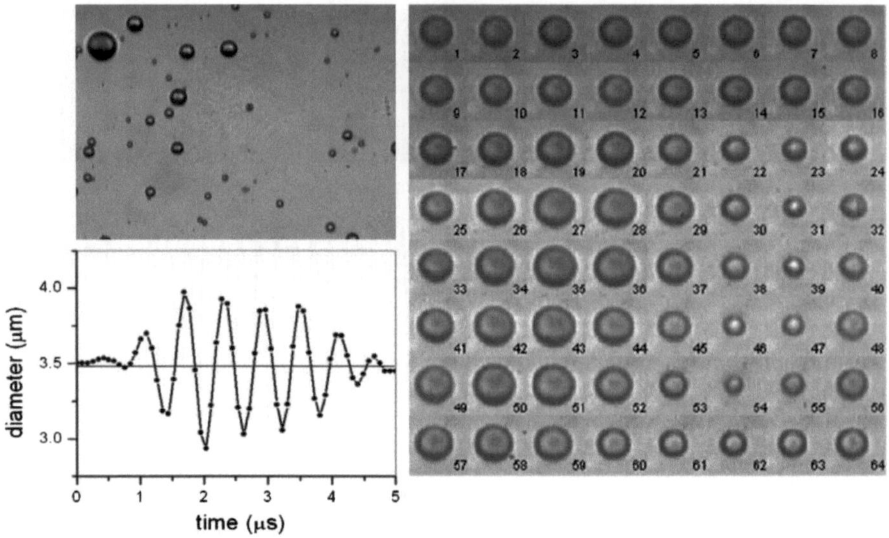

Fig. 10 *Top left* Polydisperse bubbles viewed under the microscope of the Brandaris 128 camera. The corresponding video is available online. *Right* 64 frames of a recoding of an oscillating microbubble, reprinted from N. de Jong et al., Ultrasound Med. Biol. 33 (2007), 653, with the permission of Elsevier. *Bottom left* Radius-time ($R - t$) curve of a single microbubble, reprinted from M. Emmer et al., Ultrasound Med. Biol. 33 (2007), 941, with the permission of Elsevier

medical ultrasound application domain. Figure 10(right) shows 64 frames of the recording of the oscillations of a single microbubble under the microscope taken at 10 Mfps. It nicely shows the transient behavior where the driving kicks in followed by a steady state oscillation of successive expansion and contraction of the microbubble ($\Delta R = 0.30R_0$) during the insonation with a pulse with six cycles of ultrasound. Analysis of the high-speed recording then allows to plot the size of the bubble as a function of time, giving the radius-time ($R - t$) curve, see Fig. 10 (bottom-right). Note the time scale of 5 μs on the horizontal axis.

The microbubbles are coated with a stabilizing shell to prevent gas loss, dissolution and coalescence of the microbubbles. The bubble dynamics is changed by the addition of a viscoelastic coating, typically consisting of phospholipids [22–25]; shell elasticity increases the stiffness of the system and increases its resonance frequency while shell viscosity increases the damping of the system. Figure 11(top) shows a numerical simulation of this resonance behavior. The uncoated 3.8 μm bubble resonates at 0.85 MHz, nicely following its Minnaert frequency, while the resonance curve of the same bubble coated with a viscoelastic coating has a 40% higher resonance frequency due to shell elasticity and a 5 × wider resonance curve due to shell damping. Here the resonance curves are normalized, but it is obvious that due to damping the amplitude of the coated bubble response decreases by the same amount for the same acoustic energy input.

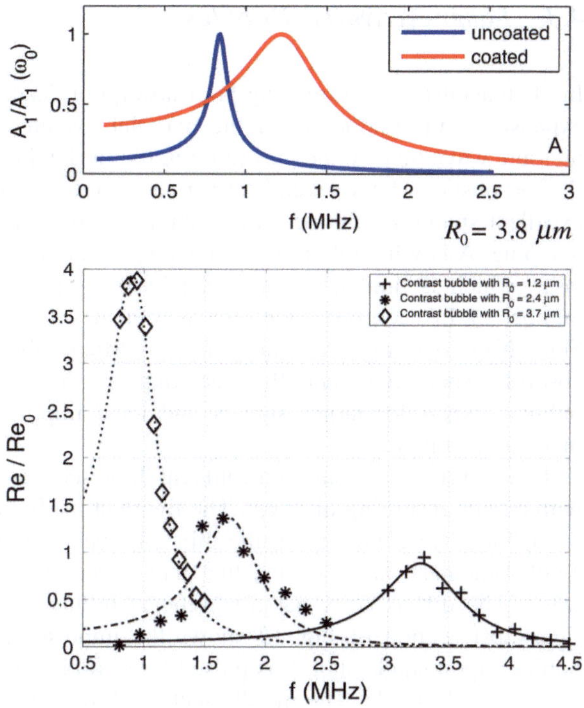

Fig. 11 *Top* Calculated resonance curves of uncoated and coated bubbles of the same size, reprinted from [24], with the permission of AIP Publishing. *Bottom* Measured resonance curves of three microbubbles of different size using the microbubble spectroscopy technique, reprinted from [5], with the permission of AIP Publishing

4.2 Non-linear Bubble Dynamics

Experimental validation of bubble shell models that incorporate a physical description of shell viscoelasticity rely on measurements of the resonance curves of single bubbles [5, 7]. To this end the response of single microbubbles to a driving ultrasound pulse is recorded for a range of ultrasound frequencies using Brandaris in segmented mode. Figure 11(bottom) shows the resonance curves of three stabilized microbubbles with a size of 1.2, 2.4 and 3.7 μm. A fit to the analytical linear resonance curve is plotted through the 11 maximum amplitude data points that were recorded in segmented mode running 12 segments of 64 frames (giving 1 ambient bubble size recoding before ultrasound arrival and 11 well-resolved responses at 11 different frequencies). Shell viscoelastic parameters can then be inferred from a direct comparison with calculated resonance curves for the uncoated bubble system. Overvelde et al. [26] operated the Brandaris 128 camera in standard mode to obtain the response of single microbubbles for 8 different pressures and 12 different frequencies. Using all 128 frames for each exposure requires a total of 16 experiments which takes approximately half an hour to record. To isolate single bubbles and keep them positioned in a fixed location for the duration of the experiment, optical tweezers were developed to trap microbubbles in the dark core of a Laguerre-Gaussian laser beam [27]. For details, see Chap. 9 (Garbin).

4.3 Laser-Activated Bubbles

In photoacoustic (PA) imaging laser absorption leads to local heating followed by expansion and contraction of the material, which results in the generation of a pressure wave that can be picked up in the far field and transformed into an image. Photoacoustics offers several attractive features in biomedical imaging including excellent spatial resolution and functional information such as blood oxygenation mapping. A key limitation of PA imaging is the penetration depth of the excitation laser, which is limited to tissue depths of 1–2 mm. To address this problem, a number of PA contrast agents have been developed, including dyes (methylene blue, ICG) and metallic nanoparticles. Microbubbles coated with an optically absorbing shell are potentially interesting for contrast-enhanced PA imaging. These bubbles may offer larger signals and have improved biocompatibility over the metallic nanoparticles.

Figure 12(top left) shows a confocal fluorescence image of a microbubble coated with a light-absorbing oil layer. The figure at the bottom left shows nicely how the flurophores are contained in the thin oil layer. Different oils were tested, namely dichloromethane and toluene, that had lower viscosities as compared to that of water. The oil viscosity has a significant influence on the bubble dynamics. The right-most section of Fig. 12 shows the impulse response of the laser-activated bubbles upon single shot 7-ns pulsed laser irradiation recorded with Brandaris 128 in bright-field mode. The middle section shows the $R - t$ curve of the two sets of bubbles and clearly displays the expansion and contraction at the eigenfrequency of the system, which can be used to fully characterize the system. More details of these experiments can be found in [28, 29] and details are also found in Chap. 9 (Stride et al.).

4.4 Bubble Shape Instabilities

Large amplitude bubble oscillations, or what is known as cavitation, can be separated in two different regimes. Inertial cavitation is a synonym of violent dynamics and represents the most chaotic form of microbubble behavior, see [30] and details in Chapter "The Bubble Challenge for High-Speed Photography" (Lauterborn/Kurz). Bubbles driven by ultrasound may also oscillate indefinitely as long as ultrasound is on in a milder regime called stable cavitation. The stability of bubbles lies in the hands of the gas diffusion in and out of the bubble due to surface tension at the gas-liquid interface, of the local dissolved gas concentration in the surrounding liquid (which may be a function of the presence of other dissolving bubbles, known as Ostwald ripening), and of bubble oscillations with changing surface area leading to the process of rectified diffusion. In some special cases a

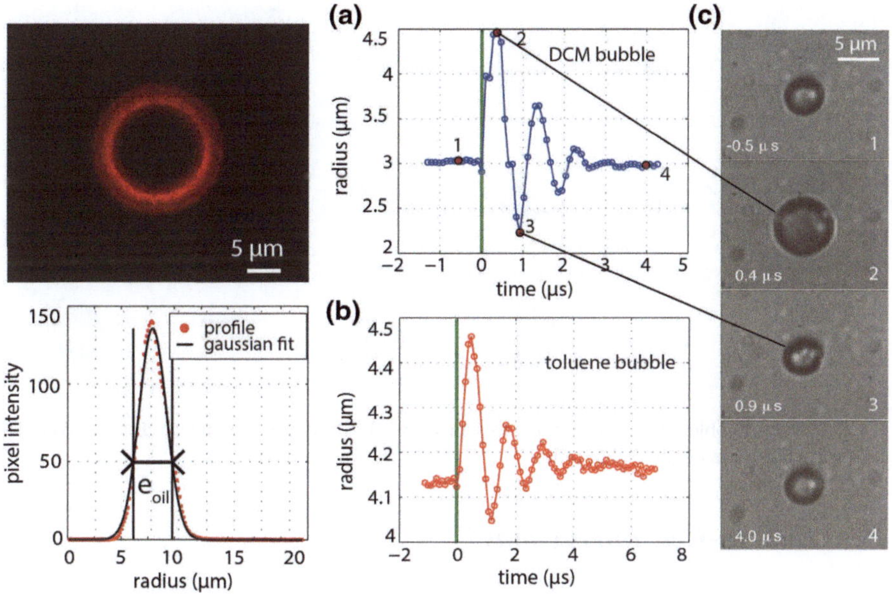

Fig. 12 Ultrafast response of laser-activated bubbles, reprinted from [28, 29], with the permission of AIP Publishing *Top left* confocal image of a light absorbing oil-coated bubble. *Bottom left* Measurement of the oil layer thickness. *Middle* $R - t$ curves demonstrating the impulse response after pulsed laser excitation for a dicholoromethane coated bubble (*top*) and a toluene-coated bubble (*bottom*) demonstrating the effect of different viscosities of the oils. *Right* images taken from a Brandaris 128 recording

small perturbation of the spherical bubble grows as a parametric instability leading to surface mode vibrations and shape instability. These shape modes are associated with a resonance threshold, which can be analyzed following a spherical stability analysis where the mode dynamics is expressed as the solution of a Mathieu equation based on a linearized bubble dynamics equation [31]. Basically we can decompose the bubble dynamics in a volumetric mode and several discrete shape modes ($n = 2 - 6$) that are described by spherical harmonics.

Using ultra high-speed imaging we observed the shape instabilities of bubbles with a size between 10 and 60 µm, see Fig. 13 for an initially spherical 36-µm bubble displaying shape mode oscillations. First, the bubble oscillates in a purely volumetric radial mode, then after 5 cycles of ultrasound the bubble becomes shape-unstable and a surface mode $n = 4$ appears. The bubble was driven by a 10-cycle ultrasound pulse at a frequency of 130 kHz and at a acoustic pressure of 120 kPa. The movie is captured with the Brandaris 128 camera at a frame rate of 1.25 Mfps. Smaller microbubbles display the same physical behavior and shape instability of ultrasound contrast bubbles has been studied in detail by Dollet et al. [32] and by Vos et al. [33].

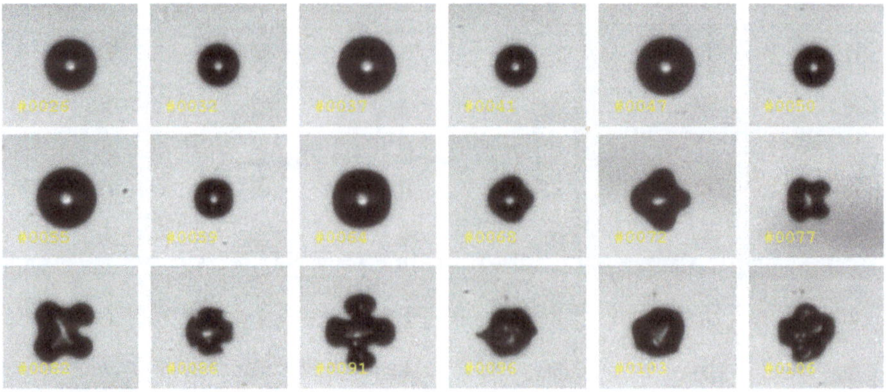

Fig. 13 Microbubble shape oscillations excited through ultrasonic parametric driving [31]. The corresponding video is available online

4.5 Bubble–Cell Interactions

The current challenges of biomedical research revolve around a potent and precise application of the biomechanical effects of targeted agents in acoustics and photoacoustics in order to bring molecular precision and highly localized therapeutic effects to a single cell level [34, 35]. Researchers in academia, as well as in industry, are spending much effort in developing agents that they see as a potential way of addressing the issues of specificity and sensitivity in early diagnostics. Amongst the 'magic bullet' candidates we identify labelled microbubbles that can be decorated with targeting ligands to bind specifically to target cells, and loaded microbubbles covered with nanoparticles, such as liposomes for local drug delivery.

Contrast microbubbles interact with the incoming pressure wave, or laser light. As discussed before, these agents can offer a lot more than contrast generation alone. While the timescale of bubble oscillations is that of microseconds and faster, a number of secondary phenomena occur on the timescale of milliseconds. These include the release of the bubble payload establishment of microstreaming, sonoprinting of functionalized materials onto cell membranes, and sonoporation and the transient disruption of these membranes [36–38]. All these effects are of primordial interest for novel therapies and highly controlled drug and gene delivery, e.g. to treat cancer or a number of genetic diseases more efficiently. This also implies that one needs to follow the subsequent biological processes occurring on a timescale of a few seconds. Following these different processes requires the combination of a range of high-speed imaging techniques [39]. First, the Brandaris 128 camera can be used both in top view and in a non-trivial side view to visualize non-spherical and non-linear features and the dynamics of bubble-induced events. Then, redirecting a fraction of the light towards a high-speed camera operating near 50,000 fps enables the visualization of these secondary effects. The light from two light sources, one pulsed and one continuous wave, are combined for this purpose.

Finally, a highly sensitive CCD camera records the slower events using the remainder of the available light. The spatial resolution of the optical microscope is sufficient to record the bubble oscillations in bright-field, but the fate of the molecules constituting the coating can only be recorded by simultaneous high-speed fluorescence microscopy. A laser is therefore coupled directly into the microscope to excite fluorescently labelled bubbles and their load with a maximum efficiency. Thus, data is collected at microscopic detail at a temporal resolution covering nine orders of magnitude.

The presence of boundaries near oscillating bubbles may pose a concern regarding the applicability of the theoretical predictions and the corresponding experimental observations to the in vivo situation. The in vitro boundary models that neighbor the buoyant bubbles are indeed known to modify significantly their behavior. Alternatively, biological boundaries such as blood vessels or soft tissues found in vivo differ greatly from the in vitro boundaries, and not only by their viscoelastic properties and geometry. In particular, the polymeric membranes that are typically used in most experimental setups, where shown to shift the resonance frequency of the microbubble by as much as 30% [40] and the oscillation amplitude by as much as 40% [27]. Additional results obtained by combining ultra high-speed imaging to advanced manipulation techniques, such as optical tweezers or acoustic radiation forces hint to a complex interaction, depending on the distance of the bubble to the wall and the physical properties of the wall itself which makes such interaction particularly delicate to model. Observations performed in ultra high-speed imaging in side-view, however, reveal how these oscillations can depart from the commonly accepted assumption of bubble sphericity and how the non-spherical character of the oscillation is enhanced by the proximity of a boundary.

Further increase in the insonifying pressure (typically a pressure above 500 kPa for a 3 μm bubble driven at a frequency of 1 MHz) gives rise to a chaotic bubble response. The bubble behavior therefore becomes not only irreproducible but also unpredictable, by definition. In this pressure range, a wide collection of phenomena can be observed, such as jetting [41]. Jetting arises from the inertial collapse of the bubble when the symmetry is broken, e.g. close to a wall or by a gradient of pressure. This particular phenomenon is known to induce damage on supporting structures and is as interesting for cleaning applications as it is concerning for cell toxicity. Ultra high-speed imaging with the Brandaris 128 camera gives direct insight into such phenomena that may also modify the bubble size, either by bubble pinch-off [42], or by gas loss due to changing surfactant concentration and composition. These effects change the acoustic properties of the bubbles and, thus, their response to the acoustic field.

Some biomedical applications require having the agents outside the vasculature, for example to porate or to kill diseased cells beyond the endothelial barrier. One way of achieving this is by injecting nanometer sized agents that are small enough to pass the intercellular spaces in the vessel walls. These agents must then be activated in order to form larger bubbles that can scatter an insonifying ultrasound wave [45, 46]. Many solutions are being investigated, that primarily involve inertial cavitation and where ultra high-speed imaging provides crucial details on the process at hand. Superheated perfluorocarbon droplets are of particular interest for this application [43, 44, 47]. The strategy here consists in injecting nanodroplets made of volatile perfluorocarbons that can then be vaporized in the target site using ultrasound. An example of an ultra high-speed recording of such acoustic droplet vaporization is given in Fig. 14. Alternatively, one can consider the use of gas entrapping nanoparticles. Mostly polymeric, the nanoparticle structure can offer host nano-sized gas pockets. Upon ultrasound exposure above a certain threshold, that depends upon the particle shape and nature, the nucleus can grow to a size 3 orders of magnitude larger, and emit a strong acoustic response [48].

It is obvious that the contrast agent, once injected, will be brought in contact with cells, before, during and after their activation. Whether they are to be used for

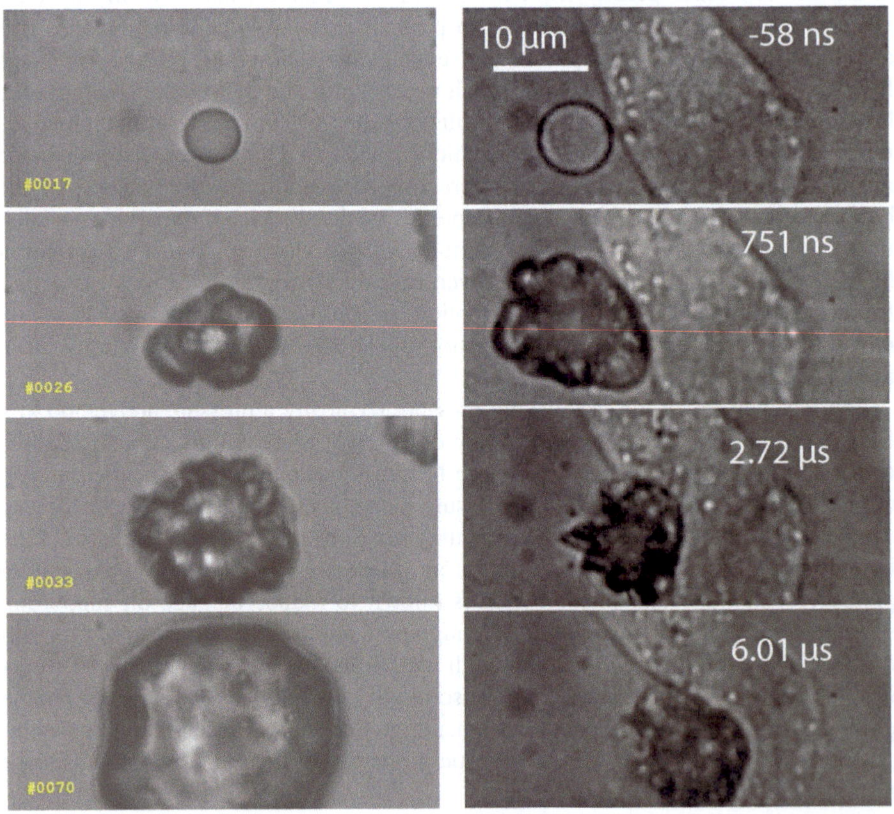

Fig. 14 Acoustic Droplet Vaporization (ADV). *Left* Brandaris 128 recording of the ultrafast vaporization of a 5-µm perfluoropentane (PFP) microdroplet [43]. The corresponding video is available online. *Right* Interaction of a vaporized PFP microdroplet with a neighboring cell. As soon as the droplet is vaporized (by superharmonic focusing [44]) the vapor becomes susceptible to acoustic radiation force and the formed bubble is pushed against the cell (ultrasound from the *left*), reprinted from G. Lajoinie et al., Biomicrofluidics 10 (2016), with the permission of AIP Publishing

therapy of for contrast enhancement, it is crucial to investigate the resulting interaction. Here again, only a combination of both ultra high-speed bright-field imaging and fluorescence microscopy can provide simultaneous information of the cell structure and functions, and on the bubble size and location. Generally, biological events also occur on longer timescales. The closure dynamics of pores for example occurs over the timescale of a second. Therefore, a successful optical observation of, for example, cell response to a bubble-induced mechanical stimulus also requires to combine fluorescence and bright-field high-speed microscopy for accessing simultaneously the long term bubble behavior and molecular information. Ultra high-speed imaging further provides complete and crucial information on the bubble behavior on a nanosecond timescale.

An example of nanosecond bubble dynamics related to cell biological response to the stress induced by US-activated microbubbles is presented in Fig. 15. Here, bubbles loaded with fluorescent nanospheres are introduced near cells cultured on a CLINIcell membrane. The seconds timescale confocal image (bottom left) shows fluorescently labeled microbubbles (green) and fluorescently labelled cells (Cell

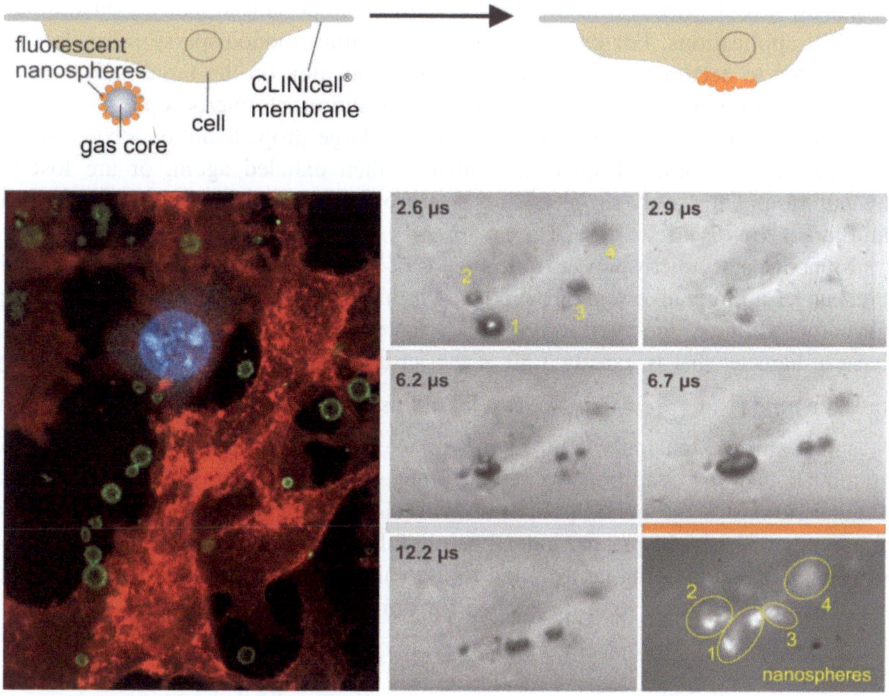

Fig. 15 Bubble-cell interactions are studied in a flow cell (*top*) where bubbles are introduced near a cell monolayer. Activation of the loaded microbubbles with ultrasound leads to deposition of the load onto the cell (sonoprinting [39]). *Bottom left* Confocal fluorescence image of cells (*red*) and bubbles (*green*), with the transfected cell nucleus in blue. *Bottom right* Brandaris 128 recording of the associated bubble dynamics at the nanoseconds timescale

mask Red) and where one cell has been porated (cytox blue). The bottom-right picture shows ultra high-speed imaging of the microbubble oscillations that lead to nanosphere release (nanosphere-loaded microbubbles, side view). Bright-field imaging was performed with the Brandaris 128 camera, operating at a frame rate of 10.5 Mfps. An ultrasound pulse of 100 cycles and a pressure of 300 kPa was sent and the first 12 μs of the experiment were recorded. A fluorescence image was acquired before and after bright-field imaging. The microbubbles are observed to exhibit non-spherical oscillations with repeated gas core coalescence and fragmentation, while acoustic radiation forces translate them towards and along the CLINIcell membrane. The fluorescence images show the localization of the fluorescent nanospheres, and indicate that the microbubbles release and deposit their nanospheres during microbubble activation, termed sonoprinting, as sketched in the top cartoon.

4.6 Diminutive Microjets

A narrow size distribution in droplet formation is important in many industrial and medical applications. For example, in inkjet printing monodisperse microdroplets are needed to accurately control impact and deposition on the target substrate. In pulmonary drug delivery a narrow size distribution of microdroplets is highly advantegeous for optimal drug delivery, as too large droplets are deposited on the trachea, while smaller droplets are inhaled, then exhaled again, or are lost by evaporation. Thus, only droplets of a size around 3 μm reach the deepest alveoli and target the disease.

Monodisperse microspray atomization can be achieved efficiently by Rayleigh breakup of a diminutive microjet produced in a 1-μm nozzle [49]. The jet breaks up into droplets due to the interfacial capillary forces [50, 51] and the droplets produced have a diameter that is exactly 1.89 × the jet diameter. The corresponding capillary breakup time $\tau_b = \sqrt{\rho_w r_j^3 / \sigma_w}$. With a jet radius of 1.25 μm, a density of 1000 kg/m ^{3}, and a surface tension σ_w of 72 mN/m we find τ_b = 160 ns. If we obey the minimum Nyquist condition we need to image at a frame rate of at least 12 Mfps. Figure 16 shows the breakup of such a liquid jet into microdroplets, taken at a frame rate of 13.76 Mfps, and indeed barely fulfulling the criterion. These results of ultra high-speed imaging contributed to the validation of one and two-dimensional numerical models that were developed for diminutive Rayleigh jets.

4.7 Drop-On-Demand Inkjet Printing

Drop formation in inkjet printing is an almost perfectly reproducible process, and single-flash photography is frequently applied to its visualization [13, 52–54]. For

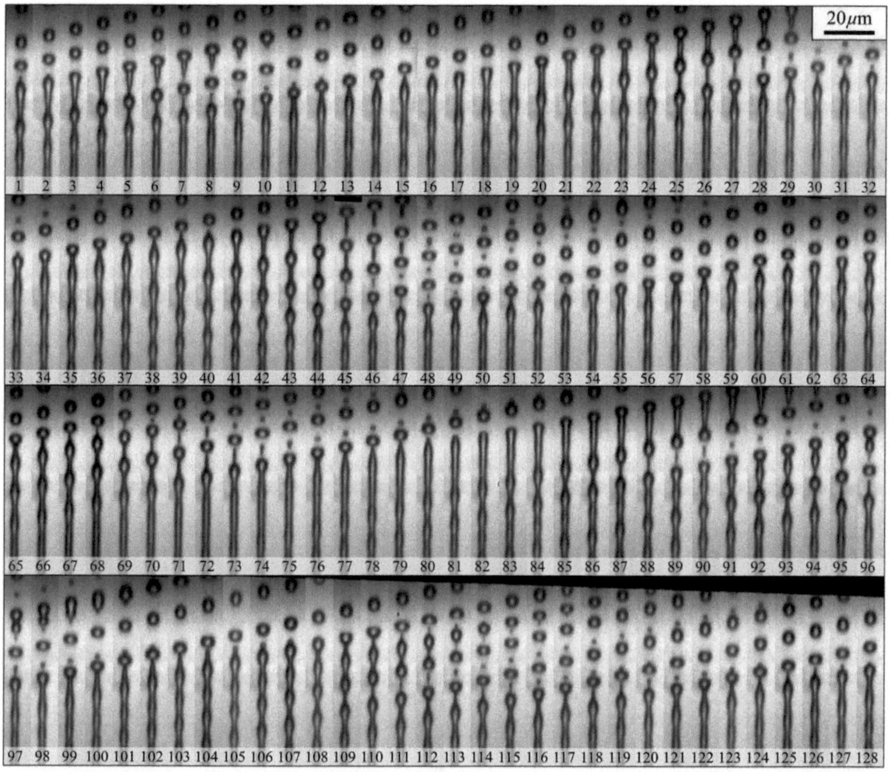

Fig. 16 Atomization following droplet breakup of a 1-μm radius micronozzle jet taken with the Brandaris 128 camera near 14 Mfps, reprinted from [49], with the permission of AIP Publishing

non-reproducible transients that cannot be captured with stroboscopic imaging we resort to real-time imaging with the Brandaris 128 camera. Figure 17 shows the formation of a pL droplet at a drop-on-demand frequency 50 kHz. Special emphasis is put on the breakup of the tail from the meniscus. Figure 17(bottom) shows the formation of a secondary tail (bottom left) that breaks up and forms a series of satellite droplets (bottom middle, right) with a volume of less than 50 fL. Preventing the formation of these satellite droplets is of utmost importance for print quality and for the reduction of aerosol contamination in inkjet machines. The physical insight in the breakup mechanism provided by the Brandaris 128 camera can lead to further optimization in a variety of nanotechnology applications.

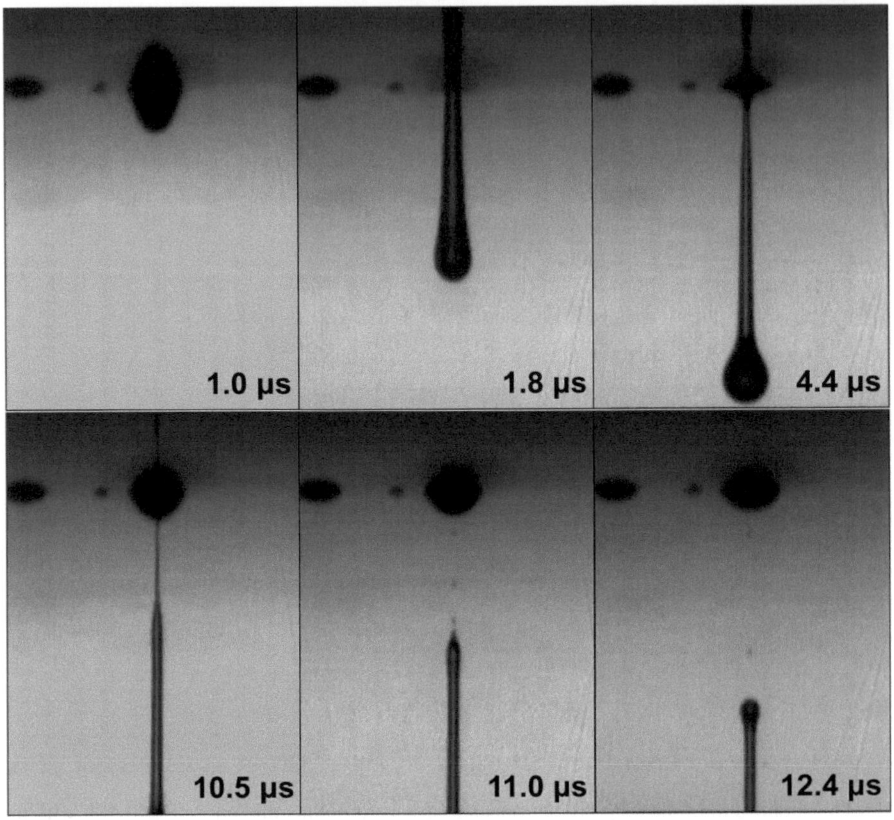

Fig. 17 Inkjet droplet formation imaged with the Brandaris 128 camera at 10 Mfps. Focus is put on the breakup of the secondary tail that connects the primary tail of the droplet with the meniscus. Reduction of satellite droplet formation after breakup improves print quality and reduces aerosol contamination

5 Conclusions

The Brandaris 128 camera has been a unique ultra high-speed imaging system, ever since its introduction in 2003. Various improvements were made in the last decade to further contribute to the versatility of the system, including a faster segmented mode, a region of interest mode boosting the number of acquisition frames to 16,000, the addition of optical tweezers, and the option to capture images in fluorescence, even combined with bright-field imaging. The reduction of the time between separate recordings has been reduced by three orders of magnitude from 80 ms initially to 17 μs now, enabling imaging at multiple timescales ranging from nanoseconds to milliseconds. Combined with its maximum frame rate of 25 Mfps, the Brandaris 128 remains an exceptional camera with great potential for future research both in medical imaging and in nanotechnology industry.

Acknowledgements We gratefully acknowledge the inspiring and indispensable assistance of Frits Mastik, on both hardware and software issues, and also that of Gert-Wim Bruggert for support on the engineering and mechanical work. The first-hour constructors of the camera, Chien Ting Chin and Charles Lancée†, thank you for your ideas and designs. We thank Detlef Lohse for stimulating discussions. We thank Cordin Company, for support in the first years of the project. We also would like to thank our collaborators from Ghent, Pittsburgh and Oxford. We also thank the collaborative support from industry: Bracco Suisse, Océ Technologies, and Philips. This work has been supported by FOM Foundation for Fundamental Research on Matter, Technology Foundation STW, Netherlands Heart Institute ICIN, Netherlands Organisation for Scientific Research NWO, European Commission Innovation Subsidies, and NanoNextNL, a micro- and nanotechnology consortium of the Government of The Netherlands and 130 partners.

References

1. M. Versluis, High-speed imaging in fluids. Exp. Fluids **54**, 1–35 (2013)
2. M. Mennaert, On musical air-bubbles and the sounds of running water. Phil. Mag. **16**, 235–248 (1933)
3. T.G. Leighton, *The Acoustic Bubble* (Academic Press, London, 1994), p. 306
4. P.N. Burns, D. Hope Simpson, M.A. Averkiou, Nonlinear imaging. Ultrasound Med. Biol. **26**, S19–S22 (2000)
5. S.M. van der Meer, B. Dollet, C.T. Chin, A. Bouakaz, M. Voormolen, N. de Jong, M. Versluis, D. Lohse, Microbubble spectroscopy of ultrasound contrast agents. J. Acoust. Soc. Am. **128**, 648–656 (2007)
6. J. Sijl, B. Dollet, M. Overvelde, V. Garbin, T. Rozendal, N. de Jong, D. Lohse, M. Versluis, Subharmonic behavior of phospholipid-coated ultrasound contrast agent microbubbles. J. Acoust. Soc. Am. **128**, 3239–3252 (2010)
7. T. Faez, M. Emmer, M. Docter, J. Sijl, M. Versluis, N. de Jong, Characterizing the subharmonic response of phospholipid-coated microbubble for carotid imaging. Ultrasound Med. Biol. **37**, 958–970 (2011)
8. E. Muybridge, AV. Mozley, in *Human and Animal Locomotion* (Dover, New York, 1887)
9. B. Brixner, J.M. Dewey, R.G. Racca (Eds.), in *20th International Congress on High Speed Photography and Photonics, Proceedings* (SPIE, Bellingham, 1992), vol. 1801, pp. 52–60
10. J. Honour, in *21st International Congress on High Speed Photography and Photonics, Proceedings* (SPIE, Bellingham, 1994), vol. 2513, pp. 28–33
11. B.R. Lawrence, Review of ULTRANAC high-speed camera: applications, results, and techniques. Proc. SPIE **2869**, 882–887 (1997)
12. A. van der Bos, A. Zijlstra, E. Gelderblom, M. Versluis, iLIF: illumination by laser-induced fluorescence for single flash imaging on a nanoseconds timescale. Exp. Fluids **51**, 1283–1289 (2011)
13. A. van der Bos, M.J. van der Meulen, T. Driessen, M. van den Berg, H. Reinten, H. Wijshoff, M. Versluis, D. Lohse, Velocity profile inside piezoacoustic inkjet droplets in flight: comparison between experiment and numerical simulation. Phys. Rev. Appl. **1**, 014004 (2014)
14. E.A. Igel, M. Kristiansen, *Rotating-mirror streak and framing cameras*, vol. PM43 (SPIE, Bellingham, 1997)
15. C.T. Chin, C. Lancée, J. Borsboom, F. Mastik, M.E. Frijlink, N. de Jong, M. Versluis, D. Lohse, Brandaris 128: a digital 25 million frames per second camera with 128 highly sensitive frames. Rev. Sci. Instr. **74**, 5026–5034 (2003)
16. E. Gelderblom, R. Vos, F. Mastik, T. Faez, T. Kokhuis, T. van der Steen, N. de Jong, D. Lohse, M. Versluis, Brandaris 128 ultra-high-speed imaging facility: 10 years of operation, updates and enhanced features. Rev. Sci. Instr. **83**, 103706 (2012)

17. X. Chen, J. Wang, M. Versluis, N. de Jong, F.S. Villanueva, Ultra-fast bright field and fluorescence imaging of the dynamics of micrometer-sized objects. Rev. Scie. Instr. **84**, 063701 (2013)
18. C.D. Miller, Half-million stationary images per second with refocused revolving beams. J. Soc. Motions Pic. Eng. **53**, 479–488 (1949)
19. V. Parker, C. Roberts, Rotating mirror and drum cameras, in *High Speed Photography and Photonics*, ed. by S.F. Ray (SPIE, Bellingham, 2002), pp. 167–180
20. A. Skinner, Versatile high speed rotating mirror cameras. J. Sci. Instrum. **39**, 336–343 (1962)
21. T. Ding, T.H. van der Meer, M. Versluis, M. Golombok, J. Hult, M. Aldén, C.F. Kaminski, Time-resolved PLIF measurements in turbulent diffusion flames, in *Third International Symposium on Turbulence, Heat and Mass Transfer, Proceedings*, ed. by Y. Nagano, K. Hanjalic, T. Tsuji (2000) pp. 857–864
22. P. Marmottant, S.M. van der Meer, M. Emmer, M. Versluis, N. de Jong, S. Hilgenfeldt, D. Lohse, A model for large amplitude oscillations of coated bubbles accounting for buckling and rupture. J. Acoutst. Soc. Am. **118**, 3499–3505 (2005)
23. N. de Jong, M. Emmer, A. van Wamel, M. Versluis, Ultrasonic characterization of ultrasound contrast agents. Med. Biol. Eng. Comput. **47**, 861–873 (2009)
24. J. Sijl, M.L.J. Overvelde, B. Dollet, V. Garbin, N. de Jong, D. Lohse, M. Versluis, "Compression-only" behavior: A second-order nonlinear response of ultrasound contrast agent microbubbles. J. Acoust. Soc. Am. **129**, 1729–1739 (2011)
25. T. Faez, M. Emmer, K. Kooiman, M. Versluis, A.F.W. van der Steen, N. de Jong, 20 years of ultrasound contrast agent modeling. IEEE Trans. Ultrason. Ferroelec. Freq. Contri. **60**, 6–20 (2013)
26. M.L.J. Overvelde, V. Garbin, J. Sijl, B. Dollet, N. de Jong, D. Lohse, M. Versluis, Nonlinear shell behavior of phospholipid-coated microbubbles. Ultrasound Med. Biol. **36**, 2080–2092 (2010)
27. V. Garbin, D. Cojoc, E. Ferrari, E. di Fabrizio, M.L.J. Overvelde, S.M. van der Meer, N. de Jong, D. Lohse, M. Versluis, Changes in microbubble dynamics near a boundary revealed by combined optical micromanipulation and high-speed imaging. Appl. Phys. Lett. **90**, 114103 (2007)
28. G. Lajoinie, E. Linnartz, P. Kruizinga, N. de Jong, E. Stride, G. van Soest, M. Versluis, Laser-driven resonance of dye-doped oil-coated microbubbles: A theoretical and numerical study. J. Acoust. Soc. Am. **141**, 2727–2745 (2017)
29. G. Lajoinie, J.-Y. Lee, J. Owen, P. Kruizinga, N. de Jong, G. van Soest, E. Stride, M. Versluis, Laser-driven resonance of dye-doped oil-coated microbubbles: experimental study, J. Acoust. Soc. Am. (in print) (2017)
30. W. Lauterborn, Kavitation durch Laserlicht (Laser-induced cavitation). Acustica **31**, 51–78 (1974)
31. M. Versluis, P. Palanchon, D.E. Goertz, I. Heitman, S.M. van der Meer, B. Dollet, N. de Jong, D. Lohse, Microbubble shape oscillations excited through an ultrasound-driven parametric instability. Phys. Rev. E **82**, 026321 (2010)
32. B. Dollet, S.M. van der Meer, V. Garbin, N. de Jong, D. Lohse, M. Versluis, Nonspherical oscillations of ultrasound contrast agent microbubbles. Ultrasound Med. Biol. **34**, 1465–1473 (2008)
33. H.J. Vos, B. Dollet, J.G. Bosch, M. Versluis, N. de Jong, Ultrasound Med. Biol. **34**, 685–688 (2008)
34. I. Lentacker, I. de Cock, R. Deckers, S.C. de Smedt, C.T. Moonen, Understanding ultrasound induced sonoporation: definitions and underlying mechanisms. Adv. Drug Deliv. Rev. **72**, 49–64 (2014)
35. K. Kooiman, H.J. Vos, M. Versluis, N. de Jong, Acoustic behavior of microbubbles and implications for drug delivery. Adv. Drug Deliv. Rev. **72**, 28–48 (2014)
36. A. van Wamel, A. Bouakaz, M. Versluis, N. de Jong, Micromanipulation of endothelial cells: Ultrasound-microbubble-cell interaction. Ultrasound Med. Biol. **30**, 1255–1258 (2004)

37. A. van Wamel, K. Kooiman, M. Harteveld, M. Emmer, F.J. ten Cate, M. Versluis, N. de Jong, Vibrating microbubbles poking individual cells: drug transfer into cells via sonoporation. J. Contr. Rel. **112**, 149–155 (2006)
38. K. Kooiman, M.R. Böhmer, M. Emmer, H.J. Vos, C. Chlon, W.T. Shi, C.S. Hall, S. de Winter, K. Schroen, M. Versluis, N. de Jong, A. van Wamel, Oil-filled polymer microcapsules for ultrasound-mediated delivery of lipophilic drugs. J. Contr. Rel. **133**, 109–118 (2009)
39. I. de Cock, G. Lajoinie, M. Versluis, S.C. de Smedt, I. Lentacker, Sonoprinting and the importance of microbubble loading for the ultrasound mediated cellular delivery of nanoparticles. Biomaterials **83**, 294–307 (2016)
40. M.L.J. Overvelde, V. Garbin, B. Dollet, N. de Jong, D. Lohse, M. Versluis, Dynamics of coated microbubbles adherent to a wall. Ultrasound Med. Biol. **37**, 1500–1508 (2011)
41. C.D. Ohl, M. Arora, R. Ikink, N. de Jong, M. Versluis, M. Delius, D. Lohse, Sonoporation from jetting cavitation bubbles. Biophys. J. **91**, 4285–4295 (2006)
42. H.J. Vos, B. Dollet, M. Versluis, N. de Jong, Nonspherical shape oscillations of coated microbubbles in contact with a wall. Ultrasound Med. Biol. **37**, 935–948 (2011)
43. O. Shpak, T.J.A. Kokhuis, Y. Luan, D. Lohse, N. de Jong, B. Fowlkes, M. Fabiilli, M. Versluis, Ultrafast dynamics of the acoustic vaporization of phase-change microdroplets. J. Acoust. Soc. Am. **134**, 1610–1621 (2013)
44. O. Shpak, M. Verweij, H.J. Vos, N. de Jong, D. Lohse, M. Versluis, Acoustic droplet vaporization is initiated by superharmonic focusing. Proc. Natl. Acad. Sci. **111**, 1697–1702 (2014)
45. N. Reznik, O. Shpak, E.C. Gelderblom, R. Williams, N. de Jong, M. Versluis, P.N. Burns, The efficiency and stability of bubble formation by acoustic vaporization of submicron perfluorocarbon droplets. Ultrasonics **53**, 1368–1376 (2013)
46. N. Reznik, G. Lajoinie, O. Shpak, E.C. Gelderblom, R. Williams, N. de Jong, M. Versluis, P. N. Burns, On the Acoustic Properties of Vaporized Submicron Perfluorocarbon Droplets. Ultrasound Med. Biol. **40**, 1379–1384 (2014)
47. O. Shpak, L. Stricker, M. Versluis, D. Lohse, The role of gas in ultrasonically driven vapor bubble growth. Phys. Med. Biol. **58**, 2523–2535 (2013)
48. J.J. Kwan, G. Lajoinie, N. de Jong, E. Stride, M. Versluis, C.C. Coussios, Ultrahigh-speed dynamics of micrometer-scale inertial cavitation from nanoparticles. Phys. Rev. Appl. **6**, 044004 (2016)
49. W. van Hoeve, S. Gekle, J.H. Snoeijer, M. Versluis, M.P. Brenner, D. Lohse, Breakup of diminutive Rayleigh jets. Phys. Fluids **22**, 122003 (2010)
50. J.A.F. Plateau, *Statique expérimentale et théorique des liquides soumis aux seules forces moléculaires* (Gauthier-Villard, Paris, 1985)
51. L. Rayleigh, On the capillary phenomena of jets. Proc. R. Soc. London **29**, 71–79 (1879)
52. H. Dong, W.W. Carr, J.F. Morris, Rev. Sci. Instr. **77**, 085101 (2006)
53. I.M. Hutchings, G.D. Martin, S.D. Hoath, High speed imaging and analysis of jet and drop formation. J. Imaging Sci. Techn. **51**, 438–444 (2007)
54. D. Cressey, 365 days: images of the year. Nature **516**, 304–309 (2014)

Part III
Cameras with CCD/CMOS Sensors

Evolution of High-Speed Image Sensors

Takeharu G. Etoh and Quang A. Nguyen

Abstract The first digital high-speed video camera was developed in 1989. Since then, the highest frame rate has exponentially increased from about 10^3 fps (frames per second) to 10^8 fps, and will reach 10^9–10^{10} fps in the very near future. The evolution has been supported by successive innovations of the technology: (1) in the middle of the 1980s, solid-state image sensors enabling parallel and partial readout were introduced, (2) in 1989, a continuous-readout digital-recording high-speed video camera was developed, (3) 1996, a burst image sensor with in-pixel SPS-CCD storage was invented, (4) in 2002, a burst image sensor with in-pixel slanted linear CCD storage achieved 1×10^6 fps (1 Mfps), (5) in 2011, a backside-illuminated burst image sensor significantly increased the sensitivity and also the frame rate to 16 Mfps, (6) in 2012, a CMOS burst image sensor with the pixel-based storage in the periphery of the chip was developed, (7) in 2015, a macro-pixel multi-framing image sensor achieved a frame interval of 5 ns (200 Mfps), and, (8) currently, the 3D-stacking technology is further increasing the frame rates. The history is reviewed with descriptions of the epoch-making achievements in each stage and works going on. When the spatial resolution of lenses approached the limit, Rayleigh discussed the theoretical spatial resolution limit. It's time to search for the temporal resolution limit of high-speed image sensors. The limit for the silicon image sensors was theoretically derived. For example, the temporal resolution limit for green light of 550 nm is 11.1 ps. Therefore, the theoretical highest frame rate is about 10^{11} fps.

Electronic supplementary material The online version of this chapter (doi:10.1007/978-3-319-61491-5_4) contains supplementary material, which is available to authorized users.

T.G. Etoh (✉) · Q.A. Nguyen
Graduate School of Science and Engineering, Ritsumeikan University, Kusatsu 525-8577, Japan
e-mail: yb6t-etu@asahi-net.or.jp

Q.A. Nguyen
e-mail: quangnabkhn@gmail.com

K. Tsuji (ed.), *The Micro-World Observed by Ultra High-Speed Cameras*,
DOI 10.1007/978-3-319-61491-5_4

1 Continuous-Readout Digital High-Speed Video Cameras

In the 1980s, high-speed video cameras utilized analogue recording in magnetic tapes. A famous one was Kodak Spin Physics SP2000. The frame rate was 2000 frames per second (fps) for 240 $\times$ 192 pixels. In 1989, the camera with digital storage replacing the magnetic tape recorder was developed and marketed as Kodak Ektapro EM. In the same year, as Etoh did not know the fact, he began development of another digital high-speed video camera to visualize motions of water. The camera was developed in early 1991, which, fortunately, achieved much higher performance than the Ektapro EM. The frame rate was 4500 fps for 256 $\times$ 256 pixels with 16-parallel readout and was increased to 40,500 fps for 64 $\times$ 64 pixels by partial readout [1]. The camera was later marketed as Kodak Ektapro HS4540 or Photron FASTCAM (the first generation), and widely expanded application fields of high-speed video imaging. For example, most automobile companies in the world employed the camera for crash tests to save human lives and for visualization of ignition and combustion processes in engines to reduce carbon dioxide in the exhaust gas and to improve the combustion efficiency.

The target frame rate of the continuous-readout high-speed video cameras in the market at this moment (2017) is about 10^6 fps for 10^4 pixels, and 10^4 fps for 10^6 fps, which is equivalent to the readout pixel rate of 10^{10} pixels per second (pps). The advanced broadcasting technology is targeting 120 fps for 1.33 $\times$ 10^8 pixels for the 8k-format video camera [2], making the readout pixel rate of 1.60 $\times$ 10^{10} pps, while the frame rate on the paper was still at 60 fps.

2 Ultra-High-Speed Video Cameras with Silicon Image Sensors

2.1 In Situ Storage Image Sensors

Soon after the development of the digital high-speed video camera in 1991, various requirements for the performance of the camera arose from users, such as a higher frame rate, a higher pixel count, a more frame count, higher sensitivity, a compact camera unit and functions enabling user-friendly operations. Therefore, in 1993, Etoh et al. started two preliminary researches:

(1) A questionnaire study on users' requirements for various high-speed imaging devices and their applications [3, 4], and
(2) A technical feasibility study on a much faster future video camera.

The questionnaire studies were also conducted in 2000 and in 2016. Figure 1 shows an example of the questionnaire study on the requirement for the frame rate. In 2000, the users simply required a much higher frame rate than in 1993. The

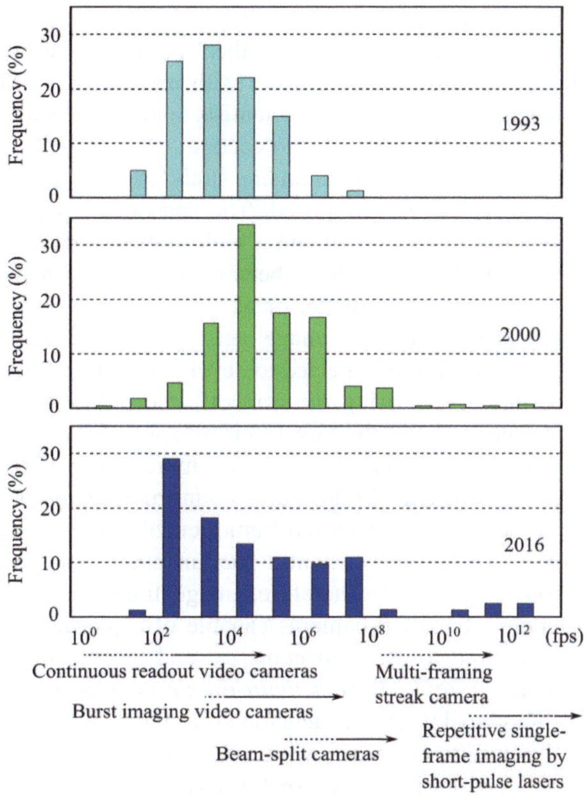

Fig. 1 Users' requirements for the frame rate. In 1993 and 2000, 1000 and 3000 questionnaires were sent via surface mail to potential users of high-speed imaging. The respondent ratio was about 15%. In 2016, the questionnaire was distributed to about 280 participants in the 31st Int. Cong. on High-speed Imaging and Photonics. The number of the answers for the frame rate was 86

required highest frame rate was increased to 10^{12} fps (frame interval, $\Delta t = 1$ ps), which was mainly for laser fusion studies.

The result of the questionnaire in 2016 is quite different.

(1) The peak of the required frame rate was lowered to between 10^2 and 10^3 fps,
(2) The requirement increased for the frame rate of 10^7–10^8 fps, and
(3) The portion requiring more than 10^{11} fps also increased.

The second result may be attributed to burst image sensors operating at more than 10^6 fps developed after 2000, and the third to advanced ultra-short-pulse lasers.

The first peculiar difference may be partly attributed to the different groups of respondents. The former two questionnaires were sent to potential users of high-speed imaging in Japan via surface mail (1000 mails in 1993 and 3000 mails in 2000; about 15% of them returned.): the latter one was distributed to about 280 participants in the 31st International Congress on High-speed Imaging and Photonics, ICHSIP-31, in 2016. The sample size for the requirement for the frame rate was 83 (much less than the previous studies). In the conference, many biological scientists participated to report applications of high-speed imaging for cell, brain, medical, sport, microbe and animal sciences. This is a new trend of high-speed imaging. The other possible reason may be common use of high-resolution image sensors with more than 10 Mpixels even in smart phones. The data suggests that the scientists tend to prefer a higher frame count to the frame rate, taking the tradeoff into account.

Based on the questionnaire study in 1993, a next-generation ultrahigh speed image sensor was proposed. The sensor is equipped with multiple image signal memories in each pixel and a built-in event-driven trigger [3]. Image sensors with the in-pixel memory have been developed and named as *burst image sensors* [5] after the operation mode, or as *in situ storage image sensors* (*ISIS*) [6] after the pixel architecture. An image sensor with the built-in event-driven trigger has not been realized. It is expected to develop an image sensor with the function as early as possible. Considering the tradeoff among the requirements for the multiple performance indices, the target performance of the next generation video camera was determined to be (1) the frame rate more than 1 Mfps, (2) the pixel count more than 100 kpixels, and (3) the frame count more than 100 frames [4].

A series of one hundred frames enables a replay of consecutive images at 10 fps for 10 s. The time constant of the fundamental saccade motion of human eyeballs is about 200 ms [7]. Therefore, image frames replayed at 10 fps look sufficiently smooth. Hereafter, cameras capable of capturing consecutive 100 frames or more are referred to as *video cameras*, and those capturing less than 100 consecutive frames are referred to as *multi-framing cameras.*

Kosonocky et al. developed a burst image sensor operating at 0.5 Mfps ("1 Mfps" in the paper) with the frame count of 30 frames [5]. In 2001, Etoh et al. developed an image sensor for a video camera almost satisfying the targets specified above, achieving the frame rate of 1 Mfps, the frame count of 103 frames, and the pixel count of 81 kpixels (slightly less than the target) [6]. The camera was marketed as Shimadzu HPV, which opened the new horizon of high-speed imaging.

The pixel architecture of the Kosonocky's burst image sensor is shown in Fig. 2. A SPS (serial-parallel-serial) CCD (charge coupled device) memory structure was

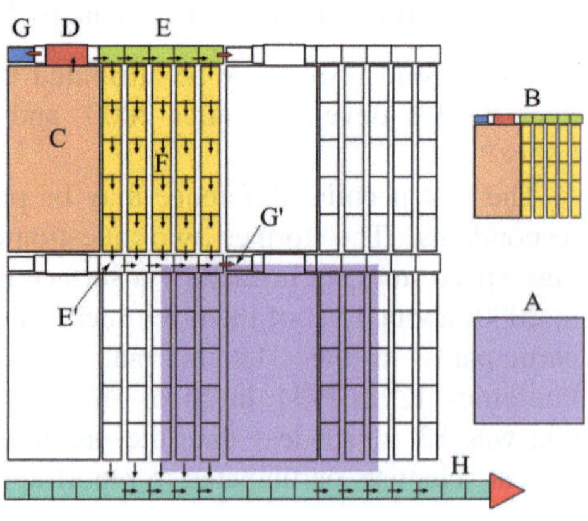

Fig. 2 Pixel architecture of a burst image sensor by Kosonocky et al. [5]. *A* an optical pixel, *B* a pixel structure unit, *C* a photodiode, *D* an input gate, *E*, *E'* a serial (horizontal memory) CCD, *F* parallel (vertical memory) CCDs, *G* a drain, *H* a horizontal readout CCD

employed for the in situ memory. Generated electrons at a photodiode (C) of a pixel are transferred to the input gate (D), serially to the horizontal CCD with five elements (E), and, then, to parallel five vertical CCDs (F).

The SPS CCD memory was also employed recently by a burst image sensor "Kirana" [8]. The frame rate of Kirana is 5 Mfps for the pixel count of 700 pixels.

Pixel architecture (B) of the in situ storage image sensor, ISIS, developed by Etoh et al., is shown in Fig. 3. A *slanted linear CCD* was employed as the in situ memory. Signal electrons captured by a photodiode (C) are straightly transferred on the linear CCD (F) without a change in the transfer direction, which enables a higher transfer rate and a more pixel count with the simplified transfer and storage structure. A drain (G) is attached at the end of each linear storage CCD, which enables continuous overwriting. Upon an occurrence of a target event, the overwriting operation stops, and image signals stored in the linear CCDs are readout through the vertical and then the horizontal readout CCDs. A video camera with the slanted linear ISIS achieved 1 Mfps for the first time in the world in 2001.

A color version of a video camera with the slanted linear ISIS operating at 1 Mfps was developed by NHK in 2013 [9]. The frame count and the pixel count were significantly increased to 144 frames and 300 kpixels, respectively. Figures 4 and 5 respectively show an experimental thunderbolt and a shock wave passing through a cylinder taken by the camera at 1 Mfps [9] and 330 kfps [10]. The shock wave was visualized with the Mach-Zehnder method.

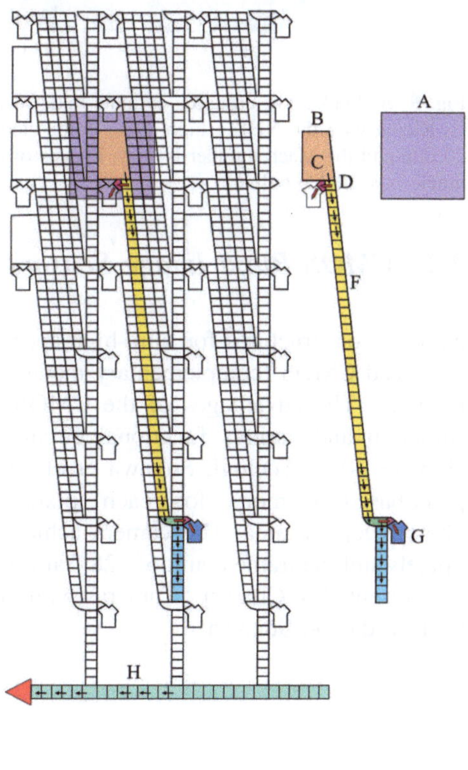

Fig. 3 Pixel architecture of ISIS developed by Etoh et al. [6]. *A* an optical pixel, *B* a pixel structure unit, *C* a photodiode, *D* an input gate, *F* a linear slanted storage CCD, *G* a drain, *H* a horizontal readout CCD

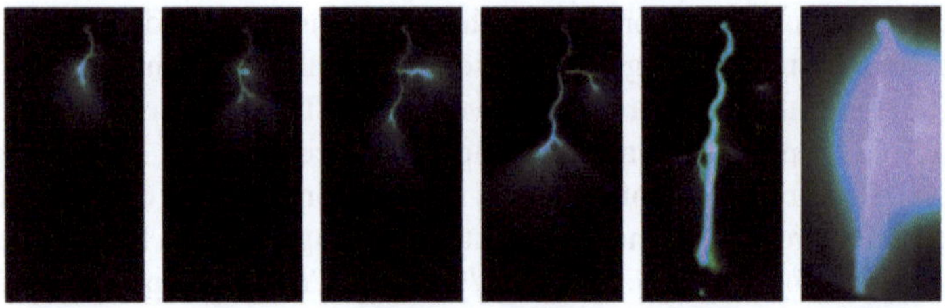

Fig. 4 An experimental thunderbolt, taken at 1 Mfps by T. Arai et al. with the NHK's high-speed color video camera [9]; brush discharges propagate downward seeking for a landing site, and, at the instance of the landing, a sudden extremely high current flows, which is the thunderbolt recognized by human eyes; images of every 20 frames are shown. The corresponding movie is available online

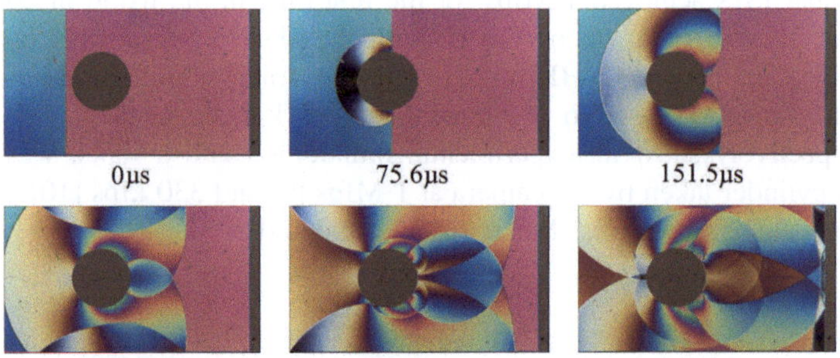

Fig. 5 A shock wave around a cylinder taken with color Mach-Zehnder imaging at 330 kfps by H. Kleine with the NHK's high-speed color video camera [10]; the diameter of the cylinder is 20 mm and the Mach number is 1.31; images of every 25 frames are shown. The corresponding movie is available online

2.2 CMOS Burst Image Sensors for Video Cameras

New sensor structures for ultra-high-speed imaging are being developed by using advanced CMOS (complementary metal oxide semiconductor) image sensor technologies. The advantages of the CMOS image sensors are the low-power consumption and various functional circuits installed in each pixel and/or on the chip. In 2012, Tochigi, Sugawa et al. developed a CMOS image sensor with a pixel-based memory for each pixel in the peripheral area outside the photo-receptive area. The camera achieved 10 Mfps with the frame rate of 100 kpixels and the frame count of 128 frames [11]. Detailed explanations of the sensor are presented in Chapter "Cameras with On-chip Memory CMOS Image Sensors" by Kuroda and Sugawa.

2.3 Backside Illuminated ISIS

W.S. Boyle and G.E. Smith received the 2009 Nobel Prize for the development of a CCD image sensor. A backside illuminated image sensor (BSI image sensor) was already proposed in their paper in 1970 [12]. For ultra-high-speed imaging, the sensitivity is the most crucial issue next to the operation rate. The fill factor of the original front-side-illuminated (FSI) ISIS [6] was only about 15% due to the light shield covering the large in situ storage area of each pixel. A BSI ISIS was then developed in 2011 [13]. The cross-section structure is shown in Fig. 6a.

Two problems had to be solved to realize the structure:

(1) Direct intrusion of incident photons remaining after absorption in the silicon layer to the memory area installed on the front side.
(2) Migration of generated electrons to the memory area.

To avoid the first problem, the sensor thickness of 30 μm was employed, which absorbs 99.9% of the incident photons of 700 nm. For the second problem, a p-well with a hole at the center was created. Then, electrons move over the p-well, gather at the center and move down to the pixel center through the center hole.

The innovative structure significantly increased the sensitivity and the frame rate as well. Additional wiring on the front side drastically reduced the RC delay in

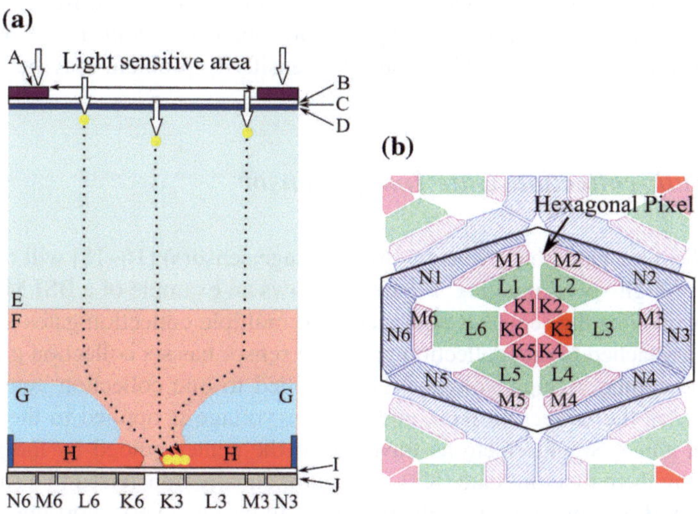

Fig. 6 A cross section and a circuit configuration of a backside-illuminated multi-collection-gate image sensor. **a** A cross-section structure of a pixel of a BSI image sensor and paths of signal electrons. *A* a light shield; *B* an oxide layer; *C* an interface between the backside oxide layer and the silicon layer; *D* a backside hole accumulation layer; *E* p^- layer, *F* n^- layer; *G* p-well; *H* memory and circuit areas; *I* an oxide layer; *J* poly-silicon electrodes. **b** Pixel architecture on the front side. *K* Collection gates; *L* storage gates; *M* barrier gates; *N* transfer gates. K3 is the collecting gate

delivery of the driving voltages without care for decrease of the fill factor and violation of the pixel uniformity. The sensor achieved 16 Mfps for 165 pixels. Later, the pixel count was increased to 300 kpixels.

A backside-illuminated CMOS burst image sensor with a built-in amplifier as well as in situ storage in each pixel was also proposed and tested [14], which achieved the frame rate of 20 Mfps.

3 Multi-framing Cameras with Silicon Image Sensors

3.1 Macro-pixel Image Sensors

A macro-pixel image sensor is defined as an image sensor with macro-pixels each consisting of several pixels (subpixels). The pixel count is the number of the macro-pixels K, and the frame count is the number of the subpixels L in each macro-pixel. The total number of the subpixels is KL. Recording image signals at a very short interval at subpixels in turn enables very fast capture of consecutive images. A major disadvantage is the lower sensitivity and the less pixel count, which are inversely proportional to L.

Mochizuki, Kagawa et al. presented a technology introducing different random shuttering sequences to the subpixels and incorporating an advanced signal process technology to reproduce consecutive images, which achieved a frame interval of 5 ns [15]. With this technology, subpixels are open for about a half of the total consecutive imaging period. Therefore, the sensitivity problem may be mitigated.

3.2 Multi-collection-Gate Image Sensor

Multi-collection-gate image sensors (MCG image sensors) [16–18] will open a new era of ultra-high-speed imaging. Figure 6b shows an example of a BSI MCG image sensor. A pixel of the sensor is equipped with multiple collection gates and storage gates each attached to one collection gate. The sensor has six collection gates (K1 to K6) at the center. A high voltage VH is applied to one collection gate, which is referred to as the *collecting gate*, and a lower voltage is applied to the other five collection gates. Signal electrons traveling to the center guided by the p-well are collected by the collecting gate (K3), and automatically transferred to a storage gate (L3). The VH is applied to the collection gates in turn at a very short interval. Then, six consecutive image signals are captured in the storage gates (L1–L6) at the very short frame interval. In the actual design, the area L6 is a drain and the gate K6 is the drain collection gate. Therefore, five consecutive image signals are captured. After the image capturing, the five stored image signals are transferred through the transfer gates (N1–N6) downwards to a readout HCCD placed outside the photo-receptive area.

When a CMOS architecture is incorporated with the BSI MCG structure, a floating diffusion with a reset gate is attached to each collection gate. Then, the drain gate can be removed and consecutive six signals can be captured.

The sensitivity and the pixel count of a macro-pixel image sensor decrease to 1/L, where L is the number of the subpixels, i.e., the frame count. The BSI-MCG image sensor provides the 100% fill factor, independent of the frame count.

The major problem of the BSI-MCG is the temporal crosstalk due to tempo-spatial dispersion of signal electrons. The average travel time of electrons from the backside layer to the front side in the cross-sectional structure shown in Fig. 6 is more than 1 ns. Therefore, when the multi-collection-gates operate at the speed of less than 1 ns, signal electrons generated by photons incident at the same instant are collected by two or more collection gates, which causes a serious temporal crosstalk [19]. A test chip of the sensor shown in Fig. 6 has been manufactured. The design frame interval is 1 ns (10^9 fps).

A light shield with the open ratio less than 50% effectively reduces the crosstalk [19], and an on-chip microlens array recovers the fill factor. A thinner sensor chip further decreases the temporal crosstalk by suppressing dispersion of the travel time of electrons. The modification reduces the temporal resolution to about 200 ps [18].

High-speed CMOS image sensors can utilize the multi-collection-gate. A test chip of a front-side illuminated MCG image sensor with eight collection gates (8-tap) is fabricated by Shirakawa et al. [20].

A combination of the macro-pixel and multi-collection-gate structures seems to be a compromised solution. For example, a macro-pixel image sensor with macro pixels each with four sub-pixels each with six collection gates can capture 24 consecutive frames at the frame interval of $\Delta t/4$, sacrificing the sensitivity (1/4), where Δt is the frame interval of a usual single collection gate image sensor.

On the other hand, the conversion rate of most photoelectron conversion layers of conventional ultra-high-speed imaging devices, such as streak tubes and MCP image intensifiers, is less than 25%. Therefore, they will be mostly replaced by ultra-high-speed imaging devices with depleted silicon layers.

4 3D-Stacked Image Sensors

4.1 BSI to 3D-Stacking

The 3D-stacking with an additional semiconductor chip on the front side of a sensor chip is a natural consequence of the BSI image sensor [21, 22]. The most promising applications of the 3D-stacked image sensors are (1) high-speed image sensors, (2) compact image sensors, and (3) functional image sensors. Their applications to high-speed image sensors have just begun [23, 24].

The 3D-stacked image sensors will further expand the applications of high-speed imaging. An example is shown in Fig. 7. Driver circuits are mounted on an IC chip

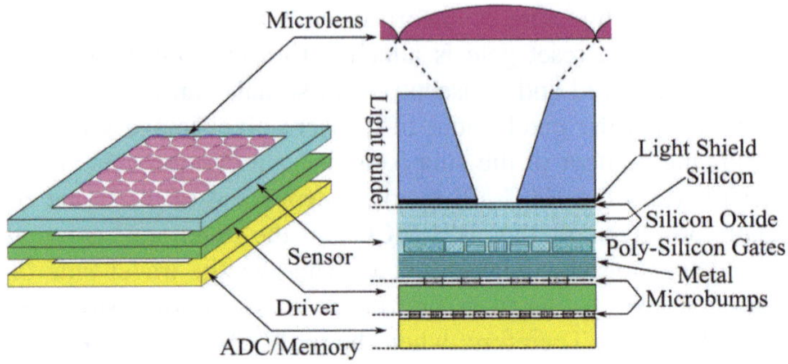

Fig. 7 Pixel architecture of a 3D-stacked BSI-MCG image sensor with an on-pixel microlens and a light guide as a focusing device. The focusing reduces dispersion of travel time due to horizontal spread of generated electrons which is the major cause for degrading the temporal resolution

attached to the front-side of a BSI-MCG image sensor. A simulation study suggested a frame interval less than 200 ps is achievable with a sensor stacked with a driver chip [18]. On the attached chip, various functional circuits can be installed, such as AD converters, in situ memories, and signal compression circuits.

For example, Kondo et al. developed an image sensor with the following dual functions:

(1) continuous imaging at 60 fps for 1.6×10^7 pixels with the global shutter to eliminate the motion-based distortion, and
(2) burst imaging of eight consecutive frames at 1×10^5 fps for 2×10^6 pixels [24].

For X-ray imaging, image sensors capable of capturing 2 or 4 frames at the frame interval of 2 ns were developed in the image sensor team of Sandia National Laboratories [25].

4.2 *A Driver Circuit for Ultra High-Speed Imaging*

An example of the driver circuit attached to a BSI MCG image sensor is presented in Fig. 8a [26].

A ring oscillator "RO" consists of multiple inverters (A) linearly connected in a ring. When a signal "0" or "1" is input to an inverter, it is inverted to "1" or "0". An XNOR circuit (C) receives a pair of input signals. If they are (0, 0) or (1, 1), the XNOR outputs "1", and otherwise "0". Inverting a signal is almost instantaneous, but; not zero. If input and output signals of an inverter are simultaneously monitored, they show (0, 0) or (1, 1) in a very short time before completion of the inversion. By using the input and output signals as a pair of input signals to an XNOR circuit, a signal "1" is output from each XNOR during the extremely short duration. We named the circuit a "ROXNOR" circuit.

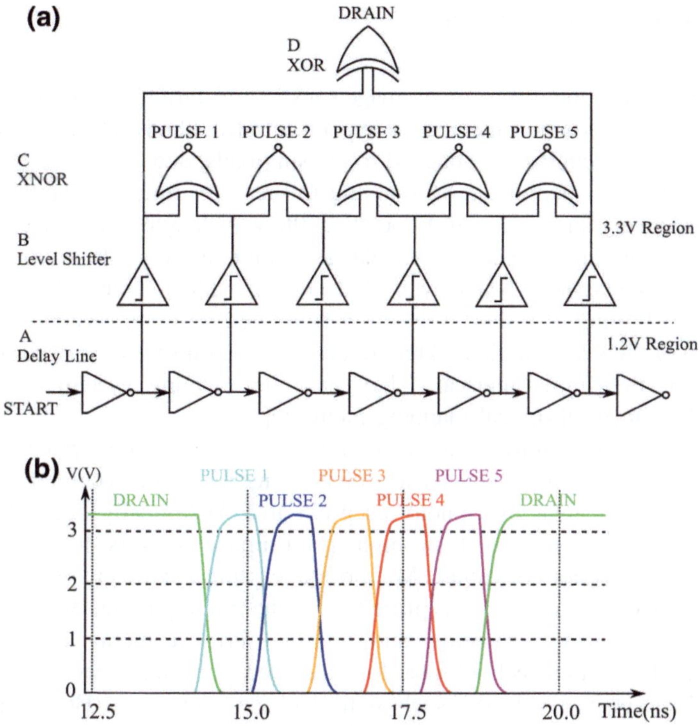

Fig. 8 a ROXNOR circuit [26]: a circuit dedicated to driving multi-collection-gate and macro-pixel image sensors, consisting of a ring oscillator (RO) and an XNOR circuit attached to each invertor of the RO; the voltage level shifter is not necessary if the sensor can be operated by a pulse height of 1.2 V. **b** A simulation result: the pulse width Δt = 1 ns

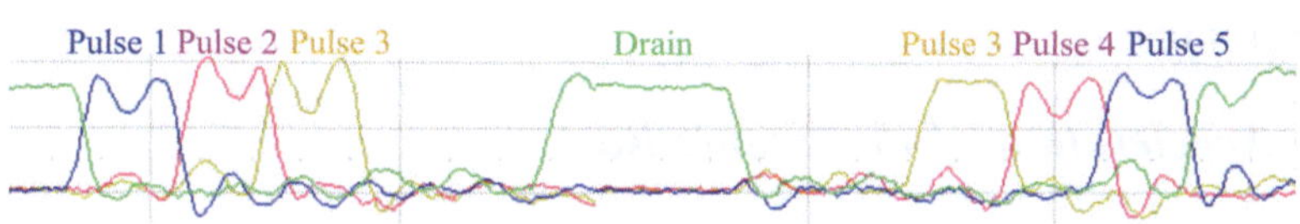

Fig. 9 Measurement results of the outputs from a test ROXNOR circuit. The target pulse width is 2.00 ns; the measured widths of pulses are (*1*) 2.11 ns, (*2*) 2.08 ns, (*3*) 2.03 ns, (*4*) 2.01 ns, (*5*) 2.00 ns

Voltage level shifters (B) to increase the voltage swing from 1.2 to 3.3 V are inserted in Fig. 8a to operate an MCG image sensor by 3.3-V pulses.

Figure 8b shows a simulation result for the pulse width Δt = 1.0 ns. Figure 9 shows pulse shapes from a test chip of a ROXNOR. Four probes can be attached to the oscillograph. Therefore, the pulses 1, 2 and 3, and 3, 4 and 5 were separately measured. The target pulse-width for the case was 2.00 ns; the average pulse width was 2.046 ns (+2.3%) and the standard deviation was 0.0422 ns (2.05%).

5 Time-Space Conversion

One application of the macro-pixel image sensors for ultra-high-speed imaging is a time-spectrum conversion method (a hyperspectral method). On-chip color filters for different wavelengths are attached to the subpixels, and, conceptually, multiple lasers with the wavelengths corresponding to the color filters are applied for illumination at a very short interval. In practice, the wavelength of an ultra-short-pulse laser is swept within the spectral (wavelength) range in a very short time. Although the concept is classical, Nakagawa, Goda et al. made it practical [27]. In their system, the chirping of a short-pulse laser is applied for compression of the pulses and sweep of the wavelength, and re-stretching of the pulse is employed for flexible selection of the frame interval. They call the imaging system STAMP, the sequentially timed all-optical mapping photography.

The time-spectral conversion method is one kind of *time-space conversion methods*, in which image signals at different instances are recorded in different pixels or different areas in a photo-receptive area. Historically, many high-speed imaging technologies other than high-speed image sensors have utilized the time-space conversion methods. Muybridge captured consecutive images of a galloping horse by many cameras placed along the running route. Beam-split mirror cameras, rotating-drum cameras, and rotating-mirror cameras are the popular examples. Many other technologies for advanced ultra-high-speed imaging technologies have been developed, such as time-phase conversion (Holography) [28], and time-dilation by a sudden decrease of a strong field applied to an electron beam in a vacuum tube [29–31].

6 Theoretical Highest Frame Rate of Silicon Image Sensors

6.1 Acceleration of the Frame Rate

Figure 10 shows the evolution of high-speed imaging devices. The blue squares and the green circles show the highest frame rate recorded by cameras with high-speed silicon image sensors. The blue squares indicate the frame rates achieved by standard high-speed cameras with one silicon image sensor [1, 5, 6, 9, 11, 13, 16–18, 25]. The green circles represent cameras with a set of multiple cameras [32, 33] and a camera with an image sensor with macro-pixels [15]. The red triangles indicate the frame rates recorded with imaging devices supported by technologies to increase the frame rates other than high-speed image sensors [27–29, 34, 35]. The size of each plot represents the frame counts. The squares drawn with the blue solid and the blue dashed lines respectively show the frame rates of silicon image sensors under process and those under design based on simulations.

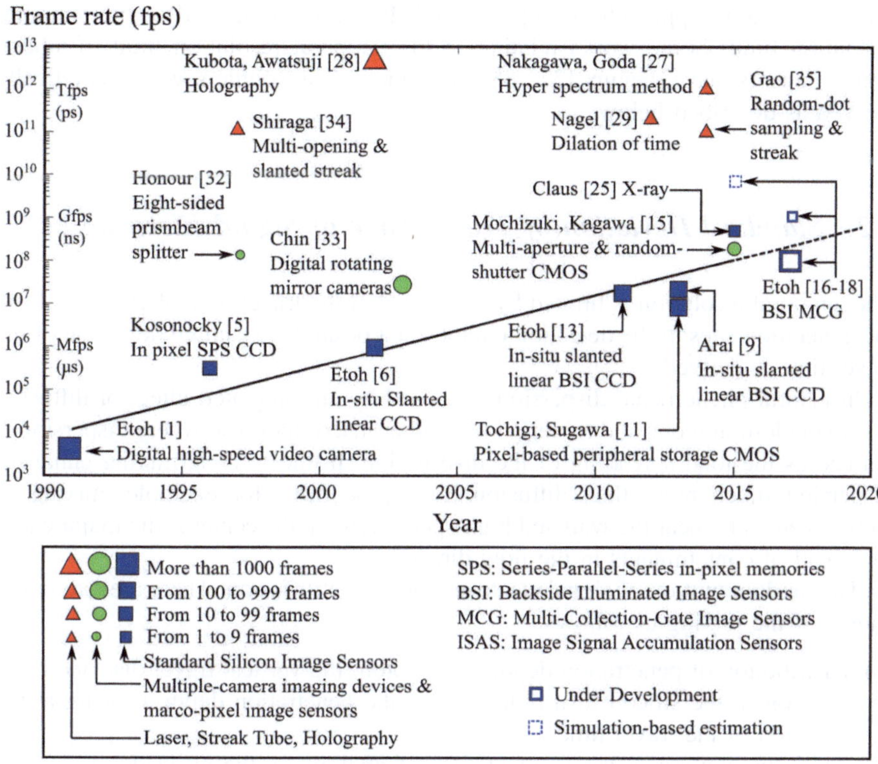

Fig. 10 Evolution of high-speed imaging devices [42]

Cameras with the silicon image sensors are compact, sensitive and user-friendly, and their performance has been rapidly improved. For example, the highest frame rate has exponentially increased (lineally increased in the semi-logarithmic scale in Fig. 10). A simulation study predicted that the frame interval less than 200 ps (about 10^{11} fps) can be achieved with an existing process technology [18]. On the other hand, the frame rates of other continuous imaging devices mostly stay between 10^{11} and 10^{12} fps.

There is no plot under the trend line of the frame rate of the silicon image sensors. Once a silicon image senor reaches a frame rate, its user-friendliness eliminates the competitors. For example, continuous recording film cameras much faster than 10,000 fps had existed before the digital high-speed video camera of 4500 fps was developed in 1991. The film was color and the resolution was much higher. Still, they completely disappeared before 2000.

The highest frame rate of the cameras with silicon image sensors has come closer to the frame rate achievable only by other high-speed imaging devices in the past. The spatial resolution limit was discussed by Abbe, Rayleigh and others, when the resolution of lenses approached the limit [36]. The frame rate of silicon image sensors seems to approach an upper bound. It is time to search for the temporal resolution limit. Some works relating to the temporal resolution limit of silicon image sensors have begun [37, 38]. Fortunately, a solution was theoretically derived as described below.

6.2 Standard Deviation of Travel Time of Signal Electrons

The temporal resolution is limited by the spread of the travel time of electrons from the generation sites to the detection circuit. First of all, the standard deviation of the travel time is derived.

In natural phenomena, dispersion is caused by an integrated effect of diffusion due to random motions and mixing due to other macro motions, where dispersion is defined as the total spreading of the distribution. In most cases, mixing enhances dispersion much more than diffusion. In a pipe flow, for example, mixing of slow-speed water near the wall and high-speed water at the center is the major cause of spread of contaminants in the flow direction.

The random motion of signal electrons causes the diffusion. There are two major causes of the mixing as follows:

(1) Distribution of penetration depth of a photon: the vertical travel distance of an electron in the silicon distributes due to the penetration depth, causing a distribution of the travel time.
(2) Areal distribution of the incident sites of photons on the backside: as seen in Fig. 6a, a photon incident to the center and that to the periphery of a pixel travel a shorter and a longer horizontal distances, and the horizontal field is much less than the vertical field, causing significantly large spread of the travel time of electrons; horizontal diffusion during the vertical travel also causes the areal distribution of electrons.

The effect of the horizontal travel of electrons can be suppressed by design efforts, such as implementing a micro-lens/a light guide, thinning the chip and optimizing the p-well shape as shown in Fig. 7. However, the diffusion due to the random motion and the mixing due to the distribution of the penetration depth are unavoidable.

The two effects are included in a simplified model on the electron travel as shown in Fig. 11. Figure 11a also includes modifications for a Monte Carlo simulation explained later. The total thickness is 30 μm. At the back and the front sides, 5 μm-thick buffer layers are excluded from the Monte Carlo simulation to avoid the boundary-layer effects. Photons are incident to the plane at the distance of 25 μm from the bottom.

Figure 11b shows the trajectories of five generated electrons, in which W, s, and δ are respectively the thickness of the chip, the penetration depth of each electron, and the average penetration depth. Therefore, the travel distance of an electron

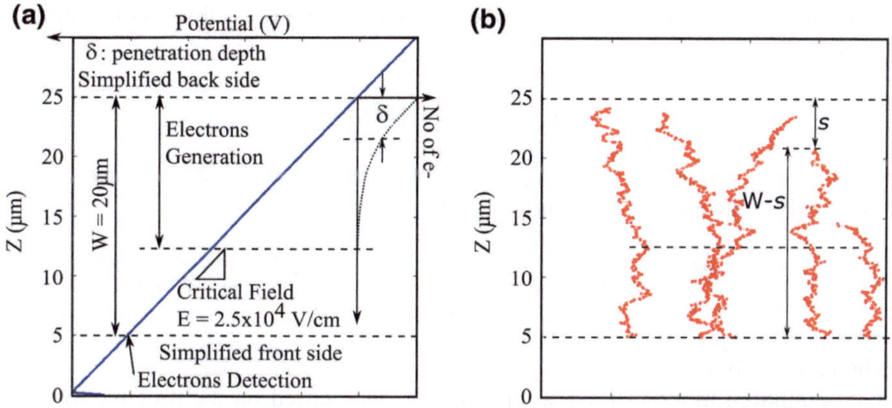

Fig. 11 Modelling the electron travel process: **a** the model for the analysis, **b** example trajectories of electrons. *W* thickness, *d* average penetration depth, *s* penetration depth of each photon. To avoid the effects at the boundaries, the depth from 5 to 25 μm is used; light is incident to the plane at 25 μm; the detection plane is changed from 24.9 to 5 μm; therefore, the active thickness *W* is 0.1–20 μm

generated at the depth s from the backside to the front side is $(\mathrm{W}-s)$, and the average travel time t_r is $(\mathrm{W}-s)/v$, where v is the drift velocity.

The probability density of the penetration depth s is expressed by the exponential distribution f(s) as follows:

$$\mathrm{f}(s) = (1/\delta)\exp(-s/\delta) \tag{1}$$

The zeroth to the second moments of the average travel time are:

$$p = \int_0^{\mathrm{W}} \mathrm{f}(s)\,\mathrm{d}s = 1 - \exp(-\mathrm{W}/\delta) \tag{2}$$

$$\mathrm{E}(t_r) = \int_0^{\mathrm{W}} \{(\mathrm{W}-s)/v\}(\mathrm{f}(s)/p)\,\mathrm{d}s = (\mathrm{W}-\delta p)/(vp) \tag{3}$$

$$\begin{aligned}\mathrm{E}(t_r^2) &= \int_0^{\mathrm{W}} \{(\mathrm{W}-s)/v\}^2(\mathrm{f}(s)/p)\,\mathrm{d}s \\ &= (\mathrm{W}^2 - 2\mathrm{W}\delta + 2\delta^2)/v^2 + \frac{\mathrm{W}^2 - 2\mathrm{W}\delta}{v^2 p}\exp(-\mathrm{W}/\delta)\end{aligned} \tag{4}$$

Then, the variance σ_{m}^2 due to the mixing effect of electrons penetrating to different depths is:

$$\sigma_m^2 = \mathrm{E}(t_r^2) - \{\mathrm{E}(t_r)\}^2 \tag{5}$$

After some cumbersome manipulation of equations, the expression for σ_{m}^2 was fortunately reduced to a very simple form:

$$\sigma_m^2 = \frac{\mathrm{W}^2}{p^2 v^2}[-\exp(-\mathrm{W}/\delta) + \frac{\delta^2}{\mathrm{W}^2}p^2] = -\frac{\delta^2 \mathrm{W}'^2}{p^2 v^2}\exp(-\mathrm{W}') + \frac{\delta^2}{v^2} \tag{6}$$

where, $\mathrm{W}' = \mathrm{W}/\delta$.

The random motion of signal electrons is superposed on the drift motion. The solution of the drift and diffusion equation shows that, when the travel time increases, the travel time distribution approaches the Gaussian distribution. The average travel time is t_r and the variance is $2\mathrm{D}/v^2$, where D is the diffusion coefficient due to the random motion. Therefore, the variance is proportional to t_r multiplied by $2\mathrm{D}/v^2$. The average travel time t_r distributes due to the distribution of the penetration depth of light. Therefore, the variance σ_d^2 due to the diffusion effect is given by integration of the variance conditional to the average travel time weighted by the distribution of the penetration depth. Then,

$$\sigma_d^2 = \frac{2\mathrm{D}}{v^2}\mathrm{E}(t_r) = \frac{2\mathrm{D}}{v^2}\int_0^{\mathrm{W}} \{(\mathrm{W} - s)/v\}(\mathrm{f}(s)/p)\,\mathrm{d}s = \frac{(\mathrm{W}' - p)}{p}\frac{2\mathrm{D}}{v^2}\frac{\delta}{v} \tag{7}$$

The mixing due to the distribution of the penetration depth and the diffusion conditioned by the average travel time are independent processes. Therefore, the total variance σ_{s}^2 is the sum of σ_m^2 and σ_d^2.

$$\sigma_s^2 = \sigma_m^2 + \sigma_d^2 \tag{8}$$

The derived expression of the standard deviation σ_s is the basis of evaluation of the temporal resolution.

A measure of the temporal resolution should be defined based on the real situation. Assume two infinitesimal light spots in time and space arrive at the same spot on the backside of the sensor with the interval of $\Delta\tau$. A batch of signal electrons generated by each light spot travels to the front side, spreading due to diffusion and mixing. The temporal distribution of the dispersion approaches the Gaussian distribution when the travel distance from the backside increases. During the travel, the difference between the average travel times of two batches of electrons is kept at $\Delta\tau$. The no-dip condition for two superposed Gaussian distributions is strictly $\Delta\tau = 2\sigma$ as shown in Fig. 12, where σ is the standard deviation of one Gaussian distribution. Therefore, the value 2σ is employed as a measure of the temporal resolution. Then, the temporal resolution limit $\Delta\tau$ is expressed as:

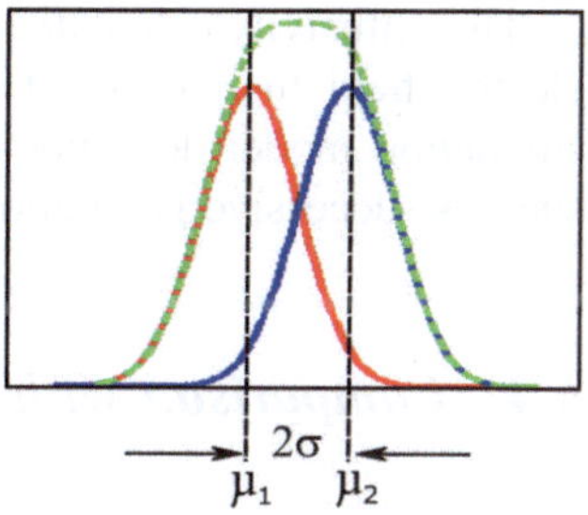

Fig. 12 Definition of separation limit (no-dip condition) for two superposed Gaussian distributions: $2\sigma = \mu_2 - \mu_1$, where μ and σ are the average and the standard deviation

$$\Delta\tau = 2\sigma_s = 2\sqrt{\sigma_m^2 + \sigma_d^2} \tag{9}$$

6.3 Identification of Parameters

The expression includes four parameters: v, D, W and δ. Another parameter, the probability p, is the ratio of the photons absorbed in the silicon layer, which is calculated from Eq. (2) for given values of W and δ, or W′. Photons of (1-p) completely penetrate the silicon chip and are lost.

Sensitivity is crucial in ultra-high-speed imaging. The quantum efficiency seriously decreases for the thickness less than the average penetration depth, since most incident photons penetrate through the chip. The temporal resolution limit becomes longer with the increase of the thickness as shown in Eqs. (6)–(9). Therefore, the thickness can be safely assumed to be equal to the average penetration depth, i.e., $W' = W/\delta = 1$. Then, $p = 1 - e^{-1} = 0.6321$, and the temporal resolution limit is:

$$\Delta\tau = 2\sqrt{0.079\frac{\delta^2}{v^2} + 0.582\frac{2D}{v^2}\frac{\delta}{v}} \tag{10}$$

Later, Monte Carlo simulations show that $\sigma_m \gg \sigma_d$ for a range of W in practical applications. Then, the diffusion effect σ_d can be neglected, and the expression of the temporal resolution limit is reduced to an extremely simple form:

$$\Delta\tau = 0.562\frac{\delta}{v} \tag{11}$$

Figure 13 shows the drift velocity v and the diffusion coefficient D with respect to the field E for the intrinsic Silicon with the crystal orientation of <111> at 300 K. When E increases, v increases and D decreases, and thus the temporal resolution $\Delta\tau$ decreases as expressed in Eq. (9). However, at $E = 2.5 \times 10^4$ V/cm, the drift velocity approaches a constant value and the diffusion coefficient takes the minimum value for the experimental data, which minimizes the temporal resolution. The authors named the field satisfying the condition as the critical field.

The diffusivity calculated by the Monte Carlo simulation further reduces for the electric field higher than the critical field. The reason is that the Monte Carlo simulation model does not include the phenomena arising in the very high field such as successive generation of the secondary electrons.

6.4 Comparison with Monte Carlo Simulations

Equation (9) is compared with the results of Monte Carlo simulations. The physical model is shown in Fig. 11a. A depleted layer with a linear field is created by the backside negative voltage. The total thickness is 30 μm. At the back and the front sides, 5 μm-thick buffer layers are excluded in the Monte Carlo simulation to avoid the boundary-layer effects. Photons are incident to the plane at the distance of 25 μm from the bottom. The detection height is changed from 5 to 24.9 μm. Therefore, the thickness is changed from 0.1 to 20 μm.

Figure 14 shows the temporal resolution $2\sigma_s$ plotted against the thickness W for blue, green and red light of 450, 550 and 650 nm for the intrinsic silicon layer under

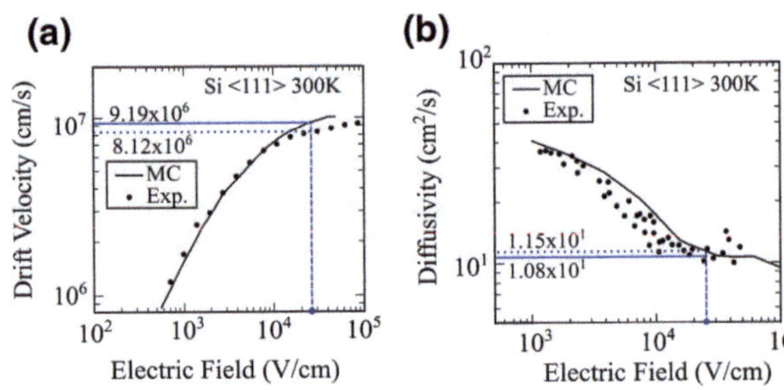

Fig. 13 Drift velocity v and diffusivity D versus electric field E. **a** Drift velocity [39, 40] **b** diffusivity [39, 41]

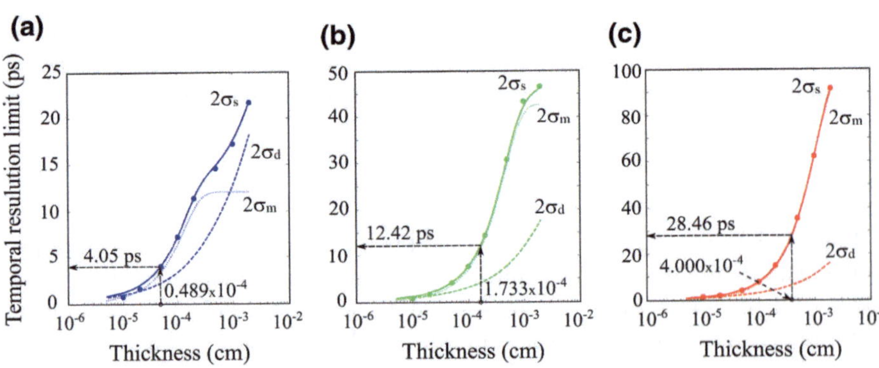

Fig. 14 Theoretical temporal resolution limits $\Delta\tau = 2\sigma_s$ [42] [Eq. (9): *solid lines*] versus Monte Carlo simulation results (*dots*); $2\sigma_m$: Mixing effect; $2\sigma_d$: Diffusion effect; Electric field: 2.5×10^4 V/cm; Wavelengths: **a** 450 nm, **b** 550 nm, **c** 650 nm

the critical field. The values of v and D are calculated by an advanced Monte Carlo simulations, and the random processes including the distributed penetration depth and the random motion of electrons are simulated. Though the field for Fig. 14 is the critical field, the agreement was confirmed in a wider range of fields.

The solid lines and the dots respectively indicate the results calculated from Eq. (9) and by the Monte Carlo simulations. They agree almost perfectly for the thickness larger than 1 μm (10^{-4} cm).

Another important observation is that the mixing effect is dominant over the diffusion effect, validating the approximation of Eq. (11). The values estimated by Eqs. (10) and (11) are respectively 12.4 and 12.0 ps for the green light.

Experimental data of the drift velocity and the diffusion coefficient at the critical field are slightly different from those estimated by the Monte Carlo simulation. The experimental data shows that the drift velocity and the diffusion coefficient at the critical field are respectively 9.19×10^6 cm/s and 10.8 cm^2/s. By substituting these values into Eq. (10), the temporal resolution limit for green light of 550 nm is calculated as 11.1 ps. The inverse, the upper-bound frame rate, is 90.1 fps (about 10^{11} fps).

By inserting the value of v for the critical field, Eq. (11) is further reduced to:

$$\Delta\tau = 6.12\,\delta \tag{12}$$

where $\Delta\tau$ and δ are expressed in ps and μm.

6.5 Generalization of the Theory

For the temporal resolution of practical image sensors, the distribution of traveling times in horizontal direction is dominant. The design target is to minimize the effect.

In the macro-pixel image sensors, electron batches in different subpixels do not mix, resulting in a much shorter theoretical temporal resolution limit. The standard deviation evaluated by Eq. (8) can be applied in the evaluation of the limit.

In X-ray imaging, many other factors are incorporated. The measure of the temporal resolution is also different.

Some discussion on these cases is presented elsewhere [42].

Now, the target is clear. It's time to search for a practical structure to achieve the temporal resolution closer to 11.1 ps.

References

1. T. Etoh, A high-speed video camera operating at 4500 fps, J. Inst. Telev. Eng. Japan (J. ITE) **46**, 543–545 (1992) (in Japanese)
2. R. Funatsu et al., 133 Mpixel 60 fps CMOS image sensor with 32- column shared high-speed column-parallel SAR ADCs, IEEE ISSCC2015. Digest of Technical Papers **58**, 1–3 (2015)

3. T.G. Etoh, K. Takehara, Needs, requirements, and new proposals for ultra-high-speed video cameras in Japan, in Proc. SPIE **2513**, *21st Int. Congress on High-Speed Imaging and Photonics*, pp. 231–242 (1995)
4. Y. Takano, T.G. Etoh, K. Takehara, Users' requirements and specification on high-speed video cameras. J. Visualization Soc. Japan **23**, 11–14 (2003). (in Japanese)
5. W.F. Kosonocky et al., 360 × 360-element very-high frame-rate burst image sensor, IEEE ISSCC1996. Digest of Technical Papers **39**, 182–183 (1996)
6. T.G. Etoh et al., A CCD image sensor of 1 Mframes/s for continuous image capturing of 103 frames, IEEE ISSCC2002. Digest of Technical Papers **45**, 46–47 (2002)
7. https://en.wikipedia.org/wiki/Saccade. (Saccade: Timing and kinematics)
8. J. Crooks et al., Kirana: A solid-state megapixel uCMOS image sensor for ultrahigh speed imaging, in *Proceedings of the SPIE **8659**, Sensors, Cameras, and Systems for Industrial and Scientific Applications XIV*, 865903 (2013)
9. T. Arai, et al., Back-side-illuminated image sensor with burst capturing speed of 5.2 Tpixel per second, in *Proceedings of the SPIE **8659**, Sensors, Cameras, and Systems for Industrial and Scientific Applications XIV*, 865904 (2013)
10. H. Kleine et al., Time-resolved Mach-Zehnder interferometry of shock waves, in *Proceedings of 28th International Symposium on Shock Waves*, vol 1, pp. 577–583 (2012)
11. Y. Tochigi et al., A Global-Shutter CMOS image sensor with readout speed of 1Tpixel/s burst and 780Mpixel/s continuous, IEEE ISSCC2012. Digest of Technical Papers **55**, 382–383 (2012)
12. W.S. Boyle, G.E. Smith, Charge coupled semiconductor devices. Bell Labs Tech. J. **49**, 587–593 (1970)
13. T.G. Etoh et al., A 16 Mfps 165 kpixel backside-illuminated CCD, IEEE ISSCC2011. Digest of Technical Papers **54**, 406–407 (2011)
14. L. Wu et al., A 20 Mfps high frame-depth CMOS burst-mode imager with low power in-pixel NMOS-only passive amplifier, in *Proceedings of the SPIE **10328**, 31st International Congress on High-Speed Imaging and Photonics*, 1032803 (2017)
15. F. Mochizuki et al., Single-shot 200Mfps 5×3-aperture compressive CMOS imager, IEEE ISSCC2015. Digest of Technical Papers **58**, 116–117 (2015)
16. T.G. Etoh, V.T.S. Dao, T. Yamada, E. Charbon, Toward one giga frames per second – evolution of in-situ storage image sensors. Sensors **13**, 4640–4658 (2013)
17. T.G. Etoh et al., Toward 1 Gfps: evolution of ultra-high-speed images sensors: ISIS, BSI, multi-collection gates, and 3D-stacking, in *Proceedings of the International Electric Devices Meeting (IEDM)*, San Francisco, CA, 10.3.1–10.3.4 (2014)
18. V.T.S. Dao et al., Toward 10 Gfps: factors limiting the frame rate of the BSI MCG image sensor, in *Proceedings of the International Image Sensor Workshop (IISW)*, Vaals, The Netherlands (2015)
19. A.Q. Nguyen, V.T.S. Dao, Y. Kamakura, K. Shimonomura, T.G. Etoh, Crosstalk analysis and a cancellation method for an image sensor operating at 1 Gfps. Mech. Eng. J. JSME **3**(6), 16-00284 (2016)
20. Y. Shirakawa, M.W. Seo, K. Yasutomi, K. Kagawa, N. Terenashi, S. Kawahito, An 8-tap CMOS Lock-In Pixel CMOS Image Sensor for Real-Time Biomedical Imaging Application. J. ITE Technical Rep. **40**, 17–18 (2016)
21. V. Suntharalingam et al., Megapixel CMOS image sensor fabricated in three-dimensional integrated circuit technology, IEEE ISSCC2005. Digest of Technical Papers **48**, 356–357 (2005)
22. J. Aoki et al., A rolling-shutter distortion-free 3D stacked image sensor with −160 dB parasitic light sensitivity in-pixel storage node, IEEE ISSCC2013. Digest of Technical Papers **56**, 482–483 (2013)
23. A. Suzuki et al., A 1/1.7-inch 20 Mpixel back-illuminated stacked CMOS image sensor for new imaging applications, IEEE ISSCC2015, Digest of Technical Papers **58**, 110–111 (2015)

24. T. Kondo et al., A 3D stacked global-shutter image sensor with pixel-level interconnection technology for high-speed image capturing, in *Proceedings of the SPIE **10328**, 31st International Congress on High-Speed Imaging and Photonics,* p. 1032804 (2017)
25. L. Claus et al., An overview of the Ultrafast X-ray Imager (UXI) program at Sandia Labs, in *Proceedings of the SPIE **9591**, Target Diagnostics Physics and Engineering for Inertial Confinement Fusion IV*, p. 95910P (2015)
26. C. Zhang et al., Pixel parallel localized driver design for a 128 × 256 pixel array 3D 1Gfps image sensor, in SPIE **10328**, *31st International Congress on High-Speed Imaging and Photonics*, 1032807 (2017)
27. K. Nakagawa, A. Iwasaki, Y. Oishi, R. Hirosaki, A. Tsukamoto, A. Nakamura, K. Hirosawa, H. Liao, T. Ushida, K. Goda, F. Kannari, I. Sakuma, Sequentially timed all-optical mapping photography (STAMP). Nat. Photonics **8**, 695–700 (2014)
28. T. Kubota, Y. Awatsuji, Observation of light propagation by holography with a picosecond pulsed laser. Opt. Lett. **27**(10), 815–817 (2002)
29. S.R. Nagel, T.J. Hilsabeck, P.M. Bell, D.K. Bradley, M.J. Ayers, K. Piston, B. Felker, J.D. Kilkenny, T. Chung, B. Sammuli, J.D. Hares, A.K. Dymoke-Bradshaw, Investigating high speed phenomena in laser plasma interactions using dilation x-ray imager. Rev. Sci. Instrum. **85**, 11E504 (2014)
30. H. Cai, X. Zhao, J. Liu, X. Xie, Y. Bai, Y. Lei, Y. Liao, H. Niu, Dilation framing camera with 4 ps resolution. APL Photonics **1**, 016101 (2016)
31. T.J. Hilsabeck et al., Picosecond Imaging of Inertial Confinement Fusion Plasmas Using Electron Pulse-Dilation, in *Proceedings of the SPIE **10328**, 31st International Congress on High-Speed Imaging and Photonics*, 103280S (2017)
32. J. Honour, Electronic camera for simultaneous framing and streak recording, in Proceedings of the SPIE **2869***, 22nd International Congress on High-Speed Photography and Photonics*, p. 668 (1997)
33. C.T. Chin, C. Lancee, J. Borsboom, F. Mastik, E. Frijlink, N. de Jong, Brandaris 128: a digital 25 million frames per second camera with 128 highly sensitive frames. Rev. Sci. Instrum. **74**, 5026–5034 (2003)
34. H. Shiraga, N. Miyanaga, M. Heya, M. Nakasuji, Y. Aoki, H. Azechi, T. Yamanaka, K. Mima, Ultrafast two-dimensional x-ray imaging with x-ray streak cameras for laser fusion research (invited). Rev. Sci. Instum. **68**, 745–749 (1997)
35. L. Gao, J. Liang, C. Li, L.V. Wang, Single-shot compressed ultrafast photography at one hundred billion frames per second. Nature **516**(7529), 74–77 (2014)
36. J. Hornak, *Encyclopedia of Imaging Science and Technology*, 1st edn. (Wiley & Sons, USA, 2002), pp. 1106–1141
37. R. Turchetta, Notes about the limits of ultra-high speed solid-state imagers, in *Proceedings of International Image Sensor Workshop (IISW)*, Vaals, The Netherlands (2015)
38. H. Mutoh, Device simulations for ultrahigh-speed and high-voltage image sensors. IEEE T. Electron Dev. **63**, 49–56 (2016)
39. T. Kunikiyo, M. Takenaka, Y. Kamakura, M. Yamaji, H. Mizuno, M. Morifuji, K. Taniguchi, C. Hamaguchi, A Monte Carlo simulation of anisotropic electron transport in silicon including full band structure and anisotropic impact ionization model. J. Appl. Phys. **75**, 297 (1994)
40. C. Canali, C. Jacoboni, F. Nava, G. Ottaviani, A. Alberigi-Quaranta, Electron drift velocity in silicon. Phys. Rev. B **12**, 2265–2284 (1975)
41. C. Canali, C. Jacoboni, G. Ottaviani, A. Alberigi-Quaranta, High-field diffusion of electrons in silicon. Appl. Phys. Lett. **27**, 278 (1975)
42. T.G. Etoh, A.Q. Nguyen, Y. Kamakura, K. Shimonomura, T.Y. Le, N. Mori, The theoretical highest frame rate of silicon image sensors. Sensors **17**(3), 483, 1–15 (2017). doi:10.3390/s17030483

Cameras with On-chip Memory CMOS Image Sensors

Rihito Kuroda and Shigetoshi Sugawa

Abstract In this article we give an overview of ultra high-speed (UHS) CMOS image sensors with on-chip analog memories placed on the periphery of pixel array and high speed cameras with the UHS CMOS image sensors for the visualization of UHS phenomena. The developed image sensors consist of $400^H \times 256^V$ pixels and 128 memories/pixel, and the readout speed of 1 T pixel/s is obtained, leading to 10 Mfps full resolution video capturing with consecutive 128 frames, and 20 Mfps half resolution video capturing with consecutive 256 frames. The first development chip was employed in the high-speed video camera and put in practical use in 2012. By the development of dedicated process technologies, photosensitivity improvement and power consumption reduction were simultaneously achieved, and the performance-improved chip has been utilized in the commercialized high-speed video camera since 2015 offering 10 Mfps with ISO 16000 photosensitivity. Due to the improved photosensitivity, clear images can be captured and analyzed even under low light condition, such as under a microscope, and capturing of UHS light emission phenomena.

1 Introduction

Ultra high-speed (UHS) image sensors with over 1 Mfps (frames per second) video capturing speed are utilized in various scientific and engineering fields for studying UHS phenomena, such as microbubbles, fluids, high speed impacts, breakdown, plasma and discharge, and others [1–10]. A high frame rate is needed especially when analyzing small moving objects under the microscope, because the field of

---

Electronic supplementary material The online version of this chapter (doi:10.1007/978-3-319-61491-5_5) contains supplementary material, which is available to authorized users.

R. Kuroda (✉) · S. Sugawa
Graduate School of Engineering, Tohoku University, 6-6-11
Aza-Aoba Aramaki, Sendai 980-8579, Japan
e-mail: rihito.kuroda.e3@tohoku.ac.jp

K. Tsuji (ed.), *The Micro-World Observed by Ultra High-Speed Cameras*,
DOI 10.1007/978-3-319-61491-5_5

view becomes small. In order to visualize these fast object, a frame rate of around 10 Mfps is required.

In order to capture the UHS phenomena precisely, a high frame rate is not the only performance that is required for imaging devices. A high resolution and a long record length are also required. There are fundamental trade-off relationships among these three performances; it is hard to improve all of the speed, the resolution and the record length simultaneously. Consequently, it is a realistic target to develop an image sensor that achieves a UHS frame rate of 10 Mfps while maintaining a sufficient resolution and record length.

Various types of UHS image sensors have been reported in recent years [10]; a back side illumination CCD (charge coupled device) image sensor with serial CCD memories near photodiode (PD) in each pixel [11], a CMOS (complementary metal oxide semiconductor) and CCD hybrid image sensor with two dimensional CCD memories near photodiode in each pixel [12, 13], a CMOS image sensor with multiple on chip analog memories per pixel separately placed from the pixel region developed by the authors' group [10, 14–17], and a multi-aperture CMOS image sensor with temporally compressive sampling [18].

The UHS CMOS image sensor developed by the authors' group has simultaneously achieved 10 Mfps (20 Mfps in development), 10 Kpixels, 128 frames, and the high-speed video camera employing the sensor chip has been utilized to capture what had not been seen before [5–10].

The following points are the main performance requirements that we targeted to achieve for the development of the UHS image sensor.

- Maximum frame rate: about 10 Mfps
- Number of pixels: 60–100 Kpixels, i.e., the signal readout rate given by the product of frame rate and number of pixels of 1 Tpixel/s
- Light sensitivity: High sensitivity due to a large number of incident photons and high photo-electron conversion efficiency
- Spectral response: High sensitivity for a wide wave band of visible to near infrared (NIR) light
- Record length: Over 100 frames with burst capturing
- Capturing mode: Both burst and continuous video capturing
- Electronic shutter: Global shutter of which shutter period is controllable independent of the frame rate
- Standby operation: Sufficiently long standby operation for UHS video capturing, i.e., over 1 min with the highest frame rate condition
- Trigger for burst capturing: Start, end and middle trigger modes with various trigger signals such as light, sound, voltage, current, image signal and so on
- Parasitic light sensitivity of the on-chip memory: No impact to the stored signal under a strong light illumination that is often employed during video capturing of UHS phenomena
- Usability/maintenance: Without complicated setup sequence and maintenance
- System size: Easy-to-handle small size and light weight without cooling system

For the UHS video camera, a CCD image sensor with on-chip CCD cells amounting to the recording frame number near the photodiode was developed and put into practical use, and the research and development have been made for further performance improvement. However, there are some difficulties to be overcome; the charge transfer characteristic degrades at a high speed, the stored signals tend to be affected due to strong light illumination, the standby operation period is limited due to the heat generation even though a cooling system is employed. Subsequently, the practical frame rate has been limited at around 1 Mfps.

On the other hand, for the CMOS image sensors due to their active pixel operation scheme, it is easier to reduce the power consumption than for CCD image sensors, as well as on-chip circuitry is flexibly implemented. Therefore, it is possible to develop a UHS CMOS image sensor with a higher performance than the CCD image sensors.

Figure 1 shows the schematic illustrations of the various high speed CMOS image sensor architectures in comparison to the general analog signal output CMOS image sensors [10]. The high-speed CMOS image sensors are classified into three groups; the parallel analog output type shown in Fig. 1b, the column parallel analog-to-digital converter (ADC) and digital signal output type shown in Fig. 1c, and the on-chip analog memory type shown in Fig. 1d. The first two types shown in Fig. 1b–c are for the continuous video capturing operation. The pixel readout rates of 1–10 Gpixel/s have been achieved.

Figure 2 summarizes the factors affecting the frame rate of CMOS image sensors [10]. These are;

(A) Photo-carrier collection time which is limited to diffusion/drift time of photo-carriers in the photodiode.
(B) Settling time of the pixel driving signals mainly limited by the RC time constant of metal wires.
(C) Readout time from the pixel to the column line mainly limited by the constant current source and RC time constant of metal wires.

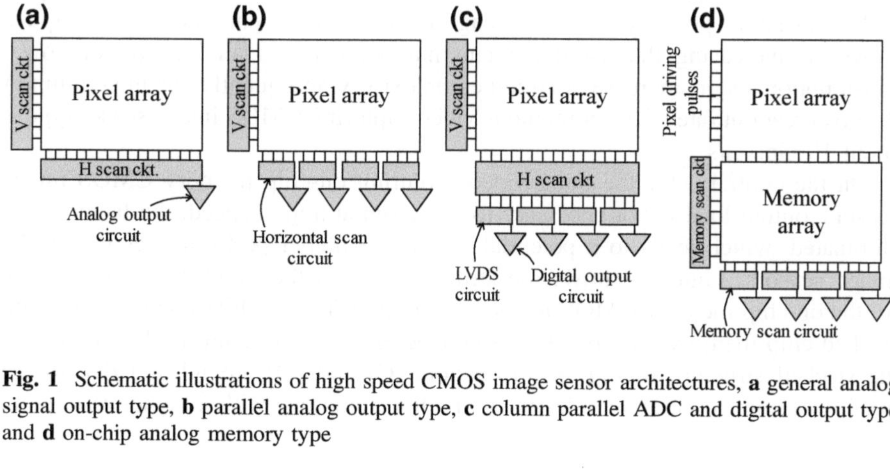

Fig. 1 Schematic illustrations of high speed CMOS image sensor architectures, **a** general analog signal output type, **b** parallel analog output type, **c** column parallel ADC and digital output type and **d** on-chip analog memory type

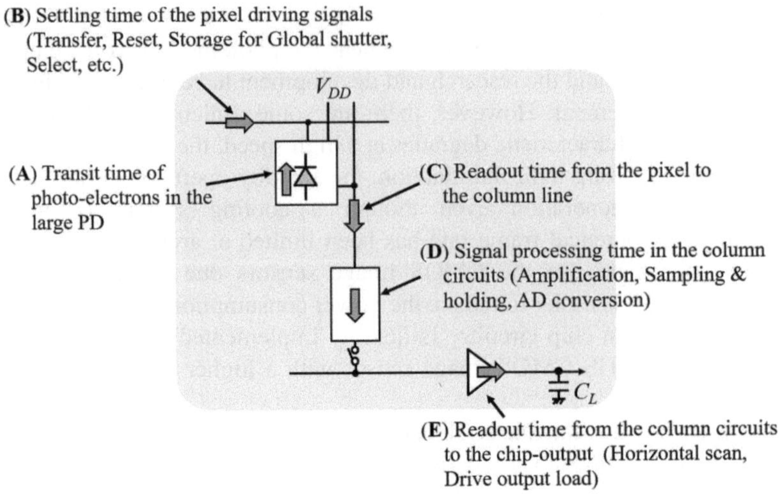

Fig. 2 Factors affecting the frame rate in a CMOS image sensor

(D) Signal processing time in the column circuits mainly due to amplification, sampling & holding and analog-to-digital conversion.
(E) Readout time from the column circuits to the horizontal line and the chip output limited by RC time constant of metal wire and output load.

If the factor (A) is the only constraint, a frame rate of over 100 Mfps is to be achieved regardless of pixel numbers. However, in reality, there are other constraints regarding the global pixel driving pulse time with a large current and electro-migration limit of applicable current. Thus, with the current large scale integrated circuit (LSI) manufacturing technology, 1–5 Tpixel/s is a practical limit, where the frame rate is dependent on the number of pixels.

For the continuous video capturing CMOS image sensor types shown in Figs. 1b–c, the major speed limiting factor is the (E) [19, 20]. Theoretically, a higher speed is possible by increasing the number of parallel signal outputs. However, increasing the parallel circuit number leads to an increase of power consumption and an increase of circuit complexity. Consequently, the upper limit of the pixel readout rate of the continuous video capturing CMOS image sensor type is about 10 Gpixel/s.

On the contrary, for the burst video capturing on-chip memory CMOS image sensors, output load is not necessary to be driven at a high speed, the factor (E) is eliminated, which leads to a potential to achieve higher performances than CCD image sensors. Figure 3 shows the structures of (a) the on-chip memory CCD, (b) the on-chip memory CMOS image sensor and (c) the CMOS image sensor with stacked-chip memory with pixel-wise connections as a structure in the future. For the stacked structure in Fig. 3c the factors (C) and (D) are relaxed due to the pixel-wise connection, which leads to a higher frame rate with a higher resolution.

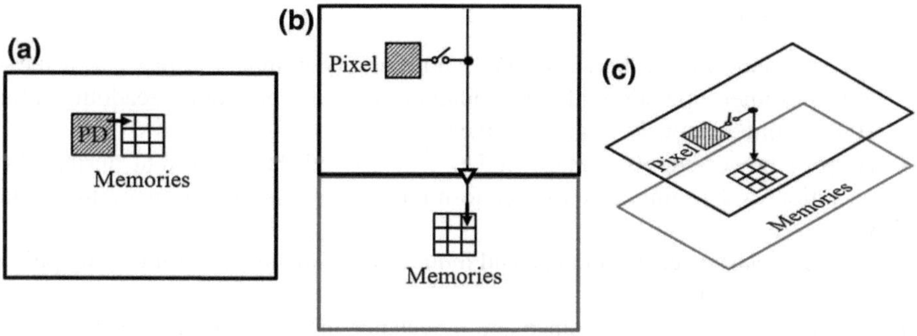

Fig. 3 UHS Image sensors with on-chip memories. **a** CCD image sensor with on-chip memories, **b** CMOS image sensor with on-chip memories, and **c** CMOS image sensor with stacked-chip memory

We have developed the on-chip memory CMOS image sensors that achieve the aforementioned requirements and put them into practical use. In this chapter, the detail of the design, key technologies and the performances of the developed on-chip memory CMOS image sensor achieving 10 Mfps and 1 T pixel/s shown in Figs. 1d and 3b, as well a the captured UHS phenomena are described.

2 Structure of the CMOS Image Sensor with On-chip Memory

Here follow the major considerations when developing the structure of the UHS CMOS image sensor with on-chip memory.

For pixel design;

- Large pixel size to receive sufficient amount of incident photons within a short integration period of less than 100 ns
- Short photo-carrier transit time due to drift and diffusion and electrical field formation inside the photodiode
- Wide spectral sensitivity response
- High resistance to smear and blooming under strong light illumination
- In-pixel circuitry (source follower, current source, amplifier, noise reduction circuit, etc.)
- Floating diffusion (FD) structure that determines the photo-electron to voltage conversion gain

For the total sensor chip;

- Die size and balance between pixel and memory numbers in order to achieve global shutter operation of reset, integration, transfer, signal readout within 100 ns simultaneously at all the pixels
- Drop of power supply voltage and rise of ground due to current during chip operation and immunities to electro-migration, stress-migration of the metal wires
- Parasitic resistance, capacitance and inductance of metal wires for driving pulses in horizontal direction
- Parasitic resistance, capacitance and inductance of pixel output wires in vertical direction and its structure

For memory design;

- Equalization of distance between pixel to memory and introduction of signal waveform shaping circuits
- Analog memory capacitor structure with a high areal efficiency and high-symmetry for minimizing the thermal noise and maximizing record length under a limited area
- Structure of analog memory with high light shield performance and low leakage current in order to avoid signal distortion during signal storing period

For signal output circuitry;

- Selection of analog or digital signal output
- Number of parallel circuits

For chip package and assembly;

- Structure to minimize impedance of metal wires especially the parasitic inductance component in the voltage supply wires
- Structure for high heat dissipation capability with high thermal conductivity in order to suppress the rise of temperature at maximum speed operation

In order to meet the requirements to the UHS video camera, the structure of the UHS CMOS image sensor was developed by taking into consideration the issues mentioned above, while all the elements are approaching near the physical limits. Under the constrain that a standard image sensor process is to be used, we decided that the CMOS image architecture with on-chip analog memory array placed on the periphery of the pixel array, where the number of memories for each pixel corresponding to the recording frames are connected to one or several pixels through a metal wire, is most adequate for this purpose. Figure 4 shows the schematic illustration of the developed UHS CMOS image sensor architecture. In this figure, an example is shown where the analog memories per pixel is connected in the pixel by pixel manner.

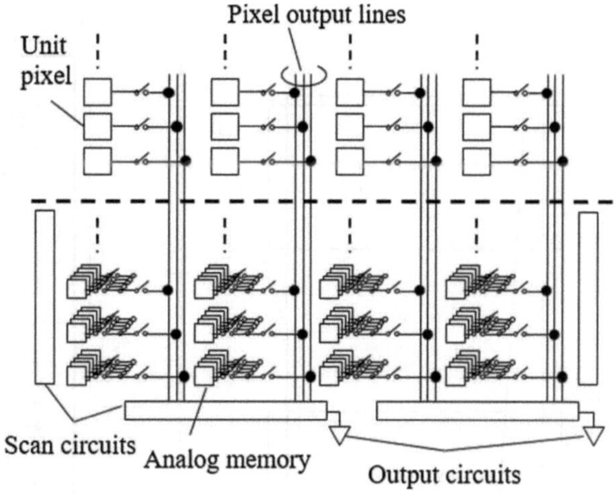

Fig. 4 Schematic illustration of the UHS CMOS image sensor architecture

It is one of the features of the developed UHS CMOS image sensor that the pixel array and memory array are separately placed on a chip, which leads to a prominent merit that the signals stored in memory are not distorted even when a strong light is used for illumination. On the other hand, in the case of UHS image sensors with on-chip CCD memory, the memories are placed close to the photodiode in pixel as shown in Fig. 3a. Due to this structure, signal distortion under a strong light illumination condition has been a challenge for the UHS CCD image sensors. In addition, for the UHS CMOS image sensor, parasitic capacitance of the wires of driving pulses for horizontal direction and vertical output wires is about one or two orders of magnitude lower than that of the CCD image sensors, as well as the supply voltage is less than about 1/3 of the CCD. By these features, a higher speed, a lower power consumption with a suppression of heat generation are achievable with the UHS CMOS image sensor.

For the optical UHS image sensors, a large pixel size is required as it is indicated by the following equation. Suppose A_{pixel} [cm^2], T_i [s], $F_{\#}$, L [lux] are pixel area, integration time, F number of lens, light illumination, and transmittance of lens T_{lens} is 100% and the wavelength λ is 555 nm, the number of incident photons per pixel in an integration time is given by the following equation [21],

$$N_{photon} = \frac{L}{683}\frac{\lambda}{hc}\frac{A_{pixel}T_{lens}T_i}{F_{\#}^2} = 4.05 \times 10^{11} \cdot \frac{LA_{pixel}T_i}{F_{\#}^2} \quad (1)$$

Here, if $L = 1000$ lx, $F_{\#} = 2$, $T_i = 100$ ns (10 Mfps), $N_{photon} = 1$ for $A_{pixel} = 10\ \mu m^2$ ($3.16 \times 3.16\ \mu m^2$) and $N_{photon} = 100$ for $A_{pixel} = 1000\ \mu m^2$ ($31.6 \times 31.6\ \mu m^2$). Since the N_{photon} is proportional to A_{pixel}, a large pixel size is advantageous to receive a sufficient number of incident photons for imaging under a short integration time less than 100 ns. For the developed UHS CMOS image

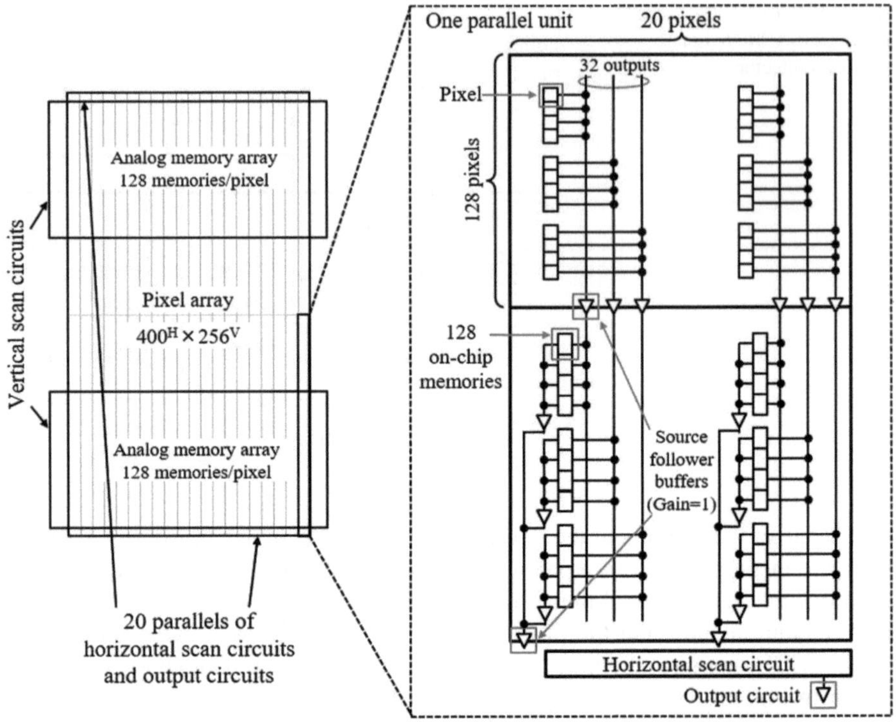

Fig. 5 Block diagram of the developed UHS CMOS image sensor chip

sensor, by considering the balance between die size, pixel number and memory number, the largest pixel pitch of 32 μm was selected.

Figure 5 shows the block diagram of the developed UHS CMOS image sensor chip [14–17]. The chip consists of a pixel array, two on-chip memory arrays, vertical scanning circuits, 20 × 2 parallel horizontal scanning circuits and 20 × 2 parallel output circuits. The number of pixels is $400^{H} \times 256^{V}$ and the pixel pitch is 32 μm. Figure 3b shows the block diagram of one parallel unit from pixel to output circuit. In order to achieve a burst video operation, each pixel has 128 on-chip memories corresponding to the number of recording frames during burst video operation. The pixel region and the on-chip memory region are spatially separated so as to allow us to independently design both the pixel and the on-chip memory layouts. With this architecture, number of pixels and memories are flexibly designed. This design also allows us to form a light shield layer on the on-chip memory region. As the memory array area is fully covered by a shield metal, signal degradation on on-chip memory due to high light illumination does not occur. This CMOS image sensor carries out both the burst video operation and the continuous video operation on the same chip. With the burst video operation, this CMOS image sensor operates at maximum speed of 10 Mfps with 128 frames in full resolution or at 20 Mfps with 256 frames in checkered-pattern half resolution.

In addition, in continuous video operation, 780 Mpixel/s, corresponding to 7.8 Kfps at full resolution is performed. During the burst video operation, the signals are readout to the on-chip memories without being output from the image sensor chip. On the other hand, during the continuous video operation, the signals are read out to the on-chip memories and are output from the image sensor chip at every frame. In order to achieve a high readout speed and a large number of pixels with a high fill factor, 32 output wires are placed in each column, thus, four pixels share one output wire. Figure 6 shows the equivalent circuit and the layout of the pixel. The pixel size is 32 $\mu m^H \times 32\ \mu m^V$. The pixel includes a pinned photodiode with a photoelectron collection time of less than 10 ns [15], a correlated double sampling (CDS) circuit to reduce the number of readout signal per frame, and current source transistors to eliminate IR-drop in output wires [22]. The on-chip memories consist of memory switches and highly area-efficient capacitors. The die size is 15 $mm^H \times 24\ mm^V$. The number of pins of package is 424.

The basic structure of this type of UHS CMOS image sensor that contains pixel region and frame memory region is similar to that of frame transfer CCD and CMOS image sensors [23, 24]. By an introduction of several key technologies such as the high-speed charge collection PD [15], multiple pixel column output lines, in-pixel CDS circuit and current sources, the first developed UHS CMOS image sensor model employing a 2-polycrystalline Si 4-metal layers (2P4M) 0.18 μm CMOS image sensor process technology has achieved signal reading out from the pixel region to the on-chip memory region within 10 ns at each frame [14]. The image signals are stored in on-chip memories and readout after the end of video capturing. Consequently, a burst video capturing over 20 Mfps has been achieved.

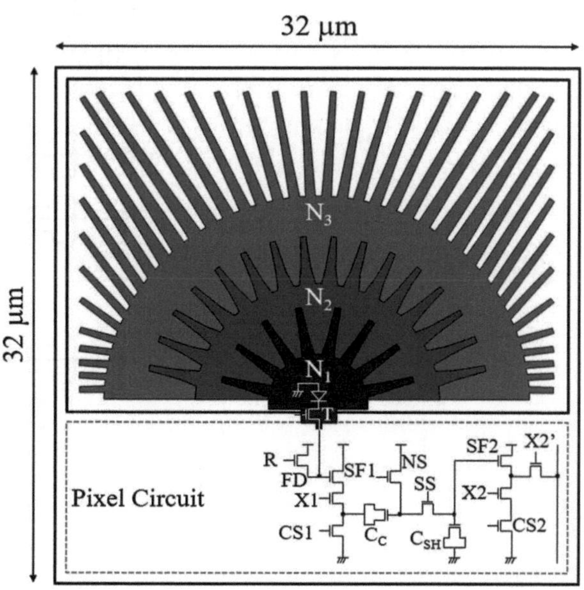

Fig. 6 Photodiode and pixel circuit

For the performance-improved chip, based on 1-polycrystalline Si 6-metal layers (1P6M) 0.18 μm CMOS image sensor process technology, several key technologies were introduced for the improvement of light sensitivity and power consumption [16, 17]. Figure 7 show the circuit schematic diagram of the analog memory. Figure 8a, b shows the cross sectional illustrations of the active and passive device components developed for the performance-improved UHS CMOS image sensor. A higher light sensitivity was achieved by the following three means; A. increasing number of photo-generated carriers to be collected under a same light illumination by increasing the fill factor and charge collection efficiency, B. increasing conversion gain by reducing the floating diffusion capacitance enabled by optimization of PD's buried n-layer, and C. increasing signal readout gain by optimizing the readout circuit. Also, the lower power consumption is to be achieved by the decrease of power supply voltage from 5.0 to 3.3 V and the reduction of source follower amplifiers' current.

Regarding point A, there are 32 pixel output lines per column, where each line is shared by 4 pixels in different rows in order to improve the signal readout speed from the pixel region to the analog memory region [14]. In the first development chip, one of the four metal layers was designated to form the 32 pixel output lines with a same pitch above the PD. For the performance-improved chip, two of the six metal layers were designated for the same usage. Consequently, the fill factor was

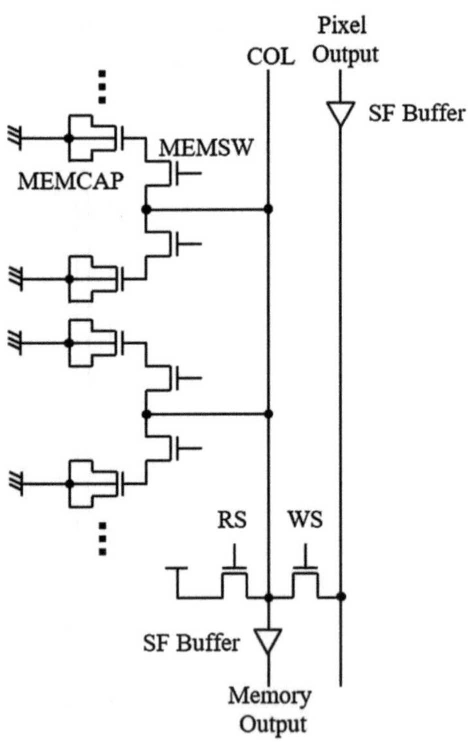

Fig. 7 Analog memory circuit schematic

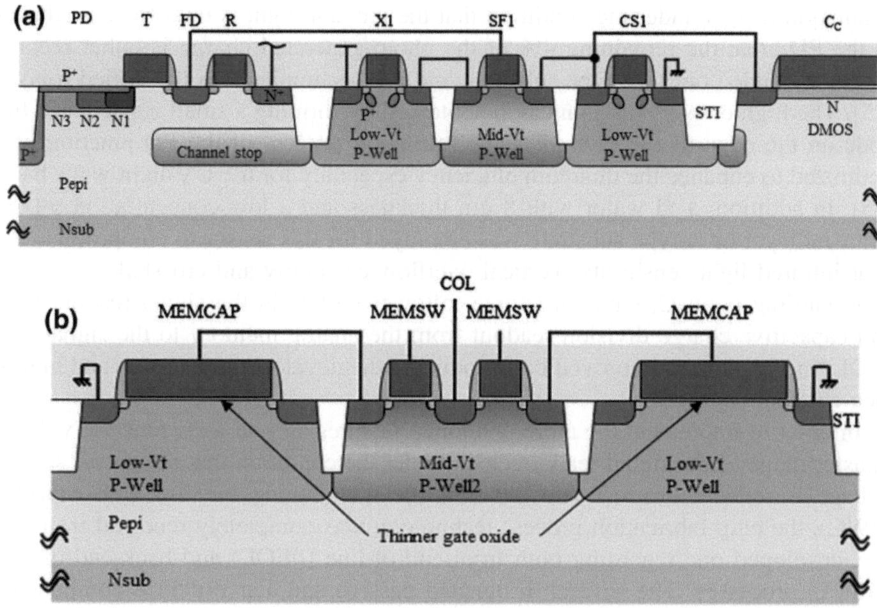

Fig. 8 Cross sectional illustrations of passive and active devices in **a** pixel and **b** analog memory developed for the performance-improved chip

improved from 37 to 55%. In addition, parasitic capacitance in output lines was also reduced due to the enlarged spacing between each line, which is also beneficial for a reduction of the current of the source follower amplifiers in pixels and in-between the pixel and memory regions under the same operation speed. In addition, by an introduction of the PD structure mentioned next, a high charge collection efficiency of 96% is to be achieved in a short charge transfer time.

Regarding point B, the PD structure achieving the full charge transfer from the large area PD within 10 ns shown in Fig. 6 was introduced [15]. In order to generate the electric field over 400 V/cm in the PD region toward the FD node to achieve high speed charge collection, two technologies were implemented; one is to form a gradient of buried pinned n-layer doping by employing the three levels of dopant concentrations and the other is to form a gradient of the width of the n-layer. The width of each sector-shaped portion of the n-layer becomes narrower from the proximal end to the distal end, making the whole shape similar to hedgehogs. By these technologies, the gradient of pinned potential of the n-layer, i.e., the electric field is efficiently and uniformly formed with a limited increase of process steps. Using process, device and optical TCAD simulations the PD design was carried out with a sufficient margin when taking into account the variation of ion implantation dosage and proximity of the lithography. By doing so, the full charge transfer within 10 ns to a single FD node with the charge collection efficiency of 96% was achieved for a large PD size of $30.00\ \mu m^H \times 21.34\ \mu m^V$ [15]. Here, with simulation results under the condition that the incident light is uniformly irradiated to the PD area, the remaining 4% of the photo-generated charge is either recombined or drained out to n-type substrate and not accumulated in the buried n-layer [15]. The high conversion gain can be obtained by forming a small capacitance FD node on the edge of the PD. Also, the formation process of the PD junction was optimized to enhance the quantum efficiency especially for the UV-light wave band [25]. In addition, a Si wafer with 8 μm thickness and a low concentration p-type epitaxial layer on n-type substrate was employed for this sensor to compromise the near infrared light sensitivity, vertical overflow capability and crosstalk.

Regarding point C, gains of source follower buffers in the signal readout path and capacitive charge division readout from the analog memory to the signal line (COL) in Fig. 7 were improved by introducing the developed transistors and highly area efficient capacitors.

In order to implement the aforementioned features as well as to tune the voltage transfer range of the signal readout chain under the condition that the power supply voltage is reduced from 5.0 to 3.3 V while maintaining the full well capacity (FWC), the chip fabrication process technology was completely renewed from the first developed one regarding both front-end-of-line (FEOL) and back-end-of-line (BEOL) processes. The several designated passive and active device components shown in Fig. 8 were developed besides the basic high threshold voltage (Vth) and low Vth transistors and they were utilized in the developed image sensor listed as follows; the low Vth nMOS with channel stop p-type implantation and without halo implantation for reset gates in pixels, the medium Vth nMOS without halo implantation designated to the first source follower driver, the high capacitance depletion-type MOS (DMOS) capacitor for in-pixel CDS circuit with a large capacitance linearity range, the medium Vth nMOS without halo implantation for analog memory select switch with small junction and subthreshold leakage currents, and the highly area efficient MOS capacitor for analog memory with thinner gate oxide than that for 3.3 V transistors. Regarding the pixel circuit schematic, the position of X1 was changed from between the reset voltage node and the SF1 in the first developed chip [14] to between the SF1 and the CS1, and the X1 source is connected to the Cc. Subsequently, the voltage drop at X1 guarantees that the Cc operates in its linear capacitance region during the whole operation condition. The total conversion gain was improved due to the smaller FD capacitance without the bootstrap effect utilized in the first developed chip [14]. Here, an increase of the capacitance value of analog memory has several merits, such that the thermal noise is reduced, the signal readout gain from analog memory to signal line due to capacitive charge division readout is increased, and the effect of leakage current during signal holding period becomes smaller. Thus, an introduction of a capacitor with a higher area efficiency is critically important to benefit from the merits mentioned above under the same usage of chip area. In order to utilize the thinner

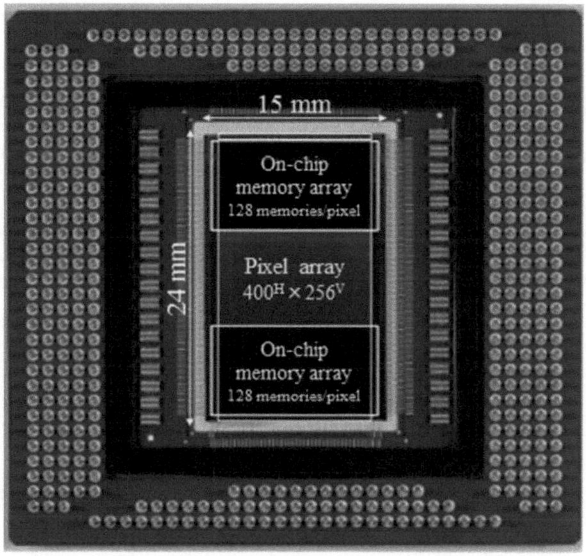

Fig. 9 Fabricated chip micrograph

gate oxide to the analog memory for this purpose, the voltage range applied to the analog memory was controlled to be below 1.8 V. In addition, a triple well process was employed for the source follower circuits in order to obtain the unity gain by removing the body bias effect.

Figure 9 shows the fabricated chip micrograph of the developed UHS CMOS image sensor with on-chip memory [16, 17]. Besides the numbers of pixels and on-chip memories, the positions of PADs and the chip package were identical as those of the first developed sensor chip. It is also beneficial for an easy substitution of the image sensor chips of the formerly developed UHS video cameras. The on-chip memory array regions are fully covered by the shield metal, so that the stored signal is not affected by incident light. The quartz glass was used as the package lid for the chip performance measurement described in the next section.

3 Performances of the Developed CMOS Image Sensor with On-chip Memory

Figure 10 shows the photoelectric conversion characteristic of the fabricated UHS CMOS image sensor chip. The input-referred FWC of 10,000 e^- was obtained. The major noise source of the developed sensor chip is the thermal noise at the on-chip memory, as it was the same case in the first developed chip [14]. By the increase of conversion gain, signal readout gain and the on-chip memory capacitance, the total input referred temporal noise of about 5 e^-_{rms} was obtained.

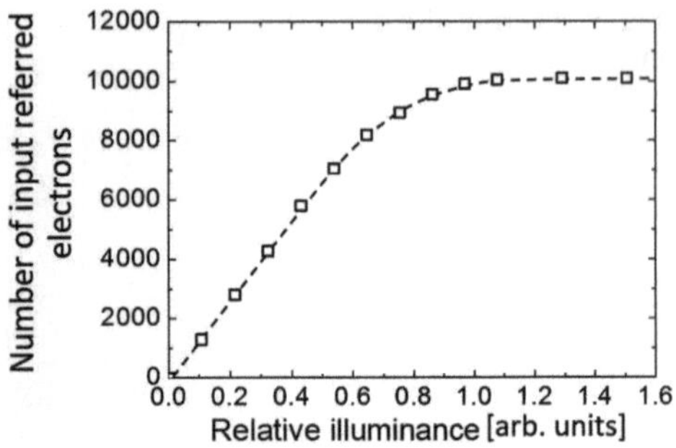

Fig. 10 Photoelectric conversion characteristic

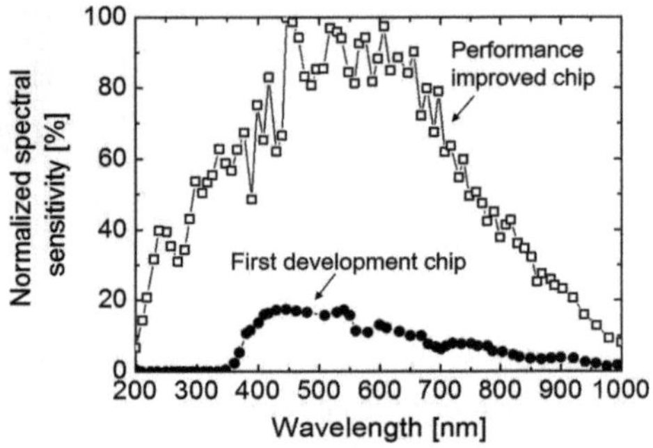

Fig. 11 Normalized output referred characteristic

Figure 11 shows the output referred spectral sensitivity characteristics of the performance-improved chip and first developed one. The improvement by factor eight of the light sensitivity was confirmed at the wavelength of around 550 nm. Here, about four-times improvement of light sensitivity is attributed to the increases of the fill factor, the charge collection efficiency and the conversion gain, and the rest is attributed to the improvement of signal readout gain, respectively. Also, the spectral sensitivity range was extended to the UV-light wave band for the performance-improved sensor chip due to the employed PD junction formation technology [25].

This is beneficial to capture ultra-high speed phenomena with UV-light, such as the dynamic behavior of plasma and flames. The spectral sensitivity of polarized light parallel and perpendicular to the pixel output lines above the PD was measured, and the results suggest that the polarization effect was negligibly small across the whole sensitive wave band.

The characteristic of image lag is shown in Fig. 12. In this experiment, under the frame period of 50 ns, the incident light from a 635 nm laser diode was cut off, just before the pulse rising timing of the transfer gate at the frame number 1, so that the incident light was kept off for the subsequent frames. The vertical axis is the ratio of output signal at each frame to that of the frame number 1, i.e., the image lag. Here, the output signal of the frame number 1 was about 3/4 of the saturation level. The output signal at frame numbers 2 and higher were the same as the dark level,

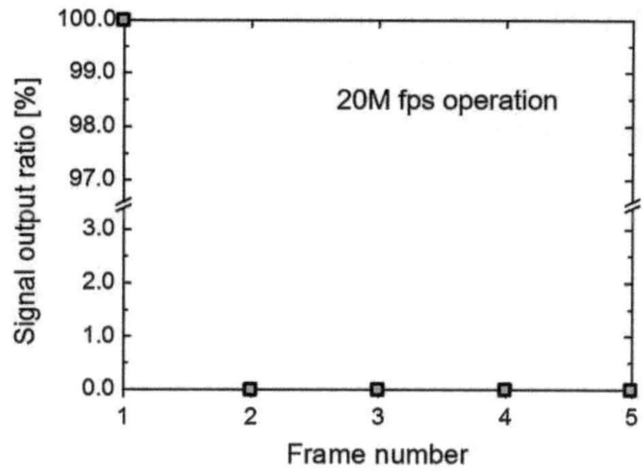

Fig. 12 Measured image lag at 20 Mfps

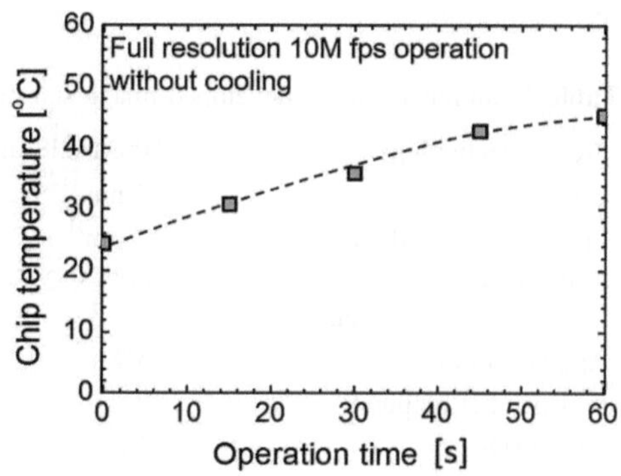

Fig. 13 Chip temperature as a function of operation time

indicating that the image lag was under the detection limit even at 20 Mfps operation [15].

In Fig. 13 the chip temperature is plotted as a function of the operation time. The temperature was measured by the on-chip thermometer during the 10 Mfps operation with burst full pixel mode under room temperature. No cooling system was used. The measured chip temperature was under 50 °C even after 60 s operation time. This result indicates that the developed image sensor chip is capable of operating at the maximum pixel readout speed while waiting for the arrival of trigger signal for the video capturing. This ability is critically important for capturing non-repeatable UHS phenomena.

Figure 14 presents the power consumption of the performance-improved image sensor chip at various frame rate with burst full and half pixel modes. Due to the reduction of the power supply voltage, the power consumption at 1 Tpixel/s operation was reduced by half, i.e., 10 W compared to 20 W of the first developed chip [14]. The performances of the developed UHS CMOS image sensor are summarized in Table 1.

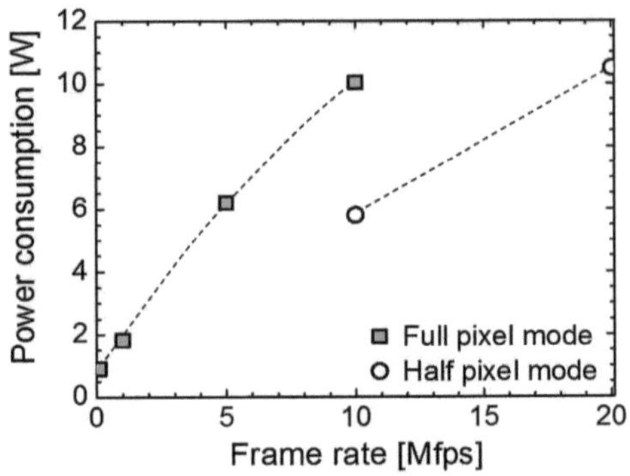

Fig. 14 Power consumption as a function of frame rate with burst full pixel and half pixel modes

Table 1 Summary of the developed image sensor performance

Process technology		1P6M 0.18 μm CMOS with pinned photodiode			
Die size		15 mm^H × 24 mm^V			
Pixel size, photodiode size		32 μm^H × 32 μm^V, 30.00 μm^H × 21.34 μm			
# of pixels	Total, effective	400^H × 256 ^{V}, 400^H × 250^V			
Aperture ratio		55%			
# of parallel outputs		40			
# of PADs		424			
# of frames in burst operation		Full	128	Half	256
Maximum frame rate	Burst		10 Mfps (1 Tpixel/s)		20 Mfps (1 Tpixel/s)
	Continuous		7.8 Kfps (780 Mpixel/s)		15 Kfps (780 Mpixel/s)
Conversion gain (input referred)		112 μV/e$^-$			
Full well capacity		10,000 e$^-$			
Power consumption		10 W at 1 Tpixel/s			
Image lag at 20 Mfps		Below the measurement limit			
Spectral sensitive range		200–1000 nm			
Full charge transfer time		≤ 10 ns			

4 Camera with On-chip Memory CMOS Image Sensor

The first developed UHS CMOS image sensor with on-chip memory was employed in the commercialized high speed video camera: HPV-X [26]. This high-speed video camera has been utilized to capture what had not been seen before [5–10]. By the development of dedicated process technologies, eight times photosensitivity improvement and 50% power consumption reduction were simultaneously achieved [16, 17], and the performance-improved chip has been utilized in the commercialized high-speed video camera: HPV-X2 since 2015 that offers 10 Mfps with ISO 16000 photosensitivity. Due to the improved photosensitivity, clear images can be taken and analyzed even under low light conditions, such as under a microscope as well as in capturing of UHS light emission phenomena [27–31].

Figures 15, 16 and 17 are some examples of images taken by this high-speed camera. Figure 15 shows a series of images of a crack generation of glass. A glass plate of 1 mm thickness was shot with a plastic bullet of 6 mm diameter at a speed of 20 m/s. The video was taken at 5 and 20 Mfps. The trigger signal was generated by the intercept of a laser by the bullet. The light intensity of the flash lamp was 76,000 W and the light-emitting time was 2 ms. The experimental set-up is illustrated in Fig. 15a–c show the series of images: at first, the crack extended radially

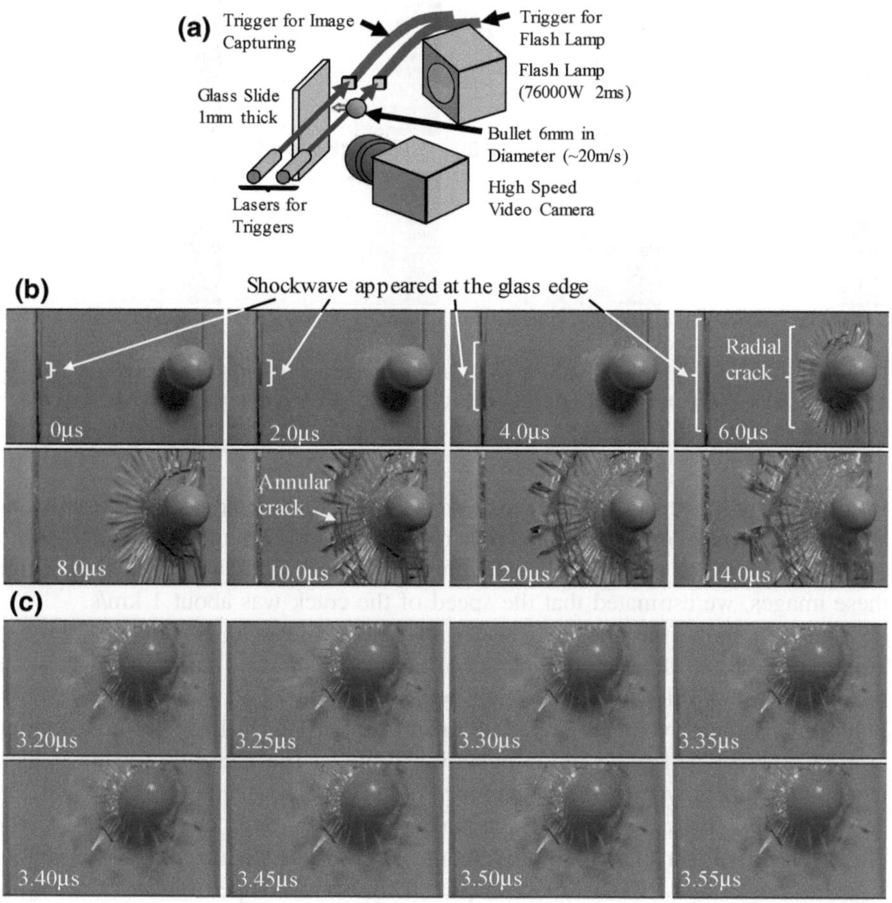

Fig. 15 **a** Schematic illustration of the video capturing of glass crack generated by bullet shooting and captured images by **b** 5 Mfps full resolution mode and **c** 20 Mfps half resolution mode, respectively. The corresponding movies are available online

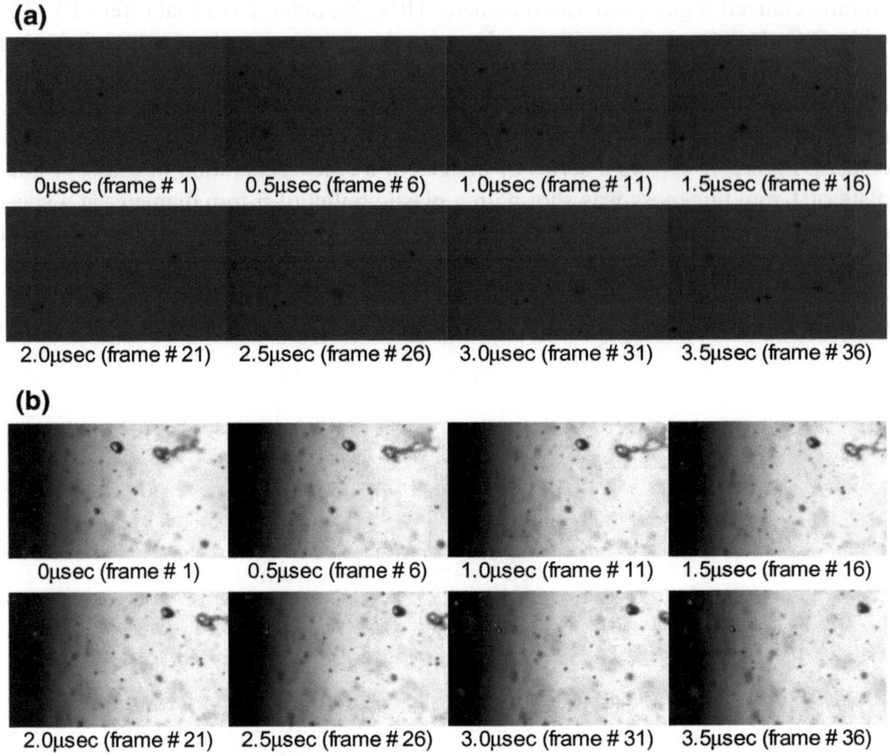

Fig. 16 Images of water mist injected from an air brush taken at 10 Mfps burst half pixel mode by **a** first developed chip and **b** performance-improved chip. The corresponding movie is available online

from the point of impact, shock waves were reflected at the edges of the glass, and after that the crack developed annularly from the position around the meeting points of reflected shockwaves. Crack branching at several spots was also seen. From these images, we estimated that the speed of the crack was about 1 km/s.

Figure 16 shows the images of water mist ejected from an air brush captured at 10 Mfps burst half pixel mode operation. Figure 17 shows a series of images of a breakdown of Al/SiO_2/n–Si capacitor with an oxide film of 100 nm thickness, induced by the gate voltage ramped up to and kept at 100 V. Images were captured at 1 Mfps burst half pixel mode operation. Figures 16a and 17a are the images taken by a camera with the first developed chip (HPV-X), while Figs. 16b and 17b are those taken by a camera with the performance-improved chip (HPV-X2). The effectiveness of sensitivity improvement is clearly recognized by the higher signal-to-noise ratio (SNR) images captured by the developed sensor. The higher sensitivity is greatly useful for capturing microscopic UHS phenomena, capturing images with a deeper depth of focus, and capturing light emitting objects.

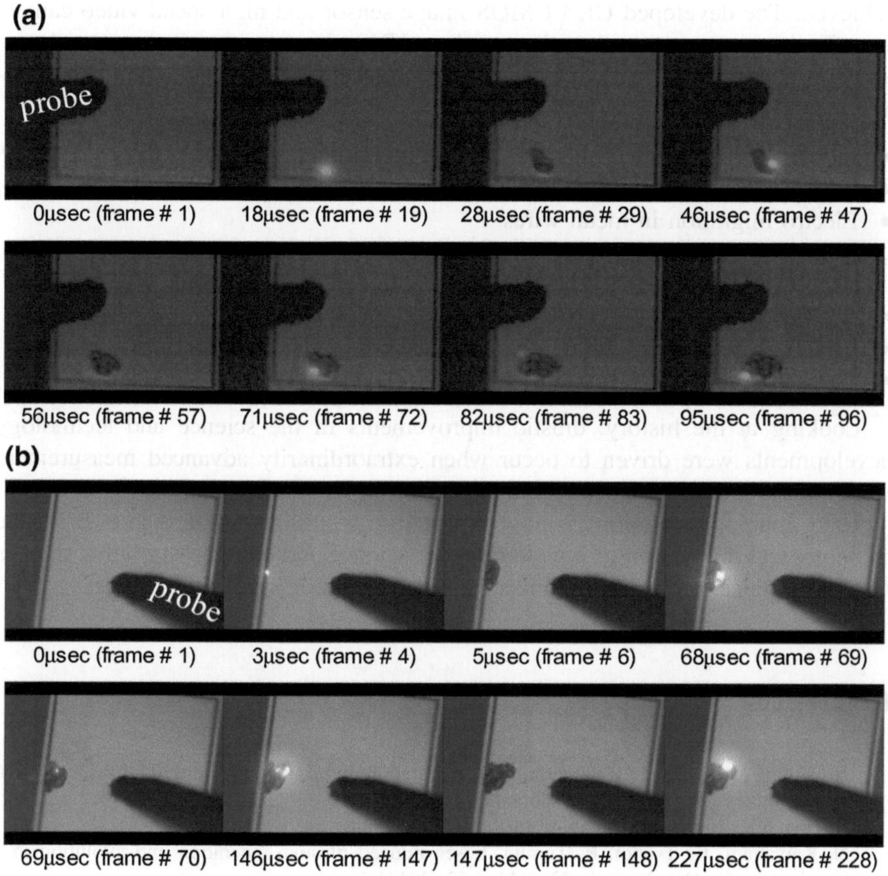

Fig. 17 Images of the electric field breakdown in Al/SiO_2 (100 nm)/n–Si MOS capacitor taken at 1 Mfps burst half pixel mode by **a** first developed chip and **b** performance-improved chip

5 Summary and Future Perspective

UHS CMOS image sensors with on-chip analog memories placed on the periphery of a pixel array for the visualization of UHS phenomena are overviewed. The developed UHS CMOS image sensors with $400^H \times 256^V$ pixels and 128 memories/pixel exhibit readout speed of 1 Tpixel/s, leading to 10 Mfps full resolution image capturing with consecutive 128 frames, and 20 Mfps half resolution image capturing with consecutive 256 frames. By the development of dedicated process technologies such as the large area high speed charge collection PD, the DMOS capacitor for in-pixel CDS circuit, the high capacitance MOS capacitor for on-chip analog memory and the designated threshold voltage MOS transistors for SF based signal readout chain with 3.3 V power supply voltage, eight-times higher photosensitivity and 50% reduction of power consumption were simultaneously achieved. The developed UHS CMOS image sensor and high speed video camera are effective to capture and analyze various UHS phenomena.

From now onward, the main factors that obstruct higher speed and higher pixel- and memory-counts are as follows:

- Inductance of package-interconnect and pad-bonding-line
- Power consumption and heat generation
- Electro-migration in metal wires

And the challenges to be tackled in order to overcome these obstacles are:

- A large number of on-chip memories with high area efficiency capacitors
- Low-heat-generation and heat-sink
- Short interconnect by stacked-chip

Looking at the history, drastic improvements in the science and technology developments were driven to occur when extraordinarily advanced measurement and analysis technologies emerged.

UHS imaging technology is now expected to be utilized more than ever for the technological innovation of a wide range of science and engineering fields such as material science, electronic engineering, nanotechnology and life sciences.

References

1. N. Kudo, K. Okada, K. Yamamoto, Sonoporation by single-shot pulsed ultrasound with microbubbles. Biophys. J. **96**, 4866–4876 (2009)
2. M. Versluis, High-speed imaging in fluids. Exp. Fluids **54**, 1–35 (2013)
3. N. Kawahara, K. Tsuboi, E. Tomita, Laser-induced plasma generation and evolution in a transient spray. Opt. Express **22**, A44–A52 (2014)
4. T. Belmonte, A. Hamdan, F. Kosior, C. Noel, G. Henrion, Interaction of discharges with electrode surface in dielectric liquids: application to nanoparticle synthesis, J. Phys. D: Appl. Phys. **47**, 224016-1-18 (2014)
5. K. Kurosawa, Y. Nagaoka, H. Senshu, K. Wada, S. Hasegawa, S. Sugita, T. Matsui, Dynamic of hypervelocity jetting during oblique impacts of spherical projectiles investigated via ultrafast imaging, J. Geophys. Res.: Planets **120**, 1237–1251 (2015)
6. N. Kawai, S. Zama, W. Takemoto, K. Moriguchi, K. Arai, S. Hasegawa, E. Sato, Stress wave and damage propagation in transparent materials subjected to hypervelocity impact. Proc. Eng. **103**, 287–293 (2015)
7. H. Zhu, F. Pierron, Exploration of Saint-Venant's principle in inertial high strain rate testing of materials. Exp. Mech. **56**, 3–23 (2016)
8. F. Shao, D. Kimoto, K. Furukawa, H. Sugo, T. Takeda, K. Miyauchi, Y. Tochigi, R. Kuroda, S. Sugawa, Ultra-high speed video capturing of time dependent dielectric breakdown of metal oxide-silicon capacitor up to 10 Mfps, in *Proceedings IS&T/SPIE Electronic Imaging*, vol 9022, pp. 902205-1-9, San Francisco, USA, 2-6 Feb 2014
9. R. Kuroda, F. Shao, D. Kimoto, K. Furukawa, H. Sugo, T. Takeda, K. Miyauchi, Y. Tochigi, A. Teramoto, and S. Sugawa, A novel analysis of oxide breakdown based on dynamic observation using ultra-high speed video capturing up to 10,000,000 frames per second, in *Proceedings of the 2014 International Reliability Physics Symposium*, pp. 3F.3.1–3F.3.4, Waikoloa, USA, 1–5 June 2014

10. S. Sugawa, Ultra-high speed imaging, in *IEEE International Solid-State CIRC. Conference on Forum, Scientific Imaging*, San Francisco, USA, 17–21 Feb 2013
11. T. G. Etoh, D.H. Nguyen, S.V.T. Dao, C.L. Vo, M. Tanaka, K. Takehara, T. Okinaka, H.V. Kuijk, W. Klassens, J. Bosiers, M. Lasser, D. Ouellette, H. Maruyama, T. Hayashida, T. Arai, A 16 Mfps 165 kpixel backside-illuminated CCD, in *IEEE International Solid-State CIRC. Conference, Dig. Tech.*, pp. 406–407, San Francisco, USA, 20–24 Feb 2011
12. J. Crooks, B. Marsh, R. Turchetta, K. Taylor, W. Chan, A. Lahav, A. Fenigstein, Kirana: a solid-state megapixel uCMOS image sensor for ultra-high speed imaging, in *Proceedings of the IS&T/SPIE Electronic Imaging*, vol 8659, 865903-1-14, Burlingame, USA, 3–7 Feb 2013
13. J. Crooks, B. Marsh, R. Turchetta, K. Taylor, W. Chan, A. Lahav, A. Fenigstein, Ultra-high speed imaging at megaframes per second with a megapixel CMOS image sensor, in *Proceedings of the International Image Sensor Workshop*, pp. 369–372, Snowbird, USA, 12–16 June 2013
14. Y. Tochigi, K. Hanzawa, Y. Kato, R. Kuroda, H. Mutoh, R. Hirose, H. Tominaga, K. Takubo, Y. Kondo, S. Sugawa, A global-shutter CMOS image sensor with readout speed of 1-Tpixel/s burst and 780-Mpixel/s continuous. IEEE J. Solid-St. Circ. **48**, 329–338 (2013)
15. K. Miyauchi, T. Takeda, K. Hanzawa, Y. Tochigi, S. Sakai, R. Kuroda, H. Tominaga, R. Hirose, K. Takubo, Y. Kondo, S. Sugawa, Pixel structure with 10 nsec fully charge transfer time for the 20 M frame per second burst CMOS image sensor, in *Proceedings of the IS&T/SPIE Electronic Imaging*, vol 9022, 902203-1-12, San Francisco, USA, 2–6 Feb 2014
16. S. Sugawa, R. Kuroda, T. Takeda, F. Shao, K. Miyauchi, and Y. Tochigi, A 20 Mfps global shutter CMOS image sensor with improved sensitivity and power consumption, in *Proceedings of the International Image Sensor Workshop*, pp. 166–169, Vaals, the Netherlands, 8–11 June 2015
17. R. Kuroda, Y. Tochigi, K. Miyauchi, T. Takeda, H. Sugo, F. Shao, S. Sugawa, A 20Mfps global shutter CMOS image sensor with improved light sensitivity and power consumption performances. ITE Trans. MTA **4**, 149–154 (2016)
18. F. Mochizuki, K. Kagawa, S-I. Okihara, W-M. Seo, B. Zhang, T. Takasawa, K. Yasutomi, S. Kawahito, Single-shot 200Mfps 5 × 3-aperture compressive CMOS imager, in *IEEE International Solid-State CIRC. Conference on Dig. Tech.*, pp. 116–117, San Francisco, USA, 22-26 Feb. 2015
19. G. Meynants, G. Lepage, J. Bogaerts, G. Vanhorebeek, X. Wang, Limitations to the frame rate of high speed image sensors, in *International Image Sensor Workshop*, P-16, Bergen, Norway, June 2009
20. G. Meynants, High-speed CMOS image sensor architectures, in *IEEE International Solid-State CIRC. Conference on Forum, High-Speed Image-Sensor Technologies, San Francisco*, USA, 7-11 Feb. 2010
21. JIS Z 8120. Glossary of optical terms
22. Y. Tochigi, K. Hanzawa, Y. Kato, N. Akahane, R. Kuroda, S. Sugawa, A prototype high-speed CMOS image sensor with 10,000,000 fps burst-frame rate and 10,000 fps continuous-frame rate, in *Proceedings of the IS&T/SPIE Electronic Imaging, Digital Photography VII*, 78760G-1-8, Barlingame, USA, 23–27 Jan 2011
23. M.F. Tompsett, G.M. Amelio, W.J. Bertram, R. Buckley, W.J. McNamara, J.C. Mikkelsen Jr., D.A. Sealer, Charge-coupled imaging devices: experimental results. IEEE T. Electron Dev. **18**, 992–996 (1971)
24. Zhou, B. Pain, E. Fossum, frame-transfer CMOS active pixel sensor with pixel binning, IEEE T. Electron Dev. **44**, 1764–1768 (1997)
25. R. Kuroda, S. Kawada, S. Nasuno, T. Nakazawa, Y. Koda, K. Hanzawa, S. Sugawa, A highly ultraviolet light sensitive and highly robust image sensor technology based on flattened Si surface. ITE Trans. MTA **2**, 123–130 (2014)
26. Y. Kondo, K. Takubo, H. Tominaga, R. Hirose, N, Tokuoka, Y. Kawaguchi, Y. Takaie, A. Ozaki, S. Nakaya, F. Yano, T. Daigen, Development of "Hyper-Vision HPV-X" high-speed video camera, Shimadzu Review, SR13_002E, 2013

27. A.M. Knisely, J.M. Austin, geometry and test-time effects on hypervelocity shock-boundary layer interaction, in *54th AIAA Aerospace Sciences Meeting, AIAA SciTech, AIAA 2016–1979*, pp. 1–15, San Diego, USA, 4–8 Jan 2016
28. T. Shibata, Y. Zama, T. Furuhata, H. Kusano, Ultra-high speed PIV measurement for gasoline spray ejected from a multiple hole DI injector, in *18th International Symposium on the Applications of Laser and Imaging Techniques to Fluid Mechanics*, Lisbon, USA, 4–7 July 2016
29. K. Johansen, J.H. Song, K. Johnston, P. Prentice, Deconvolution of acoustically detected bubble-collapse shock waves. Ultrasonics **73**, 144–153 (2017)
30. N. Kudo, N. Tokuoka, High-speed microscopic observation of ultrasonically activated micron-sized bubbles interacting with biological cells of low image contrast, in *Proceedings of the 31st International Congress on High-speed Imaging and Photonics*, pp. 783–786, Osaka, Japan, 7–10 Nov 2016
31. A. Rack, M. Olbinado, M. Scheel, C. Schmitt, O. Betz, F. Garcia-Moreno, J. Banhart, X. Just, P. Lhuissier, J.-L. Gelet, J. Morse, Real-time X-ray microimaging using hard synchrotron radiation, in *Proceedings of the 31st International Congress on High-speed Imaging and Photonics*, pp. 629–632, Osaka, Japan, 7–10 Nov 2016

Part IV
Shock Waves

High-Speed Imaging of Shock Waves and Their Flow Fields

Harald Kleine

Abstract High-speed imaging has a long history in the field of shock wave research, but only in the last two decades has it become a frequently used diagnostic tool, primarily through the availability of powerful, compact and user-friendly digital high-speed cameras. In this research field, the high-speed imaging must usually be combined with an adequate visualisation method to make the shock waves and their flow fields visible. This chapter gives an overview of the used visualisation techniques and cameras. A few examples of applications are presented that highlight the potential but also the limitations of the available techniques to obtain time-resolved visualisations of compressible flows.

1 Introduction

This chapter gives a brief overview of the methods and instrumentation used to obtain time-resolved records of shock waves and their flow fields. The two main challenges in this area of research are that shock waves typically move with speeds of the order of the speed of sound, and that they are invisible unless one applies special visualisation techniques. Both challenges have been addressed since the inception of this research field, and they have led to numerous unique developments both with respect to test and visualisation methods as well as high-speed imaging instrumentation. Thanks to the recent advances in high-speed camera technology, the time-resolved imaging of processes involving shock waves has been substantially simplified, and numerous new applications have been made possible. This chapter will first give a brief phenomenological description of shock waves and

Electronic supplementary material The online version of this chapter (doi:10.1007/978-3-319-61491-5_6) contains supplementary material, which is available to authorized users.

H. Kleine (✉)
School of Engineering and IT, University of New South Wales,
Northcott Drive, Canberra ACT 2600, Australia
e-mail: h.kleine@adfa.edu.au

K. Tsuji (ed.), *The Micro-World Observed by Ultra High-Speed Cameras*,
DOI 10.1007/978-3-319-61491-5_6

then outline test facilities and visualisation techniques. This is followed by a discussion of some features unique to this area of research and a few examples of applications. Evaluation methods for qualitative and quantitative measurements in shock wave flow fields are introduced at the end of this chapter.

2 Shock Waves

A shock wave is a thin region within a medium that has the following characteristic features: matter crosses this region, and when doing so, the properties of this matter change by typically large amounts in a near-discontinuous, non-isentropic fashion. The amount of change depends non-linearly on the speed with which the thin region is crossed. This chapter will deal exclusively with such waves in gases and—to a lesser extent—in liquids. In solids, similar processes exist, but their description and experimental verification through high-speed visualisation (the main subject of this chapter) is in most cases considerably different and will therefore not be elaborated on. More detailed descriptions of the research related to shock waves, as well as its history and the common roots of shock wave research activities in gaseous, liquid and solid media (as well as mixtures of the above), and the different paths taken later on in these different fields, are comprehensively provided in [1].

The actual thickness of the "thin" region is related to the molecular structure of the medium—in gases, this thickness is typically a multiple of the mean free path length, that is, the average distance a molecule travels before it collides with another molecule. For most gases at pressures above approximately 100 Pa, the shock wave thickness is of the order of 10^{-7} m, which is considerably smaller than any characteristic length scale of the vast majority of applications in which shock wave phenomena have so far been investigated.[1] For this reason, shock waves can typically be treated as ideal discontinuities. They are formed when a process in a medium occurs at a speed larger than the medium's characteristic speed, that is, the speed at which a disturbance can propagate isentropically through this medium (for gases and liquids, this is the speed of sound). There are two main types of processes, in which this characteristic speed is exceeded (see Fig. 1):

1. a rapid release of energy in a medium, such as in explosions and related processes, or
2. objects propagating at a speed higher than the speed of sound of the surrounding medium.

In the first scenario (Fig. 1a), the shock wave moves away from the source of the energy release. If it changes its velocity as a function of distance from the source or if it interacts with an object in its path, this propagation is always a fully unsteady process, and the majority of all observed propagation speeds is of the order of the speed of sound of the medium. In the second scenario, if the speed of the moving

[1]The issue of minuscule shock waves will be briefly discussed in Sect. 4.

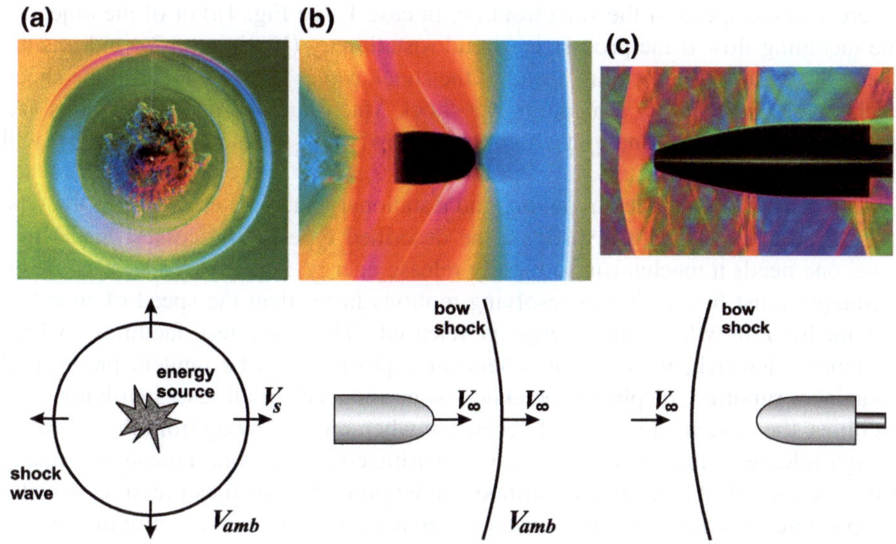

Fig. 1 Visualisations and schematic sketches of a **a** moving shock (explosion of 10 mg of silver azide, direction-indicating colour schlieren, see [2] for details), **b** pseudo-stationary shock (9 mm pistol bullet flying at $M = 1.05$, magnitude-indicating colour schlieren), **c** stationary shock (transonic flow of $M = 1.1$ over a projectile model, direction-indicating colour schlieren) Note that the visualisations show considerably more flow features that what is indicated in the sketches. V_{amb} is the speed of the ambient medium before being processed by the shock—in most applications, one has $V_{amb} = 0$

object is constant and if there are no instability processes occurring as a result of the object's movement, the distance of the shock wave with respect to the object remains constant (Fig. 1b). This process is labeled "pseudostationary" as it can be converted into a stationary one by means of a coordinate transformation, that is, by describing the process from the point of view of an observer moving with the object rather than that of a stationary observer (Fig. 1c).

It is common to all scenarios involving shock waves that the properties of the material, such as its pressure p, its density ρ, its temperature T and/or the velocity V of the material's components, change irreversibly across the shock wave by increments Δp, $\Delta \rho$, ΔT and ΔV, which are typically of at least the same order of magnitude compared to the absolute values of these properties ahead of the shock. Analytically derived equations for these changes exist, which can be found in almost all textbooks on compressible flows, e.g. [3]. These derivations indicate that the main parameters responsible for the change of thermodynamic properties across a shock wave are the angle between the flow and the shock wave, the isentropic exponent γ of the gas and primarily the shock Mach number M, defined as

$$M = V/a, \quad (1)$$

where V is the speed of the wavefront (V_s in case 1, see Fig. 1a) or of the object or the incoming flow if the process is (pseudo-)stationary (V_∞ in case 2), and a is the speed of sound of the medium ahead of the shock wave. In this chapter, if the shock itself is moving through a medium, the shock Mach number will be given as M_S, and if the shock is stationary in a high-speed flow, the associated Mach number will be labelled as M_∞.

The distinction between a moving and a stationary shock is essential for the type of test facility needed to investigate the described types of processes: in the first case, one needs a mechanism to rapidly release energy, and any visual observation technique must be capable of resolving motions faster than the speed of sound of the medium in which the energy is released. The used test facilities include chambers in which, for example, chemical explosives can be ignited, pressurised containers ruptured, or plasma breakdowns be induced. In all tests involving such facilities, the generated wave will decelerate when moving away from the source of energy release as the released amount is distributed over a continuously increasing area. As a result, the profile of thermodynamic properties such as pressure, density, temperature or induced particle velocity across the moving wavefront typically has the triangular shape indicated in Fig. 2a—the wave is actually a combination of a leading shock front immediately followed by an expansion zone. Such a combination is commonly referred to as a blast wave [4]. A pure shock front without an immediate subsequent expansion, that is, a near-ideal step function of the aforementioned thermodynamic properties of pressure, density, temperature or induced particle velocity across the moving wavefront (see Fig. 2b), can be produced in another test facility, the shock tube, which avoids the three-dimensional expansion of the flow. The shock tube consists of two sections filled with gases at different pressures and can be interpreted as a pressurised vessel, where after bursting the released gas can only move in one direction through a channel of usually constant cross section. This type of facility is commonly used to generate constant-velocity shock waves and to investigate their interaction with obstacles and area changes. As shock tubes eliminate the dependence of wavefront strength on distance from the point of energy release (and also a number of safety concerns related to this release), they have become a frequently used laboratory tool to investigate shock wave-related processes. Design criteria and performance characteristics of such facilities are extensively described in monographs and reports such as [5–7].

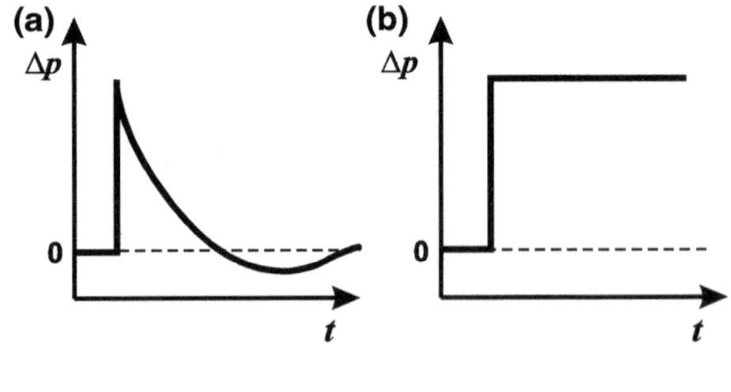

Fig. 2 **a** Blast wave profile **b** shock wave profile

The test facilities for the generation of macroscopically steady high-speed flows (scenario 2) are wind tunnels, and these facilities are typically used to investigate the behaviour of objects moving at speeds that range from fractions to multiples of the speed of sound. For such studies, only the relative velocity between object and surrounding medium is relevant, and therefore in many cases it does not matter if the object is moving through a typically quiescent medium or if this medium is flowing over a stationary object. This corresponds to the aforementioned coordinate transformation, through which an unsteady process is made steady, as indicated in Fig. 1b, c. For a macroscopically steady flow, an observation would not necessarily require high-speed imaging—but even in such cases, there usually exist areas in the flow that are marked by unsteady fluctuations, which can occur at speeds of the order of the speed of sound. While the flows generated by a rapid energy release are inherently unsteady in a wide range of temporal and spatial scales, the unsteadiness of the flows of scenario 2 is often a modulation of a macroscopically steady process. Examples for both types will be given in Sect. 5.

3 Visualisation Techniques

Shock waves in transparent gases and liquids are invisible unless one uses special techniques to make them visible. This can be done, in most cases, by exploiting the fact that a shock wave substantially changes the density of the medium it traverses, as indicated in Sect. 2. This changes the speed of light in this medium—the denser the transparent medium is, the lower is the speed of light in it (for a derivation of the equations that quantify this dependency, see, e.g., [8]). A substantial density change such as the one introduced by a shock wave must therefore affect the light that passes through it. This can, indeed, be seen without special optical systems in events with strong shock waves such as large-scale explosions: in high-speed recordings of such events one can observe that the background is distorted by the blast wave (see Fig. 3a). Such effects are also visible for less energetic everyday processes, such as layers of hot air above a surface such as the roof of a car (Fig. 3b).

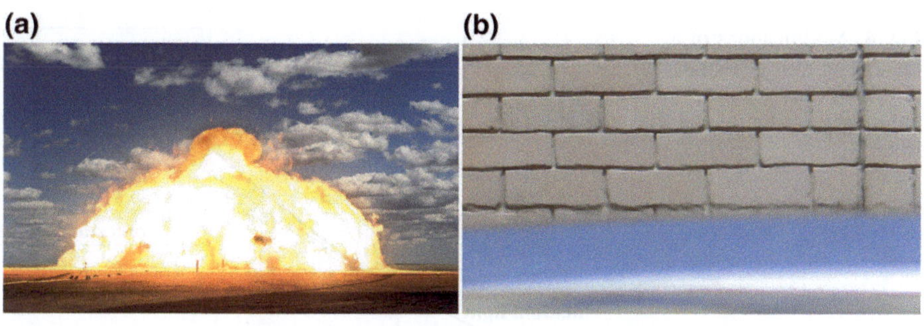

Fig. 3 Background distortion as a result of density changes **a** blast wave from a large-scale explosion (image courtesy of Dr. Fan Zhang, DRDC Suffield, Canada) **b** layer of hot air above a car roof

In most investigations of shock waves and their flows, the mentioned dependency of the refractive index on the density of the medium is the basis of the used techniques to make the flow visible. Typically, a light beam of known properties is sent through the test section where it is affected by the density changes in the flow [8]. As a result of the refractive index changes resulting from the varying density, the light beam is displaced by amounts Δx and Δy, deflected by angles ε_x and ε_y, and retarded by a time Δt as indicated in Fig. 4. Each of these modifications can be independently measured, and each measurement delivers a different derivative of the density of the medium the light beam has traversed: the displacements Δx and Δy are related to the second density derivative and are (simultaneously) recorded with a set of techniques known as shadowgraph methods, the angular deflections ε_x and ε_y are proportional to the first density derivatives ($\partial\rho/\partial x$ or $\partial\rho/\partial y$, with coordinates x and y defined as in Fig. 4) and are measured with so-called schlieren setups, while the time change Δt can be measured as a phase shift which is directly proportional to the density itself. For the latter measurement one uses so-called reference beam interferometry. Each of the mentioned sets of methods has been developed into a multitude of different systems and versions, but the basic measurement principle is the same for all techniques, namely to determine the modifications a light beam traversing the test object experiences and to relate them to the processes that caused these modifications.

A large number of monographs and review papers can be found in the literature that provide details of these methods, their operation, measurement characteristics and setup details, e.g. [8–10]. In addition to standard laboratory setups, which are used for the examples shown in this chapter, there also exist large-scale visualisation systems (retroreflective shadowgraphy, lens-and-grid schlieren, background-oriented schlieren), which typically do not require high-quality optical components (usually the main factor that limits the field of view) and are also robust enough to be used outdoors and in harsh environments. These systems are based on ideas that were already published in the middle of the 20th century (e.g. in [11]), but their actual implementation is a more recent development, facilitated by technology

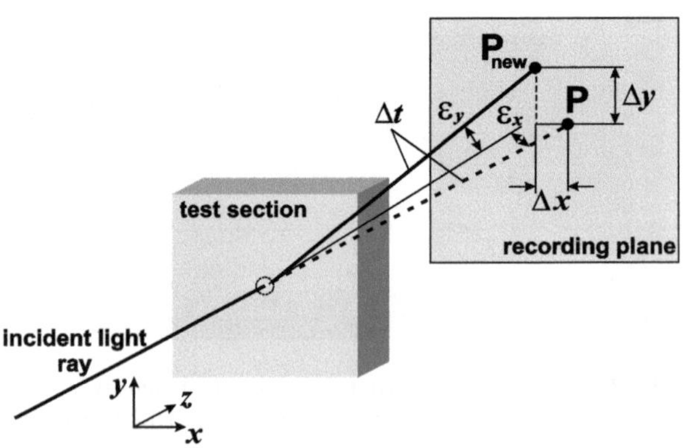

Fig. 4 Modifications of a light beam as a consequence of density changes in a transparent medium (adapted from [8])

advances and stimulated by the growing need to visualise flow fields that are larger than the regular test sections in a laboratory. There already exist a few review articles that summarise different versions and features of these setups [12–14]. As a result of the larger size of the visualised objects, time-resolved operation of these systems requires lower frame rates than in the case of the laboratory-based setups, as will be discussed in more detail in Sect. 4.

Common to all of the above mentioned systems is their line-of-sight character—they record the accumulated changes the light beam experiences during its entire passage through the test object (and the surrounding environment). A direct correlation between measured light beam modification and density variation inside the test section is therefore only possible if there are no disturbances/modifications generated in the environment and, more importantly, if the flow is highly two-dimensional in that its changes in the direction of the traversing light beam are negligible. Fully three-dimensional flows can also be readily visualised (see, e.g., Fig. 1), but it is more difficult to quantify the density changes that have generated the measured light beam modifications. Typically, such a quantification involves a tomography approach with simultaneous visualisations from different angles and/or a priori knowledge about the symmetry of the observed flow. The techniques associated with such studies are beyond the scope of this overview.

As outlined at the beginning of this section, each set of methods (shadowgraph, schlieren, interferometry) records a different modification of the light beam passing through the test object, and each modification is related to a different derivative of the density of the visualised flow. This also means that the same physical object, in this case a shock wave and its associated flow features, is differently represented in each visualisation method. In addition, there exist several variations of each set of techniques, which allow one to display the same modification of the light beam in different ways. Figure 5 illustrates this point—it shows the same flow field, visualised (in separate experiments run under identical initial conditions) by a selection of the different methods. In each case the visualisation shows a shock exiting an open tube (inner diameter 20 mm). The shock is followed by a vortex ring and a supersonic outflow from the tube. Details of the test facility and the diagnostics used are given in [15].

The largest variety of display possibilities arguably exists for the density gradient, captured with the schlieren techniques or shearing interferometry. The density gradient is a vector, and the different versions of the visualisation system capture its different properties. One can either record the vector's magnitude or its direction and display the result either in grayscale or colour. The traditional and most frequently used schlieren technique with a straight schlieren cutoff or knife edge and a monochrome light source detects the magnitude of the density gradient in a preset direction (perpendicular to the knife edge) and displays it in grayscale (Fig. 5b). This can be extended to a magnitude-indicating colour schlieren system if one modifies light source and cutoff in a way that the amount of light deflection is coded by different hues rather than intensity variations (Fig. 5c). The introduction of colour usually increases the measurement range but lowers the maximum sensitivity as a comparison of Fig. 5b, c shows. If one chooses a circular cutoff, the

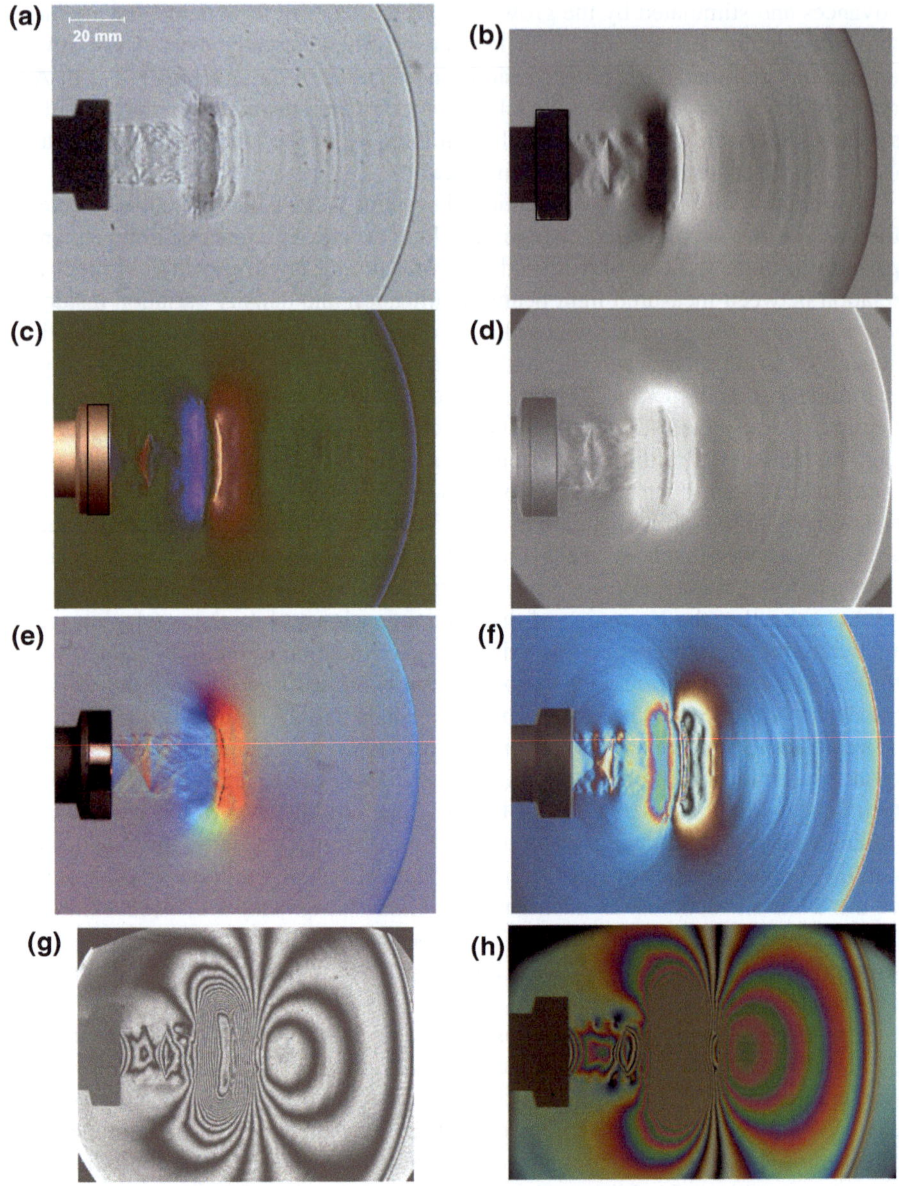

Fig. 5 Visualisation of the shock-vortex system emitted from an open shock tube, nominal shock Mach number $M_S = 1.44 \pm 0.04$; the images show the flow field approximately 280 μs after the shock has exited the tube. The images were taken with different optical systems and cameras so that the field of view is not always identical. **a** shadowgraph **b** monochrome schlieren, vertical knife edge **c** magnitude-indicating colour schlieren, vertical cutoff **d** monochrome omnidirectional schlieren (circular cutoff) **e** direction-indicating colour schlieren **f** shearing interferometry (vertical fringes) **g** monochrome Mach-Zehnder interferometry, infinite fringe mode **h** polychrome Mach-Zehnder interferometry, infinite fringe mode

overall magnitude of the gradient is recorded, irrespective of its direction (Fig. 5d). With an adequately shaped light source consisting of differently coloured segments and a matching schlieren cutoff (typically circular), one can also capture the direction of the density gradient—this is referred to as direction-indicating colour schlieren (Fig. 5e). Shearing interferometry also measures the density gradient in a preset direction (similar to the schlieren technique with a straight cutoff), but here the measurement principle is that a phase shift between two neighbouring light beams is recorded in the form of interference fringes, which can be polychrome if a white light source is used (Fig. 5f). The method depends on "comparing" two neighbouring beams separated by a given distance or shear, and as a result of this feature, discontinuities that block one of the beams are represented by a double image with the width of the shear. This includes solid boundaries and flow discontinuities such as shock waves and shear layers. On the other hand, the evaluation/measurement of the density gradient in discontinuity-free parts of the flow field is usually easier than for the schlieren methods [10].

Shadowgrams and interferograms record the distributions of scalar fields without inherent directions, which somewhat limits the number of ways the observed flow features can be displayed. Shadowgrams are typically monochrome and literally show the shadow cast by the flow field due to its refractive index variations (Fig. 5a). Their appearance is mainly influenced by the choice of the plane onto which the shadow is cast. In earlier laboratory applications the side window of a test section, with a screen or photographic film attached, served as the shadow plane. If the shadow plane itself is imaged (in this case the plane itself is virtual), the imaging process allows one to place this plane at (almost) any arbitrary position, even in front of the test object. This changes the way the shadow is displayed and also influences the sensitivity of the method.

In an interferometer operated in so-called infinite fringe mode, the interfering light beams are aligned in a way that in the absence of density variations, their phase difference is zero or a multiple of the wavelength of the used light. If the shock wave and its flow influence the propagation of light in one of the two interfering beams, fringes appear that are related to the experienced phase shift (Figs. 5g, h). In a so-called reference-beam interferometer, where one of the interfering light beams goes through the test section while the other passes through a region of constant and known reference density, these fringes then correspond, in planar flows, to so-called isopycnics or lines of constant density. In three-dimensional flows the fringes are lines of constant phase shift, which has to be converted into density contours by mathematical procedures like Abel or Radon transformations (see, e.g., [8]). A monochrome infinite fringe interferogram of a planar flow will show contours of constant density, and the increment between two adjacent contours follows from [10]

$$\Delta\rho = \frac{\lambda}{KL}, \tag{2}$$

where K is the Gladstone-Dale constant of the test gas, L is the width of the test section and λ is the wavelength of the light that has been used to generate the interferogram. Note that the sign of the increment is undetermined. Therefore, in a two-dimensional flow field visualised by such an interferogram, one needs the value of the density on at least one of the contours and some information on the direction of density change if one wants to quantify the density distribution in the flow. This problem can be avoided in polychrome reference beam interferometry, where each fringe is identifiable through its colour (Fig. 5h). A remaining ambiguity caused by the symmetry of the interference fringes around the zero-order fringe can in most cases be minimised by choosing a fringe of a different order as central fringe and background colour. Some additional information about the flow, such as clearly identifiable zones in which the sign of the density change is known, is usually required to complete the correlation between fringe colour and position and density value.

4 Special Features of High-Speed Imaging of Shock Waves

For many of the processes involving shock waves, the nature of the observed phenomenon almost inevitably requires the use of high-speed cameras. Following the ideas outlined in [16] and later refined by Versluis [17], one can estimate the required frame rate of such a camera as a function of the speed V of the to-be-imaged object and its size (or the size of the field of view, respectively). Such an estimate together with an indication of the relevant regimes of shock wave visualisation is provided in Fig. 6, which should be understood as an order-of-magnitude guideline. A similar diagram provided in [17] includes specific information on available high-speed cameras—this is not done here as the main purpose of Fig. 6 is to indicate general trends. It is readily obvious from the figure that events involving shock waves can only be adequately recorded with frame rates of 10^4 fps or below if they occur on a large scale (≥ 10 m), but that most laboratory-scale investigations require cameras capable of frame rates one or two orders of magnitude higher. If one is interested in minute details of shock patterns and their temporal development, or if one observes shocks in liquids or transparent solids, where all waves travel with speeds of the order of 1 km/s, frame rates in excess of 10^7 fps are required. These requirements have led to the development of a few high-speed camera types that have almost exclusively been used for applications in which shock waves or related phenomena were visualised. An overview of the basic high-speed imaging systems built for such purposes is given in [16].

The domains indicated in Fig. 6 are calculated in the following way:

- the size of the object and the size of the recording sensor determine the imaging factor of the optical system that is to record the event (in the calculations for Fig. 6, a sensor size of 20 mm is assumed);

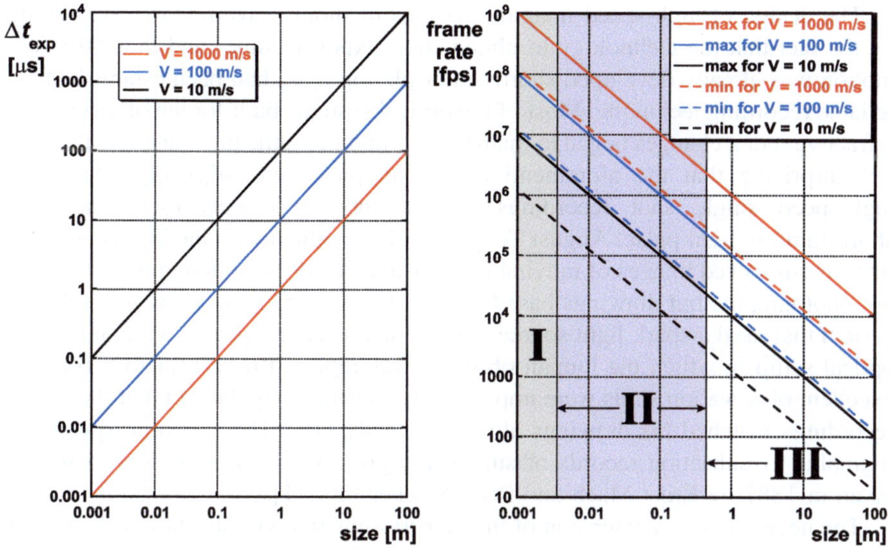

Fig. 6 Required frame exposure time (*left*) and frame rate ranges (*right*) as a function of object or test section size and object speed *V*; region I: flows with minuscule elements region II: shock tube flows region III: large-scale blast waves

- the speed of the event is then scaled down by this imaging factor, e.g. $V = 1$ km/s of a 10 m large object becomes an "imaged" speed of 2 m/s if the imaging factor is 1/500 for the assumed sensor size;
- the maximum exposure time then follows from the time the imaged object would need to travel one pixel on the recording sensor; if one assumes an average pixel size of 20 μm, this time would be 10 μs for the aforementioned example of a 10 m large object traveling at $V = 1$ km/s;
- the corresponding maximum frame then is the inverse of the maximum exposure time—any higher frame rate would not lead to a discernible movement of the object between subsequent frames; in the above example, this maximum frame rate is 10^5 fps.
- the minimum frame rate follows from the number of frames that should be recorded of the event—if Δt is the time it would take the object to leave the field of view (typically this time follows from the maximum exposure time times the number of pixels in the direction of motion), then the minimum frame rate is $nf/\Delta t$ where nf is the required number of frames; in Fig. 6, a total number of 200 pixels and a required number of frames of $nf = 25$ was assumed.

The above estimate can be refined by entering actual test and camera data and also by refining the optical resolution criterion as it was done in [17]. The resulting boundaries can then be used to determine the basic camera requirements. The diagrams of Fig. 6 indicate these requirements for the most common applications of shock wave visualisation, namely large-scale blasts, shock tube tests and flow fields with minuscule elements to be resolved.

The design of high-speed imaging systems in shock wave research was helped by the fact that the technology to obtain short exposure times and thus “freeze” a motion was already developed shortly after photography had been established as a reliable recording technique. Most of the optical systems built for the observation of refractive index changes began to mature at around the same time, and it is therefore not surprising that the aforementioned techniques were originally applied for high-speed single shot recordings of fast moving objects by means of a short-duration light pulse. August Toepler [18] is credited with having recorded the first time-resolved images of moving shock waves in 1864, although his records are not photographic but drawings based on the observations he made with a schlieren system that used a spark light source. His work was later repeated with an identical optical setup in which the human observer was replaced by a camera—Toepler’s excellent observation skills were impressively illustrated by the fact that the camera recordings matched his drawings almost perfectly [19]. The first high-speed photographic visualisation records of supersonic projectiles and supersonic jets were taken in 1887 by Ernst Mach and Peter Salcher [20, 21].

The necessity of an extension of the density-sensitive visualisation techniques to time-resolved operation was soon recognised. The main technological difficulty associated with the required rapid separation of recorded images was initially overcome by using an array of spark sources that are fired in a predetermined sequence and that create an array of images (Fig. 7). This so-called Cranz-Schardin camera [22] could achieve frame rates up to 10^6 fps as early as 1929 and became one of the first cameras frequently used for time-resolved shock wave

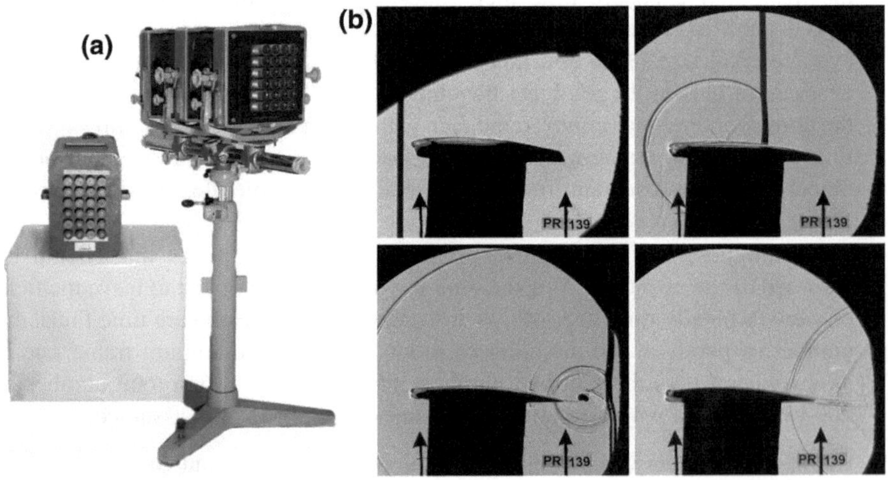

Fig. 7 Cranz-Schardin camera **a** light source array and camera **b** selected frames of a 24-frame shadowgraph series of a shock wave passing over an airfoil; the shown frames clearly demonstrate the parallax created by the different light paths. Details of the shown process and the test facility can be found in [23]. Images courtesy of Prof. H. Olivier, Stoßwellenlabor, RWTH Aachen University, Germany

visualisations. The individual images have single-image quality and resolution as the system has no moving parts. The main drawback is that the light beam emitted by each spark source passes through the test section at a slightly different angle, which gives all images a different degree of parallax. This parallax and the size of the individual spark units also limit the number of obtainable images, which is typically in the range between 4 and 24.

A parallel development that started in the early 1920s was the design of rotating mirror cameras, where the light beam used for the imaging of the flow is swept across a given length of recording material. If the light beam is pulsed or if a sequence of stops inside the camera interrupts it at certain intervals, multiple images of the observed process can be obtained [24, 25]. The built cameras could achieve frame rates that exceeded 10^6 fps (some by more than an order of magnitude). The main advantage of these cameras is that they can produce a reasonably large number (>20) of parallax-free frames. Disadvantages include image-quality deteriorating effects caused by the motion of the light beam (primarily motion blur and light diffraction) and mechanical issues related to driving the rotating mirror at very high speeds. Special turbines and mirror materials had to be used to obtain frame rates above 10^6 fps, which made the cameras not only difficult to operate but also turned them into potential health hazards when the speeds were high enough to cause structural failure of the mirror. Figure 8 shows an example of a rotating mirror camera (diameter 1.1 m, maximum speed of the mirror 640 rev/s) that has to

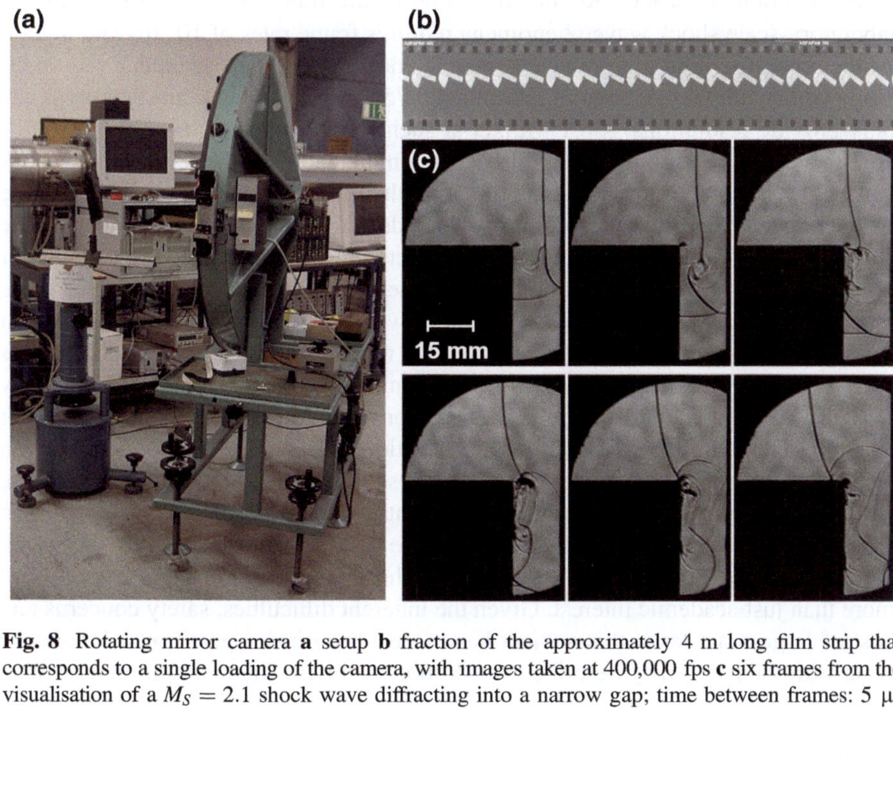

Fig. 8 Rotating mirror camera **a** setup **b** fraction of the approximately 4 m long film strip that corresponds to a single loading of the camera, with images taken at 400,000 fps **c** six frames from the visualisation of a $M_S = 2.1$ shock wave diffracting into a narrow gap; time between frames: 5 μs

be run with a stroboscopic light source, together with a representative set of frames that show the diffraction of a shock wave into a gap.

Apart from the two camera types described in the previous paragraphs, a large number of different models and variants was developed during the 20th century, both by commercial manufacturers and by individual research groups. A reasonably high level of complexity and the need for a specifically trained operator was a common feature of most of these systems, in addition to these cameras often being physically quite large and therefore not easy to implement in an optical system. Price was another factor that prevented such cameras from becoming standard equipment.

This situation was fundamentally changed (with the possible exception of the price factor) when digital recording media started to replace photographic film material in the last decade of the 20th century. The ability to perform image separation and exposure control electronically made the complex optomechanical arrangements of the earlier high-speed camera types superfluous, and this also reduced the overall size of the cameras. The fact that the cameras became increasingly user-friendly also gradually eliminated the role of the high-speed camera specialist in areas other than those specifically involved in the development of such cameras. Today's digital high-speed cameras still have a number of technological issues to overcome, but they have become much more of a standard diagnostic tool than their analogue predecessors. Other chapters in this book will give a more detailed overview of the current level of camera technology.

As mentioned earlier and shown in Fig. 6, the time-resolved visualisation of laboratory-scale shock wave phenomena requires frame rates of 10^5 fps and higher. At the time of writing this overview, no camera exists that can operate at this speed range with spatial resolutions of more than 0.75 Mpixel—in fact, most cameras can only provide resolutions ≤ 0.1 Mpixel at such frame rates.

This issue is central to an ongoing debate on the nature of the reflection of shock waves. It has been known for some time that the reflection pattern of a shock wave interacting with an obstacle primarily depends on both the Mach number of the shock wave and the angle at which it interacts with the obstacle. This behaviour highlights the previously mentioned non-linear character of shock waves. One distinguishes between a regular reflection in which one observes only the incident and the reflected shock wave, and an irregular reflection, where additional elements such as a third shock (known as the Mach stem) and a contact surface, and possibly further wavefronts, are formed, see Fig. 9a. For an overview of the various types of irregular reflections, see, e.g., [26]. The transition from one pattern to the other can be theoretically predicted, but experimental observations tend to disagree with these predictions, sometimes by substantial amounts. The nature of the reflection can significantly alter the pressure loading of the structure in the path of the shock wave, which is why a proof or disproof of the theoretical transition criteria is of more than just academic interest. Given the inherent difficulties, safety concerns and cost of large-scale tests, this subject has mostly been investigated on a laboratory scale with the previously described shock tube facilities. The main problem in such

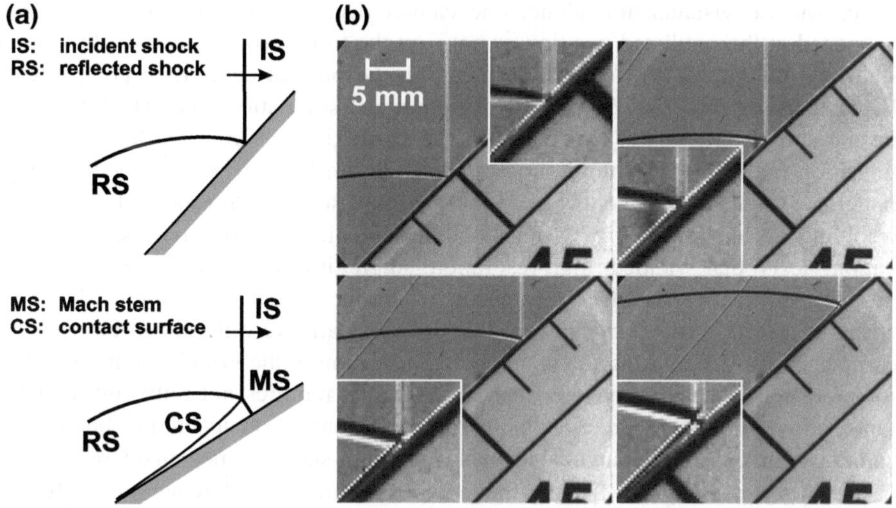

Fig. 9 Shock reflection **a** regular (*top*) and irregular (*bottom*) reflection **b** four frames from a shadowgraph series of a shock ($M_S = 1.35$) reflecting off a cylindrical arc of 450 mm radius, with 3.3 times magnified views of the reflection point area as inserts for each frame. The imaging factor of the used optical system was 0.707, the images were taken with 10^6 fps. The time interval between two successive frames shown here is 10 μs. A Mach stem is only clearly visible for polar angles larger than approx. 46° (third frame), but the emerging contact surface can already be discerned for polar angles of approx. 45°—according to theoretical predictions, however, both features should already begin to exist for polar angles above approx. 43°

tests is that close to the transition point, theory and also numerical simulations predict that the flow features indicating the presence of an irregular reflection can initially be as small as a few tens of micrometers, which presents a formidable challenge for the visualisation. A large number of experimental records exists where it is not possible to tell, whether the predicted minuscule flow features of an irregular reflection do not exist or whether they are simply too small to become discernible.

Figure 9b is an example for such a result. It shows a shock wave that is reflecting off a convex cylindrical arc of large radius (450 mm). If a shock reflects off a convex cylindrical surface, and if the initial reflection is regular, this pattern can change to an irregular one during the interaction as the angle of the cylinder wall continuously decreases. It is still an open question whether the angle at which this transition occurs is identical to the transition angles obtained for straight ramps as shown in Fig. 9a. According to the theoretical transition criteria for straight wedges, the reflection pattern should, for a shock Mach number of 1.35, have become irregular once the wave reaches the 43° mark (polar angle) on the model.[2]

[2]The exact transition angle for this shock Mach number is 42.7° polar angle or 47.3° wall angle, respectively, where wall angle and polar angle are complementary angles for 90°.

Based on the visualisation alone, one cannot decide, whether this has, indeed, happened at the predicted position or not—on the time-resolved records, the Mach stem and the subsequent contact surface only become discernible at polar angles larger than 46°, see Fig. 9b. However, numerical simulations indicate that at the instant shown in the first frame of Fig. 9b, approximately 2° past the theoretical transition point, the new third shock is only 70 μm long, which is, for the given imaging system and camera, below the spatial resolution threshold. In terms of resolving the discrepancy between theoretical prediction and experimental observation, tests that yielded high-resolution single images have largely been as unsuccessful as time-resolved sequences, where the spatial resolution of the visualisation record is inevitably lower. The limited spatial resolution can be partially compensated by increasing the imaging magnification of the visualisation setup, but for more reliable measurements such an approach requires not only substantially more powerful light sources suitable for the visualisation technique but also a camera with currently unavailable characteristics, such as megapixel resolution, frame rates exceeding 10^7 fps and frame exposure times of a few nanoseconds (see the discussion regarding Fig. 6 at the beginning of this section). Commercially available image converter cameras exist that technically meet these specifications, but the actual spatial resolution of these cameras is (significantly) lowered by the presence of a necessary image intensifier unit. Even if the technological hurdles associated with the light source and the camera could be cleared, it would still be questionable whether the described discrepancy can be satisfactorily resolved as the size of the flow features to be detected is of the same order of magnitude as the optical width of the imaged shock waves themselves. While the actual width of a shock wave is, as mentioned above, of the order of 10^{-7} m, its image obtained with density-sensitive visualisation setups is always at least two orders of magnitude larger. It is possible that only an entirely new measurement approach can provide conclusive evidence for or against the existence of the mentioned minuscule flow features.

5 Applications

The two main advantages of time-resolved high-speed imaging are substantial time savings in the investigation of reproducible processes, which would require a large number of experiments if studied by means of single-image visualisations, and the ability to obtain meaningful data for non-reproducible processes, which were in the past often simply not considered because single-shot visualisations would only have provided rather limited insight. The advent of modern high-speed camera systems has fundamentally changed this situation, and it is expected that applications of time-resolved visualisation of compressible high-speed flows will continue to increase. The following examples are meant to illustrate the potential that such high-speed imaging has in this research area.

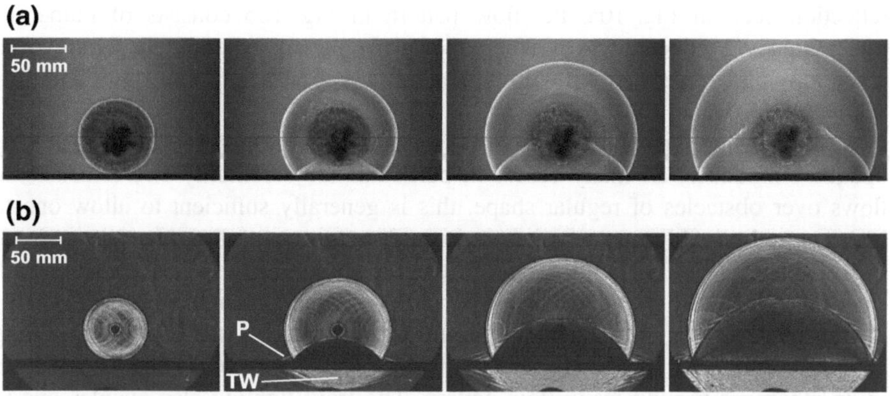

Fig. 10 Laboratory explosions visualised by omnidirectional schlieren: reflection of the blast wave generated by a 10 mg charge of silver azide for a height of burst of 35 mm **a** in air (taken with 500,000 fps, time difference between shown frames 40 μs) (The corresponding video is available online.) **b** under water (taken with 1,000,000 fps, time difference between shown frames 15 μs); *TW* transmitted wave, *P* precursor

Figure 10a shows a typical air blast wave reflection process, albeit in small scale: a 10 mg silver azide charge is ignited 35 mm above a steel plate. The initial reflection is regular which after some time turns into an irregular one (the Mach stem, which was already introduced in Fig. 9, is visible on the last frame). The mismatch in acoustic impedance, defined as the product of density and speed of sound, between air and steel is of the order of 10^5, which is why the blast wave is almost entirely reflected at the air/steel interface. This seemingly straightforward configuration is still the subject of intensive research, as the exact behaviour of the reflected shock near the transition point from regular to irregular reflection is not yet clarified (see, e.g., [4]).

The wave patterns are considerably more intricate in Fig. 10b, even though the overall initial conditions of the test are identical to the ones in Fig. 10a except that in Fig. 10b the surrounding medium is water. The wave propagation in both cases clearly shows the differences in speed of sound (approx. 1500 m/s in water and 340 m/s in air). More important is that the mismatch in acoustic impedances has been reduced to a factor of the order of magnitude ten. This means that there is now a sizeable portion of the wave transmitted into the steel plate—this results in a precursor wave running sideways with the speed of sound of the plate material (and hence faster than the wave in water) and also to a transmitted wave into the body of water below the steel plate. The transition from steel to water leads to yet another reflected wave—an expansion wave is reflected upward towards the upper side of the plate and the charge. This expansion wave is then transmitted into the water above the plate where it follows the reflected shock wave and generates cavitation in the form of numerous small-scale bubbles. This process is repeated several times, but with diminishing strength of the reflected waves. Instead of the single wave reflection seen in Fig. 10a, the flow pattern in Fig. 10b consists of numerous transmitted and reflected waves, where the latter typically experience a sign reversal when moving from a medium with high acoustic impedance to a medium with low acoustic impedance.

In typical flow visualisation records, solid objects in the test section appear as opaque silhouettes. In many cases, in particular for essentially two-dimensional flows over obstacles of regular shape, this is generally sufficient to allow one to interpret and possibly quantify the observed flow processes. For three-dimensional flows, or for the automatic inclusion of reference points in the visualisation record, simultaneous front lighting has been shown to be highly advantageous [9, 27]. This is illustrated in Fig. 11, where the front lighting reveals the (spinning) rifling grooves of a bullet flying at $M_\infty = 1.07$, which can be correlated to minute changes in an otherwise pseudosteady flow pattern. The front lighting also enables one to measure the spin rate of the projectile.

Figure 12 is an example of an unsteady flow instability within a macroscopically steady flow. Early research into so-called aerospikes—thin objects protruding from

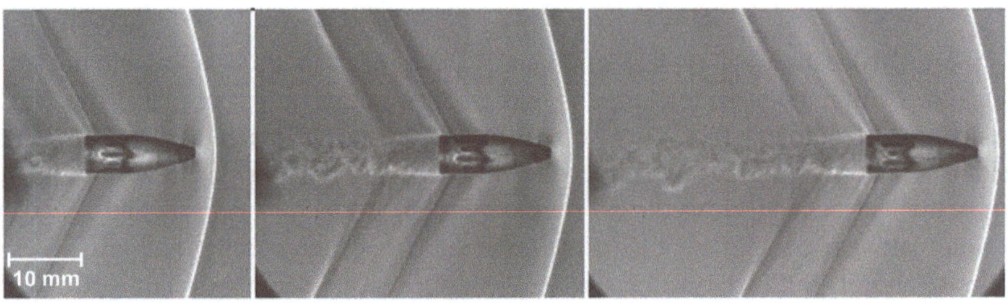

Fig. 11 Omnidirectional schlieren visualisation with front lighting of a rifle bullet flying at $M_\infty = 1.07$); frames from a visualisation taken with 500,000 fps; shown frames are 40 μs apart. The corresponding video is available online

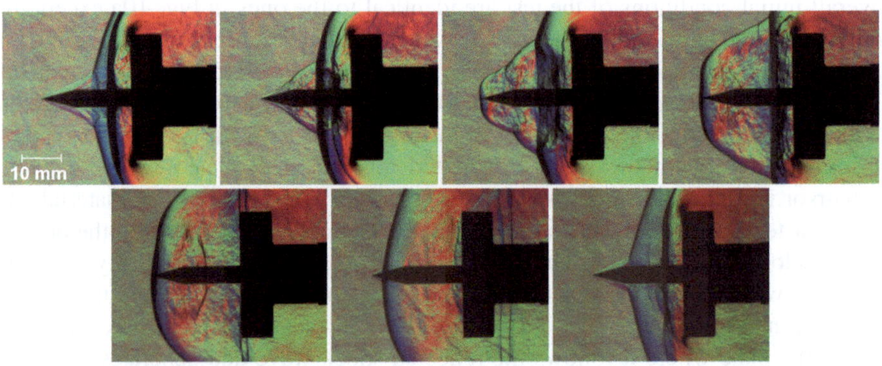

Fig. 12 Direction-indicating colour schlieren visualisation of the instability of a supersonic flow ($M_\infty = 3.0$) over a flat disk with a protruding spike; taken with 175,000 fps, time difference between shown frames 114.3 μs

a blunt body—has already shown that aerospikes of a certain length can cause the flow to pulsate [28]. The reason for this behaviour is that the gas near the blunt body has been compressed to a pressure close to the stagnation pressure of the free stream flow, which enables it to move upstream through energy-deficient flow areas that have a lower stagnation pressure. These areas are usually the boundary layers forming along the aerospike, which are being pushed outward by the high-pressure gas moving upstream (second frame in Fig. 12). This effectively forces the leading shock wave system to change its shape as the incoming flow has to be redirected. The result of this shape change is a strengthening of the leading shock system so that after a short while the compression by these shocks is able to stop and eventually reverse the upstream moving flow through the boundary layer. At the temporary equilibrium point, where the upstream moving flow has reached its maximum extension and is brought to rest, the leading bow shock has become near-normal and has therefore almost reached maximum strength (fourth and fifth frames in Fig. 12). When the upstream expansion of the gas has been reversed, the leading shock system becomes more oblique again which is equivalent to a decrease in strength. This continues until the gas near the blunt body has again been compressed to its maximum pressure, after which the described process repeats itself. From the images shown in Fig. 12 it can be concluded that the main frequency of this oscillatory behaviour is close to 1.4 kHz. While the main frequency of the process is largely constant, fluctuations in the free stream as well as in the way the compressed flow can eventually pass over the blunt body mean that higher frequency components are superimposed so that an adequate temporal resolution of the complete pulsation requires frame rates of the order of 10^5 fps.

As mentioned in Sect. 3, the shadowgraph technique records the second density derivative which makes it well suited to depict discontinuities in a flow field. The fact that this technique cannot detect gradual changes can be used advantageously in flows where some features of interest can be masked by other, more dominant structures. Typical examples are the intricate shock-vortex systems that are established when a shock exits an open shock tube where the vortex ring produces a signal that usually overrides those of the smaller flow features in its vicinity (see Fig. 5). This is also demonstrated in Fig. 13, where a shock with $M_S = 1.45$ is seen exiting from a planar channel (channel height 20 mm, channel depth 75 mm). This configuration can be considered as the planar counterpart of the axisymmetric case seen in Fig. 5, even though size limitations in the test section of the used shock tube did not allow one to let the flow develop up to the instant shown in Fig. 5 (shortly after the instant shown in the last frame of Fig. 13, reflected waves begin to enter the field of view from the top and bottom sides). The images clearly show the initial shock diffraction process with upstream moving expansion waves generated at both corners. These expansion waves further accelerate the flow that follows the incident shock until this flow becomes locally supersonic. This flow cannot negotiate the sudden area expansion at the corner and separates—the ensuing symmetric slipstreams at each corner then roll up into two vortices. A secondary shock develops on top of each slipstream in order to bring the accelerated post-shock flow back to

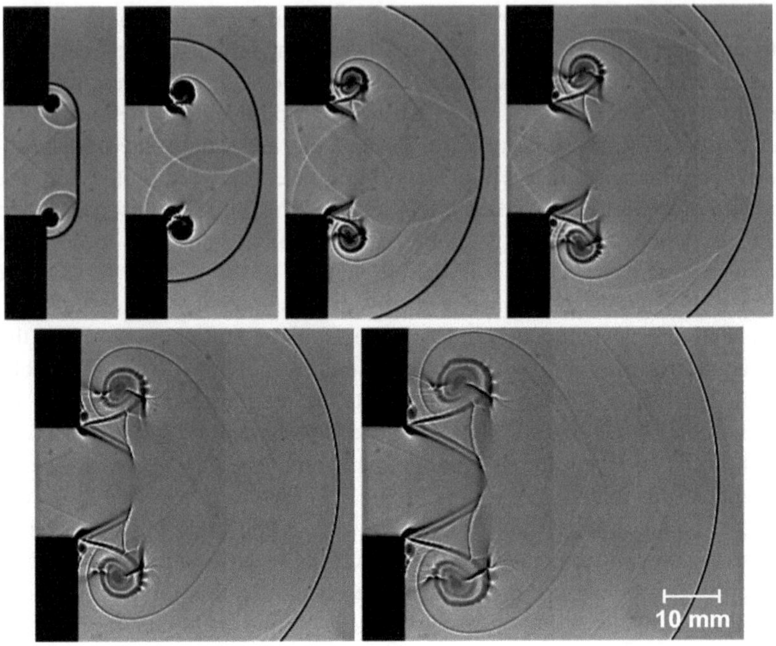

Fig. 13 Shadowgraph visualisation of a shock wave ($M_S = 1.45$) passing through a discontinuous area expansion; taken with 500,000 fps, time difference between shown frames 20 μs

subsonic levels. The secondary shock systems grow on each slipstream and eventually merge to form a single, almost straight discontinuity. An analogous process is seen in the axisymmetric case (the sequence for a corresponding Mach number and tube diameter equal to the channel height is shown in [16]), but the merging of the secondary shock systems is slower for the planar case, and the vortex cores have not moved as far from the corners as in the axisymmetric configuration. This can be explained by the fact that the mutual interaction of the two individual and physically separated diffraction patterns in the planar case is weaker than the continuous interaction along the circumference of the tube exit in the axisymmetric flow case.

The shearing interferograms of Fig. 14 show a plane shock ($M_S = 1.33$ in ambient air) passing over a diamond cylinder (a square cylinder with a side length of 30 mm tilted by 45°). These records show, similar to schlieren images, the density gradient, in this case in the direction of the shear (here: in flow direction). After reflecting off the front faces of the diamond cylinder, the shock diffracts at the top and bottom vertices of the square cylinder, which leads at each vertex to the formation of an upstream facing expansion wave, a slipstream rolling up to a vortex and a contact surface. Small secondary shock systems develop on each slip stream, which is an indication of localized patches of supersonic flow in an otherwise subsonic flow field, similar to what was seen in Fig. 13. The diffracted shocks from

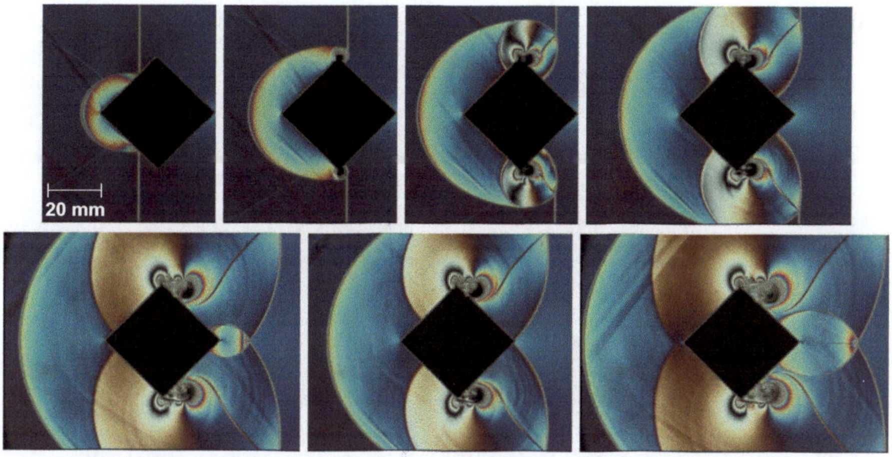

Fig. 14 Shearing interferometry visualisation of a shock wave ($M_S = 1.33$) interacting with a diamond cylinder; taken with 460,000 fps, time between shown frames $\Delta t = 21.7\,\mu s$. Adapted from [29]. The corresponding video is available online

both rear sides of the diamond cylinder intersect at the rear vertex and eventually form an irregular reflection with a rapidly growing Mach stem. The density gradients in the flow behind the Mach stem and the reflected waves concentrate near the discontinuities. The region immediately following the Mach stem is clearly seen as an area of strong gradients while the density gradient in the remaining part of the flow downstream of the square cylinder approaches zero. A multitude of new waves is generated by the interaction of the diffracted waves and the vortices. The inevitable double image (also in the direction of the shear) of all contours and discontinuities in the flow becomes obvious in the display of flow features that contain closely spaced discontinuities—for example, the rolling-up contact surface that makes up the core of the vortex generated at each vertex of the diamond cylinder, appears less clearly structured than in corresponding visualisations with other methods.

All visualisation techniques introduced in Sect. 3 have a finite measurement range. On the other hand, the thermodynamic properties in a compressible flows associated with shock waves can easily vary over more than one order of magnitude. This is illustrated in Fig. 15 for a polychrome Mach-Zehnder interferometry visualisation. In this case, the measurement range is limited by the coherence length of the used light—any phase shift exceeding this coherence length leads to increasingly weak and ultimately vanishing fringes. For normal white light operation and the given shock tube facility, the available coherence length allows one to resolve a maximum density change of $\Delta\rho_{max} \approx \pm 0.4\ \mathrm{kg/m^3}$, which corresponds to the density jump behind a normal shock with Mach number $M_S = 1.11$ or to the jump behind a normal reflected shock with initial shock Mach number $M_S = 1.06$, if these shocks are generated in ambient conditions [31]. This means that in flow

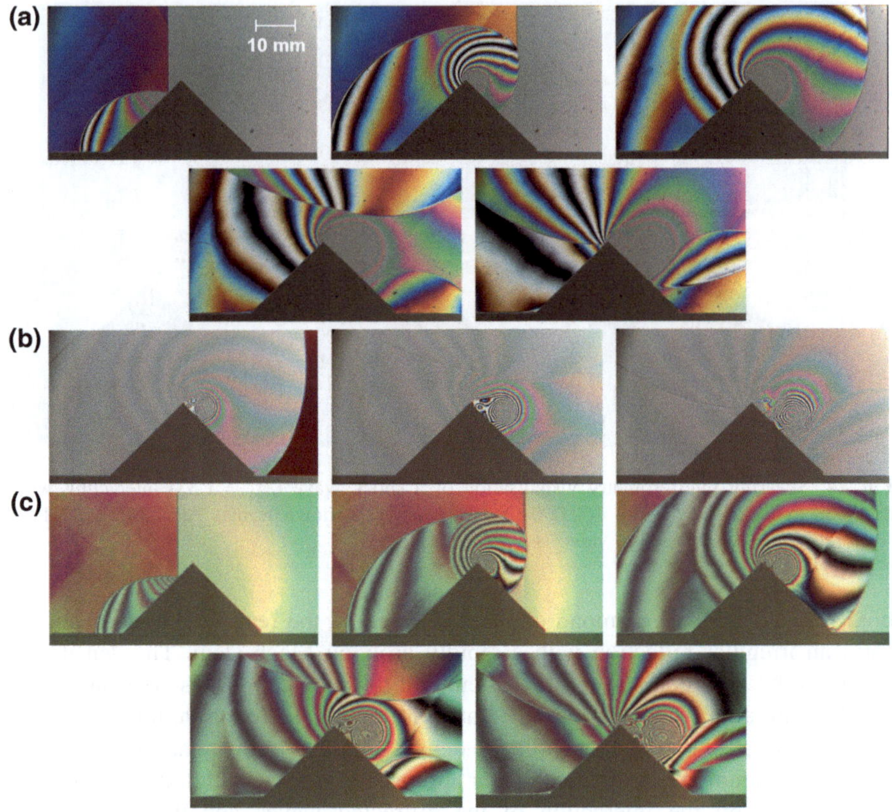

Fig. 15 Polychrome Mach-Zehnder interferometry visualisation of a shock wave passing over a triangular ramp ($M_S = 1.34 \pm 0.02$ in $p_1 = 0.48\,bar$); taken with 330,000 fps, time between shown frames $\Delta t = 45.5$ µs; **a** normal white light, negative offset to resolve the density increase on the front part of the ramp; adapted from [30] **b** normal white light, near-zero offset to resolve the density distribution in the vortex core; adapted from [30] **c** multiple-wavelength light source with extended coherence length, near-zero offset

fields generated by stronger shocks, or in flow fields that exhibit both compression and expansion zones (as in the flow field shown in Fig. 15), the measurement range would easily be exceeded. In order to visualise such flow fields, one can apply one or several of the following strategies:

- run the tests at lower initial pressure and density levels so that the absolute changes in density do not exceed $\Delta\rho_{max}$;
- if the sign of the density change is known, that is, if one expects only compressions or only expansions within a certain range but not both, or if one wants to "filter" a given flow field by only showing features within a predetermined density range, one can introduce a deliberate offset of the measurement range and choose a background fringe that is sufficiently distant from the central, zero-order fringe;

- increase the coherence length of the light by using a combination of narrow-bandwidth filters. This will eliminate some interference colours/fringes and may lead to less easily distinguishable higher-order fringes, but can extend the measurement range by a factor of up to five.

These approaches have been followed in the tests shown in Fig. 15 as the density changes induced by a shock Mach number of $M_S = 1.34$ are outside the measurement range if the shock moves into ambient air. The tests were run at a reduced initial density (48% of the ambient density) and the offset was chosen to either resolve the compression phase on the front part of the ramp (Fig. 15a) or the expansion in the vortex core on the rear side of the ramp (Fig. 15b). For an approximately three times increased coherence length of the used light, an offset was in this case not needed as the magnitude of the density increase on the front side is approximately the same as the decrease on the rear side (Fig. 15c).

6 Evaluation

Time-resolved recordings of shock wave processes are used for both qualitative descriptions of the observed process and also for quantitative evaluation of certain features. Given that shock wave interaction processes often become highly complex with a large number of new flow elements produced within short time spans (as illustrated by the examples of the previous section and Fig. 16) it is often the first main task to provide a chronological history of the events. Time-resolved records show when certain flow features evolve and can also provide the phenomenological reason for their existence. In complex processes with numerous flow features produced during an interaction, such a description would be difficult if not impossible to obtain if one only had single records from a number of repeated experiments, or time-resolved records taken with an insufficiently high frame rate. Figure 16 illustrates this in at least two different ways: the number of wave systems that evolve during the focusing of a plane shock wave in a circular reflector increases substantially, and the waves that eventually meet at the focal region have undergone numerous modifications and interactions before reaching this region (see [32] for a complete description of the process). One unexpected flow behaviour was the repeatable onset of a growing instability of the main shear layers formed during the focusing. This development, shown in in the last three frames of Fig. 16, is initiated by a second but much weaker shear layer (indicated by an arrow in the first four frames showing the focusing process). This shear layer is generated during the initial interaction of the incident shock wave with a small ramp at the inlet of the reflector (shown in magnified view in Fig. 16). This correlation would arguably have remained undetected without the availability of time-resolved records.

This highlights another important contribution that time-resolved visualisations make towards the investigation of transient events in high-speed flows: observing animated versions of these records often facilitates the physical interpretation of the

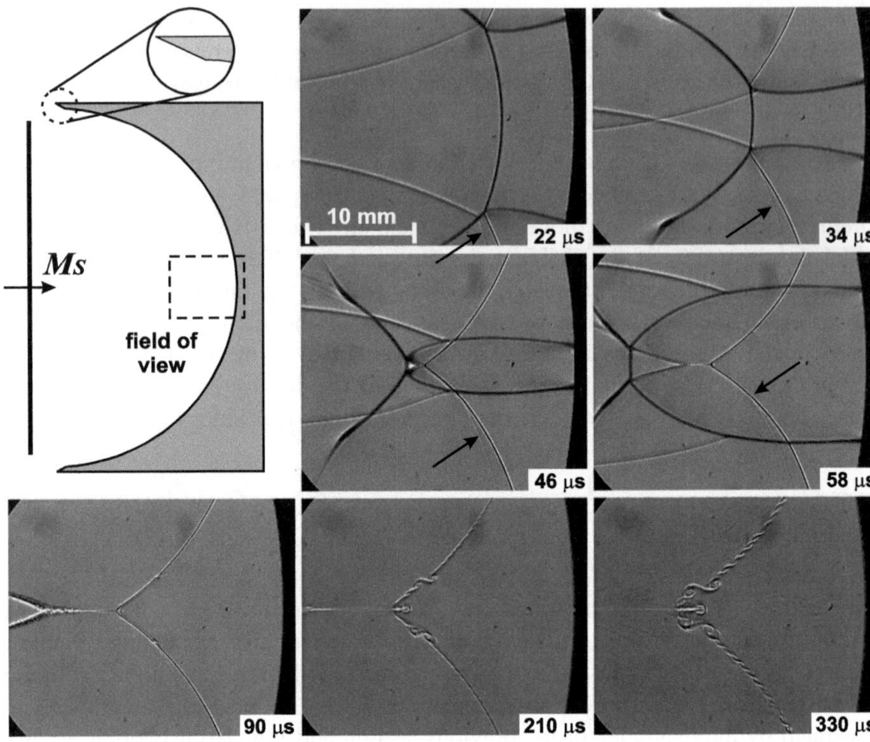

Fig. 16 Shock focusing in a circular reflector of 64 mm radius, initial shock Mach number $M_S = 1.34$; shadowgrams from a sequence taken with 250,000 fps. The first four frames illustrate the increasing complexity of the focusing process, and the last three frames show the subsequent development of a shear layer instability from a single initiation point, caused by the interaction with a second, much weaker shear layer (indicated in the first frames by an *arrow*). The indicated time t is counted from the instant the incident wave reaches the centre of the reflector. The process is described in more detail in [32]

event as the human eye is, arguably as a result of evolution, more sensitive to detect motion than to discern different signal intensity levels in a static scene. Weak waves in a flow process, for example, like the ones triggering the shear layer instability seen in Fig. 16, only generate a weak signal on each visualisation record, which can easily be masked by more prominent signals and may therefore be overlooked, but the motion of such a weak signal is much easier to detect. This effect was already discovered in the early days of high-speed imaging of compressible flows [33], and efforts were therefore made to animate the obtained visualisation sequences. The technological limitations and challenges of the pre-digital time meant, however, that such animations were not done on a regular basis.

In quantitative evaluations, tracking wavefronts and establishing their position as a function of time is arguably the most obvious approach. The close relationship between shock strength and its speed allows one to deduce therefore the thermodynamic changes induced by the shock front. In blast wave studies, for example, one can convert the distance-versus-time measurement data into an overpressure-versus-distance plot, which can be used to determine the energetic characteristics of the used explosive [2].

Such distance-versus-time measurements can be obtained by determining the position of a wavefront on each frame of the visualisation, but it is also possible to generate an equivalent $x - t$ plot directly from the images in the form of a digital streak record. Following the principle of early streak recordings, only a narrow region of the flow field is imaged as a function of time. This image then shows the trajectories of all objects visible in the selected narrow region. In the original analogue version of the technique, the narrow region is created by a slit that is either placed on the test section window or that is used as a stop in the path of the light between test section and recording medium. The digital version, which was first suggested in [34], recreates the boundaries of the narrow region by using only a given segment of the recorded full image of the test object, e.g. a row or a column of pixels. For each frame, this selected segment is then "stacked" on top of the selected segment of the preceding frame, which creates a digital streak image such as the one shown in Fig. 17. The main advantage of the digital technique over its

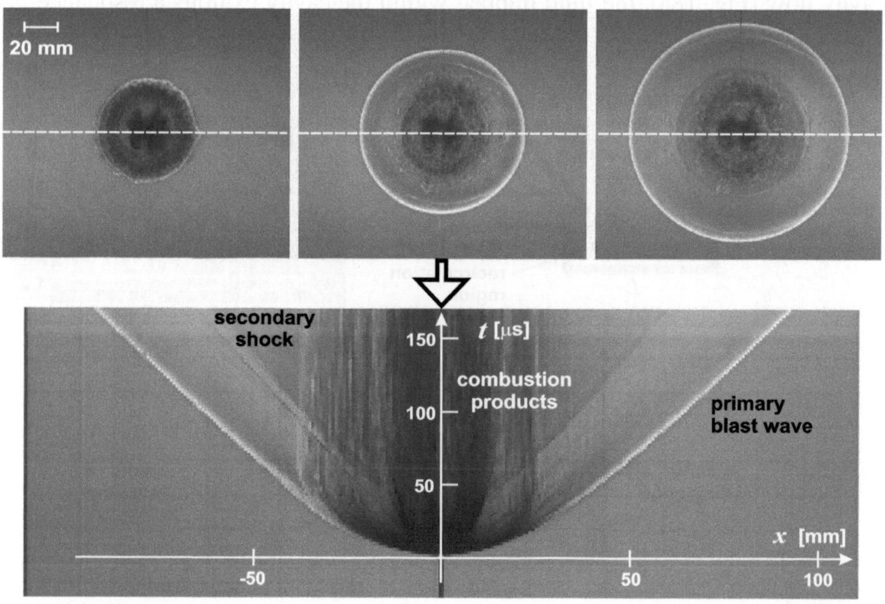

Fig. 17 Digital streak schlieren: From a sequence of schlieren visualisations of the blast wave generated by the explosion of a 10 mg charge of silver azide (*top row*), a digital streak record is created by selecting only a row of pixels from each frame (indicated by the *white dashed line* in the frames) and stacking these rows on top of each other (*bottom row*). The visualisation provided 100 frames taken with 500,000 fps, so that an overall period of approx. 200 μs is represented in the streak image. The trajectories of the expanding shock and combustion products as well as the emerging secondary shock are clearly visible

analogue predecessor is that streak size, shape and orientation can be chosen freely after the actual recording has been made. Concentrating on the features visible in the selected narrow region and discarding the remainder of the information contained in the image is equivalent to applying a narrow filter to the overall measurement represented by the full frame visualisations. This allows one to concentrate on the measurement of certain processes, such as the propagation of a flow feature in a direction given by the orientation of the slit. If this feature is easily distinguished in the visualisation record itself, the digital streak image is, in principle, highly similar to a plot of $x - t$ data, but it contains additional information that such a plot usually cannot provide. Changes in intensity (or colour) of the feature are captured as well as changes of the position, and in addition all other flow features visible in the selected segment are represented as well, as becomes obvious in the example shown in Fig. 17. This streak image therefore not only provides quantitative data such as the slope of the trajectory which usually corresponds to the propagation speed along the direction defined by the selected region, but it also facilitates a physical interpretation of the process.

In addition to graphically showing "obvious" trajectories of flow features such as the ones shown in Fig. 17, the digital streak schlieren technique can also reveal flow structures that are hidden in a more complex flow field. In a so-called open cavity flow (Fig. 18a), the fluid trapped within the cavity exhibits a resonance-like

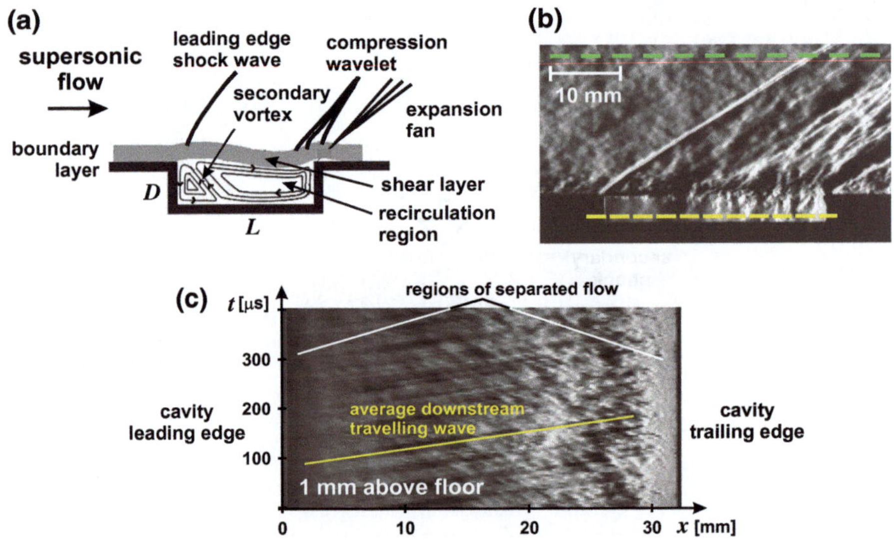

Fig. 18 Digital streak schlieren of a supersonic cavity flow. **a** schematic of an open cavity flow, typical for a length/depth ratio $L/D < 10$ (from [35]) **b** single frame of a schlieren visualisation taken with 250,000 fps, with indications of where the frame is evaluated to reveal wave structures within the cavity (*yellow line*) and the free stream (*green line*) **c** digital streak record of the flow within the cavity, evaluated 1 mm above the cavity floor. The *sloped lines* on this record correspond to periodically produced wavefronts not directly visible on the visualisation records themselves

behaviour with waves continuously passing through it at discrete frequencies [35]. These frequencies depend on the Mach number of the free stream flow outside of the cavity and the cavity dimensions (depth D and length L). The wavefronts associated with these frequencies are hard to distinguish in actual flow visualisation records such as Fig. 18b as waves with all of the different frequencies move simultaneously through the cavity. A digital streak schlieren record taken within the cavity (e.g. along the indicated yellow dashed line), however, clearly shows distinct lines that correspond to the trajectories of one set of these wavefronts (here, the ones traveling with the lowest velocity). The frequencies derived from the speed of these fronts (measurable from the slopes of the lines), correspond to the frequencies found in independently conducted pressure measurements and also theoretical and numerical predictions [35]. Similarly, the boundary layer structures on the test section windows can be used, after digital streak processing, to measure the free stream speed ahead of and behind the leading edge shock as the majority of the visible boundary layer structures propagate with a speed close to that of the free stream. The region that could be used for such a measurement is indicated by the green dashed line in Fig. 18b.

In addition to the measurement of trajectories and associated properties of flow features, the data contained in time-resolved digital images of the flow can be evaluated in many other forms, such as determining intensity (or colour) distributions/variations in the image as a function of time at specific locations. These data can be used to correlate them with certain flow features, or to detect, for example, underlying frequencies in the flow. Numerous procedures from image processing studies exist to facilitate such evaluations. One challenge that will have to be addressed is to find powerful algorithms to process the potentially enormous amounts of data provided by the high-speed visualisations.

7 Summary

This chapter has briefly outlined the application of high-speed cameras in the investigation of compressible flows including shock waves. While time-resolved visualisation of such flows has been conducted since the first half of the 20th century, a widespread use of this approach has only been seen since high-speed cameras have become more readily available and less of a specialist's tool. The use of these cameras has already clarified a number of items that previous investigations could not resolve with the instrumentation available at the time, and it can be expected that further developments and applications will continue to elucidate and explain processes in high-speed compressible flows, which so far have defied full experimental proof. The availability of modern high-speed cameras has enabled researchers to investigate processes that only two decades ago would have appeared too complicated or time-consuming to study. Future work is likely to turn to investigating small-scale flow features, for which the cameras' capabilities with respect to spatial and temporal resolution will have to be further enhanced.

Acknowledgements Most of the work presented here would not have been possible without the support by many colleagues and friends. I would like to acknowledge in particular the contributions of em. Prof. John Dewey (University of Victoria, Canada), Prof. T. Goji Etoh (Ritsumeikan University, Japan), Prof. Koju Hiraki (Kyushu Institute of Technology), Prof. Herbert Olivier (RWTH Aachen University, Germany), Prof. Gary Settles (Penn State University, USA), Prof. Beric Skews (University of the Witwatersrand, South Africa) and Prof. Kazuyoshi Takayama (Tohoku University, Japan), who all inspired and supported many of the experiments I had the privilege to conduct. I am also indebted to em. Prof. Hans Grönig (RWTH Aachen University, Germany), who introduced me to the exciting field of compressible flows and their visualisation. Last, but not least, I would like to express my thanks for the excellent technical support I have received over the last years from members of the SEIT Mechanical Workshop, in particular from Stuart Gay and Michael Jones.

References

1. P.O.K. Krehl, *History of Shock Waves, Explosions and Impact: A Chronological And Biographical Reference* (Springer, Berlin, 2009)
2. H. Kleine, J.M. Dewey, K. Ohashi, T. Mizukaki, K. Takayama, Studies of the TNT equivalence of silver azide charges. Shock Waves **13**, 123–138 (2003)
3. W. Liepmann, A. Roshko, *Elements of Gasdynamics* (Wiley, New York, 1957)
4. J.M. Dewey, Expanding spherical shocks (blast waves), in *Handbook of Shock Waves*, vol 2 (Academic, San Diego, 2001), pp. 441–481
5. B.D. Henshall, *On Some Aspects of the Use of Shock Tubes in Aerodynamic Research* (ARC RM 3044, London, 1955)
6. I.I. Glass, J.G. Hall, in *Shock Tubes*, Report NAVORD 1488, vol 6, section 18 (Bureau of Naval Weapons, Washington DC, 1959)
7. H. Oertel, *Stossrohre* (Springer, Wien, 1966)
8. W. Merzkirch, *Flow Visualization*, 2nd edn. (Academic, Orlando, 1987)
9. G.S. Settles, *Schlieren and Shadowgraph Techniques* (Springer, New York, Heidelberg, 2001)
10. H. Kleine, Flow visualization, in *Handbook of Shock Waves*, vol 1 (Academic, San Diego, 2001), pp. 683–740
11. H. Schardin, Die Schlierenverfahren und ihre Anwendungen. Erg. exakt. Naturwiss. **20**, 303–439 (1942)
12. M.J. Hargather, G.S. Settles, Retroreflective shadowgraph technique for large-scale flow visualization. Appl. Optics **48**, 4449–4457 (2009)
13. L.M. Weinstein, Review and update of lens and grid Schlieren and motion camera Schlieren. Euro. Phys. J. Special Topics **182**, 65–95 (2010)
14. M. Raffel, Background-oriented Schlieren (BOS) techniques. Exp. Fluids **56**, 60 (2015)
15. H. Kleine, C. Le Vo, K. Takehara, T.G. Etoh, Time-resolved visualization of shock-vortex systems emitted from an open shock tube. J. Visualization **13**, 33–40 (2010)
16. H. Kleine, Filming the invisible—time-resolved visualization of compressible flow. Euro. Phys. J. Special Topics **182**, 3–34 (2010)
17. M. Versluis, High-speed imaging in fluids. Exp. Fluids **54**, 1458 (2013)
18. A. Toepler, *Beobachtungen nach einer neuen optischen Methode—Ein Beitrag zur Experimentalphysik* (Max Cohen & Sohn, Bonn, 1864)
19. P. Krehl, S. Engemann, August Toepler—the first who visualized shock waves. Shock Waves **5**, 1–18 (1995)
20. E. Mach, P. Salcher, Photographische Fixierung der durch Projektile in der Luft eingeleiteten Vorgänge. Sitzungsb. Akad. Wiss. Wien **95**, 764–780 (1887)

21. E. Mach, P. Salcher, Optische Untersuchung der Luftstrahlen. Sitzungsb. Akad. Wiss. Wien **98**, 1303–1309 (1889)
22. C. Cranz, H. Schardin, Kinematographie auf ruhendem Film und mit extrem hoher Bildfrequenz. Z. Phys. **56**, 147–183 (1929)
23. H. Olivier, T. Reichel, M. Zechner, Airfoil flow visualization and pressure measurements in high-Reynolds-number transonic flow. AIAA J. **41**, 1405–1412 (2003)
24. V. Parker, C. Roberts, Rotating mirror and drum cameras, in *Scientific Photography and Applied Imaging*, ed. by S.F. Ray (Focal Press, Oxford, 1999), pp. 167–180
25. H. Schardin, The relationship between maximum frame frequency and resolution in rotating mirror framing cameras, in *Proceedings of the 3rd International Congress High-Speed Photography, Butterworths*, London, pp. 316–323 (1957)
26. G. Ben-Dor, *Shock Wave Reflection Phenomena*, 2nd edn. (Springer, Heidelberg, 2007)
27. H. Kleine, Schlieren imaging and the real world. J. Visualization **16**, 193–199 (2013)
28. S.M. Bogdonoff, I.E. Vas, Preliminary investigations of spiked bodies at hypersonic speeds. J. Aerospace Sci. **26**, 65–74 (1959)
29. H. Kleine, Time-resolved visualization of shock wave phenomena, in *Proceedings of the 29th International Congress High Speed Imaging and Photonics*, Iwate Medical University, paper C-02 (2011)
30. H. Kleine, H. Olivier, K. Tsuji, K. Etoh, K. Takehara, T.G. Etoh, Time-resolved Mach-Zehnder interferometry, in *Proceedings of the 29th International Congress High Speed Imaging and Photonics*, Iwate Medical University, paper C-07 (2011)
31. H. Kleine, H. Olivier, K. Tsuji, K. Etoh, K. Takehara, T.G. Etoh, Time-resolved Mach-Zehnder interferometry of shock waves, in *Proceedings of the 28th International Symposium*. Shock Waves, vol 1, Springer, Heidelberg, pp. 577–583 (2012)
32. B.W. Skews, H. Kleine, Flow features resulting from shock wave impact on a cylindrical cavity. J. Fluid Mech. **580**, 481–493 (2007)
33. H. Schardin, Untersuchung instationärer gasdynamischer Vorgänge als Beispiel für den zweckmässigen Einsatz der Hochfrequenzkinematographie, in *Proceedings of the 7th International Kongr. Kurzzeitphotographie*, Helwich, Darmstadt, pp. 17–23 (1967)
34. A.S. Katayama, Visualization Techniques for Temporally Acquired Sequences of Images, US Patent No. 5294978 (1994)
35. V. Sridhar, H. Kleine, S. Gai, Visualization of wave propagation within a supersonic two-dimensional cavity by digital streak schlieren. Exp. Fluids **56**, 152 (2015)

Visualization of Underwater Shock Waves

Kazuyoshi Takayama

Abstract Results of optical flow visualizations of underwater shock waves performed in the Shock Wave Research Center of the Institute of Fluid Science in Tohoku University are presented. Underwater shock waves generated in ultrasound oscillations, underwater shock waves in two-phase shock tube flows, spherical shock waves in small scale underwater explosions, their interaction with gas bubbles and interfaces, and their reflection and focusing are visualized by single and double exposure holographic interferometry. Time resolved high-speed imaging revealed the luminous emission from a contracting gas bubble during shock/bubble interactions. It is concluded that double exposure holographic interferometry can be very effective for quantitative observation of underwater shock waves, and that high-speed time resolved imaging can be useful for the underwater shock wave research.

1 Introduction

Underwater shock waves are generated not only in nature but also in the artificial world from micro-scale events such as the collapse of microscopic bubbles to the gigantic explosion of magma-water vapor during volcanic eruptions. Hence, the underwater shock wave research is one of the highlights of shock dynamics. However, unlike shock wave in gases, the optical visualization of shock waves in liquids is not as simple as those in gases, because of the higher degree of inhomogeneity of liquids. Likewise, the generation of strong shock waves in liquids is not necessarily as straightforward as that of gases, because the acoustic impedance of liquids is significantly larger than that of gases.

In this chapter, the author will demonstrate some of underwater shock wave phenomena he collected throughout forty years of his shock wave research: shock

K. Takayama (✉)
Institute of Fluid Science, Tohoku University, 2-1-1 Katahira, 980-77 Aoba Sendai, Japan
e-mail: k.takayama@mac.com

K. Tsuji (ed.), *The Micro-World Observed by Ultra High-Speed Cameras*,
DOI 10.1007/978-3-319-61491-5_7

waves generated during ultrasound oscillatory tests; underwater shock waves produced in shock tubes; underwater micro-explosions; and shock/bubble interactions associated with underwater explosions.

2 Ultrasound Oscillation

We dipped a 16 mm diameter 304 stainless steel specimen into water at 3 mm depth. The test water was ion exchanged and contained in a test chamber of 200 mm in diameter and 150 mm in width. The specimen was driven with a 500 W ultrasonic vibrator. The displacement x of the specimen surface is described as $x = \tilde{A}\sin\omega t$, where $\tilde{A}$ is the half amplitude of 17.5 μm, and ω is frequency of 17.7 kHz.

We visualized oscillatory flows by using a single exposure image holography with a light source of double pulse holographic ruby laser (Apollo Laser Inc. 22HD). The light source was coherent and had a pulse width of 25 ns. Although the holographic images were still just observation, we synchronized the images correctly with the phase angle of the vibrator at every $\pi/4$ rad and collected sequential images of flows. Although we had taken high-speed photography, still images we collected had a much finer resolution than that of high-speed photography.

We took streak recordings at the rate of 1.5 μs/mm and frame recordings at the rate of 200,000 fps (frames/s) by using Ima-Con John Hadland 790. The magnification of the images was 0.594. We used a 500 mW Argon-ion laser as a light source and synchronized its operation with a mechanical shutter of 1 ms opening duration. In the streak recordings, we put a slit of 0.75 mm in width exactly on the surface of the specimen. The slit width observable in the streak images was 0.44 mm.

Figure 1 shows the streak image. The ordinate designates the distance along the surface of the specimen in mm and the abscissa designates time in μs. When the specimen starts to pull water at $\omega t = \pi/4$, the deceleration starts and the deceleration force—$A\omega^2$ is a maximum, equivalent to −6G, which locally generates the tensile stress of water. Hence, cavitation bubble inception would take place depending on the distribution of cavitation nuclei in water.

The grey shadow in Fig. 1 indicates variation of the density of the bubble cloud. The increasing grey shadow would indicate at first the increase of number density

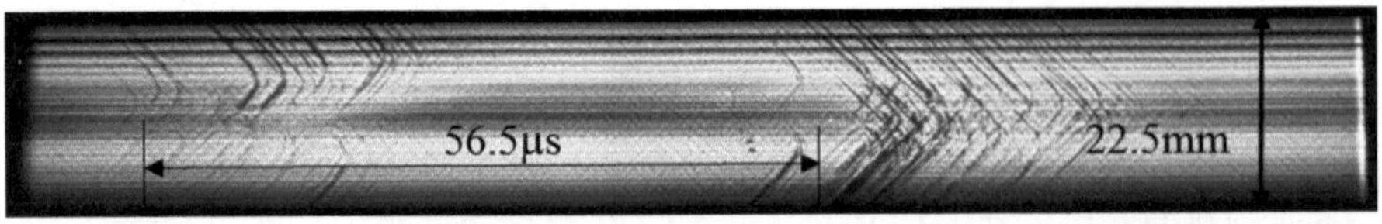

Fig. 1 Streak recording of oscillatory flows at 103 kPa, 293 K. The ordinate: the distance along the surface of the specimen in mm, the abscissa: time in μs

and at the same time growth of their size and hence the resulting grey shadow would be a maximum at around $\omega t = \pi/2$. Then the grey zone indicates the contraction of the bubbles, which were exposed to the compression waves at $\omega t > \pi/2$.

Exposed to pressure fluctuations, bubbles started to contract and eventually became a minimum volume and the pressures became highest. In the case of axially symmetric bubbles, the high pressures are induced by adiabatic compression. Then the high pressures drive so-called rebound shock waves. In the case of asymmetric bubble contractions, the bubble motion is based on the propagation and focusing of shock waves inside the bubbles. As the shock compression creates high pressure much more efficiently than the adiabatic compression, the shock wave motion would contribute to the high pressure generation.

Anyway, the higher pressures in the bubble were released into water as a train of compression waves, which instantaneously coalesced into an underwater shock wave. This may be reproducible by a fine numerical simulation.

Trajectories of shock waves view through 0.75 mm thick slit are shown as shallow V shaped lines randomly distributed in the region where the grey shadow faintly diminishing. The streak image was viewed through a width of 22.5 mm. One cycle of oscillation is 56.5 μs as described in Fig. 1. Images in the neighborhood of the diameter looked sharpest and those on the frontal and rear edges are blurred.

The creation and diminishment of bubbles are not necessarily exactly regulated by the specimen's movements. Looking at the series of streak recordings at identical initial conditions, it is found that the evolution of each bubble has its own history. The bubble collapse may belong rather to a stochastic process and is not precisely regulated by the specimen's movement. Therefore, the shock wave formations would occur slightly randomly as seen in Fig. 1.

Sanada et al. [1] measured the evolution of a sharply observable trajectory in the x-t plane and estimated that the shock Mach number Ms varied from its initial value of about 1.5 to a sound wave of Ms = 1.0. Assuming a planar underwater shock, its overpressure at Ms = 1.5 would be well over 500 MPa. Such a high pressure working on a spot would readily erode the metal surface.

In Fig. 2a–d, images of the generation and diminishment of bubbles and the formation and disappearance of shock waves at 285 K and 0.2 MPa are presented at individual phase angles. Figure 2a shows the bubble creation at a stage of $\omega t = \pi/2$, at which bubbles are created and impulsively ejected from the edge of the specimen. Columns of bubble cloud are formed at the center of the specimen, when the specimen's receding speed is a maximum. That the eight-digit figure, for example, attached to Fig. 2a, #82112502, indicates that we visualized this image in 1982, November, 25th, the test No. 2. Note that the surface deformation in reality does not show a simple mode but would show a complex surface pattern. Hence the resulting bubble column distributions are not always uniform. Then we see no bubbles being generated on the edge.

In Fig. 2a, we find a semi-circle marked by B. Measuring the radius of the semi-circle, this is a remnant of a shock wave generation which was created by a bubble collapse at 2.7 μs prior to this observation. A circle marked by A, whose

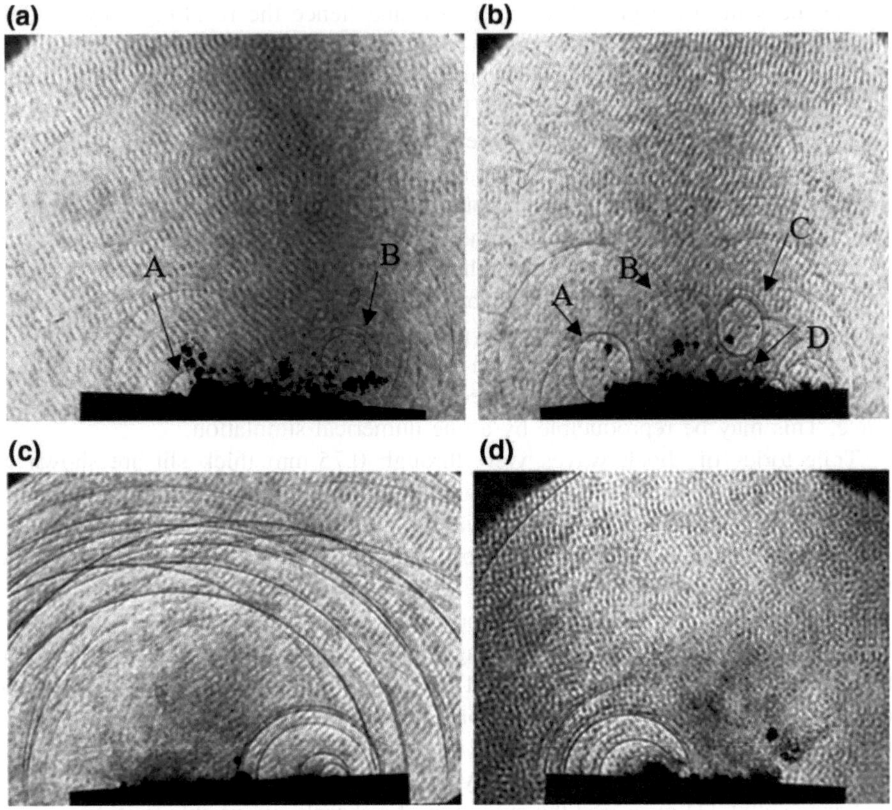

Fig. 2 Single exposure holographic images of a ultrasonic oscillation at 285 K. 0.2 MPa: **a** #82112502, at the phase angle of $\pi/2$; **b** #82112503, *at* $3\pi/4$; **c** #82112505, at $5\pi/4$; **d** #82112506, at $3\pi/2$

center is located a few mm above the specimen, is a weak shock wave created 0.8 μs prior to this observation.

The surface of the specimen moves repeatedly at a speed with amplitude of $\tilde{A}\omega$ and eventually induced convective flows in the entire flow field. Therefore, bubbles so far accumulated and surviving throughout the oscillatory motion are drifted along the center line. The grey shadow in the central region above the oscillator indicates the integrated images of bubble clouds.

Figure 2b shows the stage at $\omega t = 3\pi/4$. The bubble inception and resulting shock formation in oscillatory tests is, in principle, synchronized with the phase angle of the oscillator, but is subjected to the stochastic procedure. Hence, in Fig. 1, we identify, regardless of faintly observable shock wave trajectory or distinctly visible shock trajectories, resulting shock waves all the time. Figure 2c shows that the bubbles collapsed and the appearance of shock wave at $5\pi/4$, and Fig. 2d at $3\pi/2$.

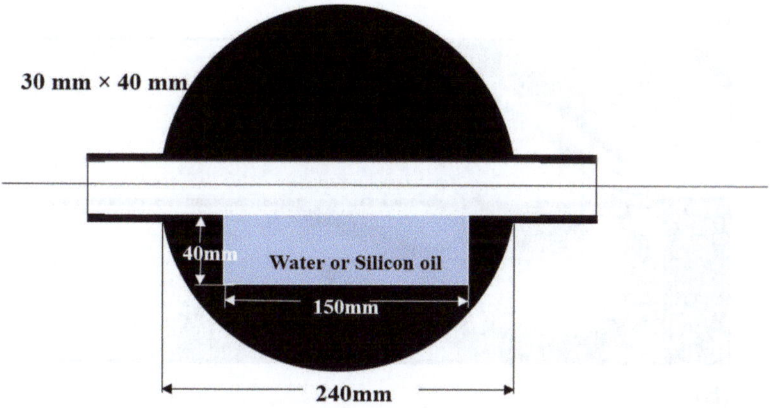

Fig. 3 A test section filling liquids in its bottom [2]

3 Shock Tube Flows

Firstly, we will observe underwater weak shock waves by means of a conventional shock tube. Figure 3 shows a test section made of stainless steel, the bottom filled with water or silicon oil TSF-10 of sound speed of 968 m/s at 293 K.

The test section was connected to a 30 mm × 40 mm shock tube. We filled silicon oil TSF-10 in the bottom of the test section. The surface of the liquid layer was very flat but was receded a few mm from the bottom of the shock tube due to the surface tension of the liquid.

In Fig. 4, we visualized the shock tube flows by using a double exposure holographic interferometry. The fringes correspond to equal density contours [3]. It should be noticed that double exposure holographic interferometry is very useful to quantitative measurements of the two-dimensional shock wave phenomena, in particular, in liquids. Individual fringes of infinite fringe interferograms correspond to changes in phase angles which occur only during double exposures. Defects of test sections are insensitive to the measured image data. Especially background noises induced by natural convections or low frequency vibrations of experimental setup can be readily eliminated, if the interval of double exposures is set very short.

In Fig. 4a, b, we can see shock waves in air propagating over TSF-10 layers: (a) shock Mach number Ms = 2.97 and its propagation speed U_s of 1127 m/s, supersonic in terms of the sound speed in silicon oil, $a_{silicon}$ = 960 m/s; and (b) Ms = 2.70 and propagation speed of 896 m/s, subsonic in terms of $a_{silicon}$ = 960 m/s.

Figure 4a shows that the shock wave being supersonic in terms of $a_{silicon}$, an oblique shock is observable in silicon oil, whose inclination angle ϕ is $\sin\phi = a_{liquid}/U_s$. Behind the oblique shock, we can see three fringes which imply gradual density decrease or pressure decrease toward the lower wall. Then the oblique shock is attenuated to be sonic wave in the neighborhood of the bottom wall.

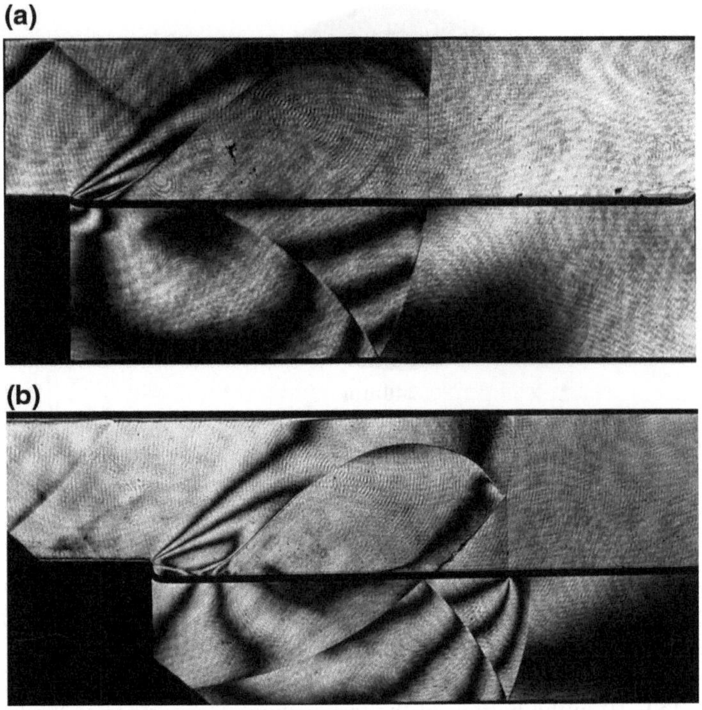

Fig. 4 Shock/silicon oil layer interaction: **a** #81052703, Ms = 2.969 at 123 hPa and 294.4 K in air; and **b** # 81051901, Ms = 2.701 at 150 hPa and 288.2 K in air

Figure 4b shows a subsonic case. A sound wave in silicon oil propagates much faster than the shock wave in air. The front of a precursory sound wave is presented to be a faint change in fringe. From the front of the precursory sound wave, the pressure gradually increases to the foot of the shock on the surface of the silicon oil, which is apparently a singular point as the fringes concentrate.

In gases, patterns of reflected shock waves over solid walls uniquely transit between regular reflection and Mach reflection, depending on the shock Mach number, reflection angles and the specific heat ratio of gases [4]. We were interested in the effect of wall conditions on the reflected shock transition. Hence, to test the transition over a liquid wedge which is perfectly flat and admit slip conditions on the wall surface, we inclined the entire 30 mm × 40 mm shock tube and the water filled test section. We eventually formed a water wedge [5] and visualized, by using double exposure holographic interferometry, shock wave reflection over water wedges. In Fig. 5a, b, we can see shock waves at Ms = 2.27 reflected from a water wedge: (a) transitional Mach reflection (TMR) from a 46.2° wedge; and (b) regular reflection (RR) from a 40.0° wedge. The speed u of the foot of the shock wave along the water wedge, $u = U_s/\cos\theta \leq a_{water}$, is subsonic but closer to $a_{water.}$, then we can see a sonic front which, although very faint, is clearly identified as a

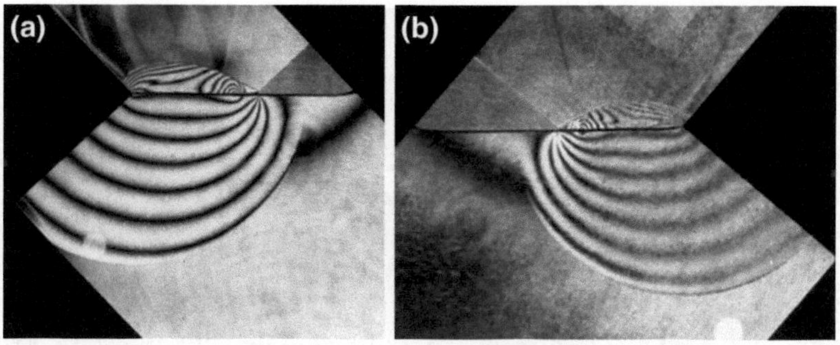

Fig. 5 Shock wave reflection over water wedges: **a** #85061701, Ms = 2.293 at 710 hPa and 294.7 K in air, 46.2°, TMR; and **b** #85061502, Ms = 2.258 at 710 hPa and 294.3 K in air, 49.0°, RR

precursory discontinuous front appearing in a finite distance. This trend is more clearly observable for the wedge angle of 49.0° as seen in Fig. 5b.

4 Underwater Explosions

The conservation equations of gas-dynamics in a continuum can be summarized by the following equation: $\delta p = a\rho\delta u$, where δp, a, ρ and δu are pressure perturbation, sound speed of the medium under discussion, density, and velocity perturbation. $a\rho$ is defined as acoustic impedance which indicates the degree of the compressibility of media. Therefore, for a given pressure perturbation, a medium having smaller acoustic impedance can induce a larger velocity perturbation than in a medium having larger acoustic impedance. As $(a\rho)_{water}/(a\rho)_{air} = 3700$, water is much less compressible than gases. Hence, it is not easy to safely generate an underwater explosion, especially, in laboratory scale experiments. In 1968, Glass and Heuckroth [6] ruptured thin wall glass spheres in water and succeeded to generate spherical underwater shock waves. He tried to visualize, by using the Schlieren method and recording with multi-frame camera, the transition of reflected underwater shock waves. As the shock waves which so far he produced were so weak, the pattern of reflected spherical shock waves was a regular reflection. Figure 6 shows sequential visualization of two underwater shock waves and their interaction.

Finishing a training of safe synthesizing lead azide, we eventually produced in house micro-pellets weighing from a few mg to 10 mg (PbN_6: density, 1.5 g/cc, detonation speed, 2.98 km/s). We had a permission from the government of handling micro-explosives in laboratory and ignited them in liquid by illuminating them with a Q-switch laser beam. Later, the Chugoku Kayaku Co. Ltd kindly provided us with 10 mg silver azide pellets, (AgN_3: density, 3.8 g/cc; detonation speed, 5.05 km/s; ignition energy, 30 $\mu J/mm^2$), for supporting our application of underwater shock wave focusing the disintegrate of urinary tract stones.

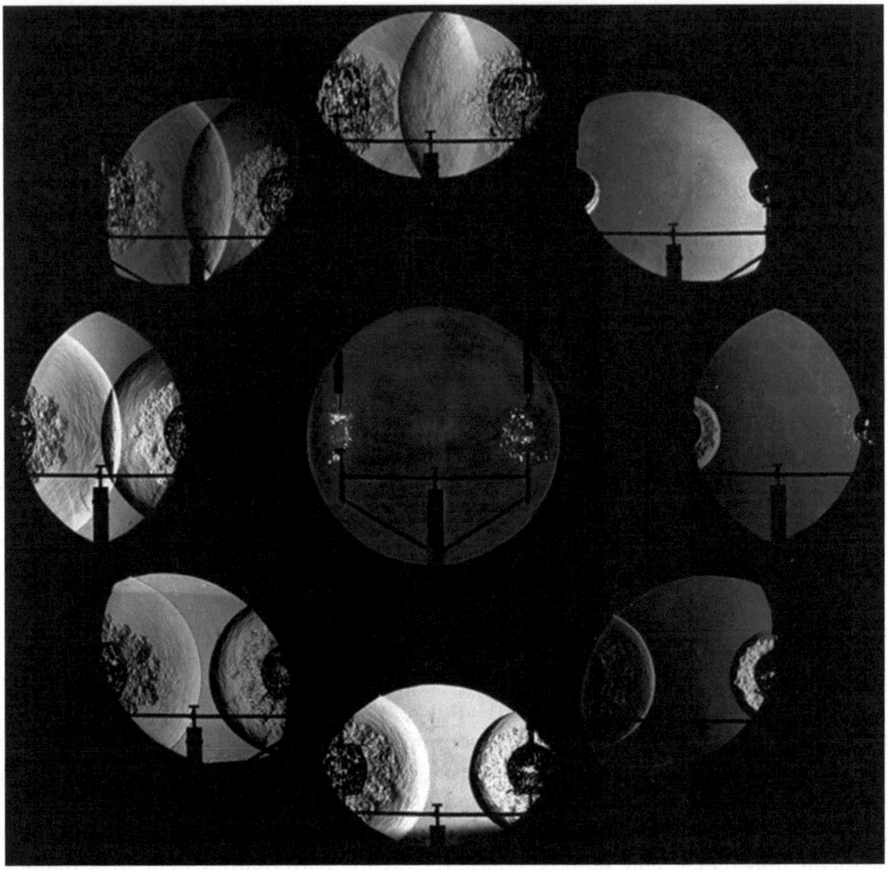

Fig. 6 Interaction of two weak shock waves reproduced from [6], with the permission of AIP Publishing

At first, we glued explosive pellets on a cotton thread and suspended it in water. We visualized underwater shock waves by using double exposure holographic interferometry. Figure 7a shows an underwater shock wave driven by 4 mg of lead azide. A small dark spot in the middle and a thin vertical line show the pellet and the cotton thread, respectively. The grey circle in the middle shows the detonation product gas. The spherical shock wave is visualized to have circular shape.

Beautifully, axial symmetrical spherical fringes appear. These were formed as integral of the light beams crossing along the path from a shock front to its reverse front, now look like co-axial fringes. The fringes are packed very densely just behind the shock front but still have spatial resolution. Assuming axial symmetric flows, we counted, from the shock front to the boundary of the detonation product gas, the order and interval of fringes on a magnified interferogram as seen in Fig. 7a. From the data, we can determine the density profile and then the pressure

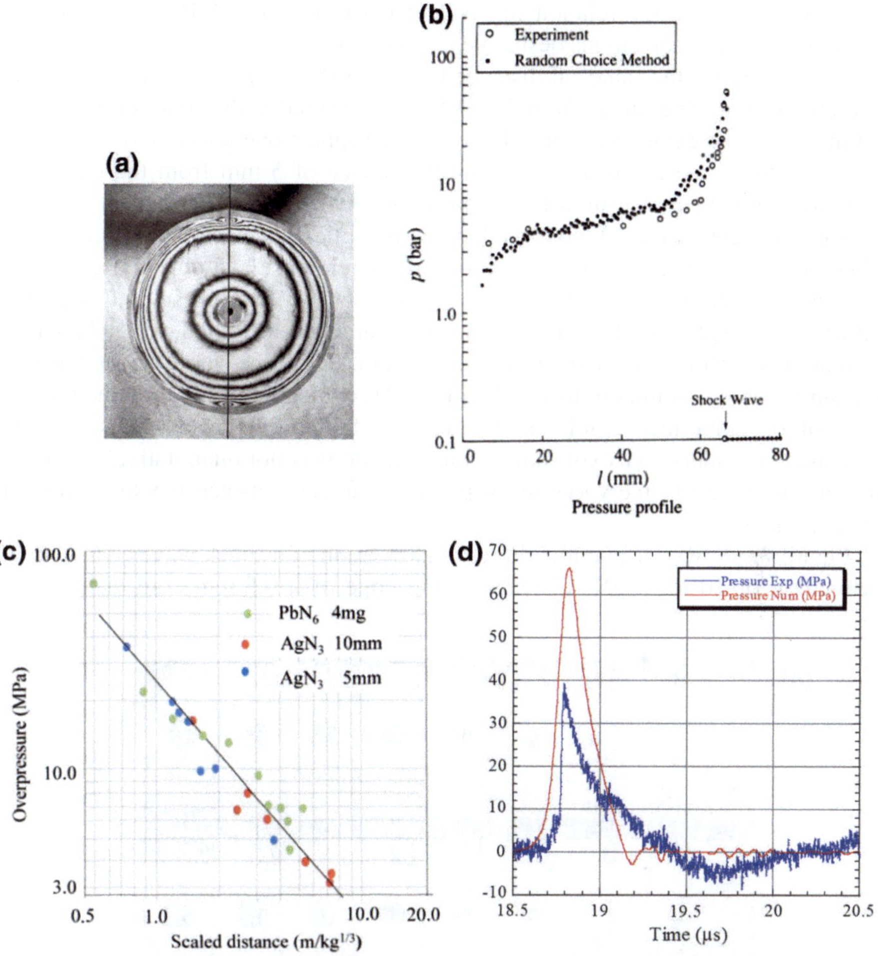

Fig. 7 Underwater explosion: **a** Interferogram, 4 mg lead azide explosion; **b** Pressure profile; **c** Overpressure-scaled distance correlation of underwater micro-explosion; **d** Comparison between the measured pressure variation and numerical simulation [9]

profile by using the Tait equation [7]. Figure 7b shows the pressure profile estimated from Fig. 7a and the comparison with the random choice numerical simulation. It should be noted that we matched the density at the shock front with the simulated value, as this value cannot be estimated from the fringe counting.

Later, our experimental skill was so refined as to attach micro-pellets on the tip of 0.6 mm diameter optical fiber. Explosives weighing even 3 μg AgN_3 were detonated by irradiating them with a Q-switched Nd:YAG laser beam of energy of 20 mJ and pulse width of 7 ns. The total energy of 3 μg AgN_3 is about 5 mJ, and such a small grain of AgN_3 still obeys the scaling law.

Figure 7c shows the relationship between overpressure (MPa) and scaled distance (m/kg$^{1/3}$). Black circles designate overpressure in MPa of 4–10 mg lead azide pellet measured at the stand-off distance from 5 to 80 mm [8]. Red circles designate overpressure of silver azide from 3 to 300 μg measured at the stand-off distance of 10 mm from the explosives, and blue circles designate overpressure of silver azide of 3 to 300 μg measured at the stand-off distance of 5 mm from the explosives. Overpressures were measured by a fibre optics hydrophone (FOPH), in which the pressure was detected not by piezo effect but by optically detecting the change in the phase angle of the spot to measure between the reference and object laser beams.

Figure 7d shows the time variation of shock wave overpressures caused by a 30 μg AgN_3 explosion. The pressure was measured by the FOPH, which had a time resolution of 10 ns and a diameter of 0.2 mm. The time variation of the overpressure profile is similar to that of 10 mg AgN_3 explosion. A red thin line indicates a result of numerical simulation by using a finite difference scheme [9]. The modeling of underwater explosion of silver azide was not quantitatively tuned yet so that the trend of time variation of pressures agreed between the simulation and the experiment.

Figure 8a shows a sequential direct shadowgraph of a shock wave created by the detonation of 10 mg AgN_3 attached at the edge of a 0.6 mm core diameter and

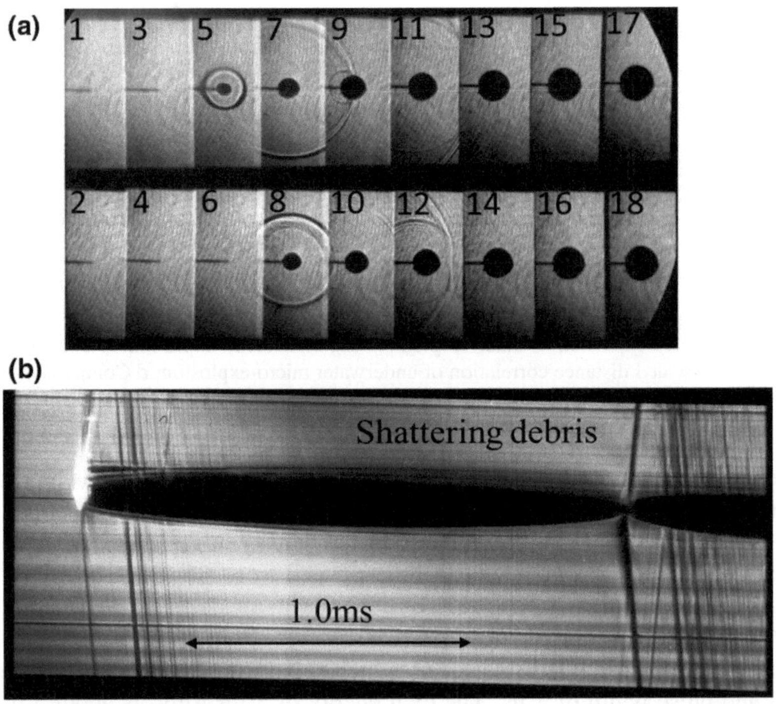

Fig. 8 Formation of shock waves by micro-explosions of 10 mg AgN_3: **a** #90070702, framing photograph 100,000 fps; **b** #85121729 streak recording

1.5 mm sheath diameter optical fiber. The images were taken by Ima-Con 970 at the framing rate of 100,000 fps. Upon the laser irradiation, AgN_3 was detonating and its detonation product gas formed a dark circle which expanded at supersonic speed and hence drove a spherical shock wave shown as a white circle. It immediately moved away from the view and about 20 μs later, was reflected from the neighboring solid wall. The reflected shock interacted with the detonation product gas. The thin dark circular curve seen in the 10th and 11th frames showed reflected expansion wave from the bubble which was induced due to the acoustic impedance mismatching at the water/gas bubble interface. On the 11th frame, the part of the reflected shock wave, which was interacted with the reflected expansion wave, was significantly deviated from the circular. That is, the reflected shock was attenuated.

Such complex interfacial reflections and long term behaviors of the detonation product gas is demonstrated in a streak recording in Fig. 8b. The ordinate designates the length scale and the abscissa designates the elapsed time. A dark shadow shows the evolution of the detonation product, the gas bubble. A bright spot at the left hand side indicates the explosion, and a resulting shock wave is shown by a dark line.

A dark line appearing from the upper part at 20 μs later was a reflected shock from a neighboring solid wall. When reaching a maximum diameter, the detonation product gas bubble started to contract. Minute fragments or debris of unburned chemical components shattered outward at nearly the detonation speed. The debris overtook the bubble surface when it started to contract. This is shown as grey clouds to be seen in Fig. 8b. When the debris shattering terminated, the bubble diameter was a minimum and started to explosively expand, driving trains of compression waves, which eventually coalesced into secondary shock waves.

5 Shock/Bubble Interaction

Wave interactions in high-speed gas flows can decrease pressures even down to vacuum, whereas wave interaction in liquids can also decrease pressures but down to their evaporation pressure. When exposing liquids to pressures lower than the evaporation pressure, a rapid phase change would be induced. This is a spontaneous bubble formation which we already saw in the ultrasonic oscillation. The bubble inception occurring in hydraulic machineries are named as cavitation. The fundamentals of cavitation research are called the bubble dynamics, which is one of the highlights of underwater shock wave research. Underwater shock wave research is very closely linked with bubble dynamics. Shima reviewed the bubble dynamic research [10]. Firstly, we decided to examine shock interaction with a single bubble. We released 1.7 mm diameter spherical air bubbles rising vertically at 0.17–0.22 m/s from the bottom of the water chamber. The bubbles were released from a thin syringe needle by keeping the release pressure difference smaller, so that we maintained spherical bubbles. For bubbles whose shapes significantly deviated from a spherical shape, the response of their motion to the shock wave would differ from that of spherical ones.

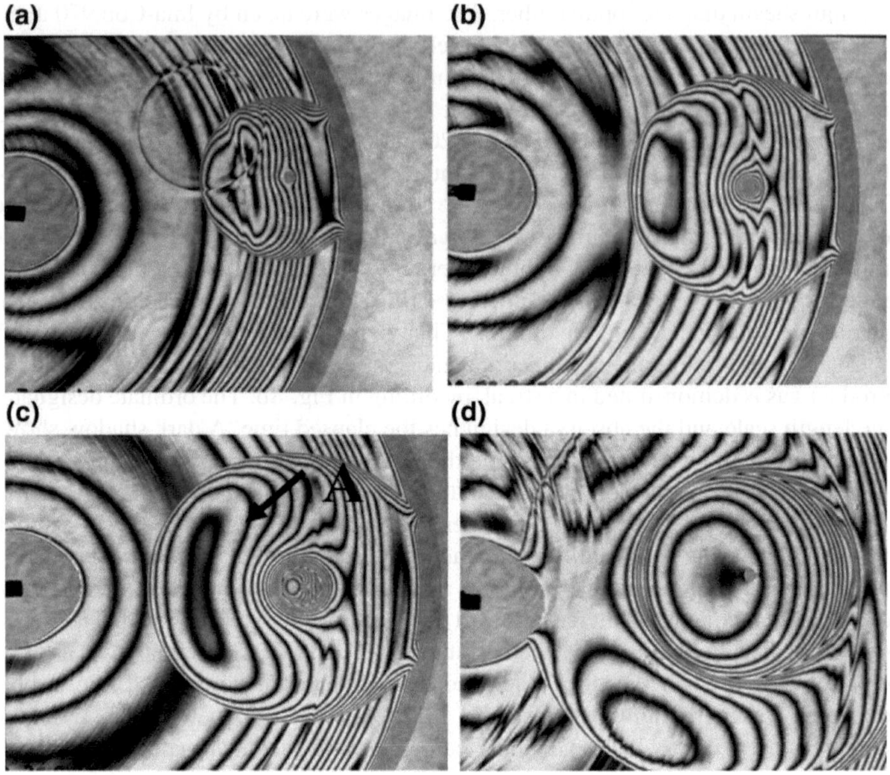

Fig. 9 Shock/bubble interaction, shock waves were generated by the detonation of 10 mg AgN_3: **a** #86082507, delay time: 18.4 μs; **b** #86082506, delay time: 20.1 μs; **c** #86082505, delay time: 22.1 μs; and **d** #86082504, delay time: 23.1 μs

We generated underwater shock waves by detonating a 10 mg AgN_3 pellet attached at the tip of optical fiber [7]. At the 20 mm stand-off distance, its overpressure would be about 50 MPa. We visualized the procedures of shock/bubble interactions by both single exposure holograms and double exposure holographic interferometry. The fringe distributions are axially symmetrical and hence, if we count correctly the fringe order and distributions, we can estimate experimentally the pressure profiles from interferograms [7].

Figure 9a–d show sequential double exposure holographic interferograms. Measuring the radius of the reflected expansion wave Fig. 9a was taken after the incident shock hit the air bubble. Figure 9a was taken at about 4 μs after the incident shock hit the air bubble. When the shock wave of overpressure of 50 MPa passed the bubble, it immediately started to contract. Its reaction, the expansion wave was released into surrounding water, which propagated in water at its sonic speed. This is the reflected expansion wave. At the same time, a transmitting shock propagates inside the bubble. This situation will be discussed later.

In Fig. 9b, we can see that the expansion wave is driven by the contraction of the air bubble. Double exposure holographic interferometric images show the phase difference of light beams which recorded phase distributions of images before and after the event. A reference image before the event and an object image that recorded the event are superimposed in a hologram. In the present case, the images of the original bubble and the contracting bubble are superimposed and hence we do not see the contracting bubble shape. We can detect these bubble deformations but clearly identify the fringe concentration at the frontal stagnation point of the bubble.

Figure 9c was taken at 6.7 μs after the incident shock impingement on the bubble: the time was calculated from the image. Note that the delay time described in the figure caption includes measurement deviation. Fringes are densely accumulated on that side, which indicates the local pressure increase. When the pressure was a maximum, a train of compression waves was released, which coalesced into a secondary shock wave. In reality, the center of the flat side was not planar but protruded toward the inside. Such a deformation was driven by the high pressure which accumulated at this side.

Eventually the formation created a water jet penetrating the bubble. Figure 9d shows interferogram at 11.2 μs (estimated from the image) after the shock impingement on the bubble. We can see the water jet penetrating the bubble. It should be noticed that the grey shadow shows the combination of the original bubble and the jet and the real recovering bubble shape was much too small.

The jet speed varied depending on the bubble diameter and the loading pressure. It is known that if a shock wave of overpressure of a few hundred MPa is loaded on a 2.0 mm air bubble, the resulting jet speed can be about 100 m/s. Hence, the resulting stagnation pressure, $p_{staganation} = a\rho u$, will be about 150 MPa.

Counting the fringe intervals and the fringe orders of the interferograms shown in Fig. 9a–d, we can calculate, by assuming the axial symmetric field, the pressure profiles over the bubbles in Fig. 10a–d, respectively. We can see the pressure decrease across the reflected expansion wave and the gradual accumulation of high pressure on the side at which the shock impinged. Then the high pressure was released as trains of compression waves, whose coalescence then formed a secondary shock wave.

For bubbles of, for example,10 μm diameter, their impingement by shock waves would follow a different scenario as described in Fig. 10. The reflected expansion wave is immediately overtaken by the secondary shock wave. Hence, the shock/bubble interaction would create secondary shock waves.

6 Shock/Bubble Cloud Interaction

Figure 11 shows the interaction of a spherical shock wave with, presumably, the densest bubble cloud observed. In Fig. 5, we already observed the bubble cloud densely migrating over the ultrasonic oscillator. At this instant, we load a strong shock wave onto the bubble cloud by detonating 10 mg PbN_6 attached on the tip of

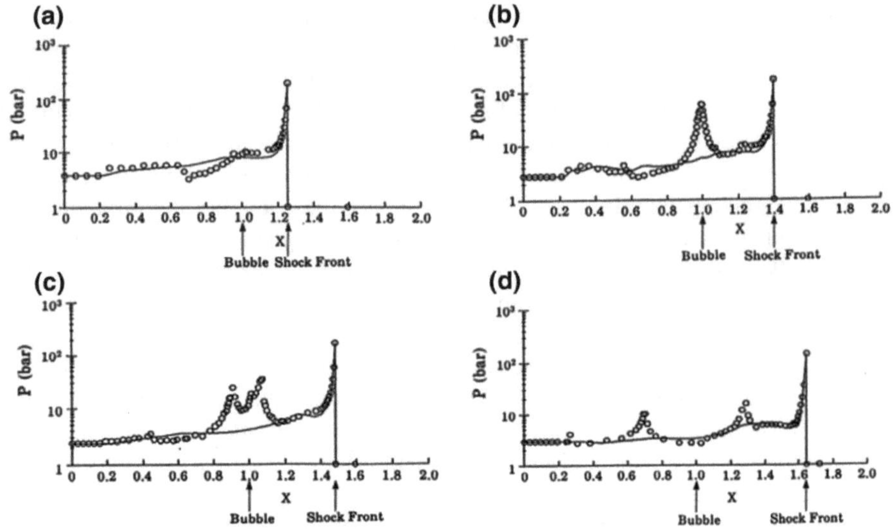

Fig. 10 Pressure profiles of shock/bubble interaction displayed in Fig. 9: **a** Fig. 9a, 18.4 μs: **b** Fig. 9b, 20.1 μs; **c** Fig. 9c, 22.1 μs; and **d** Fig. 9d, 23.1 μs

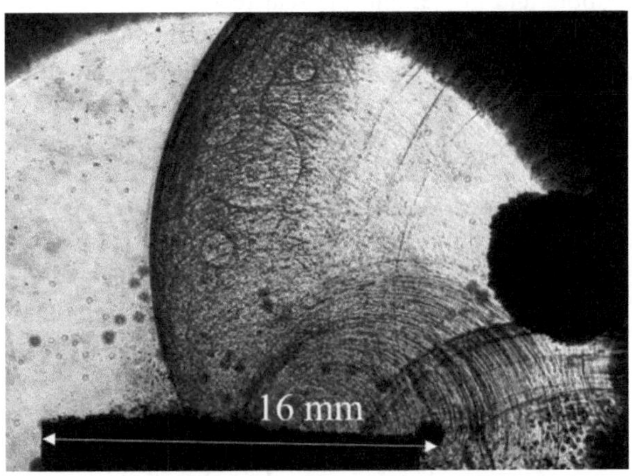

Fig. 11 Shock interaction with bubble clouds at 0.1 MPa and 301 K: #821129, delay time 73 μs

the optical fiber at about 10 mm away from the edge of the oscillator. The overpressure was about 50 MPa at a stand-off distance of 20 mm.

Upon shock loading, all the bubbles instantaneously collapsed and converted into shock waves. Circles and semi-circles in Fig. 11 are the resulting shock waves. Although we did not recognize many bubbles migrating above the oscillator, now we saw so many rings distributed behind the shock wave. We can see that the bubbles on the oscillator surface collapsed by the shock wave loading.

A bubble curtain is one of the best methods of mitigating underwater shock waves. Their diameters and spatial distributions would be important parameters to

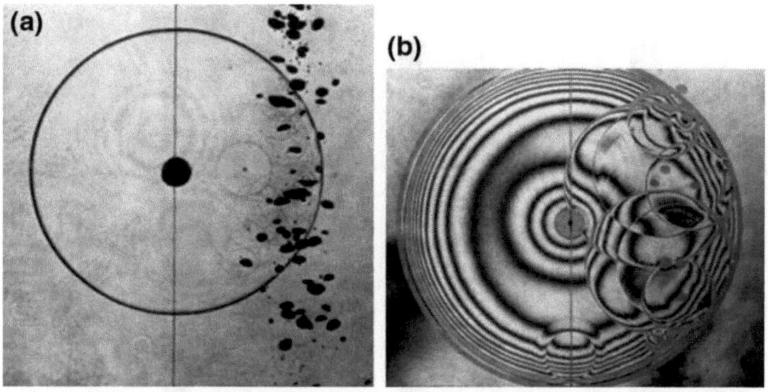

Fig. 12 Shock bubble/cloud interaction PbN_6 10 mg, stand-off distance of 20 mm: **a** #83013117, single exposure hologram; **b** #83013103, double exposure interferogram

design the bubble curtains. In Fig. 12, we visualized, by using single exposure holography and double exposure holographic interferometry. Air bubbles were ejected from a 1.0 mm diameter nozzle placed at the bottom of a water chamber. Hence bubbles had not necessarily spherical shape but randomly oriented ellipsoidal shape. Shock waves were generated by explosion of 10 mg PbN_6 and loaded at the stand-off distance of about 20 mm. The loading pressure was about 50 MPa.

Figure 12a shows a single exposure. The shock front was displayed as a broad dark ring. However, when it was caught up with the reflected expansion wave, the broad ring shape became thin. This implies that the shock front is locally attenuated. Such attenuations occurred over the bubble surface at which the shock passed the air bubbles.

Figure 12b shows a double exposure holographic interferogram. As seen in Fig. 9a–d, when the shock wave hit a large air bubble, it deforms and creates distinct reflected expansion wave. Along the part of the shock front at which the expansion wave interacted, the number of fringes drastically decreased, which implies that shock wave is attenuated.

7 Luminous Spot at Shock/Bubble Interaction

An ellipsoidal bubble of a major diameter of 2.5 mm and a minor diameter of 1.5 mm positioned at 5 mm from a solid wall started to collapse, when it was loaded at its minor diameter by a shock wave generated by explosion of a 10 mg AgN_3 pellet which was attached at the tip of an optical fiber. At the 20 mm stand-off distance, the overpressure was about 50 MPa. Figure 13a shows a sequential direct shadowgraph of 1.0 μs inter-frame time and exposure time of 10 ns by using Ima-Con 200. A 0.4 mm diameter optical fiber hydrophone was imbedded in the solid wall.

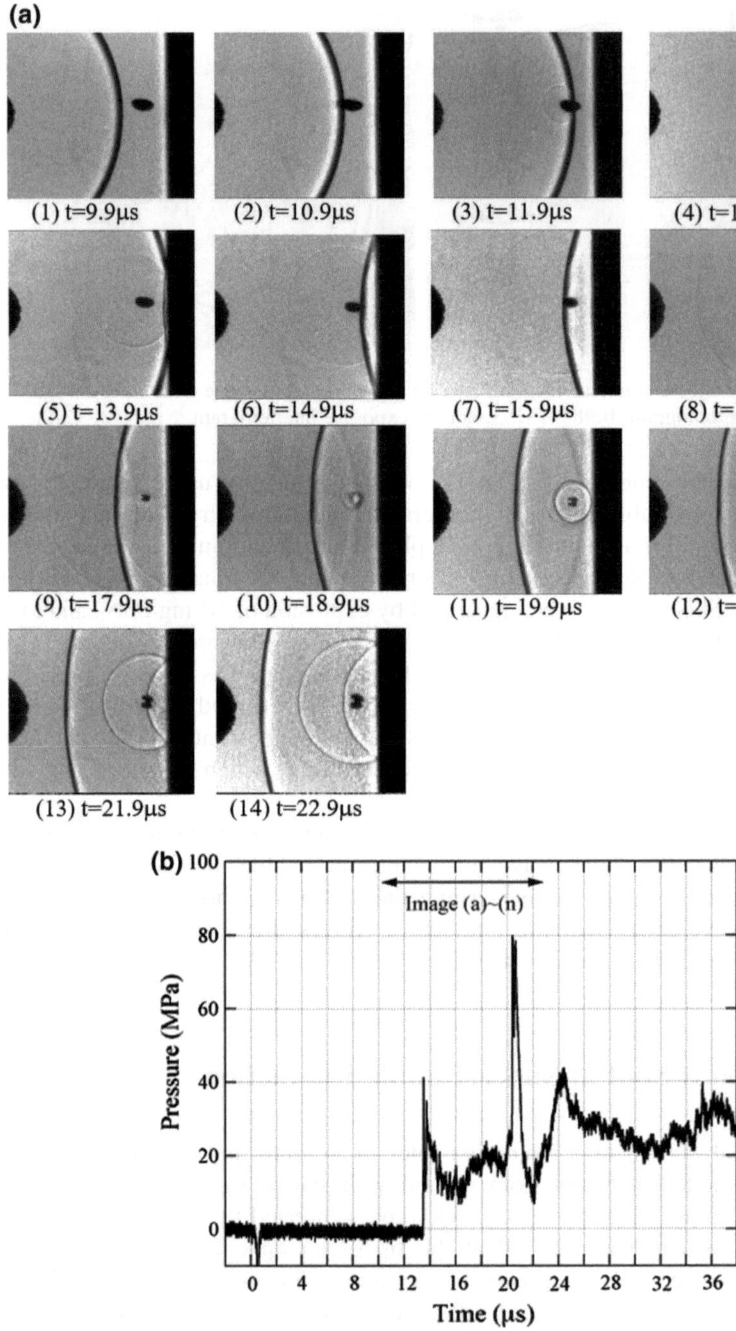

Fig. 13 Emission of luminosity at shock/bubble interaction: **a** Sequential observation; **b** pressure measurement [11]

The shock wave propagated from the left-hand side. In the second and third frames, we can see the transmission of the shock wave over the bubble. The bubble started to contract from the third frame. We can see the reflected expansion wave, which was so faint to be hardly identified. In the sixth frame, the shock wave was reflected from the wall. In the seventh and eighth frames, the bubble interacted with the reflected shock wave. The bubble contraction continued. At the ninth frame, the bubble had a minimum volume and it shape looked slightly irregularly deformed. In the tenth frame, the rear part of the bubble was shining. The bubble started to expand.

In the eleventh frame, the secondary shock wave, rather strong, was generated. So far estimated from the distance it travelled with the velocity of about Ms = 1.67. In the twelfth frame, the bubble expanded with an elapse of time and Ms = 1.49. Sanada et al. [1] estimated, from a trajectory of shock formation in streak images in ultrasonic oscillatory tests, the attenuation of a shock wave. When a bubble collapsed, the initial shock Mach number was about 1.5, which immediately attenuated to sonic speed. The Mach number of the secondary shock wave took nearly the same value.

In these frames, the bubble deformed so as for its frontal side to penetrate toward the inside of the bubble. The jet speed so far measured was in the order of magnitude of 100 m/s, which would produce the stagnation pressure of about 150 MPa, as already observed in the case of the shock/single bubble interaction in Sect. 5.

At the side of the bubble where the shock impinged, the pressure was so enhanced that it formed a jet. At the same time, the high pressure outside the bubble, say 50 MPa, is transmitted into the gas inside the bubble. When a high pressure of 50 MPa is transmitted into air or water vapor inside the bubble, very strong shock waves must be induced. When such a strong shock in gas propagates and its reflected shock wave focuses at the reverse side of the bubble, the temperature is enhanced so as to emit the luminosity.

It is well known that the oscillatory motion of a spherical bubble in a quiescent liquid creates the so-called sono-luminescence, in which at the moment of the bubble's minimum volume, the bubble emits light. However, unlike the sono-luminescence, the bubble emitted the luminosity, because of the wave interaction inside the bubble as seen in Fig. 13. The luminosity observed in Fig. 13 was hardly observable throughout holographic observation. This was based on the reasons that the holofilms did not have sensitivity to the luminous light and intense pulse laser beam always overexposed such faint luminosity.

Figure 13b shows the pressure history by embedding the FOPH pressure transducer into the wall. The ordinate designated pressure, in MPa and the abscissa designates time in μs. The first peak pressure at about 14 μs corresponds to that occurred when the incident shock wave arrived at the wall, as seen in the fifth frame of Fig. 13a. The second peak pressure at about 20 μs corresponds to which occurred when the secondary shock wave arrived at the wall, as seen in the eleventh frame. The pressure decrease at 22 μs corresponds to the pressure decrease behind the reflected secondary shock wave from the wall as seen in the fourteenth frame.

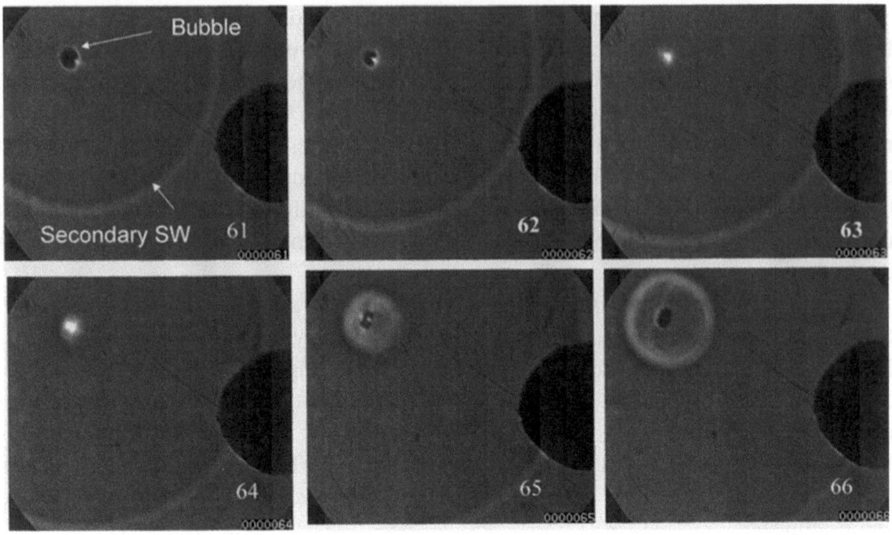

Fig. 14 Emission of luminescence at shock/bubble interaction

As a next, the case of a spherical bubble was investigated. We placed a 1.5 mm diameter spherical air bubble at 20 mm stand-off distance and impacted it with a spherical shock of overpressure of about 40 MPa, which was generated by detonation of 10 mg AgN_3 attached at the tip of an optical fiber. The visualization was carried out by using a direct shadowgraph and the images were recorded by a high-speed video camera, Shimadzu HVP-1, at the recording speed of 1,000,000 fps, the exposure time of 125 ns, and the total number of frame of 104.

Figure 14 show sequential images from the frame No. 61 to No. 66. In the frame 61, we can see the bubble on the left corner and the secondary shock wave as indicated by an arrows. The explosion product gas was shown as a dark irregular shadow at the right hand side.

In frames No. 62 and 63, the bubble contracted and the faint white spot in the frame No. 61 became brighter. In frame No. 64, the bright spot is most intense and large. The bubble is explosively enlarged, indicating the initiation of the secondary shock wave. From frame 64 to frame 65, although it is less accurate estimation, the secondary shock propagates at about Ms = 1.33, and from frame 65 to frame 66, Ms $\geq$ 1.0. It should be noted that, as we already remarked in Fig. 9, the center of the secondary shock wave is located at the frontal stagnation point of the bubble. However, in the case of shock/non-spherical bubble interaction, the center of the secondary shock was shifted from the frontal stagnation point.

In Fig. 13 the bright spot appeared only temporarily in the 10th frame. This means that the emission took about a few μs. However, in Fig. 14, the emission continued for more than 4 μs. The difference of the emission interval would be attributable to the wave propagation inside the bubble, in other words, the bubble shape.

8 Shock/Interface Interaction

The interaction of shock waves generated by underwater explosions with gaseous interfaces or water/foreign liquid interfaces is one of the intriguing research topics, in particular, related to medical application of shock waves. Analytically, the point explosion at a gas/liquid interface is a singular problem. Waves in gas and liquid propagate independently. Hence, we performed series of experiments.

Figure 15a, b show double exposure holographic interferograms of a point explosion at an air/water interface. We submerged half of a 6.4 mg PbN_6 pellet in water and ignited by irradiating a laser beam on it. In water, the spherically shaped explosion product gas drove a hemi-spherical shock wave, whereas in air the explosion product gas expanded not necessarily uniformly, but its top part ejected high speed debris and explosively vaporized water. Supersonic free flights of debris particles are clearly observable in Fig. 15b.

Hence, the top part of the explosion product gas formed a mushroom shaped shadow, which drove an irregularly shaped shock wave. The shock wave in air never looked neatly spherical. In short, the point explosion at the interface generates a hemi-spherical shock wave in water and although the shape is irregular, a little spherical shock wave is generated independently in air. As discussed, the energy transfer across a gas/liquid interface is governed by the acoustic impedances of each medium. The ratio of acoustic impedances is about 3450, so that energy in a gaseous phase is hardly transmitted into the liquid phase, and vice versa. The observation we saw in Fig. 15 would never agree with mathematical modeling. To improve the energy deposition, even if we may adopt a pulse laser beam deposition at the interface, the author would expect that the result would be much more complex. On the edge of the underwater shock front intersecting with the air, the fringe number sharply decreases, which implies that the underwater shock is locally attenuated.

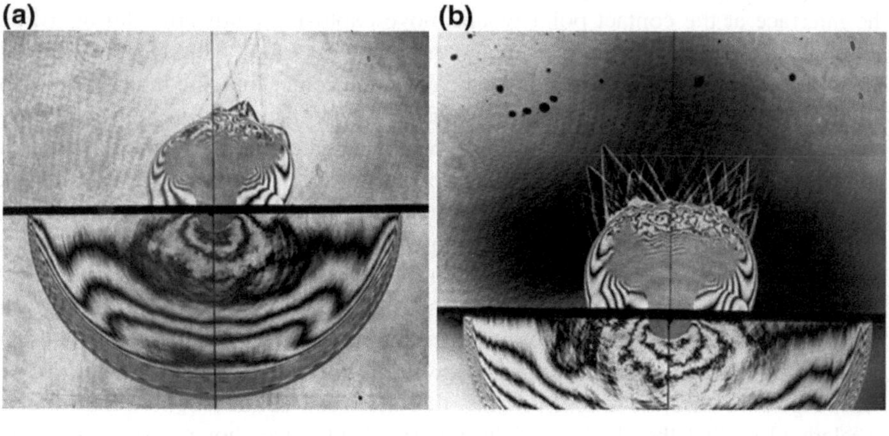

Fig. 15 Experimental verification of a singular point explosion at a gas/liquid interface, explosion of PbN_6 6.4 mg at 1013 hPa and 284 K: **a** #83011807, 30 μs; **b** #83011808, 50 μs

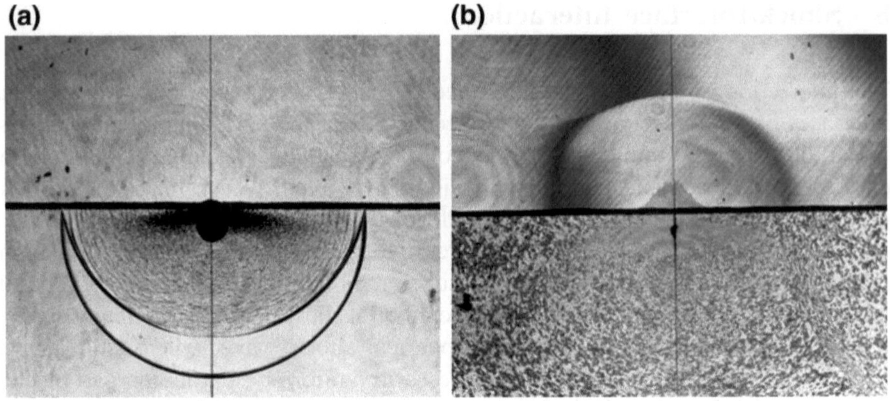

Fig. 16 Point explosion at 3 mm below air/water interface, explosion of 10 mg PbN_6 at 1013 hPa and 288 K: **a** #83020202, delay time of 30 μs, single exposure hologram; **b** #83020206, delay time of 70 μs

Figure 16a, b show the interaction of shock/air water interface visualized by double exposure holographic interferometry and a single exposure hologram. The shock waves were generated by explosion of 10 mg PbN_6 submerged at 3 mm below the interface. The experimental condition was at atmospheric pressure and 288 K.

Although Fig. 16a was taken at first with the delay time of 30 μs, we found from the distance of the shock traveled that this photo was taken at about 10 μs after the ignition. When the shock wave impinged on the air/water interface, as the ratio of acoustic impedances between water and air is about is 3.400, the most of energy is reflected from the contact point and very small amount of energy in water is transmitted to the air.

The pressure should be high at the interface at 3 mm above the explosive. Then the interface at the contact point would move explosively upward. During 10 μs, although not observable in this photo, the entire interface would move upward. Then the expansion wave was reflected. Along the center line, the interval between the reflected expansion wave and the shock wave is maintained to be 6 mm. Grey noises behind the reflected expansion wave indicate non-uniformly distributed bubble clouds. The flow around the explosion product gas is moving upward and the pressure at the center is gradually enhanced.

Due to the difference of acoustic impedance between water and air, only a fraction of energy in water is transmitted into the air. The energy is transmitted into the air, which created a visible change in fringes. The single exposure hologram missed the fringes. The shock wave created by the initial explosion attenuated with propagation and the reflected expansion wave followed just behind it.

In Fig. 16b, with the elapse of time, the interfacial deformation at the center is accelerated so that the shadow of the deformed interface shows a slightly jagged shape.

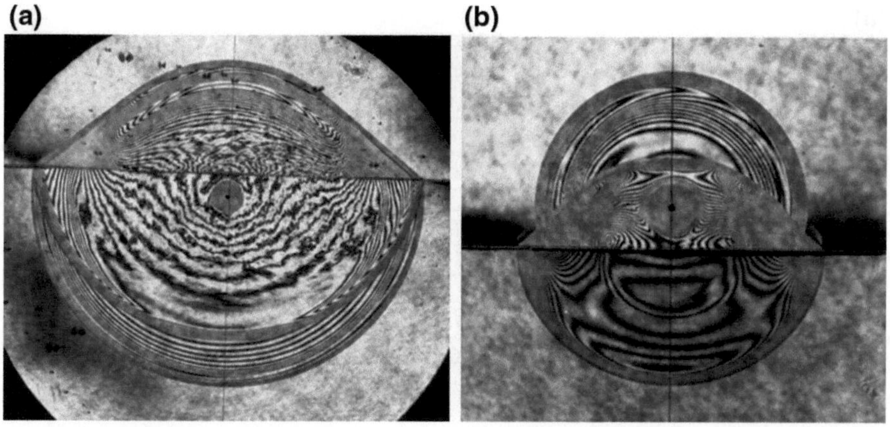

Fig. 17 Shock interaction with water/silicon oil interface, AgN_3 10 mg was detonated at 10 mm above the interface: **a** #92121502, water/10cSt silicon oil; and **b** #92123103, 10cSt silicon oil/water

We poured silicon oil on the surface of water and made a light/heavy liquid interface. Silicon oil 10cSt has sound speed of 960 m/s and is a safe liquid under normal conditions against explosion of AgN_3 10 mg. We exploded the pellet at 10 mm above the interface. When detonating in water, we have a fast/slow interface as shown in Fig. 17a. When detonating it in silicon oil, we have a slow/fast interface as shown in Fig. 17b.

Sound speeds in water and silicon oil are 1500 and 980 m/s, respectively. Hence, the ratio of the acoustic impedance of water to that of silicon oil is about 1.8. In assuming a plane shock wave reflected from this interface, about 8% of energy in water is reflected and 92% of energy is transmitted into silicon oil. Hence, a spherical shock wave is generated at the central part in silicon oil. The shock wave in water propagates much faster than that in silicon oil and transmits an oblique shock wave into silicon oil. Hence, the shock wave in silicon oil looks like a combination of a spherical shock wave with an oblique shock wave as seen in Fig. 17a.

9 Reflection of Underwater Shock Wave

In gases, the transition of reflected shock waves is visualized, in shock tubes, shock wave reflection from metal wedges. In water, however, upon shock impingements on metal walls, longitudinal waves and shear waves are generated, which are precursory to the main shock waves and precise observations of their reflection is prevented. Avoiding such a complexity, Glass and Heuckroth [6] visualized the reflection of two underwater shock waves having identical strength. However,

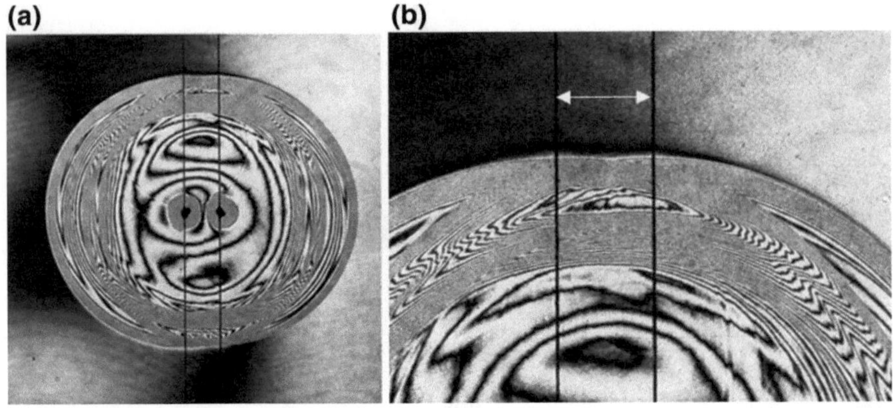

Fig. 18 Reflection of underwater shock waves, simultaneous explosion of 6.2 mg and 6.3 mg PbN_6 placed at 8 mm interval: **a** #83112805; and **b** Enlargement of (**a**)

the underwater shock waves thus created were not as strong as exhibiting Mach reflection patterns.

To achieve the generation of moderately strong underwater shock waves by explosion of PbN_6 or AgN_3, we attached 6.2 mg and 6.3 mg PbN_6 pellets each on a thin cotton thread, positioned them at an 8 mm interval and detonated them by the direct irradiation of a Q-switched ruby laser beam. Figure 18a shows a double exposure holographic interferogram. The delay time from the ignition to the visualization was 17 μs. The elapsed time, if one estimated by assuming that the direct wave propagated at sonic speed, was 20 μs. This implies that the direct wave propagated much faster than the sonic speed.

Figure 18b is enlarged image of Fig. 18a. The pattern of this intersection shows so-called von Neumann-Mach reflection (vNMR). This reflection pattern does not accompany any distinct triple points, which are a typical representation of gas-dynamic non-linearity. Therefore, vNMR appears usually when weak shock wave reflects from a shallow wedge. In other words, vNMR would appear when the flow behind the area in which the incident shock and the Mach stem merge is isentropic or nearly isentropic. Liquids are less compressible so that the particle velocity induced behind shock waves is low, even if the pressure is high. Hence, isentropic conditions are fulfilled so that the pattern of underwater shock wave reflections is mostly vNMR.

10 Underwater Shock Wave Focusing

Water is less compressible so that reflected underwater shock waves can be focused. A two-dimensional test section, 90 mm × 127.3 mm elliptical cavity (with ellipticity of $2^{1/2}$) was machined on a 10 mm thick brass plate and sandwiched with

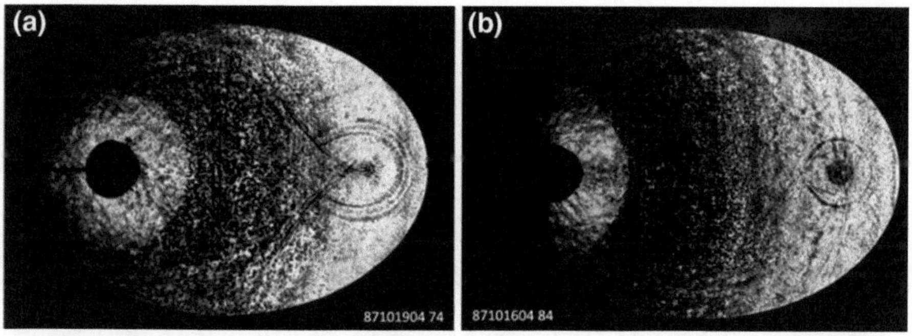

Fig. 19 Two-dimensional shock wave focusing in a 90 mm × 127.3 mm elliptical cavity: **a** #84101904, 74 μs; **b** #87101604, 84 μs

15 mm thick acrylic plates tightened with thick bolts. Filling water in the test section, we detonated a 4 mg PbN_6 pellet at one of the focal points. At first a spherical shock wave was created, whose overpressures at 5 mm distance was well over 100 MPa. The high pressure quickly diminished. After multiple reflections between the two acrylic plates, the spherical shock wave eventually turned into a two-dimensional shock wave.

Figure 19 shows sequential direct shadowgraphs. Small dark spots distributed over the field of view show cavitation bubbles, which were instantaneously created as soon as the shock wave impacted the acrylic plates. Overpressures exceeding well 40 MPa made the acrylic plates move outward. Such a deformation instantaneously incepted cavitation bubbles. We never imagined that PbN_6 4 mg pellets detonated at 5 mm distance deformed even 15 mm thick acrylic plates.

At first, we wanted to visualize test fields with a 5 mm distance from the explosive by using double exposure holographic interferometry. We then gave up to apply the idea and used a single exposure holographic method, that is, equivalence to direct shadowgraph. When a cylindrical shock was formed, it propagates and is reflected from the elliptic wall. Then, each segment of the reflected shock wave directs toward the second focal point. Numerical simulations or analytical comparisons with Fig. 19 would be rather straightforward. It would be worthwhile to study experimentally elongated elliptical cavities, in short with large ellipticity. However, it is an open question what will happen, if we try to focus reflected shock waves from a truncated ellipsoidal cavity.

We focused a reflected spherical shock wave generated at the first focal point of the ellipsoid which had the identical geometry as shown in Fig. 19 from a truncated ellipsoid of the same geometry. We wondered whether or not the reflected shock wave could focus at the second focal point. This was a part of a basic research of the shock wave application to medicine [8].

A 4 mg PbN_6 pellet was detonated at the first focal point of the ellipsoid as shown in Fig. 20. The spherical shock wave is reflected from a 50 mm diameter truncated reflector, the bottom of which is located at about 109 mm from the focal

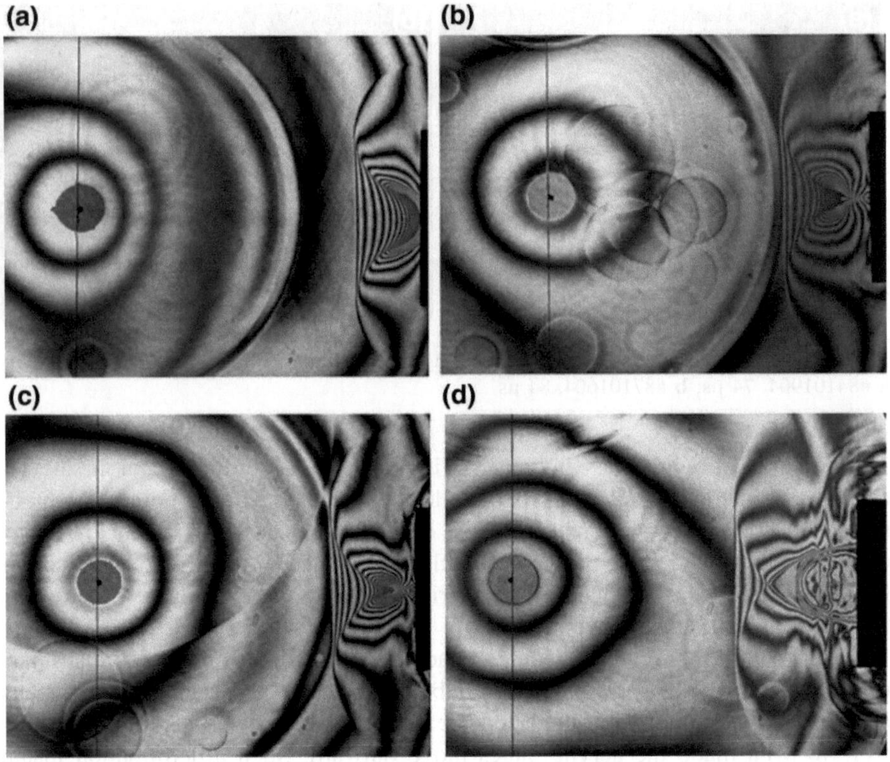

Fig. 20 Shock wave focusing from a 50 mm diameter truncated reflector, PbN_6 4 mg detonation at 1013 hPa and 285 K: **a** #83012111, delay time of 68 μs; **b** #83012012, delay time of 74 μs; **c** #83012103, delay time of 78 μs; **d** #83012128, delay time of 89 μs

point. In this arrangement, about 1.2% of the energy of the shock wave may be focused at the second focal point.

We took double exposure holographic interferograms sequentially at 2 μs intervals. Selective images are shown in Fig. 20a–d. In Fig. 20a the spherical shock wave, which was generated at the first focal point, propagated, and hit the curved wall, was reflected. The reflected shock wave directed toward the second focal point. We can see the accumulation of fringes toward the second focal point. In Fig. 20b, the fringes are going to focus at the second focal point. The density locally concentrated so high that the object beam could not pass locally through this region. Hence the image of this region became dark and the fringes were no longer resolved.

Many faint double rings are observed. These are the remnants of reflected expansion waves and secondary shock waves released from bubbles on the test section windows. The bubbles were impinged by the shock wave created by the explosion of a 4 mg PbN_6 pellet. The outer rings correspond to the reflected expansion waves and the inner rings correspond to the secondary shock waves, which are now propagating at the sonic speed. The interval of the double rings corresponds to the delay time in which the secondary shock wave was created after the shock impingement on air bubbles on the window glass. At that time we did not clean the observation glass windows. Probably air bubbles of 1 mm diameter attached on the glass windows.

In Fig. 20b, the dense fringe concentration indicates the density increase at the second focal point, in other words, the pressure increase. The density is so high that the object beam was significantly refracted and did not pass straight. Hence, the fringes were not resolved in the focal region. The region looked grey.

Once again in Fig. 20, we are visualizing an axial symmetric flow field so that, unlike two-dimensional shock convergence, the fringe concentration indicates exponential increase in density and pressure toward the focal point. Counting fringes, we estimated the peak pressure to be 40 MPa at the focal point [8].

Within the limitation of the experimental accuracy, the point at which peak pressure appeared was deviated by 2 mm to the reflector [8]. Probably this would be attributable to the fact that the reflection of shock waves with finite strength from solid walls is slightly different from that of a sound wave and shock waves accompany flows which is more or less affected by the presence of the wall boundary layer

In Fig. 20c, the focal point looked grey and the peak pressure was achieved. Figure 20d shows a flow field after about 20 μs from Fig. 20c.

We manufactured a 50 mm shallow dish having a 70 mm radius of curvature and 10.5 mm in depth and set it in a water chamber. We made a spherical shock wave reflect from the dish. We exploded a 10 mg PbN_6 pellet at the 40 mm distance from the shallow dish and 25 mm away from its center line.

11 Remarks

In order to investigate cavitation phenomena, the bubble dynamic research and the underwater shock wave research complement each other, in which optical flow visualization plays an important role. Since 1980 we became interested in applications of underwater shock wave research to medicine, then realized and collected a little bit of knowledge. Today medical applications of shock wave research made progress. However, the author realized that to support the interdisciplinary work, one has to devote more enthusiasm to underwater shock wave research.

References

1. N. Sanada, J. Ikeuchi, K. Takayama, O. Onodera, in *Generation and propagation of cavitation induced shock waves in a ultrasonic vibratory testing*, ed. by R.D. Archer, B. Milton (eds.), Proceedings of 14th International Symposium on Shock Tubes and Waves, (1983), pp. 405–412

2. K. Takayama, O. Onodera, H. Esashi, Motion of shock waves propagating over surfaces of liquids. Mem. Inst. High Speed Mech. Tohoku Univ. **48**, 419 (1982)
3. K. Takayama, Application of holographic interferometry to shock wave research. Proc. SPIE **298**, 174–180 (1983)
4. G. Ben-Dor, *Reflection of Shock Waves* (Springer, New York, 1980)
5. K. Takayama, H. Miyoshi, A. Abe, Shock wave reflection over gas/liquid interface, Rep. Inst. High Speed Mech., Tohoku Univ. **57**, 394 (1989)
6. I.I. Glass, L.E. Heuckroth, Low energy spherical underwater explosions. Phys. Fluids **11**, 2095–2107 (1968)
7. K. Takayama, Holographic interferometry study of shock wave propagation in two-phase media, in H. Groenig (ed.) Proceedings of 16th international symposium on shock tube and waves (1987), pp. 51–62
8. H. Esashi, *Study of shock wave propagation in liquids, Master Thesis* (Graduate School of Tohoku Univ, Faculty of Engineering, 1981)
9. N. Nagayasu, Study of shock waves generated by micro-explosiom and their applications. Doctoral Thesis, Graduate School of Engineering Tohoku University (2001)
10. A. Shima, Studies on bubble dynamics. SW J. **7**, 33–42 (1997)
11. K. Ohtani, K. Takayama, Spherical shock wave interaction with a single gas bubble in liquids, 14th Caviatation Sym. Sendai (2009)

Part V
Materials Research

On the Use of Digital Image Correlation for the Analysis of the Dynamic Behavior of Materials

François Hild, Amine Bouterf, Pascal Forquin and Stéphane Roux

Abstract The present chapter is devoted to the analysis of the mechanical behavior of materials subjected to dynamic loadings via digital image correlation (DIC). This measurement technique provides 2D or 3D displacement fields that can be evaluated thanks to the use of high-speed cameras. Various declinations of DIC are first presented. Uncertainty quantifications are also discussed. Last, different examples illustrate how DIC can be used to analyze and quantify deformation, damage and fracture mechanisms of brittle and ductile materials.

1 Introduction

The development of reliable measurement techniques is very important when characterizing heterogeneous mechanical manifestations at various scales. Solids, be they heterogeneous or homogeneous under complex loading, induce deformations whose multiscale analysis is needed to relate them to the underlying

---

Electronic supplementary material The online version of this chapter (doi:10.1007/978-3-319-61491-5_8) contains supplementary material, which is available to authorized users.

F. Hild (✉) · A. Bouterf · S. Roux
Laboratoire de Mécanique et Technologie (LMT), ENS Paris-Saclay/CNRS/Université Paris-Saclay, 61 avenue du Président Wilson, 94235 Cachan cedex, France
e-mail: francois.hild@ens-paris-saclay.fr

A. Bouterf
e-mail: amine.bouterf@ens-paris-saclay.fr

S. Roux
e-mail: stephane.roux@ens-paris-saclay.fr

P. Forquin
Laboratoire 3SR, Université Grenoble Alpes, Domaine Universitaire, 1270 rue de la piscine, 38400 Saint Martin d'Hères, France
e-mail: pascal.forquin@3sr-grenoble.fr

K. Tsuji (ed.), *The Micro-World Observed by Ultra High-Speed Cameras*,
DOI 10.1007/978-3-319-61491-5_8

microstructure and/or to the type of prescribed loading. From the knowledge of kinematic fields, it is possible to identify global and local material parameters, or validate material models and numerical tools [1].

Different techniques are used to measure displacement and strain *fields* under dynamic loading conditions. When utilizing optics concepts, and when coupled with mechanics, it is referred to as photomechanics [1–3]. Photoelasticity was the first technique that was used to study dynamic phenomena [4–7], in particular propagating cracks [8]. Moiré interferometry was then introduced [9, 10]. Although its practical use was more difficult than photoelasticity [11], some applications were shown in the 90s [11–13]. In contrast to photoelasticity, this technique as well as the following ones, deal with the actual material of interest and not a substitute chosen for its birefringence properties. Caustics proved to be very useful for estimating stress intensity factors and crack propagation velocities [14–17]. Digital image correlation [18], high-speed photography of laser and white light speckles [19–21] started being utilized to analyze dynamic tests in the 80s. Interferometric holography [22] and shearography [23] were subsequently introduced.

In the following, the measurement technique that is nowadays becoming very popular, namely, digital image correlation and its extensions will be presented. In Sect. 2, the general principles of image correlation and stereocorrelation are introduced. These techniques are utilized very frequently to monitor dynamic experiments with high-speed cameras. In many cases, the displacement (and strain) fields are not the final data to be assessed. They are generally post-processed to analyze the deformation and failure of various materials subjected to dynamic loadings. Different case studies dealing with such analyses of the dynamic behavior of materials are discussed in Sect. 3.

2 Digital Image Correlation

Digital image correlation consists in measuring displacement (or velocity) fields by registering (or correlating) a series of digital images (i.e., a set of pixels whose gray level is known). The first step of this and related approaches is the acquisition of the pictures during the experiment by use of cameras. It requires good quality imaging systems. It is to be emphasized that image correlation is not specific to—but can be used with—high speed imaging. High speed imaging opens up a range of new possibilities to tackle dynamic or impact mechanical behavior, but the basic principle of DIC in itself is unchanged. It is however fair to add that the usually small image definition and its potentially high noise level have motivated specific developments such as spatiotemporal regularization, which will be discussed below. Figure 1 shows two experimental configurations that were used in some of the cases discussed herein with a single high-speed camera (FASTCAM APX RS, Photron, Japan). Powerful lights (Dedolight 400D, Dedotec, Germany) were utilized to monitor experiments performed with in-house Hopkinson bars or a high-speed tension/compression testing machine (Dytesys, France).

(a)

(b)

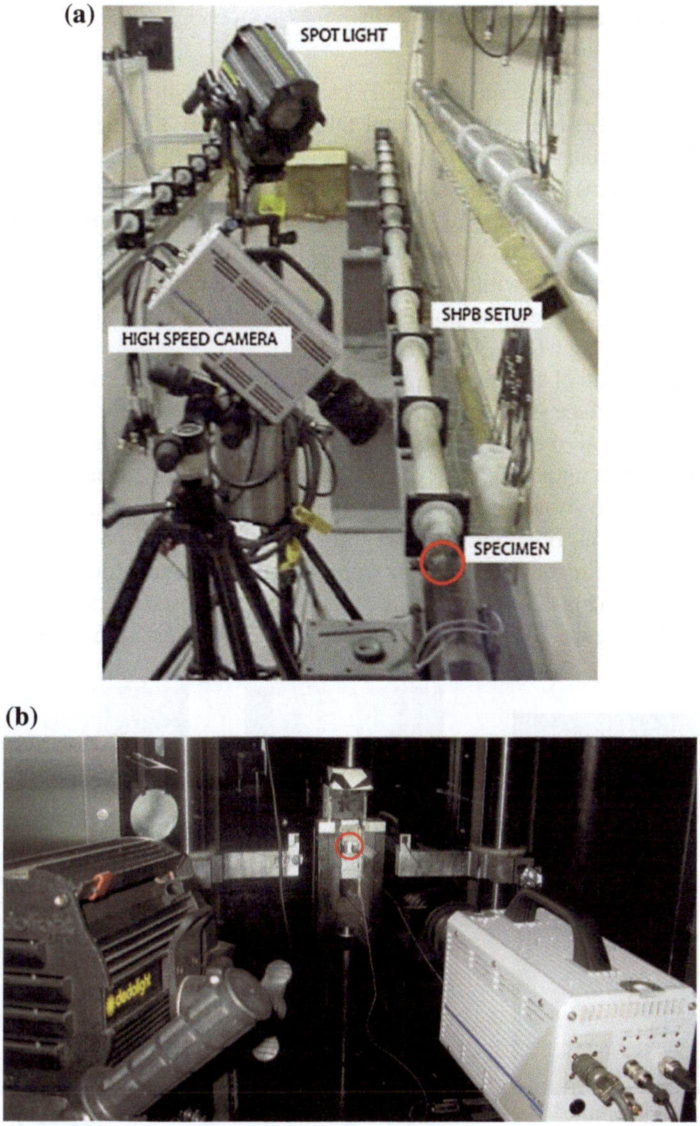

Fig. 1 Two experimental configurations using one high-speed camera (FASTCAM APX RS, Photron, Japan) and powerful lighting (Dedolight 400D, Dedotec, Germany). The monitored sample is encircled in *red*. **a** Taken from [24], **b** Courtesy of N. Swiergiel (Airbus Group Innovations)

2.1 Incremental DIC

Digital pictures are represented by matrices of (pixel) position $\boldsymbol{x}$, which are taken at different instants of time t. Incremental DIC (or 2D-DIC) is the first image correlation technique that was introduced in the early 1980s [25–27]. Only one camera (or imaging system) facing the region of interest is considered (Fig. 1) so that 2D displacements are measured. Its first use under dynamic loading conditions was reported in 1985 [18]. The stress intensity factor was evaluated and (favorably) compared to estimations with caustics. With the development of digital high-speed cameras, 2D-DIC has become popular for monitoring dynamic tests since the early 21st century (see e.g., Refs. [24, 28–33]).

The image of the observed surface in its reference configuration is denoted by I_0, and that of the surface in its deformed configuration at time t by I_t. Figure 2 shows reference pictures I_0 of experiments to be discussed hereafter. The first one (Fig. 2a) corresponds to the natural texture of an aluminum foam tested in compression (see Fig. 1a). Figure 2b shows a concrete sample fractured in dynamic spalling test. The random pattern was obtained by spraying a uniform white coating of paint and black dots. The third one (Fig. 2c) is one part of an image acquired during cylindrical expansion (Fig. 2c). In that case a sponge was used to deposit a very thin white speckle. Figure 2d shows the surface of a cylindrical sample to be tested in

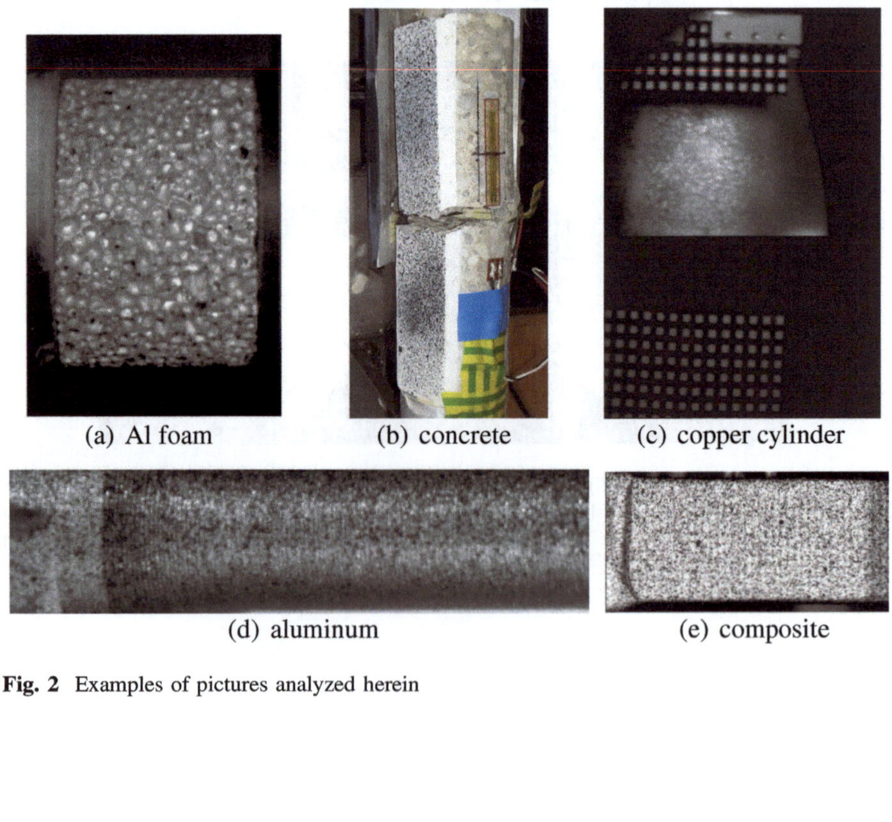

(a) Al foam (b) concrete (c) copper cylinder

(d) aluminum (e) composite

Fig. 2 Examples of pictures analyzed herein

Hopkinson bars (Fig. 5b), which was sprayed with B/W paint to create a random pattern. Similarly, Fig. 2e shows the surface of a composite sample to be tested in tension.

Pictures I_0 and I_t are related by

$$I_0(\boldsymbol{x}) = I_t(\boldsymbol{x} + \boldsymbol{u}(\boldsymbol{x}, t)) + n(\boldsymbol{x}, t), \tag{1}$$

where $\boldsymbol{u}(\boldsymbol{x}, t)$ is the displacement at pixel $\boldsymbol{x}$ and time t, and $n(\boldsymbol{x}, t)$ is a random component (e.g., photon noise, digitization when pictures are obtained by CCD or CMOS sensors, see Chapters "Evolution of High-Speed Image Sensors" and "Cameras with On-chip Memory CMOS Image Sensors"). When $n = 0$, Eq. (1) corresponds to the local brightness conservation (or gray level conservation). The determination of $\boldsymbol{u}$ is an *ill-posed* problem when no additional hypotheses are made. The information at hand (i.e., gray level of each pixel) does not allow for the evaluation of a vector field at the pixel scale. The registration is almost never performed down to the pixel level, but on a set of pixels referred to as zone of interest (ZOI), or on the region of interest (ROI) itself. Let us introduce the functional $\mathscr{R}$ of a trial displacement $\boldsymbol{u}$

$$\mathscr{R}[\boldsymbol{u}] = \|I_0 - I_t(\cdot + \boldsymbol{u})\|^2_\Omega, \tag{2}$$

where "·" denotes a dummy variable, Ω the considered domain where the registration will be performed (i.e., either any ZOI or the whole ROI). This setting corresponds to a weak formulation of the brightness conservation. When the usual $\mathscr{L}_2$ norm is considered, i.e., $\|I_0\|^2_\Omega = \sum_\Omega |I_0(\boldsymbol{x})|^2$, the determination of $\boldsymbol{u}$ consists in minimizing the sum of quadratic differences for, say, a rigid body translation [34]. The functional $\mathscr{R}_c$

$$\mathscr{R}_c[\boldsymbol{u}] = \sum_\Omega [I_0(\boldsymbol{x}) - I_t(\boldsymbol{x} + \boldsymbol{u}(\boldsymbol{x}, t))]^2 \tag{3}$$

is minimized with respect to the degrees of freedom of the sought displacement field. When the correct kinematics is found, and if $n = 0$ (see Eq. 1), $\mathscr{R}_c$ vanishes. Any offset will indicate that the registration was not perfect.

2.1.1 Local Approach

Let us assume that $\boldsymbol{u}$ is a rigid body translation. The minimization of Eq. (3) can be solved with correlation techniques that consist in maximizing the superposition of shifted copies of I_t and I_0 [25–27]. The previous minimization is equivalent to maximizing (in the limit of large ZOIs) the quantity $c(\boldsymbol{u})$

$$c(\boldsymbol{u}) = (I_0 \star I_t)(\boldsymbol{u}) \equiv \sum_{\text{ZOI}} I_0(\boldsymbol{x}) I_t(\boldsymbol{x} + \boldsymbol{u}) \tag{4}$$

where ☆ denotes the cross-correlation product. The value that maximizes the cross-correlation product is an estimate of the unknown translation vector. The evaluation of the correlation product can be performed either in the reference space [27, 35] or Fourier space [36–38]. The measured field is composed of a set of displacements assigned to the center of all analyzed ZOIs.

Let us note that the hypothesis made so far (i.e., piece-wise constant displacements) can be relaxed. Nowadays academic and commercial correlation codes use local displacement interpolations of degree one or higher. However, only the mean value is kept and assigned to the center of each ZOI. It allows more complex features to be better captured, and leads to uncertainties whose amplitude depends, among others, on the texture and its gray level interpolation, but also on the type of local interpolation, and on the ZOI size [39, 47].

2.1.2 Global Approach

The space of trial displacements $\mathscr{E}_M$ is introduced and satisfies the continuity requirement usually valid in experiments. The sought displacement is written as a linear combination in a basis of $\mathscr{E}_M$, namely, $\boldsymbol{u}(\boldsymbol{x},t) = \sum_i \upsilon_i(t)\boldsymbol{\Psi}_i(\boldsymbol{x})$. The functional to minimize now reads

$$\mathscr{R}_c[\{\upsilon\}] = \sum_{\text{ROI}} [I_0(\boldsymbol{x}) - I_t(\boldsymbol{x} + \upsilon_i(t)\boldsymbol{\Psi}_i(\boldsymbol{x}))]^2 \tag{5}$$

where $\{\upsilon\}$ is the column vector that gathers all unknown amplitudes $\upsilon_i(t)$. Moreover, the displacement field is generally evaluated by successive iterations and corrections [40]. Let us assume that I_0 and I_t be sufficiently regular at the smallest scales, and that the displacement corrections $\delta\boldsymbol{u}$ be small so that a first order Taylor expansion is possible, and $\mathscr{R}_c$ is a quadratic form of the unknown displacement corrections

$$\mathscr{R}_c[\{\delta\upsilon\}] = \sum_{\text{ROI}} [I_0(\boldsymbol{x}) - \tilde{I}_t(\boldsymbol{x}) - \delta\upsilon_i(t)\boldsymbol{\Psi}_i(\boldsymbol{x}) \cdot \nabla I_0(\boldsymbol{x})]^2 \tag{6}$$

where $\tilde{I}_t$ denotes the corrected deformed picture (i.e., $\tilde{I}_t(\boldsymbol{x}) = I_t(\boldsymbol{x} + u^{(n)}(\boldsymbol{x},t))$ for the current evaluation of the displacement field at iteration n), and $\nabla I_t(\boldsymbol{x} + u^{(n)}(\boldsymbol{x},t)) \approx \nabla I_0(\boldsymbol{x})$). The extremality condition for each amplitude $\upsilon_i(t)$ reads

$$\left[\sum_{\text{ROI}} (\nabla I_0 \cdot \boldsymbol{\Psi}_i)(\boldsymbol{x})(\nabla I_0 \cdot \boldsymbol{\Psi}_k)(\boldsymbol{x})\right] \upsilon_k(t) = \sum_{\text{ROI}} (I_0 - \tilde{I}_t)(\boldsymbol{x})(\nabla I_0 \cdot \boldsymbol{\Psi}_i)(\boldsymbol{x}) \tag{7}$$

This linear system can be written in matrix-vector form

$$M_{ik}\upsilon_k(t) = a_i \tag{8}$$

The regularity condition, which seems restrictive, can still allow non differentiable textures to be analyzed within the present framework [41]; the regularity of $\boldsymbol{u}$ allows one to bypass in many cases that of I_0 by integrations by parts and filtering. By implementing multiscale approaches to address the issue of large displacements and strains, the high frequency displacement components are progressively introduced [38, 42].

At this level of generality, various measurement spaces $\mathscr{E}_M$ can be chosen. This is one of the advantages of the present framework. For example, Fourier decompositions of the displacement field [41, 43] were considered to capture displacement fluctuations. A very often used alternative in solid mechanics is to discretize displacement fields with finite elements. The simplest choice when dealing with pictures is to consider 4-noded quadrilaterals with bilinear displacement interpolation (i.e., Q4-DIC [42]). More recently, unstructured meshes made of 3-noded triangles have also been utilized [44, 45].

2.1.3 Uncertainty Quantification

One key aspect associated with kinematic measurements is related to the estimation of displacement uncertainties [39]. In particular, the hardware associated with high-speed cameras plays a significant role in the overall performance of such equipment [46]. Figure 3 compares the measurement uncertainty for three different high-speed and ultra high-speed cameras with different Q4 element sizes. This analysis was performed by acquiring a series of pictures in the reference configuration prior to an actual experiment. All the experimental conditions [e.g., acquisition parameters, lighting (Dedocool COOLH, Dedotec, Germany)] were identical to the subsequent test. This series of images was analyzed via Q4-DIC and the standard displacement uncertainty was obtained.

The larger the element size, the lower the measurement uncertainty, which is expected. However, the three cameras do not yield the same type of result. Two cameras (i.e., B and C) lead to standard displacement uncertainties that are mostly controlled by acquisition noise [47]

$$\sigma_u = \frac{\sqrt{6}\kappa\sigma_I p}{G_I \ell} \tag{9}$$

where κ is a constant of the order of 1 that depends whether it is an inner node, edge node or corner note, p denotes the physical size of one pixel, and G_I the root mean square gradient of the reference picture I_0. It is believed that camera A hardware

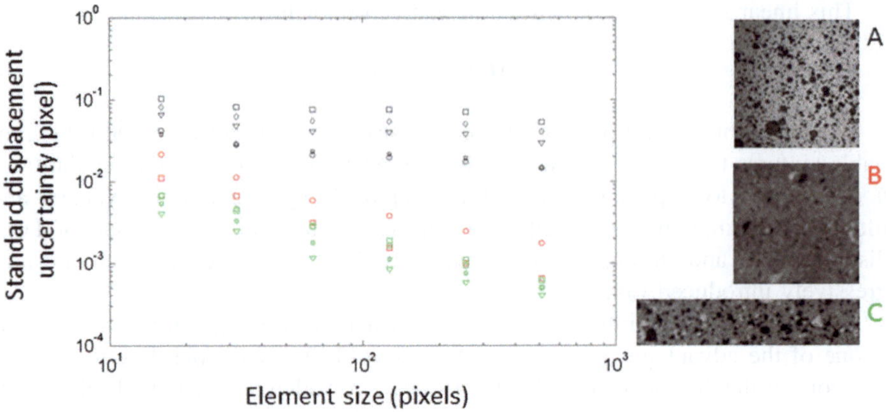

Fig. 3 Uncertainty quantification for three different high-speed cameras. For a given camera, the highest points correspond to a higher number of frames per second. Reference picture I_0 shot by the three analyzed cameras are also shown

leads to displacement fields that do not have a random signature associated with acquisition noise. Last, it is worth noting that for any camera, the higher the acquisition rate, the higher the measurement uncertainty (i.e., the lower the signal to noise ratio).

2.2 *Spatiotemporal DIC*

Virtually all applications of DIC for the investigation of dynamic experiments have not required specific algorithmic developments in comparison with quasi static tests. However, many high-speed cameras lead to image definitions that become very small as the number of frames per second increases. To compensate for this decrease in spatial definition, spatiotemporal algorithms have been introduced [48, 49] in which movies are analyzed.

An additional way of regularizing correlation procedures is to analyze a series of pictures (instead of an image pair in the previous section). This approach is referred to as spatiotemporal correlation (or 2D + t DIC) since the displacement field is now evaluated as a space/time decomposition $\boldsymbol{u}(\boldsymbol{x},t) = \sum_i \sum_j \upsilon_{ij} \boldsymbol{\Psi}_i(\boldsymbol{x}) \phi_j(t)$. The sum of squared differences

$$\mathscr{R}_c[\{\upsilon\}] = \sum_{[t_0, t_{\max}]} \sum_{\text{ROI}} [I_0(\boldsymbol{x}) - I_t(\boldsymbol{x} + \upsilon_{ij} \boldsymbol{\Psi}_i(\boldsymbol{x}) \phi_j(t))]^2 \tag{10}$$

is minimized with respect to the unknown degrees of freedom υ_{ij} over the ROI and time interval $[t_0, t_{\max}]$. The time discretization may consist of linear temporal elements extending over n frames so that a spatiotemporal analysis considers 'volumes' of size $\ell^2 n$ (expressed in voxels by equivalence with volume correlation) when the spatial discretization is based upon 4-noded quadrilaterals of edge size ℓ.

To illustrate the gain in terms of displacement uncertainty, a picture (Fig. 2e) acquired with a high-speed camera (FASTCAM APX RS, Photron, Japan) is considered. Its definition is rather small (*i.e.*, 224 × 449 pixels) and the analyzed ROI has a definition of 175 × 336 pixels. In the following a purely artificial uncertainty analysis is performed. It consists of adding to the picture of Fig. 2e a Gaussian white noise (zero mean, standard deviation σ_I equal to 3% of the dynamic range) for each picture of the artificial movie. This level of noise is quite high for illustration purposes and also to check whether a priori predictions of the standard displacement uncertainty [49] are still valid for a series of n pictures per time step in a single analysis

$$\sigma_u = \frac{\sqrt{2}\sigma_I p}{G_I \ell}\sqrt{\alpha + \frac{\beta}{n}} \tag{11}$$

where $\alpha + \beta$ is of the order of 3, and the lower α the higher the gain when considering a spatiotemporal analysis.

Figure 4 shows a comparison between the standard displacement uncertainty σ_u achieved with incremental ($n = 1$) and spatiotemporal ($n > 1$) analyses. As expected from the previous result, the larger the number of pictures per time step (i.e., the larger the size of the spatiotemporal volume), the smaller the standard displacement uncertainty. The fact that it levels off for large spatiotemporal volumes is due to the first picture that is also corrupted by noise. In the present case, a 30% gain can be expected when large spatiotemporal volumes are considered when compared with an incremental analysis.

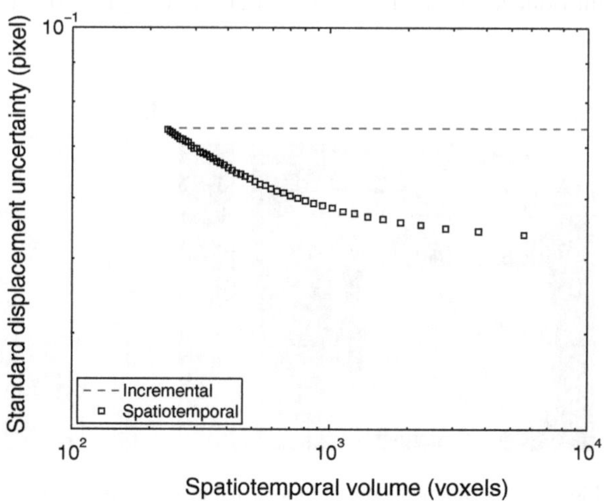

Fig. 4 Standard displacement uncertainty σ_u as a function of the spatiotemporal volume. Comparison between incremental and spatiotemporal approaches

2.3 *Stereocorrelation*

Stereovision [35, 50] consists in reconstructing 3D surfaces. By observing the latter from at least two different perspectives it is possible to determine the 3D coordinates of points. Once the initial 3D surface has been reconstructed, its 3D deformation can be measured during an experiment (i.e., 3D surface displacements). A calibrated stereoscopic setup (generally comprising two cameras, see Fig. 5b) and a DIC technique are used to register spatial and temporal sequences. This technique is widely used to monitor mechanical tests [53] for which monovision is not sufficient (because, for example, of out-of-plane motions [54]). Stereocorrelation can also be used to monitor ultra high-speed experiments (Fig. 5) as will be illustrated in the next section, provided the imaging devices are properly synchronized and characterized from a metrological standpoint [46, 51, 55, 56]. It is also possible to use a single camera and two mirrors to acquire on a single picture two view points of the sample or structure [51, 57] (Fig. 5a). The second half of the last decade has seen the first applications of stereocorrelation to the analysis of high-speed experiments [46, 55–57].

3 Applications

3.1 *Shock Enhancement of Cellular Materials*

The analysis of the deformation of different foams under shock loading is possible thanks to 2D-DIC.

An original testing procedure using a large diameter Nylon Hopkinson bar is introduced. Tests under two different configurations (i.e., pressure bar behind/ahead

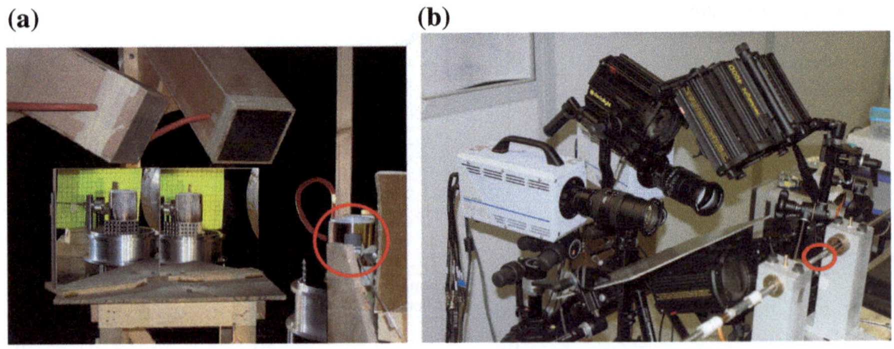

Fig. 5 Experimental configurations using either **a** one rotating mirror framing camera (LCA, France) or **b** two FASTCAM APX RS (Photron, Japan) cameras for stereocorrelation analyses. The monitored sample is encircled in *red*. **a** Taken from [51]. **b** Taken from [52]

of the assumed shock front) at the same shock speed are used to obtain the force/time histories behind and ahead of the shock front within cellular material samples [24]. Figure 1a shows one of the experimental configurations monitored by a high-speed camera (FASTCAM APX RS, Photron, Japan) for a tested aluminum foam (Fig. 2a). One of the two Dedocool 400D (Dedotec, Germany) lighting systems is also shown. A shock front is clearly revealed in the Q4-DIC analysis shown in Fig. 6 for a rather small impact velocity (here of the order of 50 m/s). From such type of data it could be shown that the experimentally observed shock enhancement was not due to the foam behavior but to a structural effect [58].

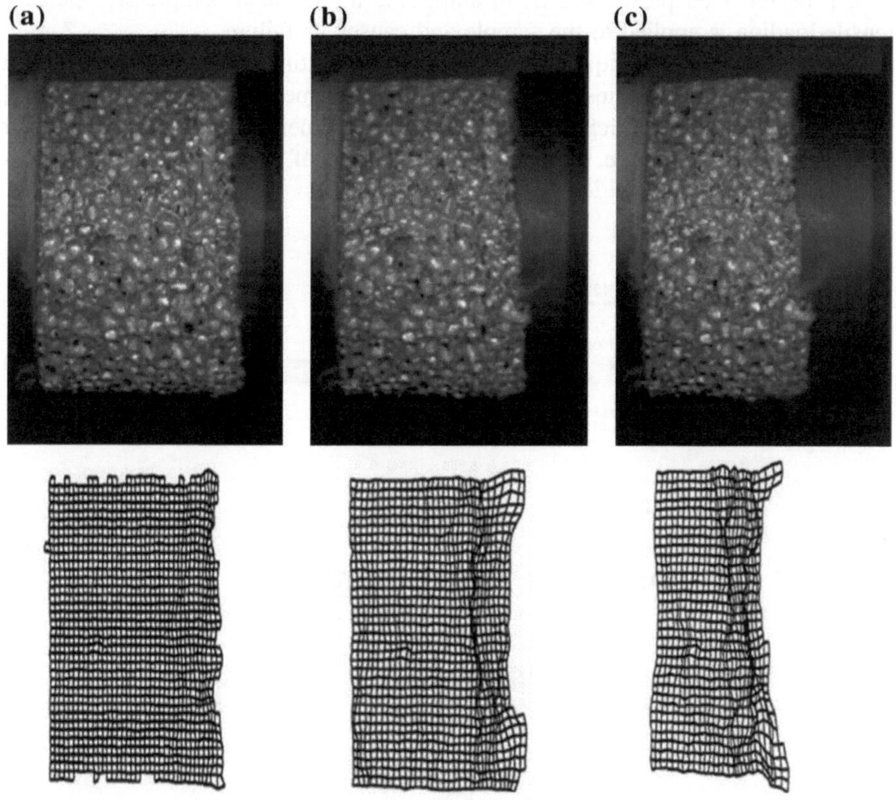

Fig. 6 Pictures of the deformed configuration (the reference configuration is shown in Fig. 2a) and corresponding mesh of Q4-DIC analyses [24]

3.2 Dynamic Tensile Damage and Cracking of Geomaterials

For the last two decades, spalling techniques have been widely employed for investigating the tensile strength of geomaterials (e.g., standard concrete, high strength concrete, mortars, rocks) at strain rates ranging from 20 to 200 s^{-1} [59–62]. The experimental setup consists of using a single Hopkinson pressure bar (i.e., input bar) that is placed in contact with the tested sample. A short compressive pulse (blue dashed line, Fig. 7a), which is generated by a small explosive charge or the impact of a short projectile, propagates through the Hopkinson bar and the sample and reflects back at its free end as a tensile pulse (red dashed line, Fig. 7a). When the reflected pulse exceeds in amplitude the incident compressive pulse, a tensile loading is applied to the sample and causes its failure.

Among different techniques used to identify the ultimate tensile strength of the sample during a spalling test the so-called Novikov method prevails [60, 62, 63]. However this method, which is based on the measurement of the velocity profile on the rear face of the sample, relies on the assumption of a linear-elastic behavior of

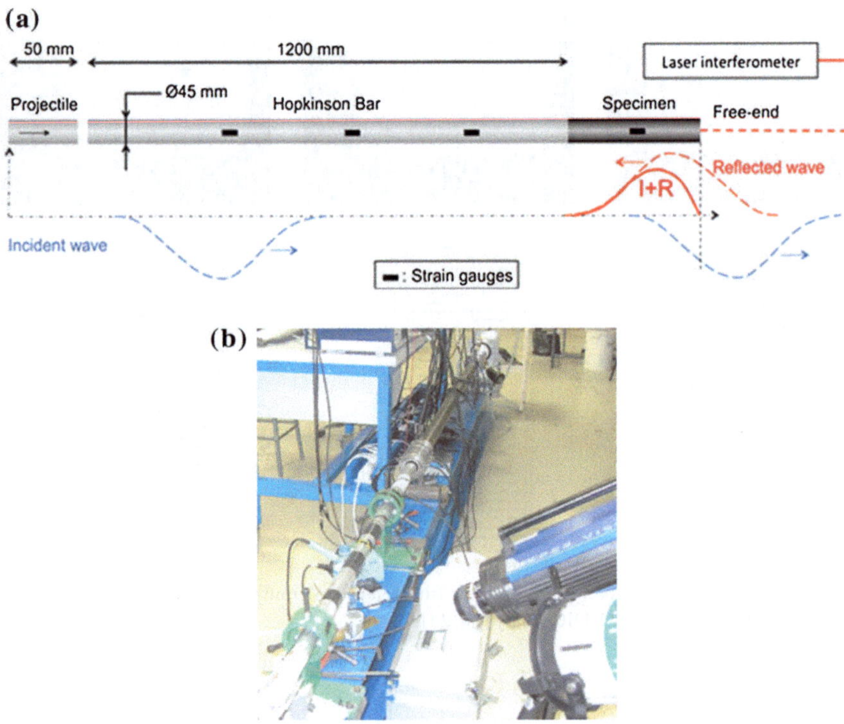

Fig. 7 **a** Principle of the spalling test. **b** View of the experimental setup [62]

the tested material until the peak-stress and only provides its spall strength with no information on its softening response. More recently, a new processing technique has been proposed [64] to evaluate the stress level in the sample and its apparent Young's modulus thanks to an ultra-high speed camera (HPV-1, Shimadzu, Japan) that is used to visualize the lateral surface of the sample (Fig. 7b). 2D displacement fields are obtained through full-field measurement techniques and DIC processing methods. The acceleration map is obtained by a double differentiation of the displacement field and a temporal fit of a second-order polynomial function over five images. Next, the mean stress in any cross-section along the length of the sample shot by the camera can be calculated. Knowing the strain level in the same cross-section, the stress-strain curve in any cross-section is obtained. The quality of the stress and strain measurements was evaluated by considering an aluminium alloy sample [65] and Q4-DIC [42].

The spalling technique provides an estimation of the tensile strength of concrete. However the tensile damage generated in the sample is composed of numerous oriented cracks that make their individual study and analysis very difficult. It is the reason why an experimental method referred to as "rocking spalling test" was proposed to investigate the dynamic propagation of a single crack in concrete samples under dynamic tensile loading [66]. The previous spalling setup is employed to apply a short compressive pulse to the sample. A parallelepiped is used instead of cylinder in which two notches are machined. The largest notch provides a "rocking" effect that produces a fast and local dynamic tensile loading in the zone near the tip of the small notch. The geometry of the specimen (size and location of notches) is shown in Fig. 8a.

This configuration allows a single unstable crack to be initiated despite the high loading rate. Experiments (Fig. 8b) were conducted on dry and water saturated concrete samples. The cracking inception and propagation can be visualized with an ultra-high speed digital camera and quantified via DIC. The results show that in most cases a single crack is indeed initiated on the small notch and propagates perpendicularly to the loading axis. Furthermore, the cracking velocity was evaluated with dry and water saturated concrete samples (i.e., $\approx$1300 m/s [66]).

3.3 Tensile Test on Composite Material

The next example corresponds to an image sequence, which was acquired by a high-speed camera (FASTCAM APX RS, Photron, Japan) at 65,000 fps, of a tensile test (Fig. 1b) on a T700 carbon fiber/M21 epoxy matrix composite (the plies are oriented at $\pm 45°$ with respect to the loading direction) with a high-speed tension/compression testing machine (Dytesys, France). The picture definition is 224×449 pixels (Fig. 2e). This is a case where the spatiotemporal analysis is

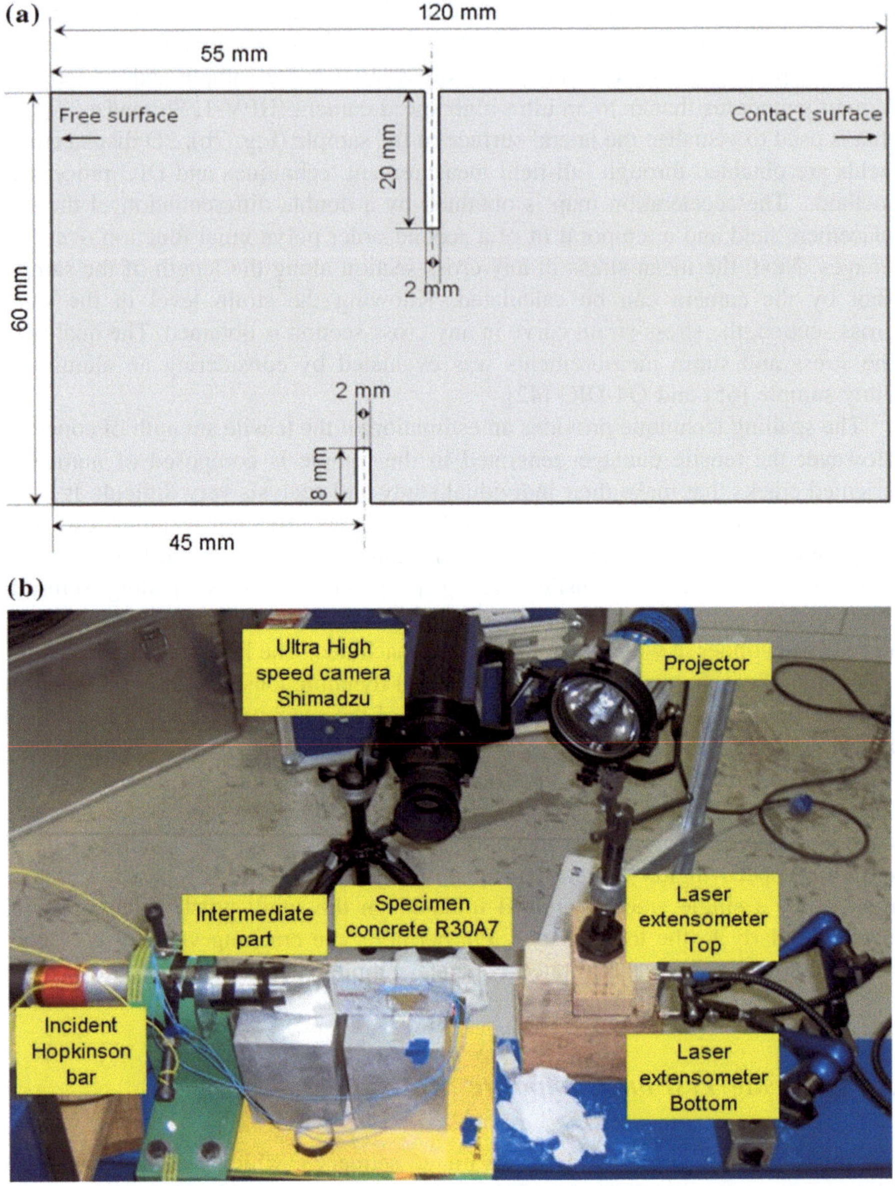

Fig. 8 **a** Specimen geometry used in rocking spalling tests. **b** Experimental setup used for dynamic crack propagation test [66]

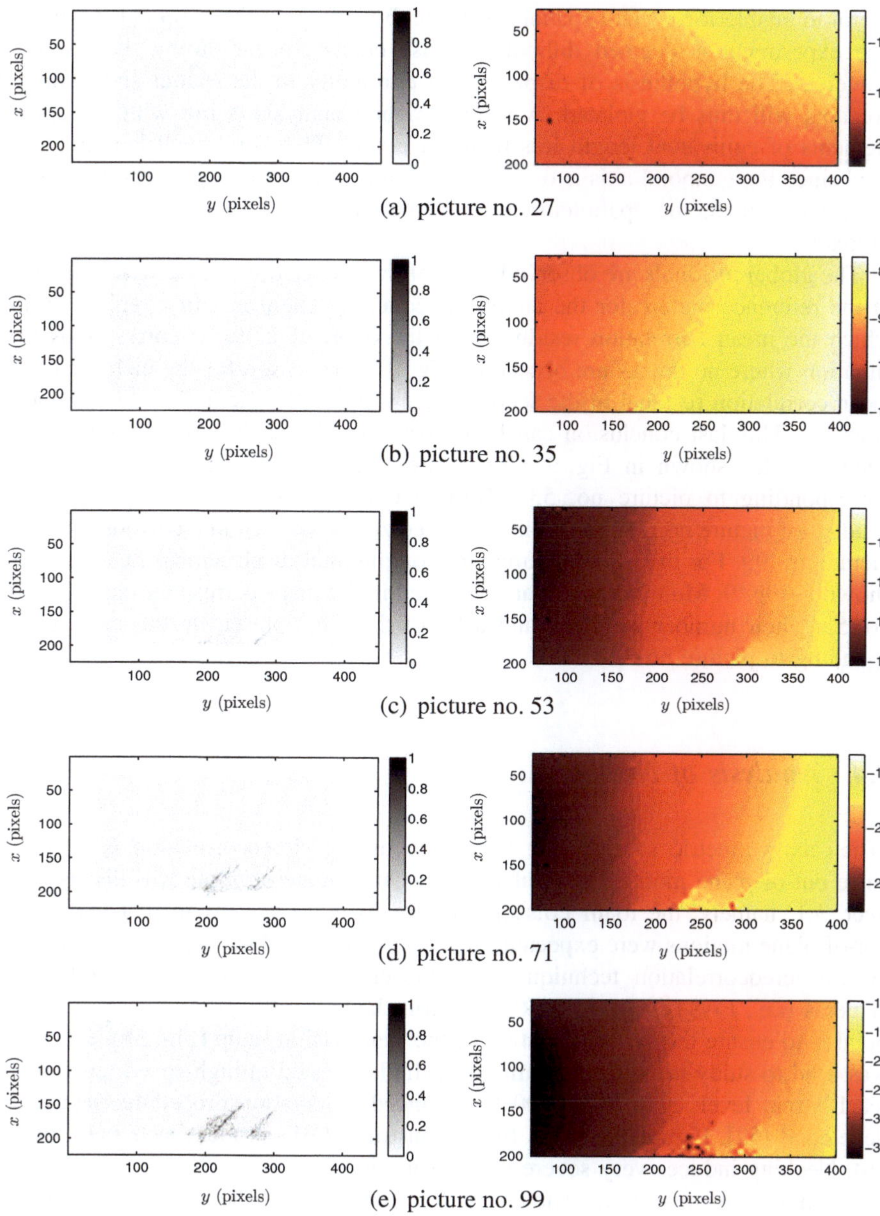

Fig. 9 Correlation residual (*left*) and horizontal (*right*) displacement maps (in pixels) at the end of the five time increments. The corresponding movie is available online

useful to enable for a small spatial resolution by enlarging the temporal resolution. The experiment is deemed difficult because cracks appear during the analyzed sequence. The hypothesis of displacement continuity in the spatial and temporal domains will thus be violated. A spatiotemporal analysis is run with Q4 spatial elements of equivalent length less than 5.4 pixels. This very small length can be used since the temporal axis is divided into 5 increments of 18 pictures each. With these parameters, the spatiotemporal volume of interest is of the order of 5.3 Mvoxels.

The global residuals are observed to be of the order of 3% of the dynamic range of the reference picture for the analyzed sequence. There is a first regime during which the mean correlation residual is of the order of 2.4%. It corresponds to a situation where no cracks are observed (Fig. 9). The reason for the increase of the mean correlation residual in the second regime (i.e., after picture no. 60) is due to cracking. This last conclusion can be drawn from the analysis of the normalized residual fields shown in Fig. 9. Cracks are clearly visible for the residual field corresponding to picture no. 53. Multiple cracking zones are observed in the residual of picture no. 71. Two fully damaged zones are seen in the residual field of picture no. 99. For the same pictures, the longitudinal displacement fields are also shown in Fig. 9. Multiple cracks are observed in the maps corresponding to picture no. 53. Their number increases and a lot of them are present in the maps corresponding to picture no. 99.

3.4 Analysis of Necking

When axisymmetric samples are tested the use of stereocorrelation is desirable since out-of-plane motions generally occur. One counterexample was discussed in Sect. 3.1, namely, the foam Poisson's ratio was close to zero so that very small out-of-plane motions were expected, thereby making 2D-DIC approaches possible. When stereocorrelation techniques are chosen with two (or more) high-speed cameras (i.e., FASTCAM APX RS, Photron, Japan, in the present case), one key point is to ensure the synchronization of the stereovision setup (Fig. 5b). The latter was used to study necking in aluminum cylinders tested in high-speed tension. To reach strain levels as high as 600%, enhanced registration procedures had to be developed [52]. Figure 10 shows two reconstructed 3D surfaces at the end of the test. Necking induces very severe radius variations.

From such results, it was shown that the formation of the major neck was not strain rate sensitive up to levels equal to 1600 s^{-1}. Conversely, the strain level at necking inception was observed to be random [52].

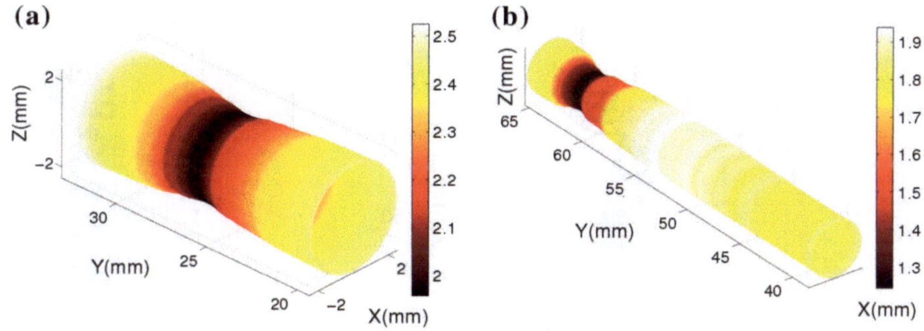

Fig. 10 Reconstructed 3D surfaces by enhanced stereocorrelation [52]. The color encodes the current radius of the sample tested in tensile Hopkinson bars

3.5 Cylinder Expansion

This last example deals with a blast experiment in which a copper cylinder is subject to rapid expansion generated by an explosive. When following microsecond events it is impossible to synchronize two ultra high-speed cameras. Consequently, two mirrors and a single high-speed rotating mirror framing camera (LCA, France) and pyrotechnic flashes (CEA-DAM, France) are used to monitor the expansion (Fig. 5a). Further, calibration, which is usually carried out prior to the experiment [35], was not possible. Consequently two calibration targets are imaged during the experiment (Fig. 2c). Last, the registration is initialized with the displacement field predicted by hydrodynamic simulations of the experiment. Any offset between the measured and predicted displacement field is an indication of model error [51].

From a sequence of pictures it is possible to reconstruct the 3D shape of the cylinder at various instances of time (Fig. 11). For the more advanced shape, it is possible to observe the inception of necking. This result shows that even under such extreme conditions (i.e., very fast phenomena, lighting with pyrotechnic flashlights) it is possible to follow the deformation of surfaces up to very large strain levels (i.e., of the order of 100%).

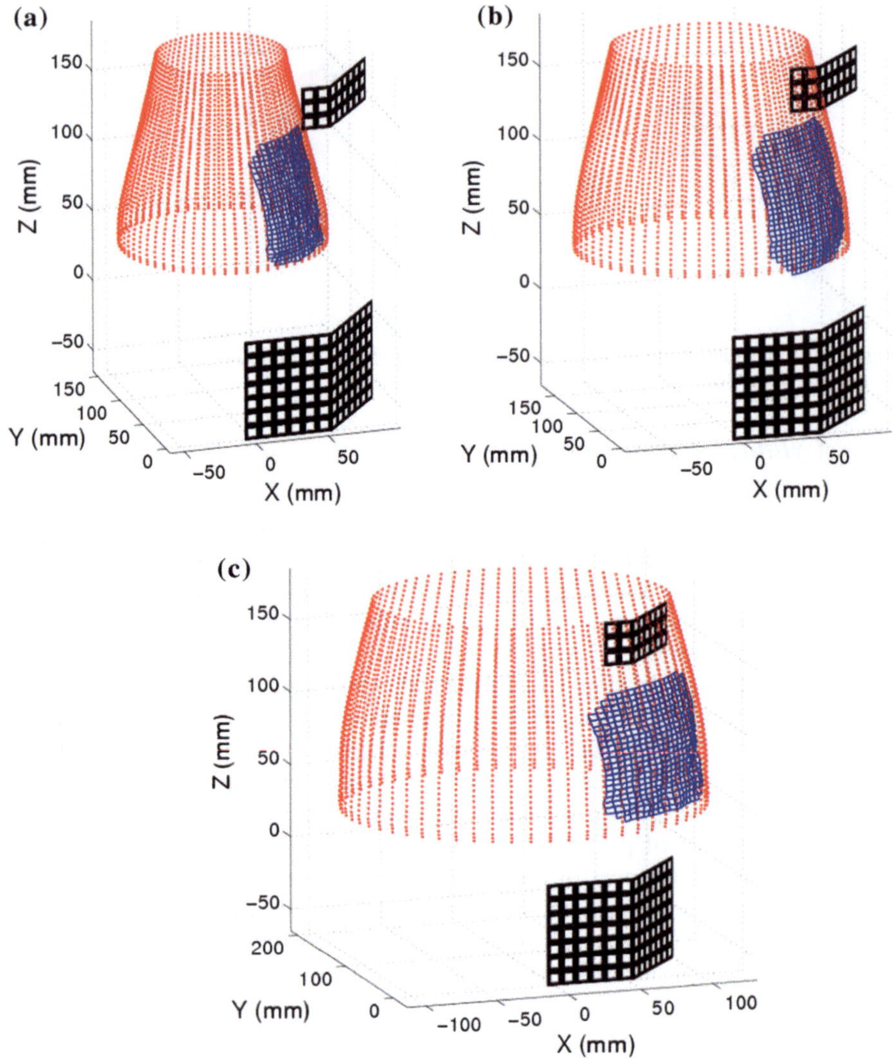

Fig. 11 3D shapes reconstructed via stereocorrelation. The *blue mesh* corresponds to the 3D reconstructed positions, and the *red dots* to the interpolated 3D surface. The corresponding movie is available online

4 Conclusions

With the development of new digital high-speed cameras with enhanced definition and higher acquisition rates, which become more affordable than in the past, their use in solid mechanics labs has increased during the last two decades. For example, one of the standard dynamic experiment (i.e., with Hopkinson bars) has been analyzed in more details and phenomena that could not be observed previously were revealed thanks to the measurement of kinematic fields via any photomechanical technique and in particular 2D-DIC or stereocorrelation techniques as above discussed. It is worth noting that one application, which was not discussed in the present chapter, is related to 4D tomography, which is also made possible thanks to nowadays high-speed cameras [67, 68]. For example, the solidification of an aluminum alloy was monitored in a nondestructive way by such means [69]. Similarly, an in situ tensile test on a metal matrix composite was studied with a frequency of 20 tomograms per second [70].

High-speed imaging allows deformation, damage and failure mechanisms to be analyzed in a qualitative and quantitative way. Such type of information is very useful when developing mechanism-based constitutive models. Further, DIC techniques allow quantitative estimates of kinematic fields to be performed. These sets of data can be compared with numerical simulations to validate the virtual models and/or calibrate parameters of constitutive models. Various identification techniques explicitly use full-field data [1]. These new approaches not only improve the way material parameters are calibrated, but they also enable the experimentalist to design new setups in which heterogeneity is no longer a limitation. Heterogeneous experiments can be imagined, not to make life more complicated, but to capture in a single test much more data so that the identification will be better conditioned.

Acknowledgements The results discussed herein have benefitted from the support of Agence Nationale de la Recherche (VULCOMP projects), CEA-DAM, CNRS and région Ile-de-France (SESAME project). The authors wish to thank G. Besnard, B. Durand, I. Elnasri, B. Erzar, Y. Girard, S. Guérard, J.-M. Guimard, J.-M. Lagrange, H. Leclerc, S. Pattofatto, D. Saletti, N. Swiergiel, H. Tsitsiris and H. Zhao for fruitful discussions.

References

1. M. Grédiac, F. Hild (eds.), *Full-Field Measurements and Identification in Solid Mechanics* (ISTE/Wiley, London, 2012)
2. A. Lagarde, *Advanced Optical Methods and Applications in Solid Mechanics, Solid Mechanics and its Applications*, vol. 82 (Kluwer, Dordrecht, 2000)
3. P.K. Rastogi (ed.), *Photomechanics, Topics in Applied Physics*, vol. 77 (Springer, Berlin, 2000)
4. A.A. Wells, D. Post, The dynamic stress distribution surrounding a running crack—a photoelastic analysis. Proc. SESA **16**, 69–92 (1958)
5. W.B. Bradley, A.S. Kobayashi, An investigation of propagating cracks by dynamic photoelasticity. Exp. Mech. **10**, 106–113 (1970)
6. J.W. Dally, Dynamic photoelastic studies of fracture. Exp. Mech. **19**, 349–361 (1979)
7. M. Ramulu, A.S. Kobayashi, D.B. Barker, Analysis of dynamic mixed-mode isochromatics. Exp. Mech. **25**, 344–353 (1985)
8. A.J. Rosakis, O. Samudrala, D. Coker, Cracks faster than the shear wave speed. Science **284**, 1337–1340 (1999)

9. F.P. Chiang, G. Jaisingh, Dynamic moiré methods for the bending of plates. Exp. Mech. **13**, 168–171 (1973)
10. J.M. Huntley, J.E. Field, High resolution moiré photography: application to dynamic stress analysis. Opt. Eng. **28**, 926–933 (1989)
11. V.A. Deason, J.S. Epstein, M. Abdallah, Dynamic diffraction moiré: Theory and applications. Optics Lasers Eng. **12**, 173–187 (1990)
12. K. Arakawa, R.H. Drinnon Jr., M. Kosai, A.S. Kobayashi, Dynamic fracture analysis by moiré interferometry. Exp. Mech. **31**, 306–309 (1991)
13. T. Bertin-Mourot, C. Denoual, G. Dehors, P.F. Louvigné, T. Thomas, High speed photography of moiré fringes—application to ceramics under impact. J. Phys. IV **Coll. C3** (Suppl. III), 311–316 (1997)
14. J.F. Kalthoff, J. Beinert, S. Winkler, Measurements of dynamic sress intensity factors for fast running and arresting cracks, in *Double-Cantilever-Beam Specimens, in Measurements of Dynamic Sress Intensity Factors for Fast Running and Arresting Cracks in Double-Cantilever-Beam Specimens, vol. STP 627*, ed. by G.T. Hahn, M.F. Kanninen (ASTM, Philadelphia, 1977), pp. 161–176
15. P. Theocaris, Dynamic propagation and arrest measurements by the method of caustics on overlapping skew-parallel cracks. Int. J. Solids Struct. **14**, 639–653 (1978)
16. K. Ravi-Chandar, W.G. Knauss, An experimental investigation into dynamic fracture: III, Steady-state crack propagation and crack branching. Int. J. Fract. **26**, 141–154 (1984)
17. C. Liu, A.J. Rosakis, L.B. Freund, The interpretation of optical caustics in the presence of dynamic non-uniform crack-tip motion histories: a study based on a higher order transient crack-tip expansion. Int. J. Solids Struct. **30**, 875–897 (1993)
18. W. Peters, W. Ranson, J. Kalthoff, S. Winkler, A study of dynamic near-crack-tip fracture parameters by digital image analysis. J. Phys. Coll. **46**, 631–638 (1985)
19. X.M. Hu, S.J.P. Palmer, J.E. Field, The application of high speed photography and white light speckle to the study of dynamic fracture. Optics Lasers Technol. **16**, 303–306 (1984)
20. J.M. Huntley, S.J.P. Palmer, J.E. Field, Automatic speckle photography fringe analysis: application to crack propagation and strength measurement. Proc. SPIE **0814**, 153–160 (1987)
21. B.W. Asay, G.W. Laabs, B.F. Henson, D.J. Funk, Speckle photography during dynamic impact of an energetic material using laser-induced fluorescence. J. Appl. Phys. **82**, 1093–1099 (1997)
22. K.-E. Fällström, H. Gustavsson, N.-E. Molin, A. Wåhlin, Transient bending waves in plates studied by hologram interferometry. Exp. Mech. **29**, 378–387 (1989)
23. H.V. Tippur, A.J. Rosakis, Quasi-static and dynamic crack growth along bimaterial interfaces: a note on crack-tip field measurements using coherent gradient sensing. Exp. Mech. **31**, 243–251 (1991)
24. I. Elnasri, S. Pattofatto, H. Zhao, H. Tsitsiris, F. Hild, Y. Girard, Shock enhancement of cellular structures under impact loading: part I Experiments. J. Mech. Phys. Solids **55**, 2652–2671 (2007)
25. P.J. Burt, C. Yen, X. Xu, Local correlation measures for motion analysis: a comparative study, in *IEEE conference on pattern recognition and image processing* (IEEE, New York, 1982), pp. 269–274
26. W.H. Peters, W.F. Ranson, Digital imaging techniques in experimental stress analysis. Opt. Eng. **21**, 427–431 (1982)
27. M.A. Sutton, W.J. Wolters, W.H. Peters, W.F. Ranson, S.R. McNeill, Determination of displacements using an improved digital correlation method. Im. Vis. Comp. **1**, 133–139 (1983)
28. F. Barthelat, Z. Wu, B.C. Prorok, H.D. Espinosa, Dynamic torsion testing of nanocrystalline coatings using high-speed Photography and digital image correlation Exp. Mech. **43**, 331–340 (2003)
29. J. Kajberg, M. Sjödahl, Optical method to study material behaviour at high strain rates, in *IUTAM symposium on field analyses for determination of material parameters—experimental*

and numerical aspects, solid mechanics and its applications, vol. 109, ed. by P. Ståhle, K.G. Sundin (Springer, The Netherlands, 2003), pp. 37–49
30. J. Kajberg, B. Wikman, Viscoplastic parameter estimation by high strain-rate experiments and inverse modelling: Speckle measurements and high-speed photography. Int. J. Solids Struct. **44**, 145–164 (2007)
31. V. Tarigopula, O.S. Hopperstad, M. Langseth, A.H. Clausen, F. Hild, A study of localisation in dual phase high-strength steels under dynamic loading using digital image correlation and FE analysis. Int. J. Solids Struct. **45**, 601–619 (2008)
32. F. Pierron, M.A. Sutton, V. Tiwari, Ultra high speed DIC and virtual fields method analysis of a three point bending impact test on an aluminium bar. Exp. Mech. **51**, 537–563 (2011)
33. D. Saletti, S. Pattofatto, H. Zhao, Measurement of phase transformation properties under moderate impact tensile loading in a NiTi alloy. Mech. Mat. **65**, 1–11 (2013)
34. L.C. Gui, W. Merzkirch, A comparative study of the MQD method and several correlation-based PIV evaluation algorithms. Exps. in Fluids **28**, 36–44 (2000)
35. M.A. Sutton, J.J. Orteu, H. Schreier, *Image correlation for shape, motion and deformation measurements: basic concepts, theory and applications* (Springer, New York, 2009)
36. D.J. Chen, F.P. Chiang, Y.S. Tan, H.S. Don, Digital speckle-displacement measurement using a complex spectrum method. Appl. Optics **32**, 1839–1849 (1993)
37. F.P. Chiang, Q. Wang, and F. Lehman, New developments in full-field strain measurements using speckles, in *Non-Traditional Methods of Sensing Stress, Strain and Damage in Materials and Structures*, vol. STP 1318 (ASTM, Philadelphia, 1997), pp. 156–169
38. F. Hild, B. Raka, M. Baudequin, S. Roux, F. Cantelaube, Multi-scale displacement field measurements of compressed mineral wool samples by digital image correlation. Appl. Optics IP **41**, 6815–6828 (2002)
39. M. Bornert, F. Brémand, P. Doumalin, J.C. Dupré, M. Fazzini, M. Grédiac, F. Hild, S. Mistou, J. Molimard, J.J. Orteu, L. Robert, Y. Surrel, P. Vacher, B. Wattrisse, Assessment of digital image correlation measurement errors: methodology and results. Exp. Mech. **49**, 353–370 (2009)
40. F. Hild, S. Roux, Digital image correlation, in *Optical methods for solid mechanics: a full-field approach*, ed. by P. Rastogi, E. Hack (Wiley-VCH, Weinheim, 2012), pp. 183–228
41. B. Wagne, S. Roux, F. Hild, Spectral approach to displacement evaluation from image analysis. Eur. Phys. J. AP **17**, 247–252 (2002)
42. G. Besnard, F. Hild, S. Roux, "Finite-element" displacement fields analysis from digital images: Application to Portevin-Le Châtelier bands. Exp. Mech. **46**, 789–803 (2006)
43. S. Roux, F. Hild, Y. Berthaud, Correlation image velocimetry: a spectral approach. Appl. Optics **41**, 108–115 (2002)
44. H. Leclerc, J.N. Périé, S. Roux, F. Hild, Integrated digital image correlation for the identification of mechanical properties, in *MIRAGE 2009, vol. LNCS 5496*, ed. by A. Gagalowicz, W. Philips (Springer, Berlin, 2009), pp. 161–171
45. Z. Tomicevic, F. Hild, S. Roux, Mechanics-aided digital image correlation. J. Strain Analysis **48**, 330–343 (2013)
46. V. Tiwari, M.A. Sutton, S.R. McNeill, Assessment of high speed imaging systems for 2D and 3D deformation measurements: methodology development and validation. Exp. Mech. **47**, 561–579 (2007)
47. F. Hild, S. Roux, Comparison of local and global approaches to digital image correlation. Exp. Mech. **52**, 1503–1519 (2012)
48. G. Besnard, S. Guérard, S. Roux, F. Hild, A space-time approach in digital image correlation: movie-DIC. Optics Lasers Eng. **49**, 71–81 (2011)
49. G. Besnard, H. Leclerc, S. Roux, F. Hild, Analysis of image series through digital image correlation. J. Strain Analysis **47**, 214–228 (2012)
50. P.F. Luo, Y.J. Chao, M.A. Sutton, W.H. Peters, Accurate measurement of three-dimensional deformations in deformable and rigid bodies using computer vision. Exp. Mech. **33**, 123–132 (1993)

51. G. Besnard, J.M. Lagrange, F. Hild, S. Roux, C. Voltz, Characterization of necking phenomena in high speed experiments by using a single camera. EURASIP J. Im. Video. Proc. **2010**(215956), 15 (2010)
52. G. Besnard, F. Hild, J.M. Lagrange, P. Martinuzzi, S. Roux, Analysis of necking in high speed experiments by stereocorrelation. Int. J. Impact Eng **49**, 179–191 (2012)
53. J.J. Orteu, 3-D computer vision in experimental mechanics. Optics Lasers Eng. **47**, 282–291 (2009)
54. M.A. Sutton, J.H. Yan, V. Tiwari, H.W. Schreier, J.J. Orteu, The effect of out of plane motion on 2D and 3D digital image correlation measurements. Optics Lasers Eng. **46**, 746–757 (2008)
55. A. Gilat, T. Schmidt, J. Tyson, Full field measurement during a tensile split Hopkinson bar experiment. J. Phys. IV **134**, 687–692 (2006)
56. P.L. Reu, T.J. Miller, The application of high-speed digital image correlation. J. Strain Anal. **43**, 673–688 (2008)
57. G. Besnard, B. Etchessahar, J.M. Lagrange, C. Voltz, F. Hild, S. Roux, Metrology and detonics: analysis of necking, in *28th International Congress High-Speed Imaging and Photonics* (Proc. SPIE 7126, 2008), p. 71261 N. doi:10.1117/12.821892
58. S. Pattofatto, I. Elnasri, H. Zhao, H. Tsitsiris, F. Hild, Y. Girard, Shock enhancement of cellular structures under impact loading: part II analysis. J. Mech. Phys. Solids **55**, 2672–2686 (2007)
59. J.R. Klepaczko, A. Brara, An experimental method for dynamic tensile testing of concrete by spalling. Int. J. Impact Eng **25**, 387–409 (2001)
60. H. Schuler, C. Mayrhofer, K. Thoma, Spall experiments for the measurement of the tensile strength and fracture energy at high strain rates. Int. J. Impact Eng **32**, 1635–1650 (2006)
61. J. Weerheijm, J.C.A.M. van Doormaal, Tensile failure of concrete at high loading rates: new test data on strength and fracture energy from instrumented spalling test. Int. J. Impact Eng **34**, 609–626 (2007)
62. B. Erzar, P. Forquin, An experimental method to determine the tensile strength of concrete at high rates of strain. Exp. Mech. **50**, 941–955 (2010)
63. S.A. Novikov, I.I. Divnov, A.G. Ivanov, The study of fracture of steel, aluminium and copper under explosive loading. Fizika Metallov i Metallovedeniye **21**, 608–615 (1966)
64. F. Pierron, M. Grédiac, *The Virtual Fields Method* (Springer, Berlin, 2012)
65. D. Saletti, P. Forquin, A comparison of DIC and grid measurements for processing spalling tests with the VFM and an 80-kpixel ultra-high speed camera. Europ. Phys. J. Special Topics **225**, 311–323 (2015)
66. P. Forquin, An optical correlation technique for characterizing the crack velocity in concrete. Eur. Phys. J. Spec. Topics **206**, 89–95 (2012)
67. P. Lhuissier, M. Scheel, M. Di Michiel, E. Boller, J. Adrien, E. Maire, L. Salvo, J.-J. Blandin, M. Suery, Ultra fast tomography: new developments for 4D studies in material science, in *1st International Conference 3D Material Science* (Wiley, USA, 2012), pp. 203–208
68. T. dos Santos Rolo, A. Ershov, T. van de Kamp, T. Baumbach, In vivo X-ray cine-tomography for tracking morphological dynamics. Proc. Nat. Acad. Sci. **111**, 3921–3926 (2014)
69. L. Salvo, M. Di Michiel, M. Scheel, P. Lhuissier, B. Mireux, M. Suéry, Ultra fast in situ X-Ray micro-tomography: application to solidification of aluminium alloys, *Materials Science Forum*, vol. 706–709 (Trans Tech Publications, 2012), pp. 1713–1718
70. E. Maire, C. Le Bourlot, J. Adrien, A. Mortensen, R. Mokso, 20 Hz X-ray tomography during an in situ tensile test. Int. J. Fract. **200**, 3–12 (2016)

Observation of Laser Materials Processing by Means of High-Speed Imaging

Christian Freitag, Thomas Arnold, Meiko Boley, Sebastian Faas, Florian Fetzer, Christian Hagenlocher, Andreas Heider, Michael Jarwitz, Rudolf Weber and Thomas Graf

Abstract High-speed imaging is a valuable tool for investigations on laser processes. The high temporal resolution of high-speed imaging allows a detailed observation of different laser processes which helps to understand process mechanisms like e.g. the formation of spatters during laser welding or the formation of a heat affected zone during laser cutting of carbon fiber reinforced plastics (CFRP). In the following the potential of high-speed imaging as a tool for laser process development is shown using different applications as an example. Initially it is described how high-quality images can be achieved during laser processing although the laser beam itself and process emissions make this a challenging task. Subsequently different applications like laser welding, laser drilling of metals with ultra-short pulsed lasers and laser processing of CFRP are introduced. During laser welding, the formation of spatters and hot cracks can be observed. Furthermore the influence of spatial beam modulation on the welding process can be investigated by means of high-speed imaging. The capillary dynamics during laser welding of metals can be studied using high-speed X-rays imaging while in transparent materials the capillary dynamics can be directly observed. During laser drilling the drilling process can be immediately seen using a suitable experimental setup. With the help of high-speed imaging it was revealed that the formation of a heat affected zone during laser processing of CFRP with ultra-short pulses is caused by heat accumulation. Furthermore the dynamics of process emissions like particles or hot vapor generated during laser processing of CFRP was investigated.

C. Freitag (✉) · T. Arnold · M. Boley · S. Faas · F. Fetzer · C. Hagenlocher · A. Heider · M. Jarwitz · R. Weber · T. Graf
Institut für Strahlwerkzeuge (IFSW), Universität Stuttgart,
Pfaffenwaldring 43, 70569 Stuttgart, Germany
e-mail: Christian.Freitag@ifsw.uni-stuttgart.de

Present Address:
A. Heider
CR/APJ1, Robert Bosch GmbH, Renningen, 70465 Stuttgart, Germany

K. Tsuji (ed.), *The Micro-World Observed by Ultra High-Speed Cameras*,
DOI 10.1007/978-3-319-61491-5_9

1 Introduction

The laser can be utilized in manufacturing engineering with great success due to its unique properties. Its most outstanding property is its highly precise temporal and spatial energy insertion at the desired location. Therefore energy losses can be minimized and the energy available for the process remains high. This results in high processing speeds and high quality of the processed parts.

The rather young history of the laser in production technologies shows that many innovative products and processes are only possible thanks to the laser. To further improve established laser processes and to investigate new fields of application, high-speed imaging is a very helpful tool. With high-speed imaging the observation of laser processes with high temporal resolution becomes possible. In the following some examples will be shown where high-speed imaging helps to understand the process mechanisms behind some typical applications. This includes laser welding of metals, laser drilling of metals and laser processing of carbon fiber reinforced plastics (CFRP).

1.1 Laser Welding

Laser welding is a joining technology where the joint between two work pieces is formed in the liquid phase. The so-called deep penetration welding is used when large welding depths, high process efficiency, and high welding speeds are required. Characteristic for deep penetration welding is the formation of a vapor capillary within the melt pool. This vapor capillary leads to a unique situation regarding energy coupling into the material, heat transport and distribution within the melt pool, and the geometry of the weld cross section. The most eminent advantage of deep penetration welding compared to conventional welding processes like arc welding is, that the energy needed for melting of the material is applied very precisely and in short times due to the high intensity of the focused laser beam. Heat losses are therefore much smaller which results in slim weld seams and a smaller thermal load of the work pieces.

To further improve deep penetration welding, the dependence of the behavior of the vapor capillary and the melt pool on different processing conditions has to be understood. High-speed imaging offers the possibility to conveniently observe fast processes like spatter formation or the movement of the vapor capillary and therefore get a better understanding of the laser welding process itself.

1.2 Laser Drilling

Laser drilling offers possibilities where mechanical drilling processes face limitations. It is possible to realize diameters down to 20 μm and also unconventional borehole geometries like negative conical drillings. To produce holes with highest quality, ultra-short laser pulses are used. The ultra-short pulses allow very high intensities and lead to evaporation of the major part of the processed material without the formation of a noticeable liquid phase. Sharp borehole outlets and inlets with virtually no melt formation are the result.

1.3 Laser Processing of CFRP

Carbon fiber reinforced plastics (CFRP) have a great potential in lightweight applications due to its unique mechanical properties. However, processing of this innovative material is challenging. Mechanical cutting, milling or drilling suffer from very strong tool wear, abrasive waterjet treatment from complicated environment and effects like delamination, and therefore, abrasive processing becomes even more difficult. The laser as a well automatable, noncontact tool without wear could overcome these restrictions.

Both components of the compound, carbon fibers as well as the thermoset matrix material, have no liquid phase and are therefore sublimated by the focused laser beam. However, the different thermal properties of the two components can lead to the formation of matrix damage caused by the heat load of the material introduced by the laser beam. Perfect quality of the processed CFRP parts is achieved by using high laser intensities and by avoiding heat accumulation effects. High-speed imaging is a useful tool to understand such basic process mechanisms e.g. the formation of the heat affected zone during laser processing.

In the following a recap is given on how to achieve high-quality images with high-speed cameras during laser material processing by the example of a laser welding process.

2 Obtaining High-Quality Images During Laser Processing

High-speed imaging is a common method of in-process monitoring during laser material processing. However, the laser beam and process emissions, like the radiation of the excited metal vapor above the work piece, can prevent the observation of features such as capillary dynamics, melt pool dynamics or particles flow. Since high spatial (range of interest, ROI < 1 mm) and temporal resolution ($\ll$0.1 ms) of the resulting image are required, additional illumination in combination with optical filtering is essential to achieve high quality images. There are two aspects of additional illumination: first to increase the required illumination level and second to avoid overexposure and disturbances by process emissions.

As a result of detailed research an optimum illumination by an additional laser with a wavelength of 808 nm in combination with a narrow bandpass filter with a central wave length of 808 ± 10 nm in front of the camera was identified [1, 2]. This allows simultaneous observation of the capillary area and the melt pool dynamics during the process with high quality images, at least in welding. As an example images of a laser welding process illuminated with different methods are shown in Fig. 1. The high-speed speed camera used during the experiments was a Photron Fastcam SA5. The laser system used was a TRUMPF TruDisk 5001.

Direct illumination of the interaction zone leads to speckle noise in the resulting image (as can be seen in Fig. 1c) due to the coherence of the illumination laser. Therefore diffuse illumination is favorable. Using a sand blasted copper plate as a diffusor results in a homogeneous illumination of the complete range of interest. Figure 1a shows single frames of a high-speed video for a weld with diffuse illumination. A simultaneous observation of the capillary, melt pool and solidified weld is possible with homogeneous illumination. Without illumination (Fig. 1b) only the capillary is visible but no information from the weld seam and the melt pool is given. Using direct illumination (Fig. 1c) bright spots are reflected and speckle noise in the images occurs and observation of detailed features becomes challenging.

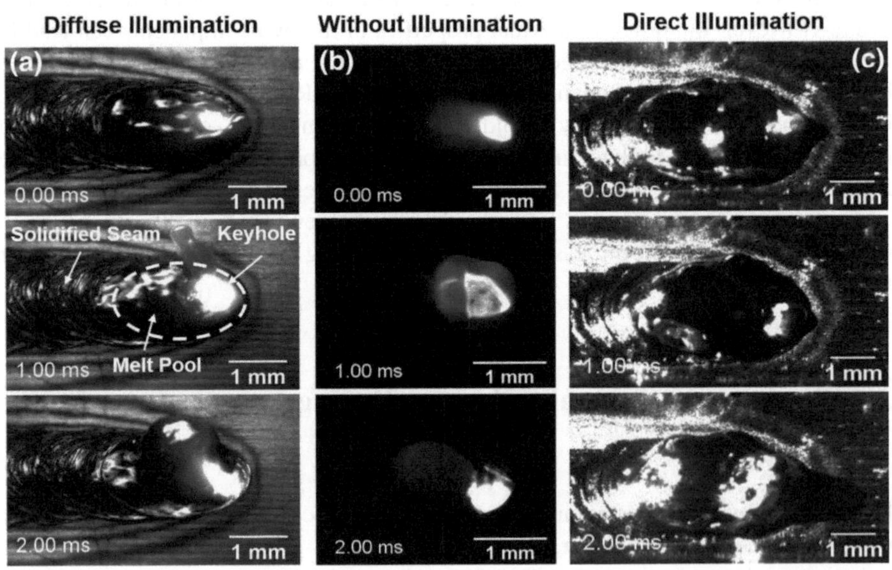

Fig. 1 Series of single frames from a high-speed video of a welding process with different illumination conditions and band pass filter (808 ± 10 nm) in front of camera. **a** With diffuse illumination by laser (808 nm). Laser is reflected on a sand blasted copper plate as diffuser, **b** without illumination, **c** with direct illumination by laser (808 nm)

3 Observation of the Laser Welding Process

3.1 Spatter Formation During Laser Welding

Efficient and reliable welding processes combined with a high quality of the resulting welds are required for industrial applications. Unfortunately, spatter formation during laser welding using 1 μm wavelength lasers is still a not satisfyingly solved [2–4]. Hence, high-speed imaging to analyze spatter formation is used to investigate the formation mechanism in order to improve the resulting weld quality.

In the following all welds were observed and recorded with a high-speed camera at frame rates of up to 100 kHz. The surface of the welded samples was illuminated by a laser with a wavelength of 808 nm, and the camera was equipped with the corresponding band pass filter.

As an example of spattering in laser welding Fig. 2 shows a series of single frames from a high-speed video of a welding process in steel. The video was recorded with a framerate of 10 kHz. From this video it can be seen that molten material is emerging from the interaction zone (0.50 ms) forming a "finger-like" structure. This can be explained by friction between the melt and the escaping metal vapor. When the energy of the finger-like structure exceeds the surface tension of the melt (1.00 ms) a droplet is formed and spatter formation occurs (1.50–4.00 ms).

One of the most critical weld defects for welds of pure Copper (e.g. Cu-ETP, which is important for electro-mobility) using 1 μm wavelength lasers are the so called "melt ejections". Melt ejections are big spatters where the complete molten

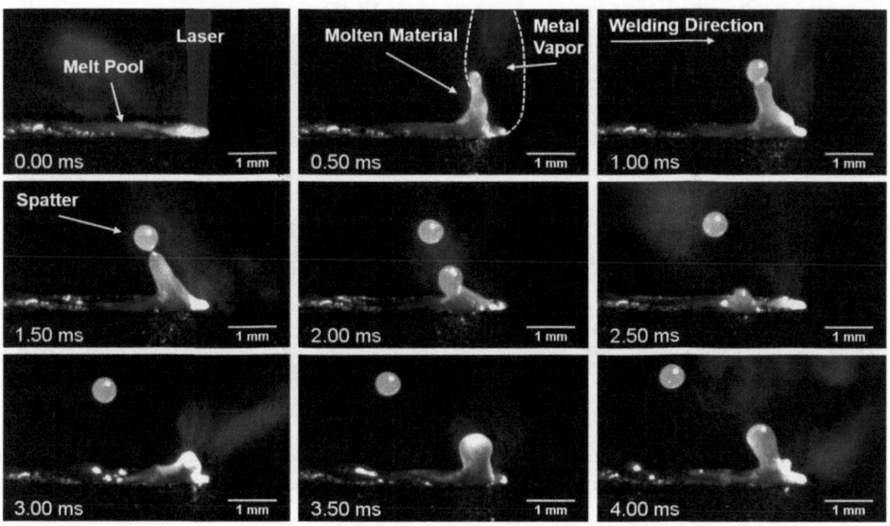

Fig. 2 Series of single frames from a high-speed video of a welding process in Cu-ETP showing the formation of a melt ejection

material is thrown explosively out of the melt pool while welding. These melt ejections lead to a hole in the weld seam and often extend to the complete penetration depth [5–8].

To investigate the formation of melt ejection high-speed imaging was performed recording the interaction zone between laser and work piece with framerates of up to 50 kHz, adapted to the range of interest and the welding speed. Figure 3 shows the formation of a typical melt ejection.

Starting from the regular process (0.00 ms) the molten material inside the melt pool is explosively blown out of the interaction zone within less than 0.10 ms. As this creates a hole in the weld seam which is surrounded by solid material (0.20 ms) the capillary and the melt pool subsequently have to be reformed as in the beginning of a welding process (0.40–0.50 ms).

High-speed imaging of spatter formation and melt ejection allow to determine the critical time constants involved in the process. This knowledge is essential for analyzing and interpreting the fundamental mechanisms behind these weld defects.

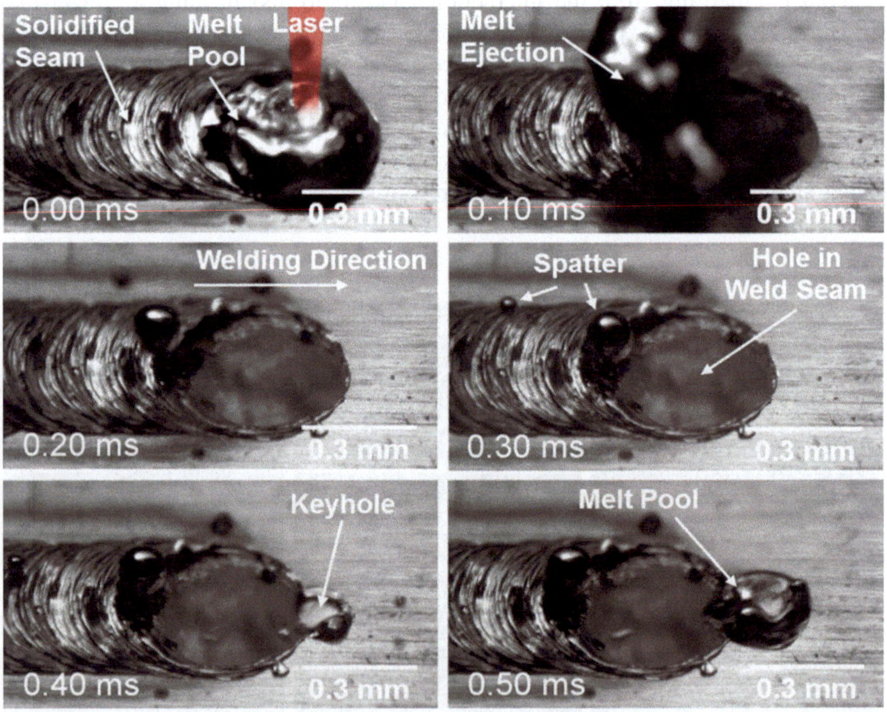

Fig. 3 Series of single frames from a high-speed video of a welding process in Cu-ETP showing the formation of a melt ejection. Laser Power P = 3 kW, velocity v = 6 m/min, focal diameter d_f = 100 μm

3.2 Hot Crack Formation During Laser Welding

During welding of alloys metallurgical and thermomechanical effects may lead to hot cracks. The formation of hot cracks happens during the solidification of the molten material between liquidus and solidus temperature of the alloy. In fact, the residual melt between the growing grains gets ruptured. The hot cracks propagate along the grain boundaries [9].

Figure 4 shows single frames of high-speed videos of welding of an AlMgSi alloy. The videos were recorded with a frame rate of 1 kHz. The observed area was illuminated by a diode laser with a wavelength of about 808 nm. In Fig. 4a the propagation of a hot crack (red arrows) was observed. The propagation starts outside the liquidus isotherm in the solidification zone. Figure 4b shows no hot crack formation on the weld seam surface. Figure 4c is the microscope photograph of the resulting grain structure in the weld seam of the process shown in Fig. 4a.

The use of a high-speed camera enables observation of initiation and propagation phenomena of hot cracks with a high temporal resolution. In addition, the shape of the melt pool can be analyzed (red dashed line). It gives an impression of the shape of the liquidus isotherm. As grains are growing in direction of the temperature gradient [10]—i.e. perpendicular to the isotherms (white arrows in Fig. 4c), the shape of the melt pool is an indicator for the orientation of the grains. In case of a sharp peak at the end of the weld pool (Fig. 4a), the grains are growing straight in direction to the centerline of the weld. This results in a continuous grain boundary along the centerline, which is highly sensitive to hot cracking [9]. In contrast, a circular shape of the weld pool (Fig. 4b) leads to a curved growth of the grains. In this case the grain boundary at the weld's centerline is discontinuous and less sensitive to hot cracking.

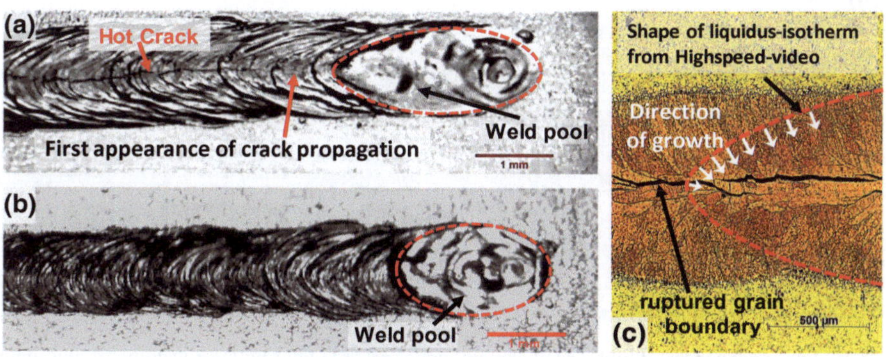

Fig. 4 Single frames of high-speed videos of a weld with a sharp peak at the end of the weld pool and propagating hot crack (**a**) and a weld with a *circular shaped* end of the weld pool and no hot crack formation (**b**) and the resulting grain structure after solidification (**c**)

3.3 Spatial Beam Modulation During Laser Welding

Another example of the application of high-speed imaging for process observation is the investigation of spatial beam modulation during laser welding. This technique is applied, e.g. in welding of dissimilar metals [11, 12]. In this case, the aim of the spatial modulation of the laser beam is to directly influence and control the mixing behavior and mixing ratio of the two joining partners, and to control the ratio of weld seam width to depth. For this purpose, different one- and two-dimensional oscillation figures (Lissajous curves) are used with typical oscillation frequencies of up to 1000 Hz and oscillation amplitudes of up to 3 mm. Respective high temporal and spatial resolutions of the high-speed camera imaging are required in order to investigate the influence of the oscillation parameters on the formation and dynamics of the weld pool.

Figure 5 shows two image sequences of high-speed camera videos recorded during laser welding of pure aluminum (Al99,5) to pure copper (Cu-OF).

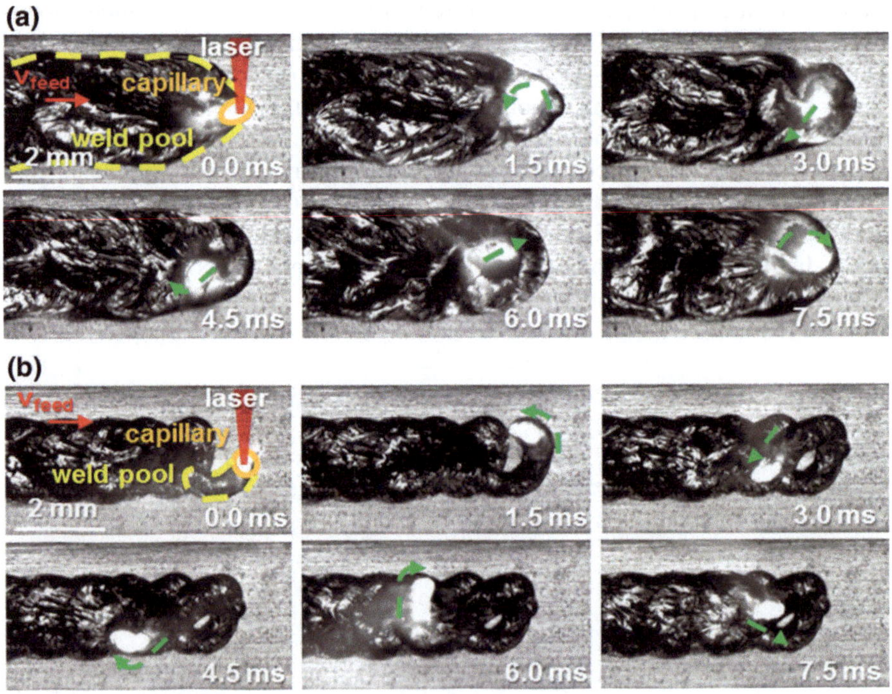

Fig. 5 Image sequences of high-speed camera videos recorded during laser welding of aluminum to copper with spatial modulation of the laser beam. *Green arrows* indicate the moving direction of the laser beam. Parameters: Laser power: P = 3.25 kW, welding speed: v_{feed} = 6 m/min, focal diameter: d_f = 280 μm. **a** oscillation frequency x: f_x = 100 Hz, oscillation amplitude x: a_x = 0.5 mm, oscillation frequency y: f_y = 200 Hz, oscillation amplitude y: a_y = 0.25 mm. **b** oscillation frequency x: f_x = 100 Hz, oscillation amplitude x: a_x = 1.5 mm, oscillation frequency y: f_y = 200 Hz, oscillation amplitude y: a_y = 0.75 mm

The two-dimensional oscillation figure used was a "lying eight", oriented in feed direction. The main differences in process parameters are the oscillation amplitudes in x- and y-direction (a_x, a_y). The process was recorded with a frame rate of 10 kHz, illuminated diffusely by an illumination laser at a wavelength of 808 nm and the high-speed camera was equipped with a respective band pass filter (808 ± 10 nm). In the case shown in Fig. 5a a rather large weld pool forms out, following the global feed direction, where the laser beam is "stirring" in the liquid weld pool. Whereas in the case shown in Fig. 5b a quite small weld pool is trailing behind the capillary, following the spatial modulation of the laser beam.

3.4 *High-Speed X-ray Imaging of the Capillary*

To investigate the inner processes of laser welding, a high-speed X-ray imaging system can be used [6, 7]. A sample is moved in-between an X-ray source and an imaging system. The welding is done from the top side of the sample as can be seen in Fig. 6.

In deep penetration mode, the laser creates a vapor filled channel, called the capillary or keyhole, which can be detected by the X-ray shadow projection. Since the vapor has a very small density it's nearly transparent and appears as a brighter area in the image, as can be seen in Fig. 7. The sample is moved under the stationary laser beam, which creates a process direction to the right.

It is used to analyze the shape and behavior of the capillary at different process parameters, to get a better understanding of their influences. For example welding aluminum is considered difficult to weld, in comparison to steel. The X-ray imaging

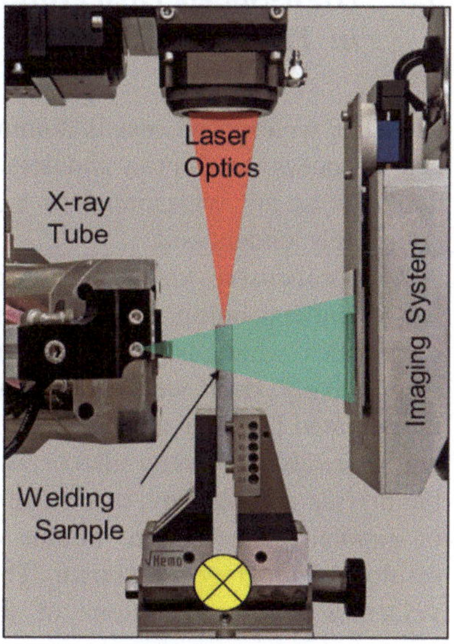

Fig. 6 High-speed X-ray imaging setup. The upright standing sample is welded from the *top side*, while it is shine through by an X-ray tube from *left to right* and imaged by the imaging system on the *right*

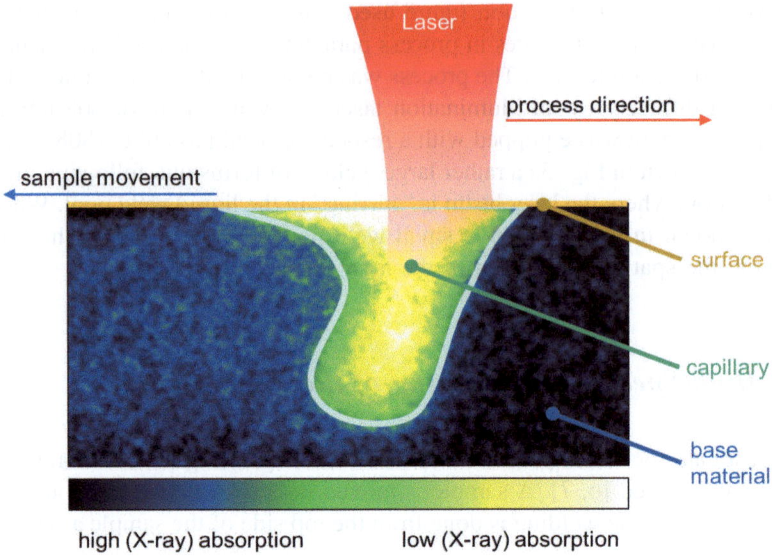

Fig. 7 Sample image of the high-speed X-ray imaging system. *Brighter colors* represent less X-ray absorption. In this case the *brighter area* is the capillary, resulting from the laser-beam vaporizing the base material

system reveals, that the capillary is very instable and collapses frequently, whereas steel is most often very stable and the capillary looks like in Fig. 7.

3.5 High-Speed Observation of Capillary Dynamics in Transparent Media

The laser welding process is dominated by highly dynamic phenomena, which can cause process instabilities and defects in the resultant weld. Although laser welding applications are widespread in industry, many of these phenomena are not yet completely understood.

Comprehensive visualizations of the process dynamics are required to gain sound understanding of the underlying processes. This can be done by X-ray diagnostics which is, however, limited in spatial and temporal resolution [13]. Alternatively laser welding of transparent media, such as water or ice can be investigated [14]. Given the appropriate illumination, framerates of several 10 kHz are achievable. Thereby effects like the collapse mechanisms of the capillary and the driving mechanisms of the fluid dynamics can be analyzed. In Fig. 8 the cycle of capillary constriction, bubble creation, the collapse of this bubble and the resultant fluid flow are shown. The collapse time of the capillary, as well as the velocities and accelerations of the resultant flow can be measured. Thereby,

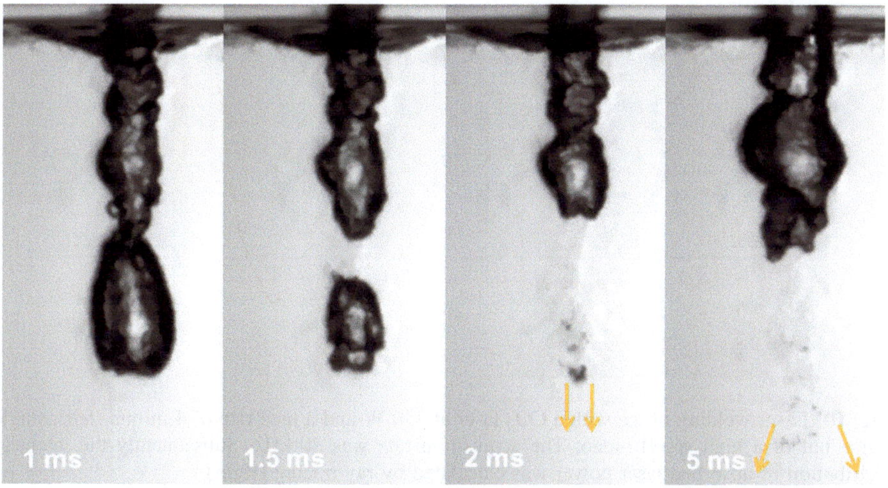

Fig. 8 Laser welding in water with a CO_2 laser at 500 W. The focal diameter at the water surface was 0.75 mm. The four images represent the denoted times after the laser was switched on

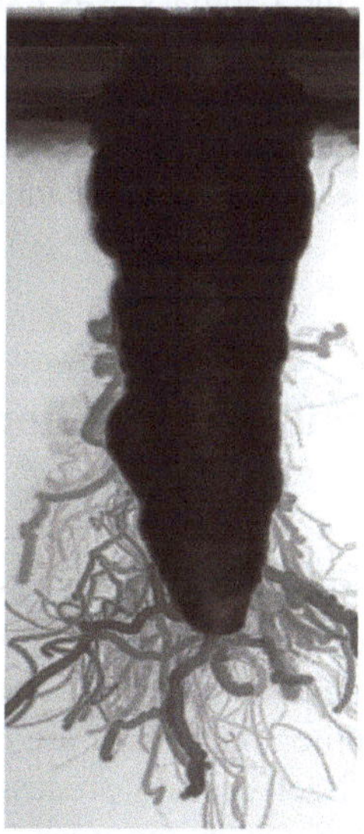

Fig. 9 Superposed minimum values over 200 ms of the process shown in Fig. 8

velocities in the range of several meters per second can be found. A better representation of the total scenery can be given by subsequent image processing. In Fig. 9 superposed minimum values of the first 200 ms of this high-speed video is given, which shows the maximum extension of the capillary, as well as the flow patterns, visualized by the remaining bubbles. Additionally, these measurements by means of high speed imaging can be combined with simulations.

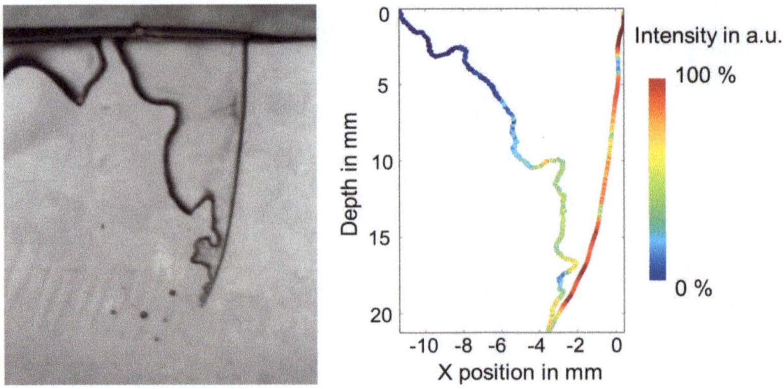

Fig. 10 Laser welding of ice with a CO_2 laser at 750 W and a feed rate of 4 m/min (*left*), single frame out of a high speed video. The acquisition rate was 20 kHz. Subsequently the intensity distribution of absorbed laser power was calculated by ray-tracing (*right*)

To investigate the drivers of process dynamics in laser welding, laser welding of ice was recorded. In Fig. 10 single frame out of a high speed video is shown (left). Subsequently the geometry of the capillary was extracted and the intensity distribution of absorbed laser power at the capillary was calculated with a ray tracing approach as given in [15].

Thereby in extension to the qualitative understanding of the process, quantitative conclusions regarding the causes for instabilities and fluctuations in welding depth can be found.

4 Laser Drilling of Metals—Ultra-Short Pulsed Laser Drilling Processes

In ultra-short pulsed laser drilling a high-speed camera is used to observe the formation of the borehole which takes typically a few seconds. Within this process the formation itself, the movement of the particles, the formation and movement of melt, and the metal vapour plume and plasma are of great interest.

To get an insight into the laser drilling process materials are used which are transparent for specific wavelengths in the near infrared (e.g. silicon) or in the visible spectrum [16]. Additionally, a new method was introduced to observe the formation of the borehole in non-transparent materials. This method is called drilling behind glass. Here, a thin glass plate is placed in front of the non-transparent sample. The sample is pressed against the glass plate and the laser is focussed onto the interface of these two pieces, Fig. 11.

To illuminate the process area a cw-diode laser with a maximum output power of 75 W at a wavelength of 808 nm was used. Figure 12 shows a series of five single

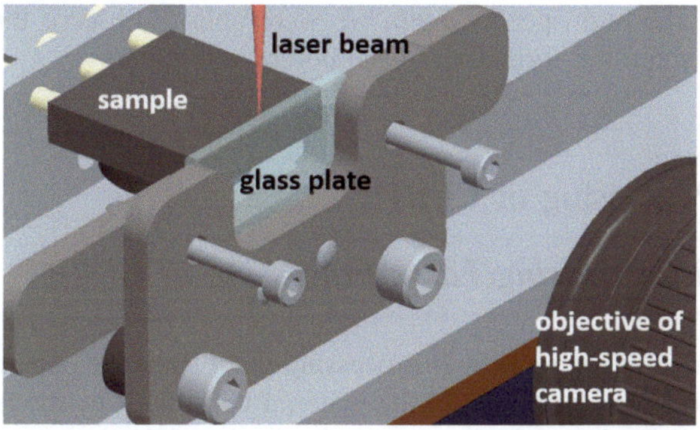

Fig. 11 Experimental setup for drilling behind glass. The sample is pressed against the glass plate. The laser is focused on the interface of the sample and the glass plate. With the high-speed camera the whole process can be observed through the glass plate

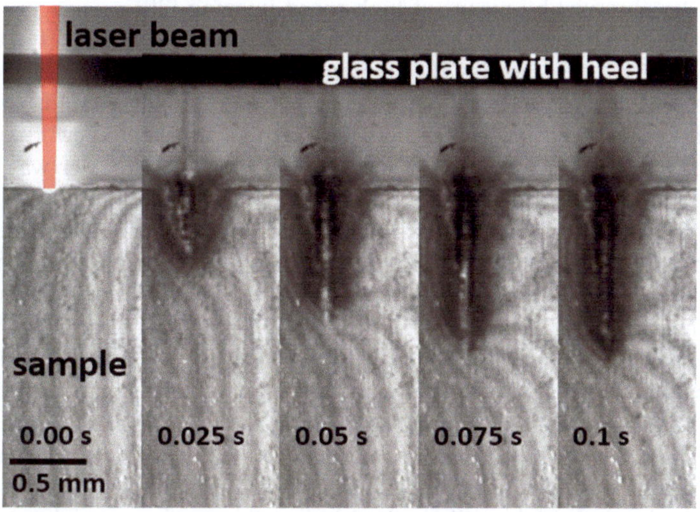

Fig. 12 The first 0.1 s of a drilling process. The repetition rate of the laser was set to 4 kHz. The frame rate was set to 28 kHz. For illumination a cw-diode laser was used with a maximum output power of 75 W

frames, the first 0.1 s of the drilling process, of a high-speed video recorded with a framerate of 28 kHz. The drilling process lasted 1.53 s and a repetition rate of 4 kHz was used. In the first picture, at the time 0 s, the drilling process starts and a reflection of the laser beam in the glass plate can be seen. After 0.025 s the borehole is already 0.46 mm deep and an additional 0.025 s later the depth of the borehole has nearly doubled. After 0.1 s the hole is 1.12 mm deep and has therefore reached half of the final depth which is 2.27 mm after 1.53 s.

Because of the high-speed camera we were able to observe this fast ongoing drilling process and the formation of the borehole in detail.

5 Laser Processing of CFRP

5.1 Heat Flow During Laser Processing of CFRP

One of the main challenges of laser processing of CFRP is the potential large matrix damage caused by the heat load of the material introduced by the laser beam [17]. High-speed imaging is a useful tool to understand such basic process mechanisms. Using high intensities, matrix damage is mostly caused by heat accumulation effects [18]. Every pulse deposits a certain amount of heat in the material. This heat accumulates with each subsequent laser pulse [19]. Eventually this heat accumulation leads to the formation of a heat affected zone. The extent of the heat affected zone can be observed by means of high-speed imaging [20].

The high-speed imaging system recorded the process with 3000 fps and a resolution of 1064 × 624 pixels. The illumination time was 24.8 μs. The scene was illuminated using an 808 nm diode laser and a corresponding band pass filter in front of the high-speed imaging system. The experiments were performed with an ultra-short pulsed laser system. The maximum average laser power was 22.2 W.

In Fig. 13 the surface of the CFRP work piece is shown 100 ms after the percussion drilling process started for different laser parameters. The process shown in the top frame was performed with an average power of 22.2 W. Below this frame on the left side, the repetition rate was kept constant and only the pulse energy was lowered resulting in a reduced average power. On the right side of the figure, the pulse energy was kept constant and only the repetition rate was lowered.

The carbon fibers of the CFRP work piece get heated when laser processing starts. The heat flows mainly along the carbon fiber due to their good heat conductivity. When evaporation temperature of the matrix material is reached, the matrix material evaporates leaving blank carbon fibers because their evaporation temperature is much higher. Since the layer of the matrix material on top of the carbon fibers is usually very thin, it is removed very quickly when the evaporation temperature is reached. For the highest average laser power of 22.2 W the extent of the heat affected zone is already about 2 mm after 100 ms. Generally speaking the extent of the heat affected zone is reduced by lowering the average laser power. However, there is a significant difference depending on whether the reduction of the average power is achieved by lowering the pulse energy or by reducing the repetition rate. At the same average power, the heat affected zone is significantly smaller using a reduced repetition rate as compared to the result for reduced pulse energies. By using high-speed imaging, it was revealed that the repetition rate is a major influencing factor defining the heat accumulation effect.

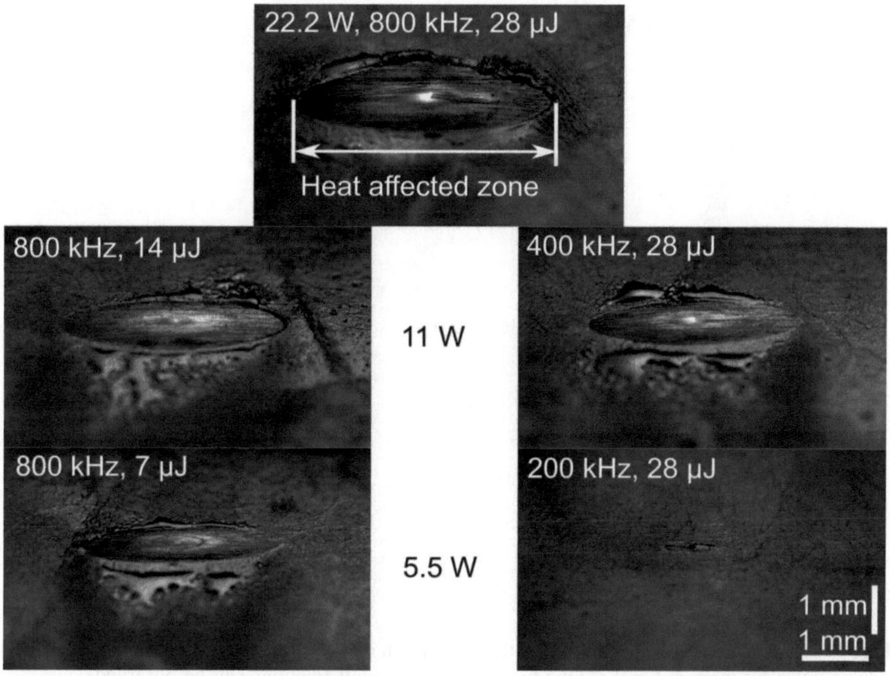

Fig. 13 Single frames of a high-speed recording that show the extent of the heat affected zone when laser processing CFRP with different process parameters

5.2 Cracking of Carbon Fibers and Particle Generation

With the help of high-speed imaging not only the heat affected zone can be observed, but also the ablation process of carbon fibers and CFRP can be investigated in detail [21]. For a high temporal resolution a high speed imaging system with four full-frame cameras was used that allows ultra-fast recording with 1 ns interframing time. To observe the ablation process, an interframing time of 100 ns was chosen.

Two carbon fibers were restrained and subjected to the laser pulse. The experiment was performed with a 5 picoseconds laser system with an average power of 600 mW, a wavelength of 800 nm and a pulse repetition rate of 1 kHz. The ablation process was illuminated by an incoherent illumination laser source so that no speckles occur. In Fig. 14 a sequence of four images can be seen. At t = 0 ns the laser pulse hits the two carbon fibers. At t = 100 ns the appearance of several small and one big fractions of the carbon fibers can be observed. The small fractions quickly move out of the focal plane of the camera and therefore cannot be seen on the other images. The bigger fraction can be followed over time. It moves with a speed of 200 m/s away from the interaction zone.

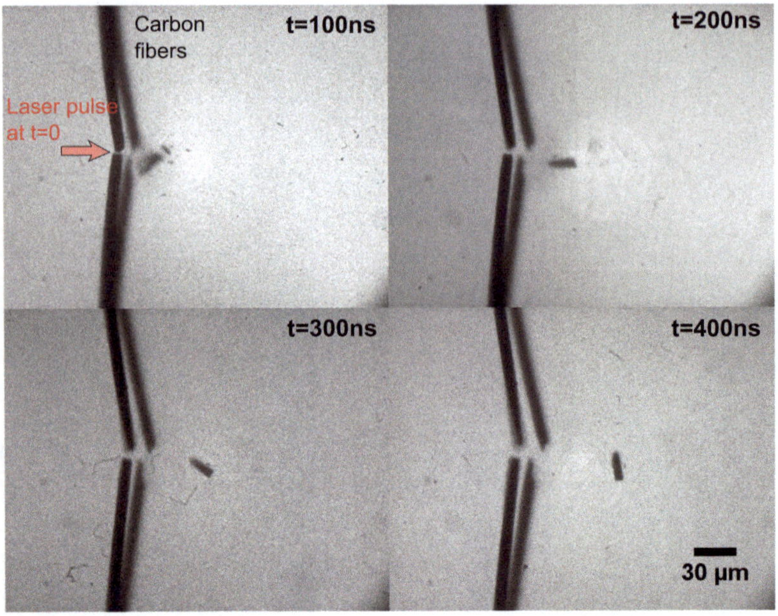

Fig. 14 Sequence of four images with a temporal pitch of 100 ns showing two carbon fibers after a laser pulse hit them at t = 0 s. One big and several small fractions can be observed

Particles are also generated during laser cutting of CFRP. These particles can be generated through cracking of the carbon fibers or through a decomposition of the compound due to evaporation of the matrix material which results in the loss of fixation of the carbon fibers. Particle generation during the cutting process was observed with the same high-speed camera as before. The time-shift between two frames was set to 2.5 µs and the exposure time of one frame was 3 ns. Because of the short exposure time the process area was illuminated in the axis of the camera view with an incoherent but monochromatic light source.

In Fig. 15 a series of four images taken during the cutting process can be seen. The red circle marks the same particle in all four images. The blue line in the first and last image shows the distance of the particle to the processing area. By means of high speed imaging we were able to determine the number of particles generated during laser processing of CFRP with a continuous wave laser. Additionally, the speed of the particles can be evaluated. These information are especially valuable for the design of an adequate exhaustion system for this process.

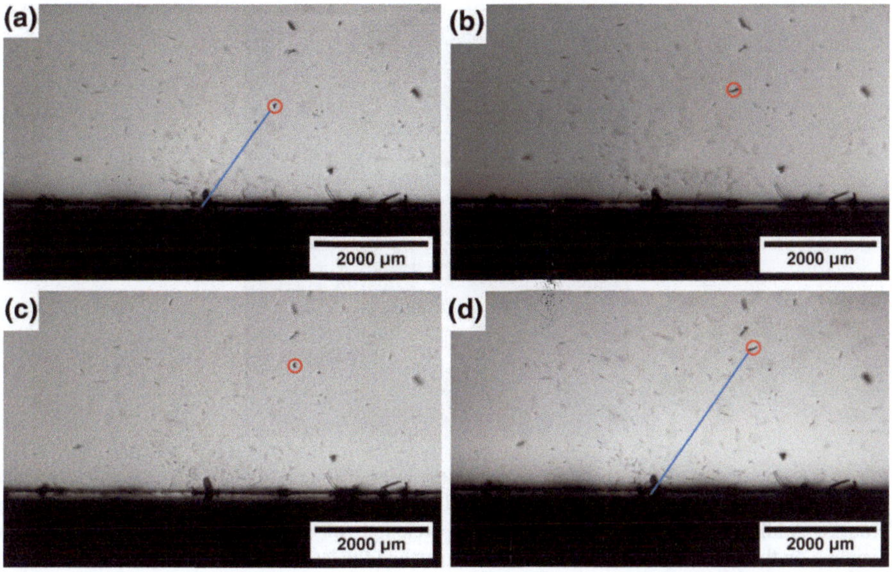

Fig. 15 Side view of a high-speed recording of laser cutting of CFRP with a continuous wave laser. The average laser power was 2.0 kW and the feed rate was 1 m/s. The frames were taken **a** 0.0 μs, **b** 2.5 μs, **c** 5.0 μs, and **d** 7.5 μs after start of the recording. Additional process parameters: the wavelength of the laser was 1030 nm, the focal diameter on the surface of the sample about 300 μm, and the laser beam was not polarized

5.3 Flow Speeds of the Ablation Products Generated During Laser Processing of CFRP with a Continuous Wave Laser System

In the following the ablation products generated during laser drilling of CFRP with a continuous wave laser system were observed. The high-speed camera was mounted on the side of the workpiece. No light source for illumination of the process was used. The process was recorded with a frame rate of 7500 fps, the exposure time was set to 133.1 μs, and the resolution was 1024 × 1000.

A single frame of the plume of the ablation products during laser drilling CFRP is shown in Fig. 16. The hot ablation products flow out of the entry of the borehole and the formation of a compression shock can be seen. Compression shocks indicate a flow speed well above the local speed of sound. The flow speed can be calculated by knowing the diameter of the borehole and the distance from it to the first compression shock, see [22].

The use of high-speed imaging enabled us to detect that the ablation products generated during processing of CFRP with a continuous wave laser flow with supersonic speed.

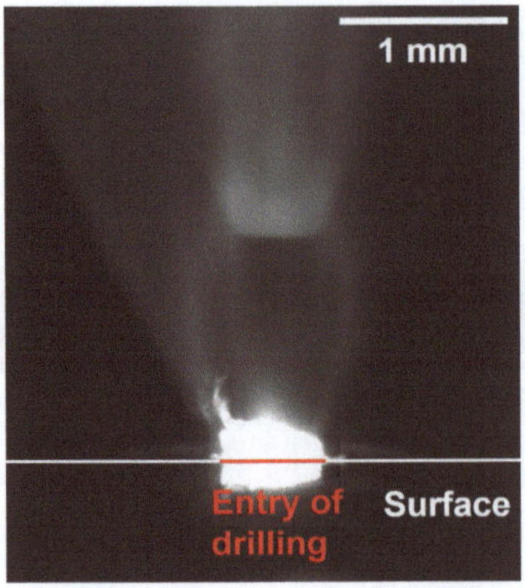

Fig. 16 Side view of a high-speed recording of the laser drilling of CFRP with a continuous wave laser. The average laser power was 3.5 kW and it was recorded 3.6 ms after start of the drilling process. Additional process parameters: the wavelength of the laser was 1030 nm, the focal diameter on the surface of the sample about 300 μm and the laser beam was not polarized. Camera parameters: 7500 frames per second, and 133.1 μs exposure time

6 Summary

It has been shown, that high-speed imaging is a very useful tool for the observation of laser processes. With the help of high-speed cameras basic process mechanisms during laser welding like spatter formation, the formation of hot cracks, the dynamics of the vapor capillary, and the influence of spatial beam modulation were investigated. High-speed imaging also enabled the investigation of ultra-fast processes like drilling with ultra-short laser pulses and laser processing of CFRP. Valuable insights like the fact, that hot vapor generated during laser drilling and cutting of CFRP with a continuous wave laser system flows with supersonic speed, can only be obtained with high-speed imaging.

References

1. J. Mueller-Borhanian, C. Deininger, F. Dausinger, H. Huegel, Spatially resolved on-line monitoring during laser beam welding of steel and aluminum. Paper presented at the international congress on applications of lasers & electro-optics, San Francisco, 4–7 Oct 2004
2. A. Heider, Final report of DVS project ZeSAR (DVS, Düsseldorf, 2015)
3. J.-P. Weberpals, Dissertation, University of Stuttgart, 2010
4. A.F.H. Kaplan, J. Powell, Spatter in laser welding. J. Laser Appl. **23**, 032005 (2011)
5. A. Heider, R. Weber, D. Herrmann, P. Herzog, T. Graf, Power modulation to stabilize laser welding of copper. J. Laser Appl. **27**, 022003 (2015)

6. A. Heider, M. Boley, R. Weber, T. Graf, Investigation of spatter formation in laser welding of copper using high-speed online X-ray imaging. Paper presented at the 6th international congress on laser advanced materials processing, Niigata, 23–26 July 2013
7. A. Heider, M. Boley, R. Weber, T. Graf, High-speed X-ray analysis of spatter formation in laser welding of copper. Paper presented at the 66th IIW annual assembly international conference on "Automation in Welding", Essen, 16–17 Sept 2013
8. A. Heider, P. Stritt, R. Weber, T. Graf, High-power laser sources enable high-quality laser welding of copper. Paper presented at the international congress on applications of lasers and electro-optics, San Diego, 20–23 Oct 2014
9. J. Schuster, Habilitation, Technische Universität Chemnitz, 2009
10. J.F. Lancaster, *Metallurgy of welding* (Abington Publ, Cambridge, 1999)
11. F. Fetzer, M. Jarwitz, P. Stritt, R. Weber, T. Graf, Fine-tuned remote laser welding of aluminum to copper with local beam oscillation. Paper presented at the 9th international conference on photonic technologies—LANE, Fürth, 19–22 Sept 2016
12. J. Gedicke, A. Olowinsky, J. Artal, A. Gillner, Influence of temporal and spatial laser power modulation on melt pool dynamics. Paper presented at the international congress on applications of lasers & electro-optics, Orlando, 29 Oct–1 Nov 2007
13. F. Abt, M. Boley, R. Weber, T. Graf, G. Popko, S. Nau, Novel X-ray system for in-situ diagnostics of laser based processes—first experimental results. Phys. Procedia **12**, 761–770 (2011)
14. P. Berger, Reduced velocity of sound in the weld pool in the presence of bubbles. Paper presented at the international congress on applications of lasers and electro-optics, San Diego, 20–23 Oct 2014
15. Y. Qin, A. Michalowski, R. Weber, S. Yang, T. Graf, Comparison between ray-tracing and physical optics for the computation of light absorption in capillaries—the influence of diffraction and interference. Opt. Express **20**, 26606–26617 (2012)
16. S. Döring, T. Ullsperger, F. Heisler, S. Richter, A. Tünnermann, S. Nolte, Hole formation process in ultrashort pulse laser percussion drilling. Phys. Procedia **41**, 431–440 (2013)
17. R. Weber, M. Hafner, A. Michalowski, T. Graf, Minimum damage in CFRP laser processing. Phys. Procedia **12**, 302–307 (2011)
18. T.V. Kononenko, C. Freitag, M.S. Komlenok, V. Onuseit, R. Weber, T. Graf, V.I. Konov, Heat accumulation effects in short-pulse multi-pass cutting of carbon fiber reinforced plastics. J. Appl. Phys. **118**, 103105 (2015)
19. R. Weber, T. Graf, P. Berger, V. Onuseit, M. Wiedenmann, C. Freitag, A. Feuer, Heat accumulation during pulsed laser materials processing. Opt. Express **22**, 11312–11324 (2014)
20. C. Freitag, V. Onuseit, R. Weber, T. Graf, High-speed observation of the heat flow in CFRP during laser processing. Phys. Procedia **39**, 171–178 (2012)
21. C. Freitag, M. Hafner, V. Onuseit, A. Michalowski, R. Weber, T. Graf, Diagnostics of basic effects in laser processing of CFRP. Paper presented at the international symposium on laser processing of CFRP and composites, Yokohama, 26–27 Apr 2012
22. D. Rist, *Dynamik realer Gase*, 1st edn. (Springer, Berlin, 1996)

Real-Time Hard X-ray Imaging

Alexander Rack, Margie Olbinado, Mario Scheel, Benjamin Jodar and John Morse

Abstract Using hard X-rays for high-speed and ultra high-speed imaging has enormous potential to visualize the interior of opaque systems as they change with time. Exposure times below one nanosecond for ultra high-speed imaging are accessible when synchrotron light sources are employed and this provides a non-destructive method of *in-motion* radiography. The polychromatic radiation of insertion devices in combination with X-ray phase contrast has proven to be suited for acquisition rates up to the MHz range. This chapter outlines the basic principles of indirect hard X-ray imaging detectors for real-time imaging, and other detection schemes and sources of radiation are briefly discussed. The potential of using hard X-rays for high-speed imaging is demonstrated with application examples from soft matter physics and materials processing.

Electronic supplementary material The online version of this chapter (doi:10.1007/978-3-319-61491-5_10) contains supplementary material, which is available to authorized users.

A. Rack (✉) · M. Olbinado · J. Morse
European Synchrotron Radiation Facility (ESRF), Grenoble, France
e-mail: rack@esrf.fr

M. Olbinado
e-mail: margie.olbinado@esrf.fr

J. Morse
e-mail: morse@esrf.fr

M. Scheel
Synchrotron Soleil, Gif-Sur-Yvette, France
e-mail: mario.scheel@synchrotron-soleil.fr

B. Jodar
UMR 6251—CNRS/Université de Rennes 1, Rennes, France
e-mail: benjamin.jodar@univ-rennes1.fr

K. Tsuji (ed.), *The Micro-World Observed by Ultra High-Speed Cameras*,
DOI 10.1007/978-3-319-61491-5_10

1 Hard X-ray Imaging

Hard X-ray imaging is a valuable tool to depict the interior of an object in a widely non-destructive manner. It is simple to implement and hence, nowadays wide spread. Radiography commonly is used when two-dimensional projection images are sufficient or other constraints are present. The latter can be simply technical limitations as it can often occur in non-destructive testing (such as inspection of pipes) or dose-considerations (medical applications). If projection images from different angles of view are accessible then X-ray imaging can be extended into three dimensions by the use of tomographic algorithms which enable reconstruction of the mass distribution within an object from the set of two-dimensional radiographs [1, 2]. For real-time studies, two-dimensional radiography has an obvious advantage in speed as compared to tomography, as the latter frequently needs at least several hundred projection images for a good reconstruction, which reduces the effective time resolution by up to three orders of magnitude with respect to two-dimensional X-ray imaging.

2 Indirect Detection

The ‘indirect’ X-ray imaging detection scheme is frequently used at synchrotron light sources. In this scheme, the luminescence image of a scintillating screen is relayed by visible light optics onto the sensor of a camera, cf. Fig. 1. Such a detector offers a high level of resistance to radiation damage arising from the intense synchrotron X-ray beam, as with a suitable ‘folded’ geometry, the only element subject to the intense radiation is the scintillator screen. Furthermore, spatial resolutions down to the sub-micrometre range are accessible, ultimately limited by optical diffraction limit set by the wavelength of maximum emission of the scintillator and the numerical aperture (NA) of the lens system (Abbe theorem). Indirect detectors may be assembled using components which are commercially available, and they allow for easy modification and hence are a versatile solution for hard X-ray imaging [3, 4]. High temporal resolution (short exposure times and/or high image acquisition rates) can be reached when fast cameras are combined with fast-decay scintillator materials and high NA optics [5–7].

Instead of using a single high-speed camera, several time sequential images may be recorded using gate-able image-intensified cameras which view the scintillator screen via an optical multiplexer [10]. The advantage of this multiplexing arrangement is the high level of flexibility that it offers in terms of synchronization as each frame can be triggered individually with arbitrary delays. The number of frames which can be acquired in one shot is simply the number of cameras used: this is limited by available photon statistics which are reduced on a per camera basis.

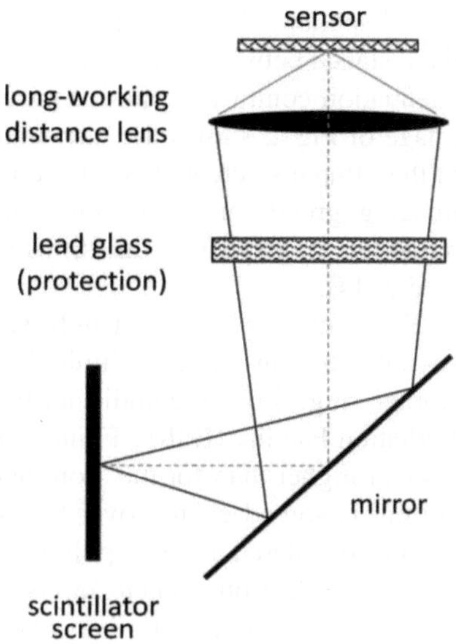

Fig. 1 Layout of an indirect X-ray imaging detector with folded optical beam path to reduce the number of components subjected to the intense radiation. Lead glass protection is frequently required to reduce X-ray scattering from the mirror (following the concepts of Hartmann et al. and Koch [8, 9])

Direct detectors, i.e. in which the X-rays are detected by a silicon imaging sensor have been developed which reach up to MHz acquisition rates. These systems offer a limited spatial resolution [11, 12] due to the practical sensor pixel sizes, typically ~20 μm and more: they are also inefficient for hard X-ray imaging due to the poor absorption of X-rays in silicon above ~20 keV, and they suffer more rapidly from radiation damage.

3 Contrast Modes and Synchrotron Light Sources

The classical method for providing contrast in an X-ray image is by absorption, i.e. the varying attenuation of an X-ray beam which passes through a specimen results in an intensity modulation which can be recorded as picture. This contrast is mainly related to the (local) mass density of the specimen. In many cases, higher imaging sensitivity can be reached using X-ray phase contrast techniques. Here, the local X-ray refraction of the specimen is used to provide contrast. Phase contrast is related to the local electron density of the specimen [13].

Synchrotron light sources offer excellent properties for phase contrast imaging. Electrons accelerated close to the speed of light are injected into a storage ring: as their trajectory is deviated by passing through magnetic fields, these electrons emit radition with X-ray photon energies. Due to the relativistic speeds, the radiation is emitted in a narrow cone, providing an almost parallel beam which enables a large source-experiment distance of more than 100 m [14]. This long distance reduces the influence of the finite source size on the image resolution and introduces significant coherence properties to the X-ray beam wavefront at the experiment position:

partial spatial coherence [13]. As a consequence, at a synchrotron light source, as the distance between the sample and the detector is increased we move from simple attenuation contrast to phase contrast imaging by means of edge enhancement: the image of Fig. 2 shows that the dramatic 'fringe enhancement' of the edge features in the sample so created is robust against noise: in consequence, this phase contrast imaging greatly increases the sensitivity of feature detection and thus enables reduction in exposure times, which is highly important for fast time-resolved studies [4].

Frequently, synchrotron light sources are operated in different modes. In these so-called timing modes, a limited number of electron bunches circulate around the storage ring: 4 or 16 equidistant bunches in the case of the European Synchrotron Radiation Facility (ESRF, France). In these modes the electron density per bunch is also far higher than for the 'continuous fill' mode in which the total electron bunch current is spread evenly over 992 bunches. In the timing modes, the synchrotron can be considered as MHz pulsed source (1.4 or 5.6 MHz in the case of the ESRF), as every electron bunch emits an X-ray flash of duration around 100 ps. A high-speed detector, i.e. one which is capable of resolving the bunch time-structure of the synchrotron source also can be operated with the 100 ps X-ray exposure time determined by the length of an individual X-ray flash, and not the (usually far longer) integrating window of the detector itself [10, 15]: this we refer to as single-bunch imaging. For the sake of completeness we must mention that special pulsed laboratory-based X-ray sources can offer X-ray flashes as short as 20 ns, but such short flashes are available only in a single-shot manner or with a far lower repetition rates than synchrotron sources.

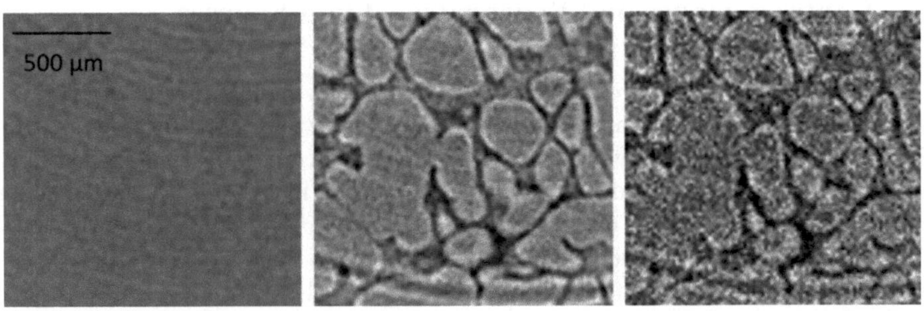

Fig. 2 Different contrast modalities for hard X-ray imaging shown here for tomographic slices of an AlSiMg alloy. Due to the similar mass densities within the sample, the attenuation contrast does not make visible the various material phases (*left*), but refraction at interfaces increases the contrast by means of edge enhancements: so-called propagation-based phase contrast (*middle*). The resultant edge contrast is also very robust against noise (*right*) [4, 13, 16]

4 Time-Resolved Microtomography

The ability of acquiring hard X-ray images at high framing rates enables us to acquire complete tomographic data sets within less than a second. Frequently, the frame acquisition rate in tomography is limited more by technical constraints such as the rotation speed of the sample stage than the image acquisition frame rate of the detector. So-called 4D tomography with acquisition rates up to 20 Hz achieved in a routine manner has already been reported in the literature. Nevertheless, these rates are far below what we describe here as high-speed or ultra high-speed and are not further considered in this chapter, and we mention 4D microtomography only for the sake of completeness (for further details the reader is referred to the literature [17, 18]).

5 Applications

Two applications have been chosen to exemplify the potential of hard X-ray imaging with ultra high-speed acquisition schemes in combination with X-ray phase contrast. The first is from the soft matter research field: the dynamics of aqueous foam. Foams are of high interest due to their wide-spread use in the food industry [19], also they can act as macroscopic model system for the study of dynamics between atoms within a crystal lattice [20]. The second example is the modification of a surface coating due to intense laser irradiation. Laser processing of materials is of high interest for industrial applications such as additive manufacturing, and MHz radioscopy may enable a detailed understanding of surface dynamic processes which occur when a high-power laser pulse interacts with a material [21, 22].

5.1 Aqueous Foam

In order to study dynamics in aqueous foams, a simple system was chosen: tap water mixed with commercial soap, foamed with the help of air. This imaging experiment was carried out at the microtomography beamline ID19 of the ESRF. For this experiment, the ESRF storage ring was operated in the 4-bunch mode, i.e. with a repetitive 704 ns spacing of the 100 ps X-ray flashes. In order to reach a sufficient photon flux density in each X-ray flash, ID19 was operated without any energy selecting monochromator. Two type u32 undulator insertion devices both set to their minimum gap to provide maximum X-ray flux were chosen as photon source. The radiation was filtered by 1.4 mm-thick diamond and 2.8 mm-thick aluminium absorbers and a 0.5 mm-thick beryllium window. The X-ray energy spectrum consequently had a broad peak at around 30 keV. The indirect detector arrangement consisted of two Hasselblad objectives (each of 100 mm focal length, in tandem geometry), resulting in an effective 1.1× magnification with 0.2 NA. A 250 μm-thick single-crystal scintillator screen of LuAG:Ce (Ce-doped $Lu_3Al_5O_{12}$, Crytur, Czech Republic) with 58 ns decay time for its dominant emission was used in order to provide high spatial resolution. The high-speed camera used was a pco.dimax (PCO AG, Germany) with 2016 × 2016 pixels, 11 μm pixel size, and 36 Gb onboard image memory.

In order to reach higher image acquisition rates, the camera was operated with a reduced region-of-interest of 450 × 450 pixels: this enabled images to be acquired every 25 μs, and an exposure time of 1.28 μs was used. A sample-detector distance of 14 m was ensured high imaging sensitivity resulting from propagation-based phase contrast.

Example images from the movie acquired are shown in Fig. 3, and the complete animated movie is available online (see Appendix). The pictures of Fig. 3 show an excellent contrast of the lamella of the pore structure, i.e. objects which have virtually zero X-ray absorption at the photon energy used. Despite the moderate frame acquisition rate, two cell wall collapses can be followed and the subsequent rearrangement of the pore neighborhood. In the first frame, labeled "500 μs" an intact cell wall is marked with a red arrow. In the next frame labeled "625 μs" this cell wall almost disappeared. A neighboring cell wall is marked with an arrow which again disappears in the following frame ("1000 μs"): a cascade of cell wall collapses. The following frames show how remains of the cell walls fade away.

5.2 *Laser Processing*

The second application sample shown originates from materials research: laser processing. A polystyrene foam was chosen as a sample as it is excellently suited for phase contrast imaging. The surface of the foam was coated with a thin aluminium layer.

For this experiment, beamline ID19 of the ESRF was again chosen due to its excellent capabilities in terms of phase-contrast imaging. The storage ring was operated in 16 bunch mode, giving equally spaced electron bunches spaced at 176 ns. In this case the indirect detector arrangement consisted of a custom 4× magnification lens of 0.2 NA (OptiquePeter, France). A 250 μm-thick LYSO: Ce (Ce-doped $(Lu_{2-x}Y_x)SiO_5$, Hilger Crystals, UK, 40 ns decay time) single-crystal scintillator screen was used. For this work an ultra high-speed camera was installed: the HPV-X2 (Shimadzu, Japan, 400 × 250 pixels, 32 μm pixel size, 128 frames recording length). Note that due to the 4× optical magnification, the detector operated with an 8 μm effective pixel size. A propagation distance between sample and detector of 6 m was used. The camera was set to an exposure time of 200 ns in order to ensure that only the X-rays from one flash contributed to teach successive image (single-bunch imaging). The camera acquired a frame every 710 ns (i.e. a 1.4 MHz frame acquisition rate). The configuration of the beamline

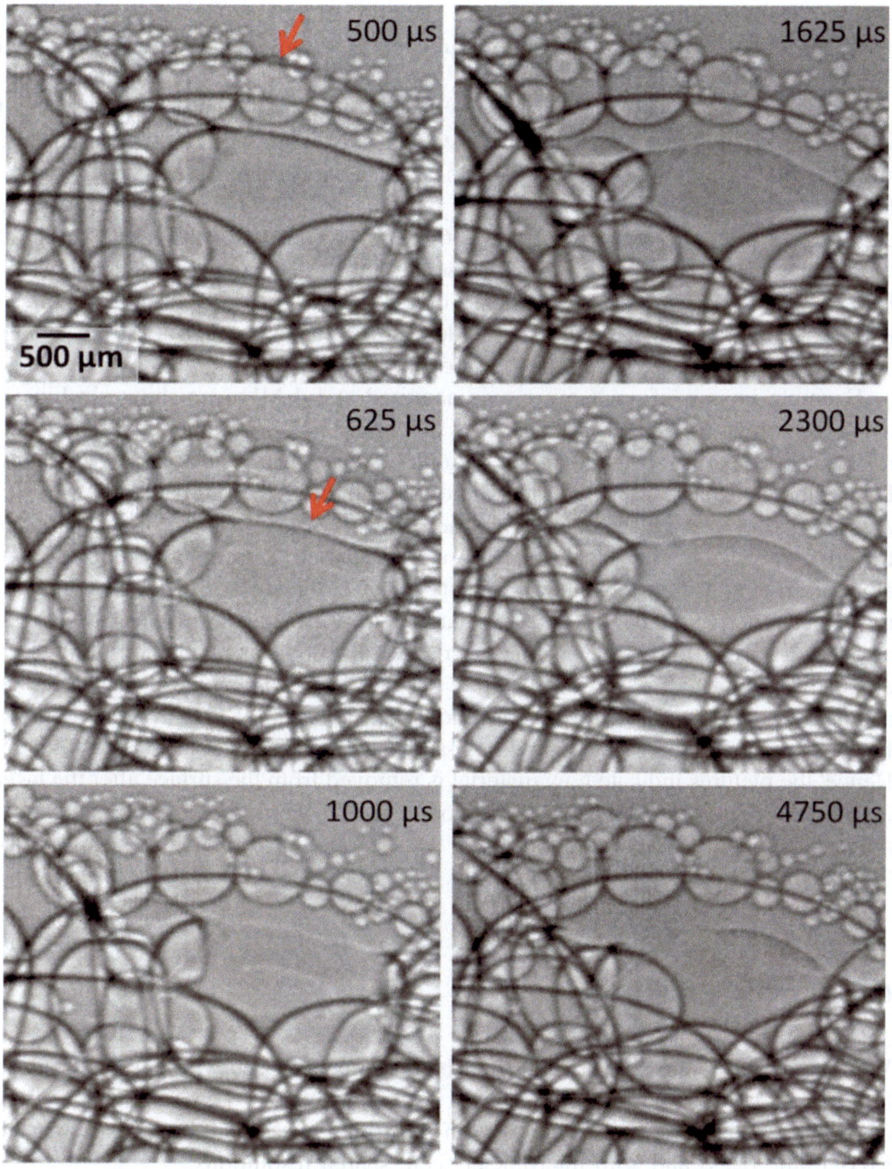

Fig. 3 Series of images from a movie (available online) showing dynamics in an aqueous foam obtained by means of high-speed phase contrast radioscopy. The collapse of two cell walls (*red arrow*) can be followed as well as the rearrangement of the pores in the immediate neighborhood

was similar as in the previous section, i.e. both undulators at minimal gap, no monochromator, 1.4 mm-thick aluminium and 2.8 mm-thick diamond filter and 0.5 mm-thick beryllium window.

In order to ensure that an X-ray image was taken when the laser beam interacted with the specimen, synchronization of the laser pulse with the time-structure of the storage ring and the camera acquisition was required. The laser chosen was a BrilliantB (Quantel, France): this is a pulsed Nd:YAG-laser with active q-switching. For this experiment the 1064 nm wavelength of the laser was used (nominal: 800 mJ per pulse, 20 ns pulse duration). The radio frequency of the ESRF synchrotron was integer divided down and used to synchronize the flash lamp (operated at 10 Hz), and coincidence of the laser impact during image acquisition was realised by a fine-delay of the q-switch.

Example images of an X-ray movie acquired at 1.4 MHz are shown in Fig. 4: the first frame at 7.10 μs shows the specimen before irradiation. The right, homogeneous part is air while on the left one can see the pore structure of the foam. Due to the size and amount of pores, hardly any individual pore is distinguishable. A bright hallow can be seen at the interface between foam and air: a clear indicator for edge-enhancement by means of propagation-based X-ray phase contrast. The aluminium coating is not visible, probably because the surface of the specimen is not exactly perpendicular to detector plane. The following frame marks with an arrow the start of the impact of the laser (the full movie is available online—see Appendix). Due to the slow pulse-rate of the laser with respect to the image acquisition rate, the following frames show the evolution of the surface without a further laser interaction. A detailed post-analysis of the specimen is not available. Hence, the pictures indicate how material is removed from the surface after the laser irradiation. The movie underlines the potential of MHz radioscopy to study fast processes such as laser ablation with the benefits of hard X-rays in a truly time-resolved manner.

6 Summary

We have briefly summarized real-time hard X-ray radioscopy using synchrotron radiation in combination with an indirect detection scheme. This enables the study of processes inside opaque samples with ultra high-speed acquisition rates (up to MHz frame rates) and with micrometer-scale spatial resolutions.

Currently, a world-wide race is in progress to build or upgrade synchrotron light sources, aimed at achieving near X-ray diffraction limited sources providing a dramatic increase in the brilliance of the X-ray beams [15]. If the necessary parallel development of fast cameras is made, then GHz frame rate hard X-ray imaging could be possible within the next decade [23, 24].

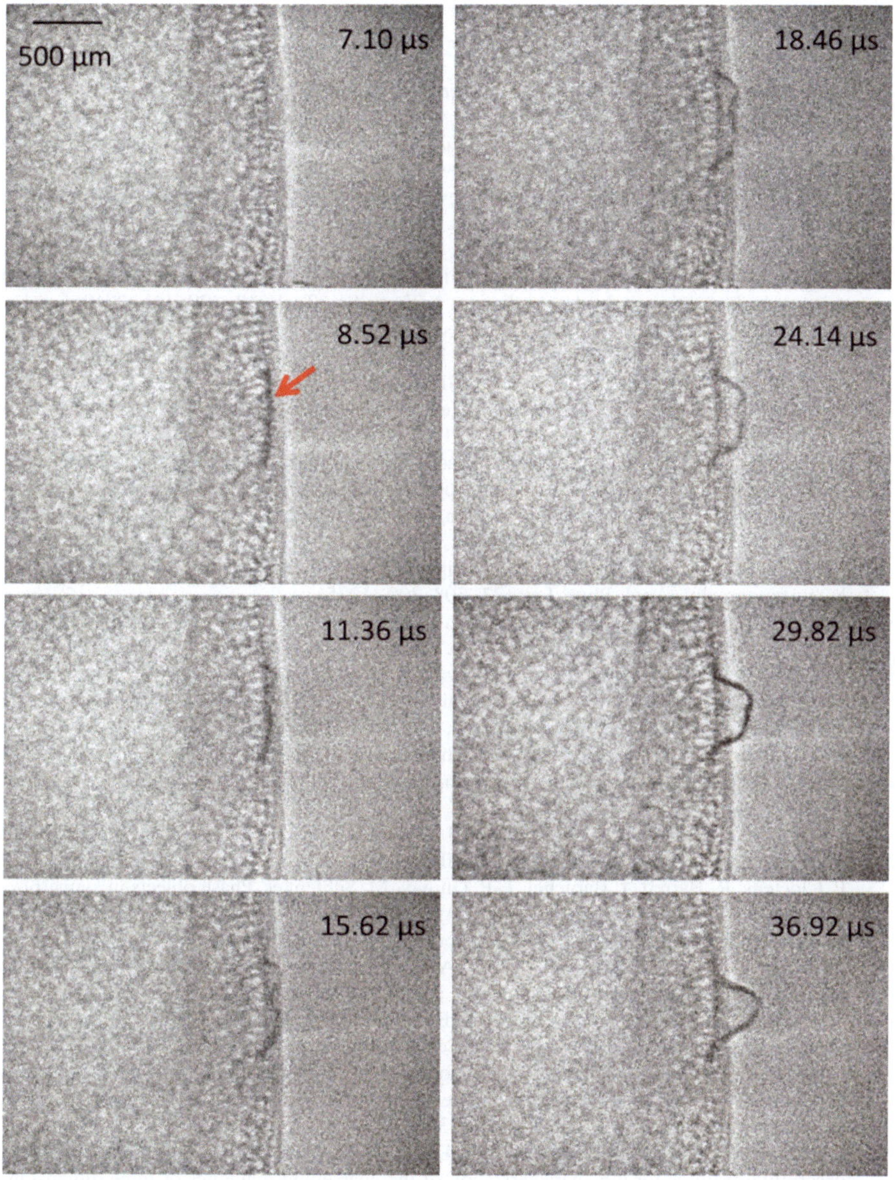

Fig. 4 Selected images of a single-bunch X-ray movie showing the interaction of an isolated laser irradiation (800 mJ, 20 ns pulse) with the aluminium-coated surface of a polystyrene foam. The complete movie is available online. The frame acquisition rate was 1.4 MHz, the integration time of the camera 200 ns (effective exposure time of 165 ps (fwhm) obtained by means of single-bunch illumination). Contrast in the pictures is dominated by X-ray phase contrast. The actual impact is the region marked with an arrow in frame "8.52 μs" (images taken afterwards do not overlap with laser irradiation due to the slow pulse rate of the laser with respect to the image acquisition rate)

References

1. H.E. Johns, J.R. Cunningham, *The Physics of Radiology*, 4th edn. (Charles C Thomas, Springfield, 1983)
2. A.C. Kak, M. Slaney, *Principles of Computerized Tomographic Imaging* (IEEE Press, New York, 1988)
3. A. Koch, C. Raven, P. Spanne, A. Snigirev, X-ray imaging with submicrometer resolution employing transparent luminescent screens. J. Opt. Soc. Am. **15**, 1940–1951 (1998)
4. A. Rack, F. Garcia-Moreno, C. Schmitt, O. Betz, A. Cecilia, A. Ershov, T. Rack, J. Banhart, S. Zabler, On the possibilities of hard X-ray imaging with high spatio-temporal resolution using polychromatic synchrotron radiation. J. X-ray Sci. Tech. **18**, 429–441 (2010)
5. M.P. Olbinado, X. Just, J.-L. Gelet, P. Lhuissier, M. Scheel, P. Vagovic, T. Sato, R. Graceffa, J. Morse, A. Rack, MHz frame rate hard X-ray phase-contrast imaging using synchrotron radiation. Opt. Expr. **25**, 13857–13871 (2017)
6. A. Rack, M. Scheel, A.N. Danilewsky, Real-time direct and diffraction X-ray imaging of irregular silicon wafer breakage. IUCrJ **3**, 108–114 (2016)
7. M. Hudspeth, B. Claus, S. Dubelman, J. Black, A. Mondal, N. Parab, C. Funnell, F. Hai, M.L. Qi, K. Fezzaa, S.N. Luo, W. Chen, High speed synchrotron X-ray phase contrast imaging of dynamic material response to split Hopkinson bar loading. Rev. Sci. Instrum. **84**, 025102 (2013)
8. W. Hartmann, G. Markewitz, U. Rettenmaier, H.J. Queisser, High resolution direct-display X-ray topography. Appl. Phys. Lett. **27**, 308–309 (1975)
9. A. Koch, Lens coupled scintillating screen-CCD X-ray area detector with a high quantum efficiency, Nucl. Instrum. Meth. Phys. Res. A **348**, 654–658 (1994)
10. S.N. Luo, B.J. Jensen, D.E. Hooks, K. Fezzaa, K.J. Ramos, J.D. Yeager, K. Kwiatkowski, T. Shimada, Gas gun shock experiments with single-pulse X-ray phase contrast imaging and diffraction at the Advanced Photon Source. Rev. Sci. Instrum. **83**, 073903 (2012)
11. H.T. Philipp, M.W. Tate, P. Purohit, K.S. Shanks, J.T. Weiss, S.M. Gruner, High-speed X-ray imaging pixel array detector for synchrotron bunch isolation. J. Synchrotron Rad. **23**, 395–403 (2016)
12. J. Schwandt, E. Fretwurst, R. Klanner, J. Zhang, Design of the AGIPD sensor for the European XFEL. J. Instrum. **8**, C01015 (2013)
13. P. Cloetens, R. Barrett, J. Baruchel, J.-P. Guigay, M. Schlenker, Phase objects in synchrotron radiation hard X-ray imaging. J. phys. D Appl. Phys. **29**, 133–146 (1996)
14. H. Wiedemann, *Synchrotron Radiation* (Springer, Berlin, 2002)
15. A. Rack, M. Scheel, L. Hardy, C. Curfs, A. Bonnin, H. Reichert, Exploiting coherence for real-time studies by single-bunch imaging. J. Synchrotron Rad. **21**, 815–818 (2014)
16. S. Zabler, P. Cloetens, J.-P. Guigay, J. Baruchel, M. Schlenker, Optimization of phase contrast imaging using hard x rays. Rev. Sci. Instrum. **76**, 073705 (2005)
17. E. Maire, P.J. Withers, Quantitative X-ray tomography. Internat, Mater. Rev. **59**, 1–43 (2014)
18. E. Maire, C. Le Bourlot, J. Adrien, A. Mortensen, R. Mokso, 20 Hz X-ray tomography during an in situ tensile test. Int. J. Fracture **200**, 3–12 (2016)
19. J. Dittmann, A. Eggert, M. Lambertus, J. Dombrowski, A. Rack, S. Zabler, Finding robust descriptive features for the characterization of the coarsening dynamics of three dimensional whey protein foams. J. Coll. Interface Sci. **467**, 148–157 (2016)
20. A. Meagher, F. García-Moreno, J. Banhart, A. Mughal, S. Hutzler, An experimental study of columnar crystals using monodisperse microbubbles. Coll. Surfaces A **473**, 55–59 (2015)
21. R. Acharya, J.A. Sharon, A. Staroselsky, Prediction of microstructure in laser powder bed fusion process. Acta Mater. **124**, 360–371 (2017)

22. S. Besner, M. Meunier, *Chapter 7: Laser precision microfabrication, in Springer Series in Materials Science*, vol. 135 (Springer, Heidelberg, 2010), pp. 163–187
23. Z. Wang, C.L. Morris, J.S. Kapustinsky, K. Kwiatkowski, S.-N. Luo, Towards hard X-ray imaging at GHz frame rate. Rev. Sci. Instrum. **83**, 10E510 (2012)
24. T.G. Etoh, A.Q. Nguyen, Y. Kamakura, K. Shimonomura, T.Y. Le, N. Mori, The theoretical highest frame rate of silicon image sensors. Sensors **17**, 483 (2017)

Part VI
Combustion

Development and Application of High-Speed Laser Visualization Techniques in Combustion Research

Marcus Aldén and Mattias Richter

Abstract In order to fulfil the requirements of available energy resources, there is a great need to obtain a sustainable and environmentally friendly energy utilization using combustion processes. In order to do this, it is of utmost importance to utilize non-intrusive diagnostic techniques with high spatial and temporal resolution which can characterize the combustion process and also validate combustion models. During the last decades different laser techniques have proven to fulfil these requirements. A special requirement in practical applications when highly turbulent flames are to be investigated, is to be able to follow the phenomena in time, i.e. it is important to develop and apply high speed laser diagnostics. In the present chapter we are describing the use of high speed lasers together with high speed detectors which make it possible to probe in two dimensions even the fastest combustion phenomena in real time. The chapter is describing the use of a so called Multi YAG laser which together with a framing camera is able to record up to eight images. Also the use of a high repetition rate laser and a high power burst laser together with CMOS cameras and their application for studies of turbulent combustion phenomena are described. The examples are mainly taken from the author's laboratory and include more academic studies of turbulent flames but also practical applications in engines.

Electronic supplementary material The online version of this chapter (doi:10.1007/978-3-319-61491-5_11) contains supplementary material, which is available to authorized users.

M. Aldén (✉) · M. Richter
Division of Combustion Physics, Lund University, Lund, Sweden
e-mail: marcus.alden@forbrf.lth.se

M. Richter
e-mail: mattias.richter@forbrf.lth.se

K. Tsuji (ed.), *The Micro-World Observed by Ultra High-Speed Cameras*,
DOI 10.1007/978-3-319-61491-5_11

1 Introduction

Thermochemical conversion processes, mainly combustion, constitute very important phenomena in our society. A quantitative description of flames and combustion phenomena is still lacking due to the extreme complexity of these thermal-mechanical-chemical phenomena involving hundreds or even thousands of elementary chemical reactions, turbulent flows, heat transfer and multi-phase interactions. Scientifically, combustion thus involves a wide range of basic and applied disciplines. The continued development of these processes requires inter-disciplinary scientific research and professional personnel with wide and detailed knowledge in various fields. Specifically, the main reasons for doing scientific research on combustion and related processes are:

- Combustion processes provide a major source of energy for our society (~>90%).
- Emissions from combustion processes constitute a major part of the air pollutants of today, >90% of NO_x and >50% of SO_x originate from combustion devices. Soot, polycyclic aromatic hydrocarbons and particulates are other major air pollutants emerging from combustion. Another environmental aspect of combustion of fossil fuels is the substantial emission of greenhouse gases, e.g. carbon dioxide, which put emphasis on combustion of new, e.g. bio-based renewable fuels.
- Knowledge in combustion science and technology is crucial for main branches of industry, e.g. the transportation industry, the heat and power industry and the manufacturing industry.
- Combustion processes have important effects on our society, for example in the areas of domestic heating, waste incineration, and prevention and extinguishing of unwanted fires.

With the outstanding scientific/technical challenges in combustion science there is a great need for development of new instrumentation tools as well as appropriate models to face and approach several research barriers e.g.:

- Turbulent combustion, with extreme temporal and spatial scales requiring adequate measurement and modeling resolution
- Combustion of renewable fuels with unknown composition and thus requiring adequate chemical modeling and new diagnostics
- Control, understanding and avoidance of combustion instabilities/oscillations, requiring for example ultra-high temporal resolution, both in modeling and experiments
- High pressure combustion; the chemistry become much more challenging when going from atmospheric pressure to tens or hundreds of bars, i.e. pressures relevant for industrial conditions
- Understanding of multi-phase conversion phenomena

In order to optimize and control combustion processes, it is of utmost importance to apply diagnostic techniques capable of measuring critical parameters. The challenges in this environment can be substantial; high temperature, high pressure, high contents of particles/droplets and a highly turbulent flow field. During the last decades the most important development for a deepened understanding of combustion processes in terms of experimental achievements, has been the introduction of various laser diagnostic techniques. The advantages with these techniques are their inherent non-intrusiveness in combination with high temporal and spatial resolution. Various techniques have been developed for measurements of species concentrations, temperatures and velocities, see e.g. [1–5] yielding possibilities for instantaneous information in two, three or even four dimensions (space + time).

Until the late nineties more or less all combustion studies yielding two-dimensional information, were made using lasers with a repetition rate of ~10–100 Hz. For this purpose different techniques, e.g. Mie scattering, planar laser-induced fluorescence (PLIF), Laser-induced incandescence (LII), were applied for measurements of particle/droplets, atoms and molecules and soot characteristics, respectively. Clearly, at these frequencies there are no way to follow the turbulent structures at the length scales prevailing in most applied combustion environments, e.g. turbulent combustion taking place in engines and gas turbines. More or less the only commercially available high speed laser at that time used for combustion diagnostics was the copper vapor laser. These lasers typically run at a repetition rates over 10 kHz and produces radiation at 511 and 578 nm. However, the practical applications have been scares due to a rather low pulse energy, ~couple of mJ, and a low efficiency when pumping dye lasers. Thus, the applicability in combustion has been limited to Mie scattering and particle image velocimetry (PIV) in practical combustion applications such as Internal Combustion (IC) engines [6, 7].

In the present chapter we will review the development and application of high speed visualization using advanced lasers/detectors for studies of combustion phenomena. Most of the examples will be given from the activities at Lund University. The first section will describe a specially designed laser/detector system for high speed visualization using a limited number of high energy pulses. In the next section application of a continuously operating pulsed high speed laser, ~10 kHz, will be described. This equipment complements the specially designed laser in the sense that it provides an endless pulse train, but it also suffers from significantly lower pulse energies. Finally, a newly available commercial laser, based on a burst system with 100–1000 pulses and still high pulse energies, will be described.

2 A Tailor Made Laser/Detector System for High Speed Visualization: A Multi-YAG Laser and Framing Cameramen Description

2.1 Instrument Description

As described above, there has been a strong need for a versatile and useful laser/detector system for studies of turbulent combustion phenomena. In order to

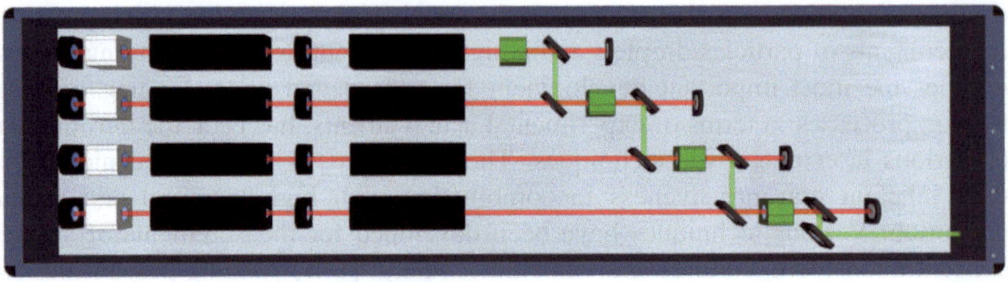

Fig. 1 Schematic overview of the Nd:YAG laser cluster, consisting of 4 Nd:YAG lasers

make high speed visualization of combustion processes a specially designed laser/detector system was developed in Lund in the late nineties. The system consists of a Nd:YAG laser cluster and a high speed camera. It is capable of producing a rapid sequence of up to 8 laser pulses with a temporal separation ranging from microseconds to milliseconds. The laser source of the high speed laser diagnostic system is a cluster consisting of 4 flash lamp pumped Nd:YAG lasers. Each laser in the cluster consists of a Q-switched Nd:YAG oscillator and a single amplifier, emitting laser pulses with a duration of 7 ns. After each oscillator there is a manual attenuator, consisting of two polarizers with a rotatable half-wave plate in between, which allows continuous adjustment of the laser pulse energy without changing the thermal load on the Nd:YAG crystals. The beams from the four lasers are combined using a scheme illustrated in Fig. 1. The laser beam from the first laser is converted from 1064 to 532 nm, using a second harmonic generation crystal. This beam is then combined with the 1064 nm beam from the second laser using a dichroic mirror, reflecting 532 nm and transmitting 1064 nm. The combined beams then pass through the second harmonic generation crystal of the second laser, the 532 nm beam passes unaffected while the 1064 nm beam is converted to 532 nm. The same procedure is then repeated two more times until all four beams have been combined. The large number of optical components in the beam path of lasers 1 and 2 introduces some energy losses and beam profile degradation. The energy losses can, however, be compensated for by increasing the flash lamp pump intensity of these lasers. It is also possible to use an alternative beam combining scheme, where beams having different polarization orientations are combined. The advantage of the latter scheme is that fewer optical components are needed, minimizing energy losses and beam profile degradation. On the other hand beam alignment is more critical and difficult to achieve.

A single fourth harmonic generation crystal, converting the 532 nm pulse train to 266 nm can optionally be inserted near the output of the unit. The four individual lasers can be fired in series with time delays, τ_1, ranging from 0 up to 100 ms, where the upper limit is given by the overall repetition frequency of the system (10 Hz). By switching the Q-switch twice during the flash-lamp discharge duration, each of the Nd:YAG lasers can be fired two times with a short separation, τ_2, between pulses. τ_2 can be adjusted from 25 to 145 μs, limited by the length of the flash lamp discharge and the gain build up time in the oscillator. The pulse timing is

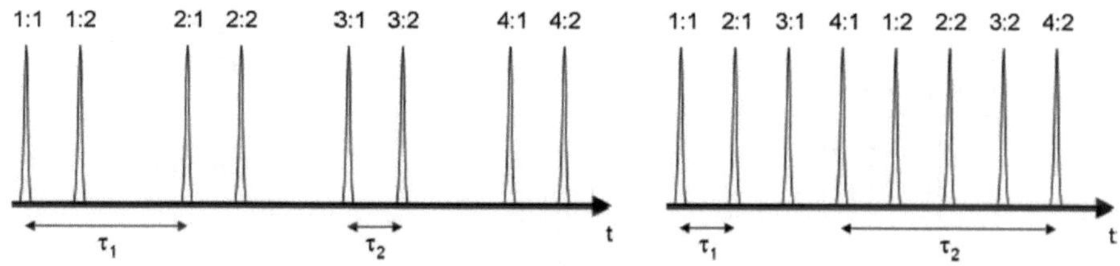

Fig. 2 Timing diagram for a pulse burst from the Nd:YAG laser cluster, *left* the double pulses from each laser are fired in series, *right* the double pulses are interleaved

Table 1 Nd:YAG laser cluster output energies

	Multi YAG energy output in each pulse (mJ)
	Single/double pulse operation
Laser energy at 532 nm	880/325
Laser energy at 355 nm	360/220
Laser energy at 266 nm	180/70

illustrated in Fig. 2. By interleaving the double pulses from the four lasers the time separation between pulses can be reduced down to 6.25 μs (=25 μs/4), as shown to the right in Fig. 2. It should be noted that the laser pulses do not have to be equally spaced as shown, as the timing of each laser is individually controllable.

The multi-YAG laser system can also generate 355 nm radiation. As 355 nm generation requires mixing 532 and 1064 nm this cannot be done using a single third-harmonic-generating (THG) crystal in a manner similar to 266 nm generation. Instead, each laser beam is individually doubled to 532 nm and tripled to 355 nm using a THG crystal. The 355 nm beam is separated from the 1064 and 532 nm beams using a dielectric mirror (coated high reflective—355 nm, high transmission—532 nm and high transmission—1064 nm). The beams at 355/532 nm from the individual lasers are then combined in a scheme similar to the 532/1064 nm combining illustrated in Fig. 1.

Maximum pulse energies of the Multi-YAG laser system are given in Table 1.

To be able to excite specific species using for example laser-induced fluorescence, it is required to convert the primary laser beams from the YAG cluster to tunable radiation. This can be made using a single dye laser, providing the dye solution is flowing in a sufficient speed to replace the liquid solution between two consecutive laser pulses [8] or to pump several dye lasers [9]. A third alternative is to use an optical parametric oscillator, OPO. Very simplistically, an OPO splits one photon into two with lower frequency. As the energy in the each of the two photons is dependent on the angle of the active crystal, the output wavelength can be scanned by turning the crystal [10].

When using the multi-YAG laser system it is clearly very important to use a high-speed camera which can match the repetition rate of the laser system. The requirement of the camera speed is thus in the MHz regime. Furthermore, in order

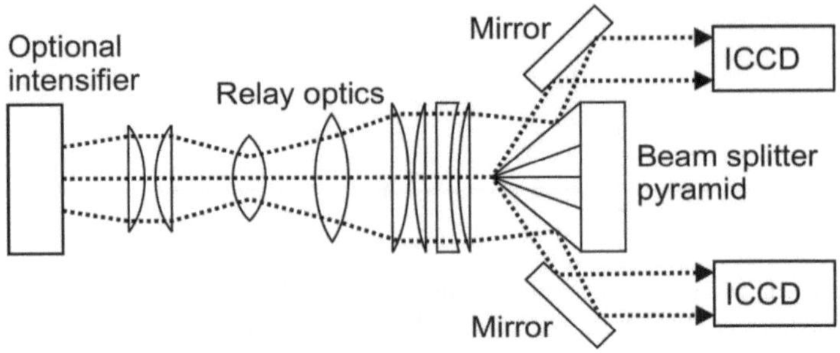

Fig. 3 A sketch of a Framing camera used in the high-speed measurements together with the Multi-YAG cluster. (Only two of the eight ICCDs around the pyramid beam splitter are shown)

to be able to measure low level signals, the efficiency of the camera is a very important parameter. In the experiments in Lund two framing cameras (Imacon 468 and SIM 8) have been used. The cameras have a similar general structure but they have a slightly different construction. Both cameras use eight individual ICCD's (intensified CCD) that can be triggered individually. This allows for frame rates up to 100 MHz but limits the number of frames to eight per event. When using the Framing cameras a single optical input lens is used so that all eight ICCDs see and image the same measurement object. In order to both amplify the signal and to convert the incoming light to be optimum for the optics and intensifiers in the camera, an optional intensifier can be placed after the initial lens. For the SIM 8 (Specialised Imaging, Tring, United Kingdom) camera, two intensifiers with varying phosphor decay time can be used: one with P46 phosphor (300 ns decay time) and one with FS phosphor (12 μs decay time). These decay times determine the minimum time separation between exposures if ghosting is to be avoided. A sketch of a framing camera Imacon 468 (Hadland Photonics, Hertfordshire, United Kingdom) is shown in Fig. 3.

2.2 *Examples from Multi-YAG Systems*

The first demonstration using the system described above was in experiments performed in a non-premixed methane/air flame consisting of two concentric tubes [11]. Air was passed through the inner tube and fuel was passed through the outer tube. The hydroxyl radical, OH, was excited using the $Q_1(8)$ transition near 282 nm in the $v'' = 0$, $v' = 1$ band of the $X^2\prod^+ \rightarrow A\ ^2\sum$system. Subsequent fluorescence in the (1, 1) and (0, 0) bands near 309 nm was observed using appropriate optical filters. The laser beams were formed into a sheet 5 cm in height and about 80 μm in width in the interaction region. The laser beam profiles were monitored on line by simultaneously imaging fluorescence from a water solution containing fluorescing dye. Each image was normalized by this profile.

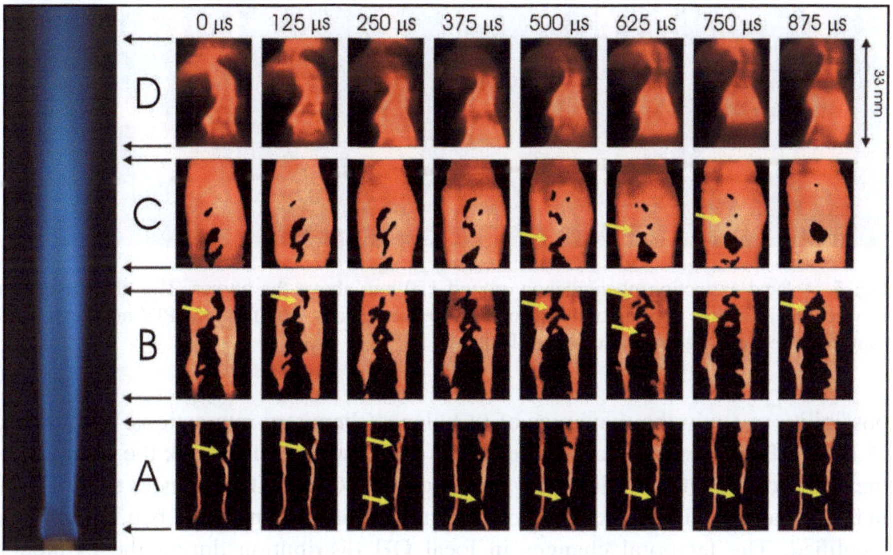

Fig. 4 Temporal PLIF sequences of OH at different heights in the turbulent, non-premixed CH_4/air-flame: A direct emission photograph of the flame is included to the *left*. Each PLIF-image shown corresponds to a single laser shot, time increases from *left to right* in 125 μs steps. The four measurement regions *A*, *B*, *C*, *D* correspond to the indicated positions in the flame [11], with permission of the Springer

Figure 4 shows typical results obtained from the flame. In the left part a direct emission photograph of the flame is included to identify the regions imaged by the OH-PLIF displayed to the right. Each temporal OH sequence (labelled A to D in the figure) corresponds to an imaged height of 33 mm or about 7 inner nozzle diameters. The time separation between successive images was 125 μs translating into an overall repetition rate of 8 kHz. In region A of Fig. 4, close to the exit nozzle where flow is still laminar, the typical 'braids' of OH can be observed marking regions where fuel and oxidizer diffuse to, and mix in, the reaction zone. Local flame extinction is clearly seen in this sequence (highlighted by arrows). Most likely this is caused by convection building up strong local gradients which effect losses by molecular transport. This leads to a slowing down of chemical reactions to the point where the flame extinguishes. The sequence clearly highlights the capabilities of the technique.

Local extinction phenomena have a major influence on flame stability and a better understanding of the mechanisms leading to extinction is both of fundamental and practical importance. To study the importance of large scale structures for local flame extinction two-dimensional measurements of relevant quantities with high time resolution are desirable. Simultaneous measurements of both flow and flame quantities, allowing correlations between the two to be studied, are also useful in this respect. Simultaneous time resolved measurements of the evolution of the OH radical distribution and the instantaneous velocity field in turbulent flames were presented in Ref. [12]. The combined measurement technique thus allowed the

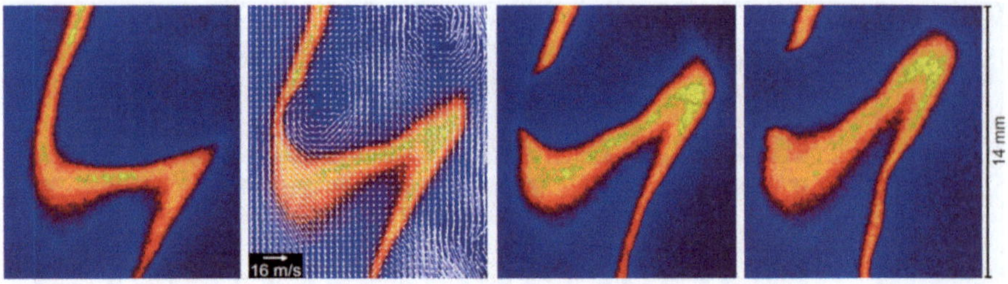

Fig. 5 A local extinction phenomenon around 150 mm above the burner. The time separation between the images is 125 μs. In the second image the velocity field from the PIV measurement is shown in the same figure as the OH-PLIF image [12]

possibility to study the dynamics of turbulence/chemistry interactions. High speed PLIF of OH at repetition rates exceeding 13 kHz was used to track the response of the flame front to the turbulent flow field. From the OH-PLIF sequences the position, in both space and time, where the local extinction of the flame front begins, could be identified. The temporal changes in local OH distribution during the extinction process could also be studied and the time scale of extinction events estimated. Instantaneous velocity field measurements were performed simultaneously using PIV. In the velocity field the flow structures associated with the identified extinction events were visualized. A result from these experiments is shown in Fig. 5. In the first image a flame front is captured. The imaged region corresponds to 14 × 16 mm and the bottom part is positioned 144 mm over the burner nozzle.

In the second image a vortex is seen, approaching the flame front and forcing it to bulge out slightly, a decrease in the OH concentration can also be seen at the top of the bulge. In the next frame the flame has just extinguished as indicated by the sharpness of the two tips of the now separated flame fronts. The large-scale wrinkling of the flame front in this downstream part of the flame is clearly seen. Three relatively large regions with high and uniformly directed velocities act to twist the already s-shaped flame front even further.

Clearly, the possibility to achieve detailed time resolved information of turbulent combustion phenomena is of instrumental importance for a deepened understanding of these phenomena. In a joint collaboration with University of Stuttgart and NEC Europe, the growth of spark ignited flame kernels in turbulent premixed gases was studied [13]. The growth of the flame kernel starts with an initially laminar kernel moving in the turbulent flow field, it subsequently grows with increasing degrees of wrinkling; and finally develops into a fully turbulent flame. A cylindrical constant-volume combustion cell, equipped with four high speed rotors, was used to study the turbulent flame kernel propagation. The cell was filled with CH_4/air or H_2/air mixtures, which were ignited by a spark in the center of the cell. The ignition system was designed to ensure highly reproducible ignition events. Controlled and variable amounts of turbulence could be generated in the mixture by the four rotors. The flow field in the center was investigated using a Laser Doppler Velocimetry system and the turbulence field was found to be homogeneous and isotropic.

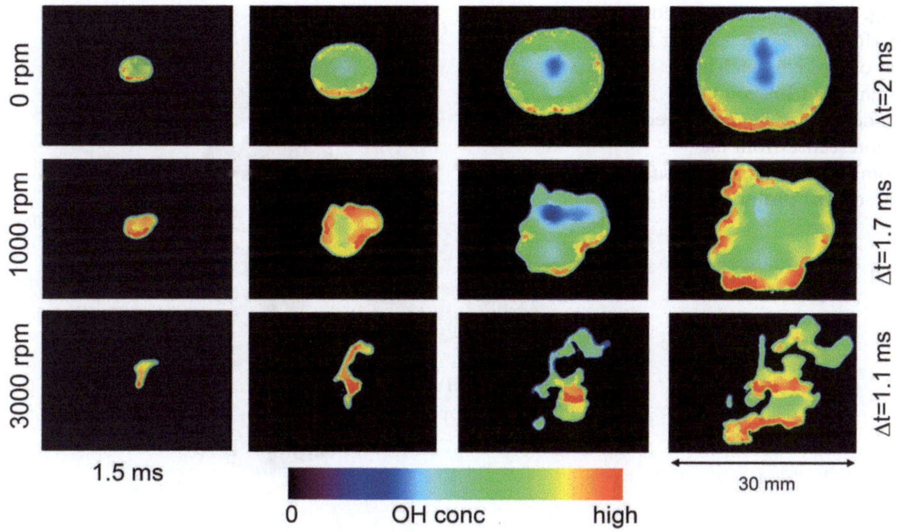

Fig. 6 *Top* OH laminar flame propagation in a stoichiometric CH_4/air mixture with the time separation between images are 2 ms. *Middle* OH-PLIF sequence of turbulent flame propagation (rotor speed = 1000 rpm) in a stoichiometric CH_4/air mixture with a time separation of 1.7 ms. *Bottom* OH-PLIF sequence of turbulent flame propagation (rotor speed = 3000 rpm) in a stoichiometric CH_4/air mixture. Time separation between images is 1.1 ms [13], with the permission of Elsevier

OH-PLIF sequences, consisting of 4 images separated by 250 μs to 2 ms, were recorded, which show a cut through the flame kernel. Parameters like fuel, stoichiometry and turbulence intensity were varied, and a large data set was recorded. In Fig. 6, the interaction between turbulence and chemical reactions is seen in action, as the flame is wrinkled by turbulent eddies. This leads to an increased reaction surface area which increases the overall reaction rate, as indicated by the shorter time between images compared to the laminar flame in the top row. When the rotor speed was increased further (3000 rpm) the flame propagation speed increases as well as the flame front wrinkling, and isolated structures appear in the PLIF images, see Fig. 6, bottom row.

In addition to the more fundamental issues of turbulent combustion phenomena, high speed laser visualization techniques add considerable sharp tools for studying applied combustion process, e.g. taking place in an internal combustion engines. The fuel/air mixing and combustion processes within the cylinder of an IC engine do to a large extent determine the engine efficiency and emissions. Simulations of the two processes can be performed using computational fluid dynamics (CFD), but the current precision of this tool makes in-cylinder diagnostics necessary in order to fully understand fuel/air mixing and combustion. By using the laser/detector system described above cycle-resolved engine measurements can be studied, where single combustion events can be followed in time. Cycle-to-cycle variations of complex events can thus be studied, which is impossible from single shot images captured in

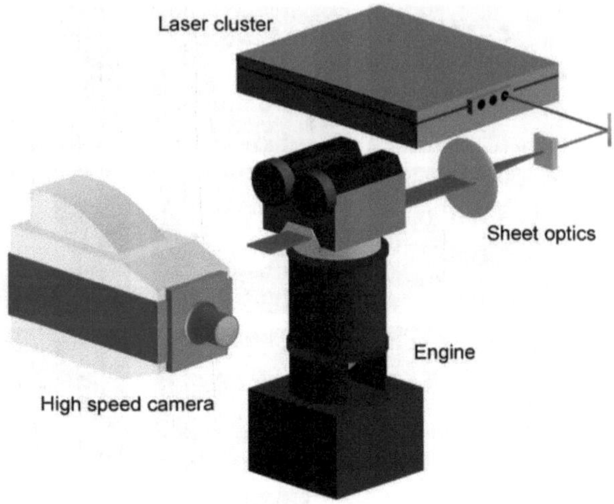

Fig. 7 Schematic setup for performing PLIF imaging in an optical engine. The high-speed Imacon 468 camera is arranged to detect fluorescence from the combustion chamber through the optical piston

different cycles. As individual structures can be followed in time, the large-scale development of a fuel injection, an ignition event, or a flame propagation can be visualized. Time scales of such phenomena can be estimated from the time resolved data.

In Fig. 7, a typical set-up for high speed visualization in an engine with optical access is shown. Isooctane was used as fuel, and 10% of 3-pentanone was added as fluorescent tracer [8]. In experiments in a spark ignited engine, the early flame propagation was captured by rapidly sequenced fuel tracer PLIF. An example of fuel tracer PLIF is shown in Fig. 8 (top) where eight consecutive images recorded in the same combustion cycle are presented [8]. The combustion chamber is viewed from above with the spark plug located to the upper-right, see photo to the left in Fig. 8 (top). The intensity in the PLIF images is proportional to fuel concentration, which means that bright regions correspond to unburned gases and dark regions to burnt gases. The turbulent flame propagation is illustrated by the expansion of the burnt regions through the sequence. The time separation between consecutive images is 100 μs, which here corresponds to 0.7 crank angle degree (CAD). The first image is acquired at 5 CAD after top dead center (ATDC). The appearance of isolated flame islands in the sequence is an effect of three-dimensional flame wrinkling, in and out of the two-dimensional laser sheet.

In order to compare different engine concepts also a homogenous-compression charge-ignition (HCCI) engine was investigated [8]. In an HCCI engine the premixed fuel/air mixture auto-ignites by compression. In Fig. 8 (bottom) a sequence covering the major part of a combustion event in the HCCI engine is shown, covering 6.5 CAD. The fuel distribution is viewed from below through the transparent piston. In the first image (1 CAD ATDC) a homogeneous fuel distribution is seen. In the next image several small dark regions can be found to the left, to the right and in the upper part of the image, indicating that the reactions consuming fuel have started. These chemical reactions are initiated by the temperature rise

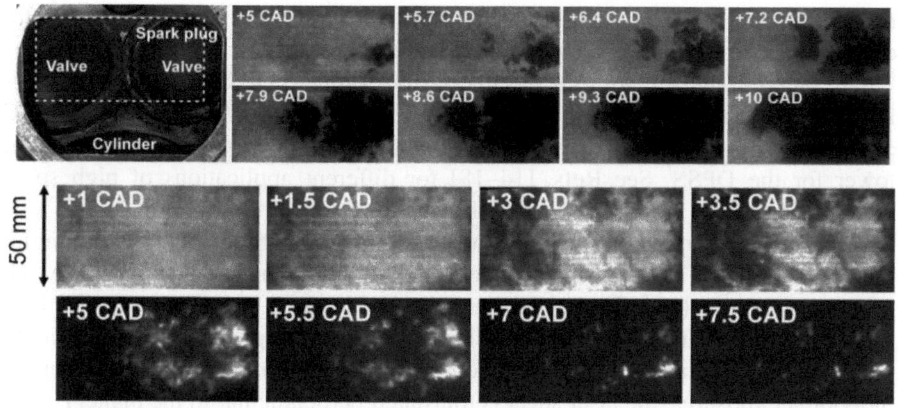

Fig. 8 *Top* Cycle resolved fuel tracer PLIF sequence recorded in a spark ignited (SI) engine. The time separation between consecutive images corresponds to 100 μs (0.72 CAD). The imaged region was 50 × 25 mm^2, and is shown to the *left*. *Bottom* Fuel tracer PLIF sequence recorded in an HCCI engine. The fuel distribution in a plane in the center of the combustion chamber was imaged through the piston, the imaged region corresponds to 95 × 50 mm^2 [8], with permission of the Optical Society

associated with the compression of the gas. The fuel in these regions is then consumed further in the next images, the regions spread and new regions of fuel consumption also appear. The HCCI combustion is rapid, in 3.5 crank angle degrees most of the fuel is consumed, and, in the end (7.5 CAD ATDC) only small separated islands of fuel remain. The use of unequal time spacing between images throughout the series allows a long event to be captured, while short term changes are still captured between image pairs.

3 Continuous KHz Repetition Rate Laser Applications

During the last decade, Diode Pumped Solid State (DPSS) lasers have gradually become more powerful and can nowadays be a viable alternative in high repetition rate two-dimensional (2D) imaging in reacting flows. Instead of employing flash lamps as pump source, as found in most 10-Hz systems and laser clusters such as the one described earlier, the gain medium is here pumped by radiation from a diode array. Compared to gas discharge lamps, the spectral profile of the emission is more narrow for the diodes. This results in a more efficient pumping, with less waste heat generated. The main advantage with the DPSS laser systems is their ability to operate continuously at relatively high (~20 kHz) repetition rates. This feature brings several distinct benefits to the field of combustion diagnostics. One is the ability to follow a relatively slow transition with overlaid rapid changes. Another advantage is the possibility to utilize the post-triggering capability of modern CMOS detectors. This can be used to capture rare events such as flame extinction or re-ignition processes where accurate pre-triggering is very hard.

The main drawback with the commercially available DPSS-systems is the limited pulse energy compared to the conventional 10 Hz systems or the Multi-YAG cluster mentioned earlier. Typically, the pulse energy is two orders of magnitude lower for the DPSS. See Refs. [14–18] for different applications of high speed lasers/detectors for combustion applications.

One further example of an application where the continuous high repetition rate output is utilized is in temporally resolved three-dimensions imaging, i.e. 4D imaging, of fluids [19]. The technique is based on rapid scanning of a thin laser sheet across the volume of interest. Each exposure of the signal generated within the laser sheet represents a 2D distribution of the species of interest. Between the individual exposures, the laser sheet is translated perpendicular to the plane. Hence, the spatial position of the probe volume is slightly altered. Through interpolation the 2D data in different cross sections can be combined to show 3D information. It is important to perform each sweep across the investigated volume fast enough to freeze the fluid motion. If there is a small but still significant motion of the fluid during the scan there are measures that can be applied to compensate for this in post-processing. More details on the practical implementation can be found in [19]. In Fig. 9, a schematic setup for performing 4D imaging of OH radicals in a Bunsen flame is shown. The DPSS YAG-laser (Edgewave HD40IV-E, Edgewave GmbH, Würselen, Germany) is operated close to its maximum repletion rate of 20 kHz. A high repetition rate dye laser (Sirah Credo, Sirah Lasertechnik GmbH, Grevenbroich, Germany) is used to convert the 532 nm radiation from the YAG to 609 nm. After frequency doubling and fine tuning of the wavelength output,

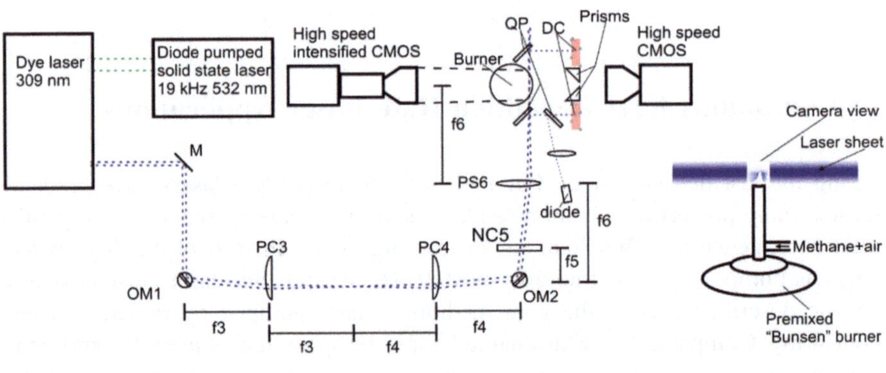

Fig. 9 Schematic setup for acquisition of 4D data through application of the rapid scanning technique. Note the use of dual oscillating mirrors to make the sheet separation more equidistant than what can be achieved with a single oscillating mirror. *M* Mirror, *f* focal length, *DC* Dye Quvette, *QP* Quartz plate, *OM* Oscillating mirror, *PC* Pos. Cyl. Lens, *NC* Neg. Cyl. lens, *PS* Pos. Spherical lens reprinted from R. Wellander, Exp. Fluids 55 (2014), with permission of Springer

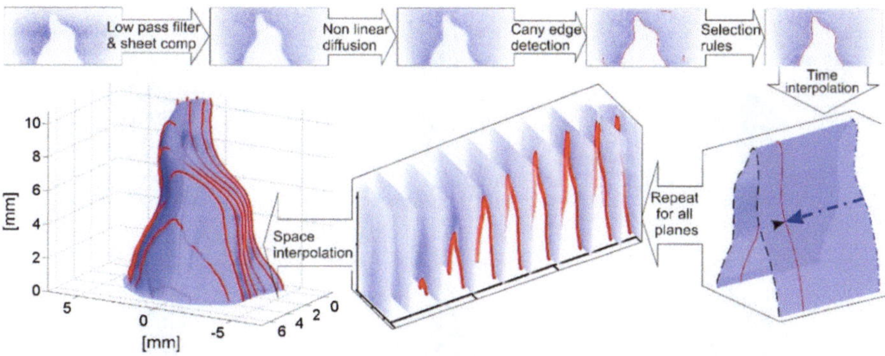

Fig. 10 Illustration of the image processing steps employed when creating 3D images from the 2D images recorded in a rapid sequence in different spatial positions across the probed volume reprinted from R. Wellander, Exp. Fluids 55 (2014), with permission of Springer

309 nm radiation matching a transition in the OH radical, is generated. The translation of the laser sheet in between exposures is achieved through implementation of two oscillating mirrors. In principle the deflection of the beam could be realized with one oscillating mirror. The sheet displacement would then be oscillating around the center point in a sinusoidal manner. However, by using two mirrors positioned according to the sketch and forcing the second mirror to oscillate at three times the frequency of the first mirror, the displacement will approach a triangular pattern. For more details on the mirror system please refer to [19, 20].

In conventional 2D PLIF imaging as described earlier, often the weak fluorescence signals are maximized through use of low f-number optics. The resulting restriction in the depth of field is then not an issue since the imaged volume, that is defined by the laser sheet, is generally very thin. However, in 4D imaging performed with a scanning light sheet the required depth of field is governed by the extension of the scanned volume. In order to achieve a sufficient depth of field the f-number must here most often be increased. This is obviously a limiting factor and the resulting tradeoff between signal-to-noise ratio and depth of field must be considered when probing species with weak fluorescence signals.

The image processing steps applied to the originally recorded 2D images and the following interpolation to achieve the 3D images are illustrated in Fig. 10. With this set-up it is possible to generate 3D images, reconstructed from ten 2D images, at a repetition rate of 2 kHz. In Fig. 11, a series of OH iso-concentration surfaces, recorded with a Photron SA-Z, are illustrated.

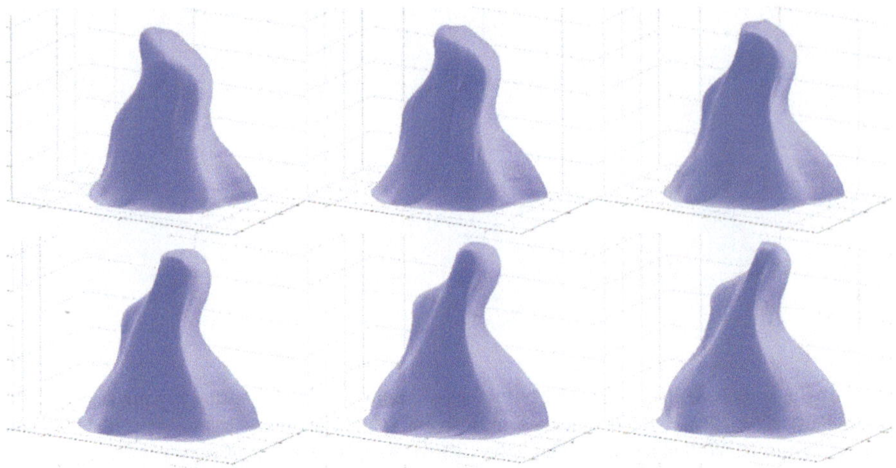

Fig. 11 Example of temporally resolved 3D iso-concentration surfaces of OH radicals in a laboratory flame. The 3D images are recorded at a repetition rate of 2 kHz reprinted from R. Wellander, Exp. Fluids 55 (2014), with permission of Springer. The corresponding movie is available online

4 High Power Burst Laser Application

Burst-mode lasers for combustion diagnostics were introduced in the late 1990s [21, 22]. These systems are based on sequential amplification of a high repetition rate seed laser. Until recently, burst mode lasers were quite rare and not commercially available. Fortunately, this has now changed and in this section results from the application of a state-of-the-art ultra-high speed burst-laser (QuasiModo, Spectral Energies LLC, Dayton, Ohio, USA) will be shown. This laser is a hybrid-pumped burst-mode laser with two diode-pumped amplifiers followed by six flash-lamp pumped amplifiers. Two 6 mm flash-lamp pumped amplifiers serve as a double-pass for the laser beam in order to increase the gain. The beam is then further amplified by passing through four additional flash-lamp pumped amplifiers, see Fig. 12.

The duration of each burst is adjustable up to 10 ms. Also the repetition rate is adjustable from 10 to several hundred kHz. There is, of course, a tradeoff between pulse energies and rep rate. As an example, if the laser is operated at 10 kHz with a burst duration of 10 ms, the resulting pulse train will consist of 100 pulses. In terms of individual pulse energies, peak values for the available wavelengths will be approximately >1200 mJ at 1064 nm, >600 mJ at 532, >300 at 355 nm, >70 mJ at 266 nm. These numbers are typically about 50 times higher than the corresponding pulse energies from a continuously operating DPSS YAG. On the detector side these systems rely on conventional CMOS technology.

Figure 13 illustrates the experimental setup for simultaneous dual-species visualization of a premixed turbulent jet flame by using ultra-high speed PLIF imaging (50–100 kHz). The burner is a hybrid porous-plug/jet type burner,

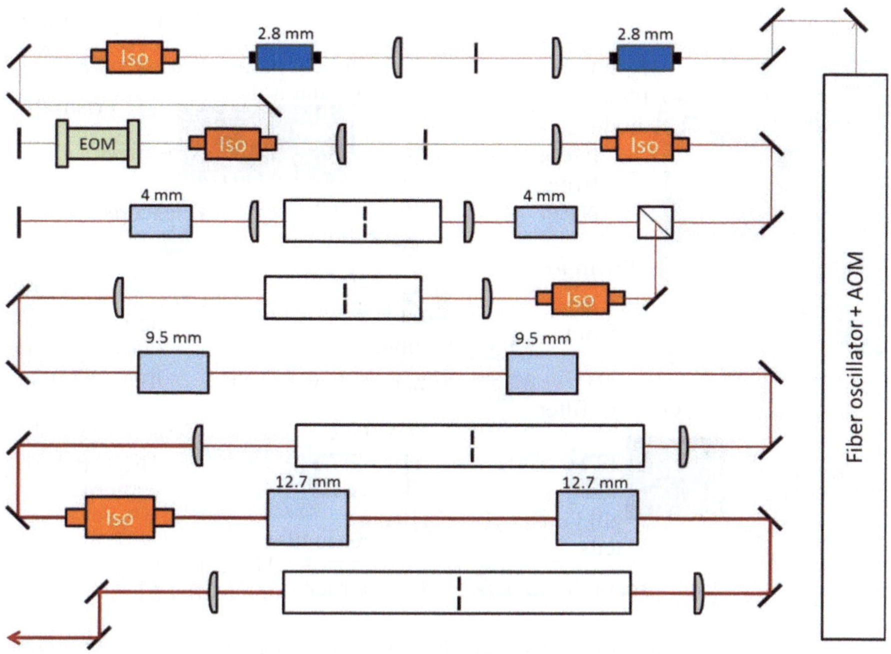

Fig. 12 Layout of the Quasimodo Burst-laser. The output from a cw fiber laser is shopped by an acoustic-optical modulator. The resulting high-rep rate pulse train is then sequentially amplified in the YAG-rods (labeled 2.8, 4, 9.5, 12.7 mm)

described in more detail in [23]. Its main parts are a porous sintered stainless steel plug with a diameter of 61 mm and a 1.5 mm-diameter nozzle in the center for the creation of a jet flame. Premixed CH_4 and air mixture was fed through the center nozzle to create the jet flame which was stabilized by a premixed CH_4-air flame above the porous plug referred to as co-flow. In the examples presented here the jet velocity was set to 66 m/s.

The OPO was pumped by 355 nm radiation from the burst system and produced excitation radiation for OH near 284 nm. The excitation of formaldehyde (CH_2O) can conveniently be performed at 355 nm. Thus, the residual 355 nm remaining after pumping the OPO was used for this purpose. The two beams were then formed into thin sheets and overlapped in the center of the jet-flame. The gate time, i.e. exposure time, of both high-speed intensifiers was set to 100 ns. At 100 ns exposure time, all the generated fluorescence signals were captured while the background signals caused by the flame chemiluminescence were sufficiently suppressed.

Experimental results generated with this setup are presented in Figs. 14 and 15. The burst system is capable of providing excitation energies comparable or even exceeding the ones from Multi-YAG lasers but for a significantly longer pulse train. This facilitates the following of interesting phenomena such as variations in

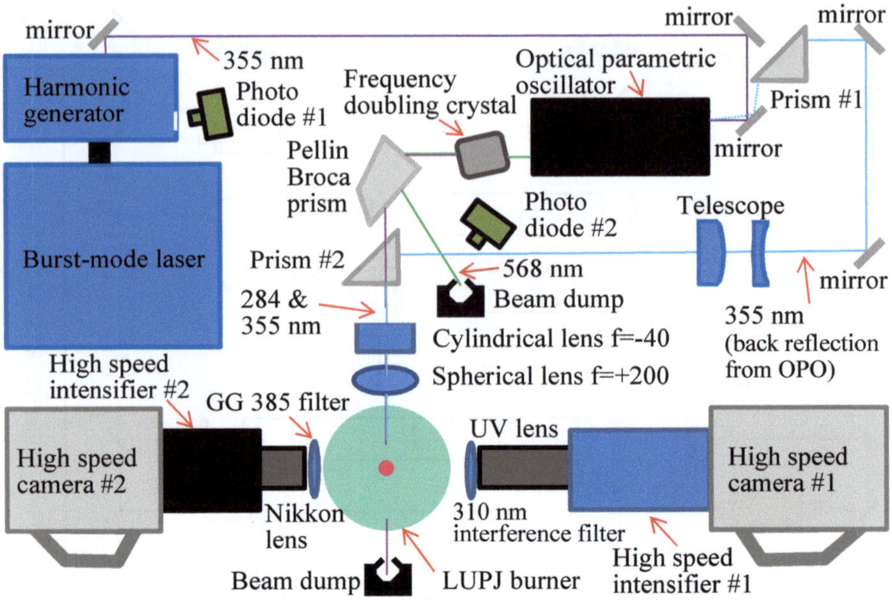

Fig. 13 Experimental setup for simultaneous PLIF imaging of OH and CH_2O at 50 kHz in a turbulent jet-flame. Two high-speed cameras in combination with two high-speed intensifiers are mounted opposite to each other in order to capture OH and CH_2O fluorescence simultaneously. For detecting the former species, a CMOS Photron Fastcam SA-Z (camera #1) was used in combination with a high-speed intensified relay optics. For the CH_2O imaging, a CMOS Photron Fastcam SA-X2 (camera #2) was employed which was also combined with a high-speed image intensifier (Lambert HS) (Courtesy of Zhenkan Wang)

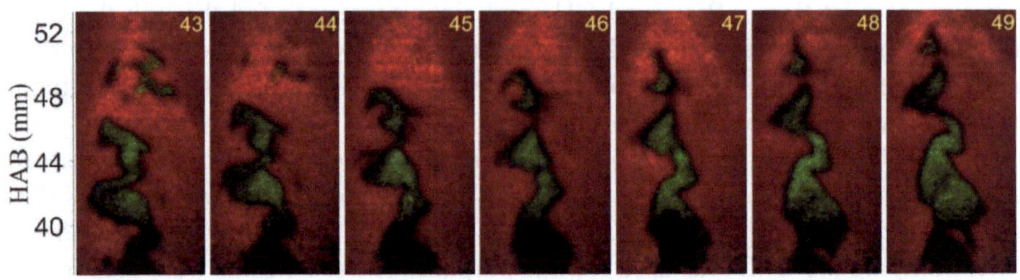

Fig. 14 Simultaneous PLIF recordings of OH (*red*) and CH_2O (*green*) performed at 50 kHz in a turbulent flame (Courtesy of Zhenkan Wang)

turbulent length scales, extinction and re-ignition over longer periods of time. In Fig. 14, the repetition rate was set to 50 kHz. To further demonstrate the capability of this unique system the repetition rate was increased to 100 kHz for the formaldehyde distribution sequence presented in Fig. 15. As indicated by the still sufficient signal-to-noise ratio, the frame rate could be increased further. The laser and detector allow for up 500 kHz with reduced pixel number.

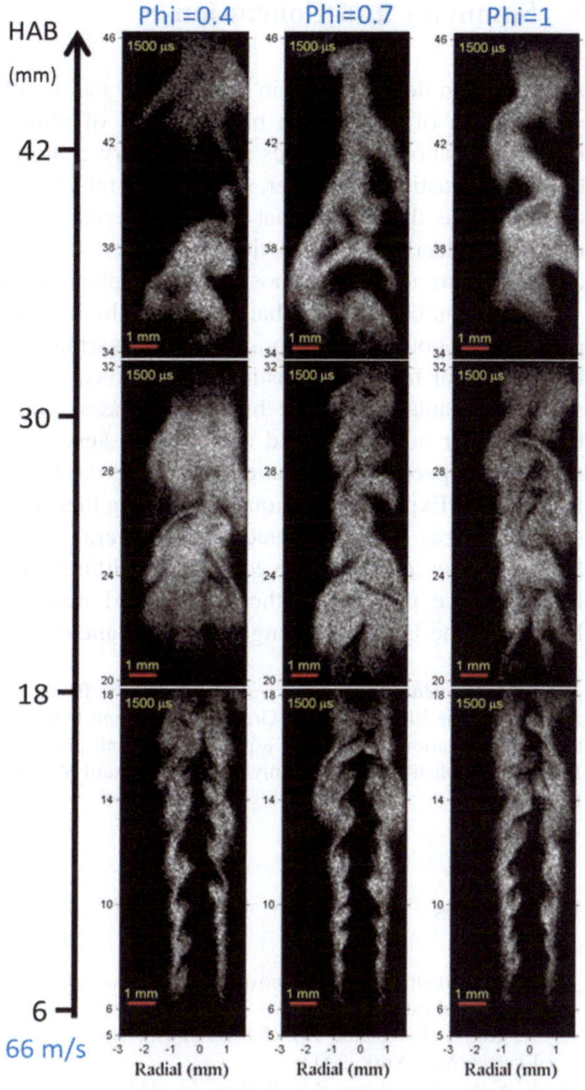

Fig. 15 Formaldehyde distributions in a turbulent jet-flame recorded at different heights above the burner for different air-fuel ratios. The burst laser excitation source and the intensified CMOS camera were operated at 100 kHz (Courtesy of Zhenkan Wang)

Another possibility with the burst system is to perform PIV measurements at extreme speeds. Since the pulse train is originally generated from a continuous wave laser, the pulse separation can be adjusted arbitrary. For high repetition rate PIV measurements the pulses can be arranged in pairs with short separation between the two pulses in each pair and longer separation between the pairs. The energies available indicate that PIV at 100 kHz should be fully feasible.

5 Summary and Conclusion

As have been demonstrated in this Chapter the use of high speed laser imaging for visualization of combustion parameters is of utmost importance in gaining new information about the physics and chemistry controlling phenomena related to turbulent combustion. Moreover, since industrial applications more or less always rely on turbulence, the benefits that can be achieved using these new diagnostic tools are multifold. Clearly, the interaction with expertise in computational fluid dynamics is also crucial to address the most important phenomena for model validation.

It can thus be foreseen that the use of high speed visualization of combustion phenomena, both in the laboratory and in various industrial applications, will be a major tool for future combustion applications.

A disadvantage with the high-speed measurements as described above is the high cost for both lasers and detectors. A very new and promising technique for some high speed/3D applications is to use the Frequency Recognition Algorithm for Multiple Exposure technique [24]. Using this approach simultaneous imaging of several images with one detector and several structured laser sheets, originating from one laser, can be achieved [24]. The different images are then separated in the Fourier space providing either high speed imaging with a framing speed only limited by the laser pulse length or instantaneous 3D imaging.

Acknowledgements The authors acknowledge the financial support from the Swedish Energy Agency and the ERC Advanced Grants TUCLA and DALDECS. The authors also acknowledge the work by various colleagues, especially past and present PhD students working in the field; Johan Hult, Johan Sjöholm, Jimmy Olofsson, Rikard Wellander, Zhenkan Wang and Panagiota Stamatoglou.

References

1. A.C. Eckbreth, *Laser Diagnostics for Combustion Temperature and Species* (Gordon and Breach Publisher, 1996)
2. K. Kohse-Höinghaus, J.B. Jeffries (eds.), *Applied Combustion Diagnostics* (Taylor and Francis, New York, 2002)
3. K. Kohse-Höinghaus, R.S. Barlow, M. Aldén, J. Wolfrum, Combustion at the focus: laser diagnostics and control. P. Combust. Inst. **30**, 89–123 (2005)
4. M. Aldén, J. Bood, Z. Li, M. Richter, Visualization and understanding of combustion processes using spatially and temporally resolved laser diagnostic techniques. P. Combust. Inst. **33**, 69–97 (2011)
5. M. Aldén, A. Omrane, M. Richter, G. Särner, Thermographic phosphors for thermometry: a survey of combustion applications. Prog. Energ. Combust. **37**, 422–461 (2011)
6. I. Duwel, M.C. Drake, T.D. Fansler, High-speed, high-resolution imaging of multihole fuel sprays in a firing spray-guided direct-injection gasoline engine. Paper presented at ILASS-Europe 19th annual conference, Nottingham, England, 2004
7. M. Reeves, D.P. Towers, B. Tavender, C.H. Buckberry, A high-speed all digital technique for cycle-resolved 2-D flow measurement and flow visualization within SI engine cylinders. Opt. Laser. Eng. **31**, 247–261 (1999)

8. J. Hult, M. Richter, J. Nygren, M. Alden, A. Hultqvist, M. Christensen, B. Johansson, Application of a high-repetition-rate laser diagnostic system for single-cycle-resolved imaging in internal combustion engines. Appl. Optics **41**, 5002–5014 (2002)
9. J. Olofsson, *Laser Diagnostic Techniques with Ultra-High Repetition Rate for Studies in Combustion Environments* (Lund University, Dissertaton, 2007)
10. J. Sjöholm, E. Kristensson, M. Richter, M. Alden, G. Goritz, K. Knebel, Ultra high speed pumping of an OPO laser, for high speed laser-induced fluorescence measurements. Meas. Sci. Technol. **20**, 025306 (2009)
11. C.F. Kaminski, J. Hult, M. Aldén, High repetition rate planar laser induced fluorescence of OH in a turbulent non-premixed flame. Appl. Phys. B **68**, 757–760 (1999)
12. J. Hult, G. Josefsson, M. Aldén, C.F. Kaminski, Flame front tracking and simultaneous flow field visualisation in turbulent combustion, in *Proceedings of the 10th International Symposium on Applications of Laser Techniques to Fluid Mechanics*, Instituto Superior Técnico, Lisbon, 2000, Paper No. 26-2
13. C.F. Kaminski, J. Hult, M. Aldén, S. Lindenmaier, A. Dreizler, U. Maas, M. Baum, Spark ignition of turbulent methane/air mixtures revealed by time-resolved planar laser-induced fluorescence and direct numerical simulations. P. Combust. Inst. **28**, 399–405 (2000)
14. C. Heeger, B. Böhm, S.F. Ahmed, R. Gordon, I. Boxx, W. Meier, A. Dreizler, E. Mastorakos, Statistics of relative and absolute velocities of turbulent non-premixed edge flames following spark ignition. P. Combust. Inst. **32**, 2957–2964 (2009)
15. B. Peterson, V. Sick, Simultaneous flow field and fuel concentration imaging at 4.8 kHz in an operating engine. Appl. Phys. B **97**, 887–895 (2009)
16. N. Soulopoulos, J. Kerl, T. Sponfelder, F. Beyrau, Y. Hardalupas, A.M.K.P. Taylor, J.C. Vassilicos, Turbulent premixed flames on fractal-grid-generated turbulence. Fluid Dyn. Res. **45**, 061404 (2013)
17. I. Boxx, M. Stöhr, C. Carter, W. Meier, Sustained multi-kHz flame front and 3-component velocity-field measurements for the study of turbulent flames. Appl. Phys. B **95**, 23–29 (2009)
18. M. Juddoo, A.R. Masri, High-speed OH-PLIF imaging of extinction and re-ignition in non-premixed flames with various levels of oxidation. Combust. Flame **158**, 902–914 (2011)
19. R. Wellander, M. Richter, M. Aldén, Time-resolved (kHz) 3D imaging of OH PLIF in a flame. Exp. Fluids **55**, 1–12 (2014)
20. R. Wellander, M. Richter, M. Aldén, Time resolved, 3D imaging (4D) of two phase ow at a repetition rate of 1 kHz. Opt. Express **19**, 21508–21514 (2011)
21. W.R. Lempert, P.F. Wu, B. Zhang, R.B. Miles, J.L. Lowrance, V.J. Mastocola, W.F. Kosonocky, *Pulse Burst Laser System for High Speed Flow Diagnostics*, AIAA-96-0179 (1996)
22. W.R. Lempert, P.F. Wu, B. Zhang, and R.B. Miles, *Filtered Rayleigh Scattering Measurements Using a MHz Rate Pulse-Burst Laser System*, AIAA-97-0500 (1997)
23. B. Zhou, C. Brackmann, Z. Li, M. Aldén, X.-S. Bai, Simultaneous multi-species and temperature visualization of premixed flames in the distributed reaction zone regime. P. Combust. Inst. **35**, 1409–1416 (2014)
24. E. Kristensson, Z. Li, E. Berrocal, M. Richter, M. Aldén, *Instantaneous 3D imaging of flame species by means of coded laser illumination* (P. Combust, Inst, 2016). in press

Visualization of Combustion Processes of Internal Combustion Engines

Nobuyuki Kawahara

Abstract Very fast visualization of fuel injection and spray impingement on the wall, the laser ignition process, and auto-ignition of the end-gas region due to flame propagation during the engine knock cycle was demonstrated using an ultra-high-speed camera for a detailed understanding of the combustion process inside an internal combustion (IC) engine. The combustion process of an IC engine is complex, involving several thermal fluid phenomena. The high-speed video camera was an essential tool for visualizing and recording these high-speed phenomena. When more magnified images are needed, the ultra-high-speed camera and a long-distance microscope with a Barlow lens were used. Here, high-speed visualization techniques were shown to understand the combustion process inside an IC engine.

1 Introduction

It is important to improve the thermal efficiency and reduce exhaust emissions of internal combustion (IC) engines, to save energy resources and reduce problems associated with exhausting nitrogen oxide, unburned hydrocarbons, carbon monoxide, particulate matter, and carbon dioxide [1, 2]. Thus, decreasing the exhaust emissions and fuel consumption of IC engines simultaneously are important matters for improving the global environment. Reducing pumping and heat losses can improve the fuel consumption of IC engines. Additionally, lean-burn combustion systems have been developed, e.g., spark-ignition (SI) engines operating with a stratified lean mixture at low loads and direct-injection spark-ignition

Electronic supplementary material The online version of this chapter (doi:10.1007/978-3-319-61491-5_12) contains supplementary material, which is available to authorized users.

N. Kawahara (✉)
Department of Energy Systems Engineering, Okayama University,
3-1-1 Tsushima-Naka, Kita, Okayama 700-8530, Japan
e-mail: kawahara@okayama-u.ac.jp

K. Tsuji (ed.), *The Micro-World Observed by Ultra High-Speed Cameras*,
DOI 10.1007/978-3-319-61491-5_12

(DISI) engines operating unthrottled under ultra-lean conditions by stratifying the charge and preparing a fuel-rich mixture around the spark plug [3].

The combustion process of an IC engine is complex, involving several thermal fluid phenomena. A reciprocating engine operates with a four-stroke cycle: intake, compression, expansion (power), and exhaust. During the intake stroke, a fresh mixture is drawn into the cylinder. During the compression stroke, the mixture inside the cylinder is compressed to a small fraction of its initial volume. Before 'top dead center', combustion is initiated from a spark in a SI engine or by direct fuel injection in a diesel engine. During the power or expansion stroke, high-temperature high-pressure gases push the piston down creating power from the IC engine. An engine may run at 1000 rpm or more; thus, each stroke is of millisecond order or shorter duration. As an example, combustion is initiated from the electrical spark discharge produced between the spark plug electrodes close to the end of the compression stroke in a gasoline SI engine. A high-temperature plasma kernel forms from the spark discharge for a microsecond and energy transfer from the high-temperature plasma to the combustible mixture occurs around the spark plug for a millisecond. The initial flame kernel created by the spark discharge becomes a propagating flame.

A detailed understanding of the combustion process is required in an IC engine, including fast phenomena such as fuel injection with its atomization process, ignition, flame propagation, and end-gas auto-ignition during a knocking cycle, to advance the performance of these engines and further reduce exhaust emissions. Many researchers have attempted to visualize fast, complex phenomena in IC engines [4]. For example, an optically accessible engine with optical windows for the top of the piston or windows in the cylinder head is an effective way to view the dynamics inside the cylinder. Moreover, studying the fuel injection and SI processes in a constant volume vessel also provides information on fast phenomena without piston motion.

Here, very fast visualization of fuel injection and spray impingement on the wall, the laser ignition process, and auto-ignition of the end-gas region due to flame propagation during the engine knock cycle was captured using a high-speed video camera or an ultra-high-speed camera.

2 Visualization Technique for Combustion Process of Internal Combustion Engine

The combustion process should be visualized under higher temperature and higher pressure conditions. There are two kinds of methods to visualize the combustion process of an IC engine. The first method is the constant volume vessel or open-air experiments with a device, such as the injector, the spark plug and so on. The second one is optical access through the sapphire or the quartz window to visualize the inside combustion chamber. There are several methods to access the inside the combustion chamber, optical access through the cylinder head, the piston and the use of an endoscope [4].

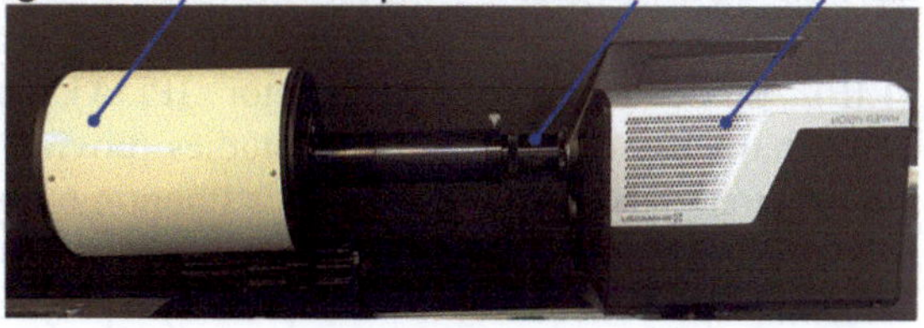

Fig. 1 Ultra-high-speed microscopic recording settings using high-speed video camera with long distance microscope

An ultra-high-speed video camera (maximum speed: 10 million frames/s (Mfps); Shimadzu Hyper Vision HPV-X2) was used to visualize the fuel break-up process, spray impingement on the wall, and auto-ignition of the end-gas region due to flame propagation during the engine knock cycle. When more magnified images are needed, the ultra-high-speed camera and a long-distance microscope with a Barlow lens could be used (Fig. 1). The high-speed video camera was an essential tool for visualizing and recording these high-speed phenomena. The maximum frame rate of the camera was 10 Mfps, allowing maximum continuous recording of 256 images. Additionally, when the frame rate was under 2 Mfps, it was possible to change the exposure time in 10-ns units. The frame size of this camera was 400 × 250 pixels. It is quite important to select the lens system with higher spatial resolution. For example, if the droplet diameter and the velocity are required to estimate the Weber number, it is crucial to be able to visualize the droplets clearly at an appropriate magnification. As we knew the threshold for droplet detection to be reached when the droplet diameter occupied eight or more pixels of the image, the level of magnification greatly influenced the diameter measurement. Droplets with a diameter of less than eight pixels in the image were excluded from image processing. The long-distance microscope used was a Maksutov–Cassegrain catadioptric microscope, consisting of a convex front lens, a rear concave mirror, and a second mirror. The microscope was originally designed and manufactured for high-speed visualization of spray impingement and atomization processes. Usually, high-speed visualization requires strong illumination due to the short exposure time. To address this problem, a photographic flash light source (Metz mecablitz 44 AF-4i) was used. Backlighting from a strong light source made illumination on enabled the spray impingement process in the near wall-region to be illuminated.

3 Observation of Liquid Break-Up Process and Spray Impingement on the Wall with Transient Spray

The gasoline DISI engine has great potential to achieve higher thermal efficiency and lower exhaust emissions. Fuel injection systems for DISI engines provide both late injection for stratified charge combustion at part load, as well as early injection during the intake stroke for homogeneous charge combustion at full load. To understand the spray characteristics resulting from a multi-hole injector, investigations of gasoline direct injection sprays using several measurement techniques, such as the laser sheet method with laser-induced (exciplex) fluorescence (LIF), phase Doppler anemometor (PDA), and particle image velocimetry (PIV) have been reported. These studies provide detailed information about spray tip penetration, spray corn angle, distribution of the liquid/vapor phase, and droplet velocity and diameter due to secondary break-up processes. However, a key process affecting spray behavior is the primary spray break-up, because it defines the starting conditions for the spray distribution, the evaporation process, and mixture formation. Experimental investigations of the atomization process have been limited, because these are very-high-speed processes that occur in a very small region of space. Although scale-up models have been used to study the primary spray structure, it has not been possible to match Reynolds, Weber, cavitation numbers, and time scales simultaneously with a practical high-pressure multi-hole injector. Novel microscopic and very fast investigations of the primary spray structure of a practical multi-hole injector are required.

In this section, experimental investigations of the atomization process and the fuel spray impingement on the wall of a practical multi-hole injector were made using a high-speed video camera. The initial state and development of the spray were investigated under variable injection pressure conditions. During the injection period, the primary spray angle was measured. Fluctuations in the surface waves at the spray boundary were assessed. Moreover, the fuel spray impingement on the wall was visualized.

The gasoline engine multi-hole injector used in this study was a prototype, fabricated specifically for research purpose. In this research, to focus on the primary spray structure of one spray plume, a two-hole injector was used. This two-hole injector had a spray angle of 90°. Safety gasoline (dry solvent, density: 770 kg/m^3, surface tension: 2.45×10^{-2} N/m) was used as the fuel instead of gasoline. The injector had a needle inwardly opening. The diameter of the nozzle exit, ϕ was 0.21 mm. A backlighting method was used to allow detection of the primary spray plume close to the nozzle exit. The development of one of the spray plumes was obtained using the video camera at a 45° incline. Two types of lighting methods were used: a scattering light method and a backlighting method. To obtain whole images of the spray plume, the scattering light method with a metal-halide lamp was used. The backlighting method was used to focus on the primary spray image close to the injection hole. Typical captured images using these lighting methods are shown in Fig. 2 (left image: scattering light image, right image: backlighting image). The image obtained using the scattering method indicates the spray plume in a white color. In this case, the captured image was $\sim 55 \times 66$ mm at a 32 kfps camera speed (exposure time: 8 μs) to obtain the entire image of one spray plume. In contrast, the spray plume close to the nozzle exit was colored black using the backlighting method. The obtained image was scaled at 10×12 mm using a macro lens at 1 Mfps camera speed (exposure time: 0.25 μs). The primary spray structure is a fast phenomenon at >150 m/s spray velocity at an injection pressure of 10 MPa. To obtain the primary spray structure, a fast camera speed is needed. However, a fast camera speed corresponds to a short exposure time, such as

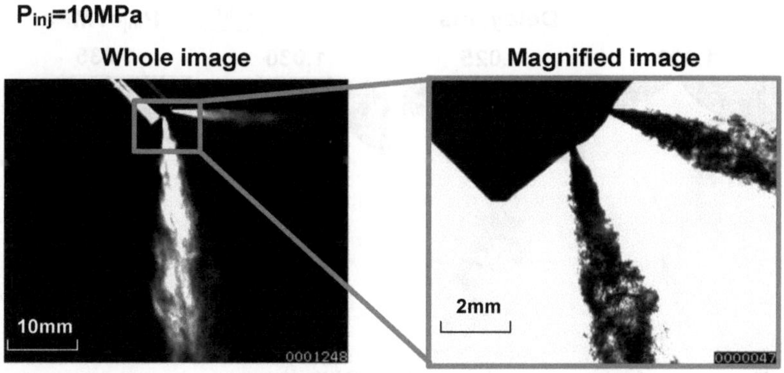

Fig. 2 Effects of lighting method, scattering light (*left*) or back-lighting (*right*), for visualization of initial spray plume

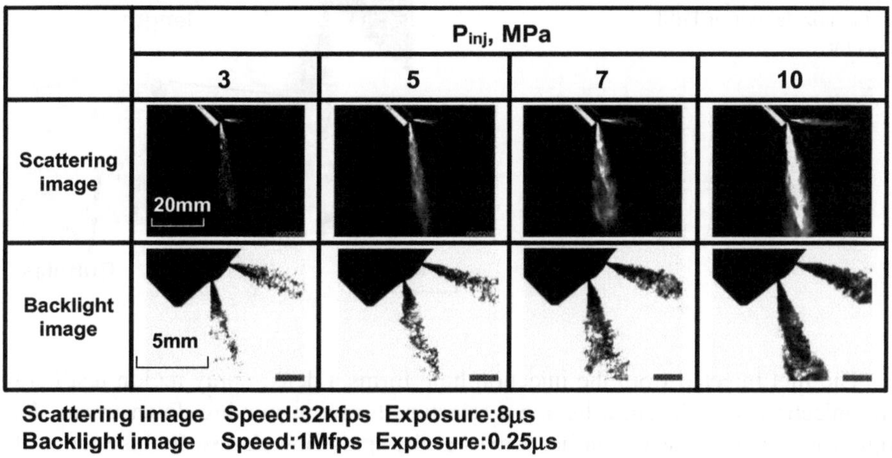

Fig. 3 Effects of injection pressure on spray image injected from DISI injector

0.25 μs; thus, a strong, bright light source is needed. Using the backlighting method, the detailed structure of the primary spray plume can be obtained.

Figure 3 indicates the effects of injection pressure on entire spray images and the primary spray, near the nozzle exit. Several injection pressures were examined: 3, 5, 7, and 10 MPa. Injection durations under each pressure condition were determined to obtain the same total mass of fuel. It can be seen that the volume of one spray plume increased with higher injection pressure. A higher injection pressure causes a larger mass of spray plume during a given exposure time. In this study, time series of spray images were obtained using the high-speed video camera; 100 images could be obtained. The time series of primary spray images under 7 MPa of injection pressure and 1 Mfps camera speed (exposure time: 0.25 μs) is shown in Fig. 4.

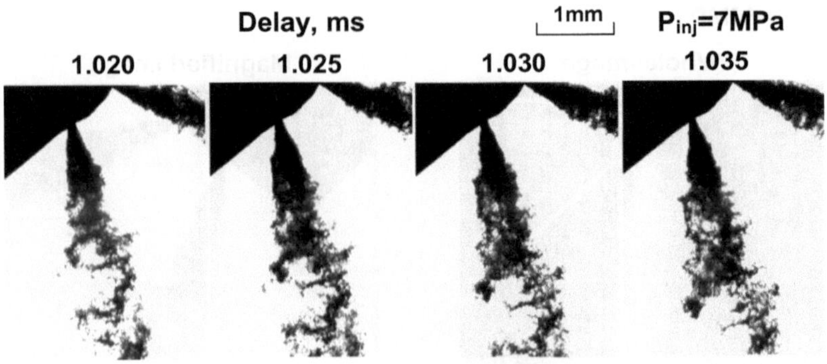

Fig. 4 Time-series of primary spray injected from DISI injector

Fig. 5 Definition of spray core of initial spray very close to the nozzle exit of DISI injector

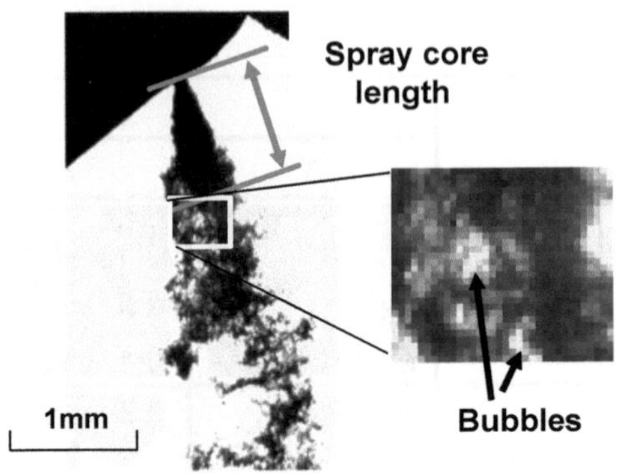

The fuel injected from the injection hole forms a dense spray region just below the injection hole. Then, a ligament structure forms in moving downstream. This ligament structure causes surface waves at the spray boundaries. In this section the dense spray region is referred to as the "spray core" (Fig. 5). The length of the spray core region is defined as the distance from the injection hole to the bubble generation point inside the dense spray core. Fuel is injected at very high velocity, and the spray core exhibits the Kelvin–Helmholtz instability, caused by the large slip velocity between the spray core and the ambient air. It is generally accepted that such aerodynamic instability causes the spray core to break up into ligaments. Here, the fluctuation in the length of the spray core was assessed using a time series of primary spray images. Figure 6 indicates the fluctuation in the length of the spray core as a function of time under 10-MPa injection pressure conditions. It can be seen that the fluctuation in the length of the spray core showed a saw-shaped fluctuation. The longer length of the spray core was shortened suddenly, and then the spray core extended in the downstream direction.

Fuel impingement behaviors were observed under atmospheric conditions. As an injector, a six-hole DI injector was used. Fuel injection pressure was varied from 1 to 13 MPa. One of the six sprays impinged on the wall at 50°. Frosted glass was used as the impinging wall in order to consider the effects of surface roughness on the fuel impingement process. Figure 7 shows time series images of deposit and splash with a fuel injection pressure of 5 MPa; both impinging behaviors, deposit and splash, can be seen clearly. The upper images show the deposit. The fuel droplet sticks and spreads on the wall after impingement. The lower images show splash. Several smaller droplets come out after droplet impingement on the wall. Thus, fuel spray impingement on the wall can be visualized using an ultra-high-speed camera with a long-distance microscope.

Direct microscopic images obtained with the ultra-high-speed video camera and a long-distance microscope were used to investigate the primary spray structure close to the injection hole of a practical high-pressure multi-hole injector used in a DISI engine and fuel spray impingement on the wall. Using the ultra-high-speed camera, fuel break-up of the primary spray and spray impingement on the wall can be visualized with high temporal resolution. The results here demonstrate that the direct microscopic imaging technique can provide valuable dynamic information about atomization and fuel impingement processes.

Fig. 6 Fluctuation of spray core length of initial spray injected from DISI injector

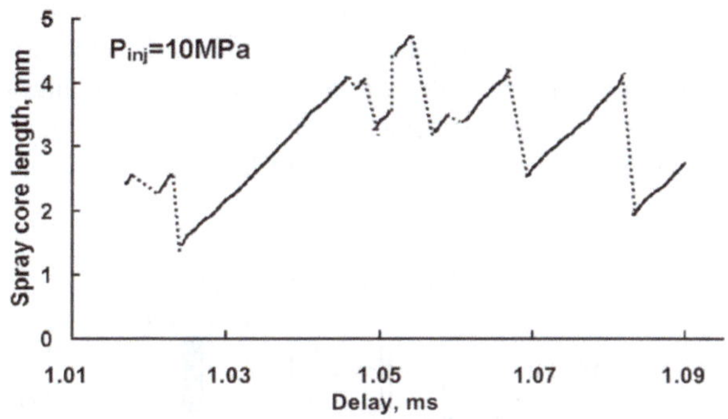

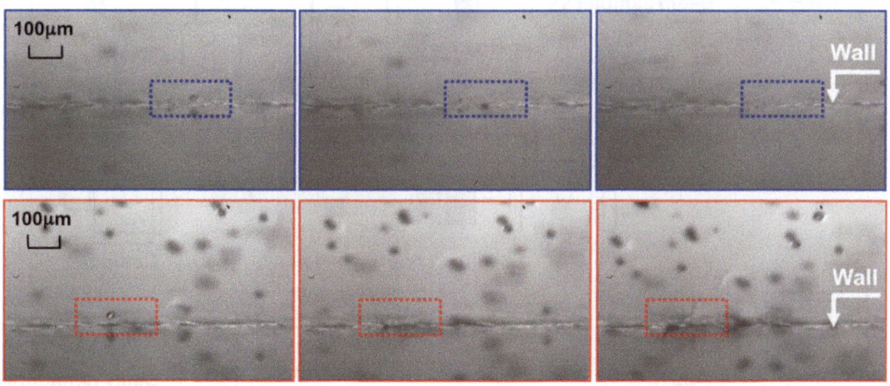

Fig. 7 Example of droplet 'deposit' and 'splash' (*Upper* Deposit, *Lower* Splash) in the spray impingement on the piston surface

4 Laser-Induced Plasma Generation in a Transient Spray

Ignition characteristics fundamentally influence the combustion process in SI engines, especially in DISI engines. The position of the spark plug can affect combustion characteristics and mixture formation in DISI engines. Additionally, heat loss occurs from the spark plug to the cylinder head of an SI engine. To overcome these drawbacks, research has focused on laser-induced plasma ignition [5, 6]. Laser-ignition devices do not affect the fuel spray, and variable ignition positions can be selected by adjusting the optical settings. These advantages make laser ignition appropriate for ignition systems using a fuel spray.

In this section, the behavior of laser-induced plasma and fuel spray was investigated by visualizing images with an ultra-high-speed camera. Isooctane and ethanol were used as fuels. Time series images of laser-induced plasma in a transient spray were visualized using a high-speed color camera. The effects of the shockwave generated from laser-induced plasma on evaporated spray behavior were assessed.

Figure 8 shows the experimental apparatus [7]. A two-hole fuel injector was set in a constant-volume vessel, with four optical windows. One of the two fuel sprays was injected vertically downwards into the vessel. Isooctane and ethanol were used as fuels. Laser-induced plasma was generated in the injected fuel spray by focusing a nanosecond pulse of the second harmonic (wavelength λ = 532 nm) of a Q-switched Nd:YAG laser. Laser incident energy was controlled by a half-wavelength plate and a polarizing beam splitter, and the laser beam was focused with a convex lens of

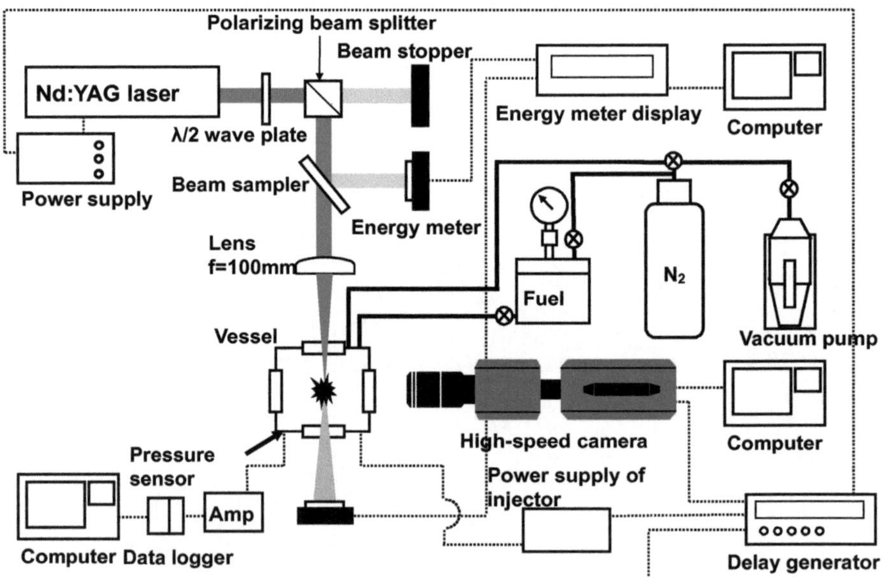

Fig. 8 Experimental apparatus for the visualization of laser-induced plasma in a transient spray

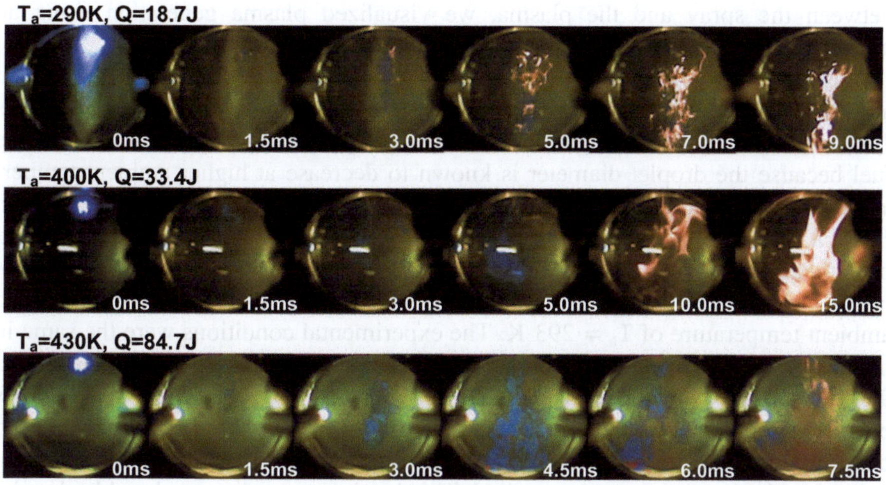

Fig. 9 Time-series of laser ignited spray flame image under each ambient temperature

100-mm focal length. The state of the plasma was visualized with an ultra-high-speed color camera, which can record images up to 1.25 Mfps (minimum exposure time of 100 ns). The injection pressure, P_i, was varied to observe the effects of flow velocity and droplet diameter on plasma and fuel formation. The ambient gas pressure, P_a, was 0.1 MPa in all experiments, and the ambient temperature was controlled using an electric heater that surrounded the vessel.

Figure 9 shows time series images of a laser-ignited flame of isooctane spray under various ambient temperature conditions. These images were used to investigate the combustion behavior of laser-ignited fuel spray at ambient temperatures T_a of 290, 400, and 430 K. The incident energy was constant at E_{in} = 90 mJ. The value Q, above the image, is the heat production, estimated from the pressure history inside the vessel. At T_a = 290 K, the flame was small and luminous. At T_a = 400 K, the flame size and heat release were larger than those at T_a = 290 K, but the flame was still luminous. At T_a = 430 K, the heat production was larger than that at T_a = 400 K, and the flame became blue. These results showed that the ambient temperature was an important parameter affecting combustion. Temperature influences the air–fuel mixture formation; it is useful to determine, whether there is a combustible mixture at each focal position.

Plasma in a fuel spray containing liquid and gas phases differs from plasma containing only a gas phase. Thus, it is important to investigate the laser-induced plasma formation process within a fuel spray. Displacement of the plasma position is a problem, because the ignition position is important for spray combustion. Additionally, it was observed that the shockwave generated from laser-induced plasma dispersed the fuel droplets. Thus, it is necessary to consider the presence of fuel droplets when selecting the ignition point. To understand the interaction between the spray and the plasma, we visualized plasma generation using an ultra-high-speed camera.

Laser-induced plasma generation within a fuel spray was recorded with an ultra-high-speed camera. The frame speed was 1 Mfps and the exposure time was 100 or 200 ns. Ethanol, which has a relatively low boiling point, was used as the fuel because the droplet diameter is known to decrease at higher ambient temperatures due to evaporation. The images show an area around the plasma generation (10.0 × 8.7 mm). The injection pressure, P_i, was set at $P_i = 1$ and 7 MPa to investigate the effects of flow velocity and droplet diameter on the formation of laser-induced plasma. Figure 10 shows plasma images at $P_i = 1$ MPa and an ambient temperature of $T_a = 293$ K. The experimental conditions were the same in images #1 and #2. The laser beam was applied from the right side of the image, and the focal point of the convex lens was along the dashed line. Figure 10 (#1) shows plasma generated around the focal point of the lens. This is similar to laser-induced plasma generation in the gas phase. The shockwave was detected in the image at 4 μs after the laser energy was applied. The droplets were dispersed and broken by the shockwave. In contrast, the images in #2 (Fig. 10) captured discrete regions of plasma, even though #1 and #2 were under the same conditions. Each region of the plasma emitted a shockwave and caused droplet dispersion, differing from the images shown in #1. Compared with #1, the plasma decay was faster when the plasma was generated at the locations shown in #2. If the same laser energy was applied, the incident energy was partitioned when the plasma generation was discrete, as seen in #2. As a result, the energy consumption per area of plasma was lower than that in #1, causing faster decay of laser-induced plasma.

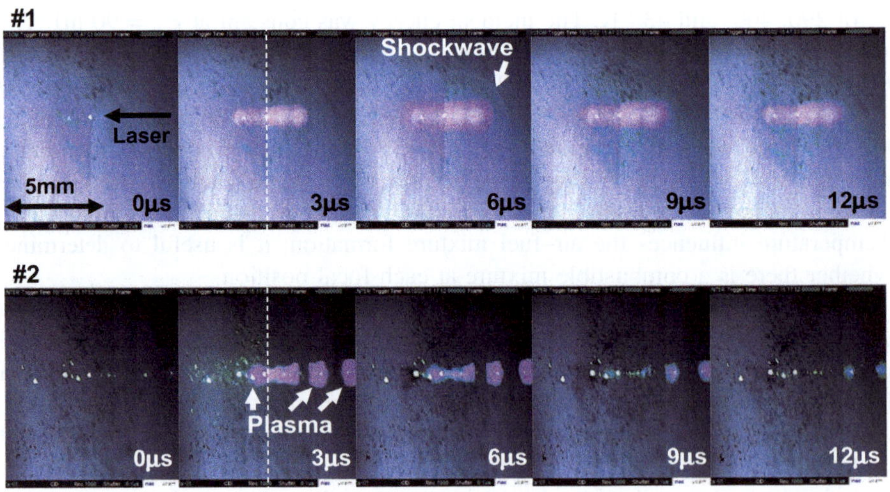

Fig. 10 Time evolution of laser-induced plasma in fuel spray at $P_i = 1$ MPa and $T_a = 293$ K

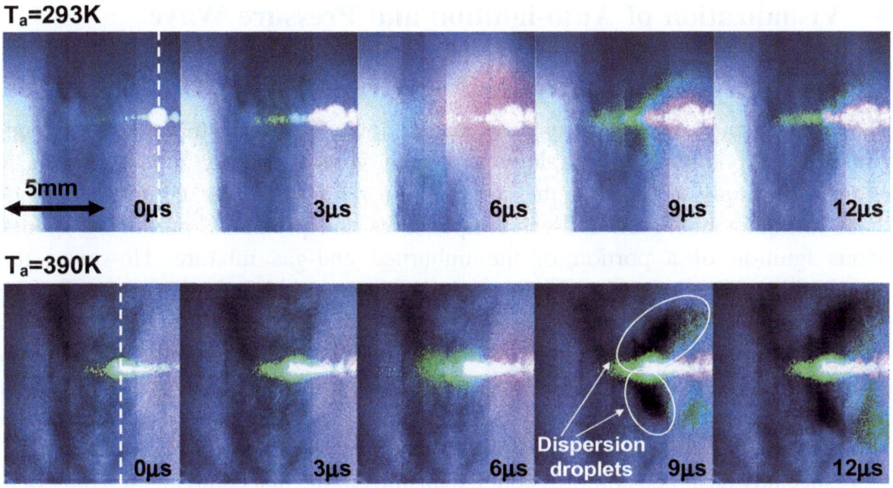

Fig. 11 Time evolution of laser-induced plasma in fuel spray at P_i = 7 MPa, T_a = 293 K and 390 K

Figure 11 shows plasma images at P_i = 7 MPa and T_a = 293 and 390 K. Comparing the images in Fig. 11 with those in Fig. 10 at T_a = 293 K shows that the droplet size decreased with increasing injection pressure and ambient temperature. In Fig. 11, the point of plasma generation shifted to the laser side, compared with Fig. 10. The plasma position shifted because the spray angle expanded with increasing injection pressure, and the contact points of the laser and the droplets moved towards the laser side. In Fig. 11, the plasma was generated at points similar to those in Fig. 10, but the dispersion was greater than that in Fig. 10. When the ambient temperature was increased to 390 K, dispersion of the fuel spray became significant. Because the ambient temperature (T_a = 390 K) was higher than the boiling point of ethanol (351 K), the fuel evaporated after injection and the diameter and weight of the droplets decreased. As a result, the droplets became sensitive to the shockwave produced by the laser-induced plasma.

The behaviors of laser-induced plasma and fuel spray were shown by visualizing images with an ultra-high-speed camera. The effects of the shockwave from the laser-induced plasma on evaporated spray behavior were assessed. Interaction between the laser-induced plasma (and the shockwave) and the fuel spray can be observed using ultra-high-speed visualization. Laser-induced plasma was generated at the surface of larger droplets under poor atomization conditions. The fuel droplets in the spray were dispersed by the shockwave generated from the laser-induced plasma under improved atomization conditions. Thus, ultra-high-speed visualization is a powerful tool for investigating ignition processes in a laser-ignited spray flame.

5 Visualization of Auto-ignition and Pressure Wave During Knocking in a Spark-Ignition Engine

The requirements of future SI engines include lower fuel consumption and reduced emission of pollutants. A strategy to obtain greater thermal efficiencies in SI engines is simply to increase the compression ratio. A greater compression ratio leads to higher unburned end-gas temperatures and pressures, producing spontaneous ignition of a portion of the unburned end-gas mixture. However, this auto-ignition of end-gas also produces shockwaves and high-frequency oscillations in in-cylinder pressure, which causes engine knocking. Knocking combustion is a major limitation on the thermal efficiency of SI engines. In this section, the auto-ignition of the end gas, due to flame propagation and intense oscillations from the resulting pressure waves, was visualized using high-speed cameras in a hydrogen SI engine to understand the initiation of engine knocking.

A specially designed compression–expansion engine that could be fired only once was used for these experiments [8, 9]. The engine had a bore and stroke of 78 and 85 mm, respectively, along with a pancake combustion chamber and a compression ratio of 9.0:1. The engine was operated at 600 rpm. The cylinder and mixture tank were charged initially with a homogeneous H_2 + 0.5 O_2 + 1.9Ar mixture to obtain a higher unburned gas temperature at the spark timing. The equivalence ratio (mass ratio of fuel to air), ϕ, was 1.0. When nitrogen was used instead of argon, knocking could not be obtained due to the lower compression temperature. The engine was driven by an electric motor while the valve was kept open. After a given time, the valve was closed, when the piston was located at the bottom dead center (BDC). The gas was then compressed and ignited by an electric spark at a selected crank angle, θ_{IT}. The spark timing was changed from 330° to 370° to obtain several knock intensities, ranging from normal combustion to severe knocking conditions.

The compression–expansion engine provided optical access via a quartz window; the window was set in the cylinder head beside the cylinder wall to obtain knocking images as the end gas was compressed by the propagating flame. The diameter of the visualized image was 32 mm. The position of the quartz window was confirmed by changing the position of the spark electrode to visualize the auto-ignited kernel clearly during knocking combustion. By gating the camera so that it was in synchronization with the engine, we were able to acquire images at specific crank angles.

Engine knocking results from the spontaneous ignition of a portion of the end-gas mixture ahead of the propagating flame. A hydrogen–air mixture has a fast burning velocity, and thus the end-gas mixture is compressed rapidly by a regular propagating flame. In the process leading to the generation of strong pressure oscillations, the pressure wave induced by the initial auto-ignition propagates towards the regular flame front.

The high-speed color camera was used to obtain knocking images as the end gas was compressed by the propagating flame. Figure 12 shows time series images at a specific crank angle for both normal and knocking engine cycles, together with the

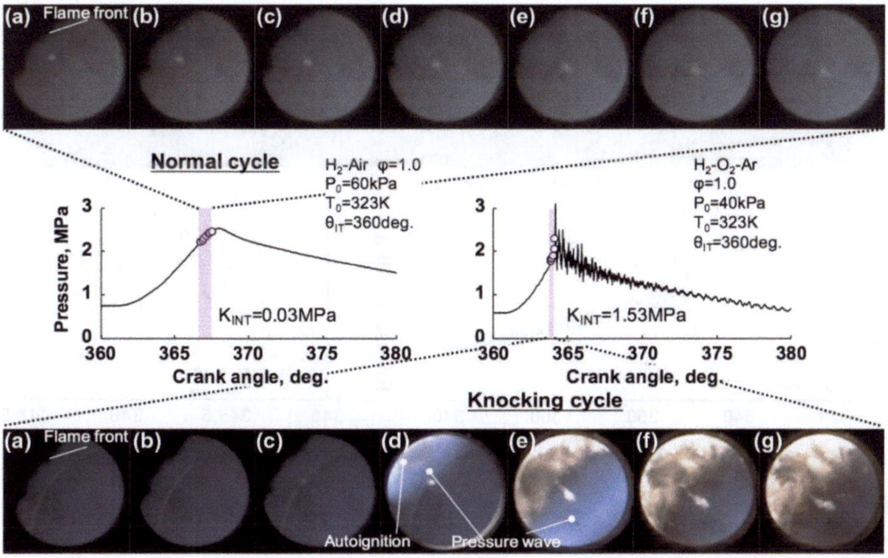

Fig. 12 Time series of high-speed direct images and related in-cylinder pressure histories for both normal and knocking engine cycles [8], with permission of IOP Publishing. The corresponding movie is available online

in-cylinder pressure histories. These images were obtained with a camera speed of 60 kfps (16.7 μs between frames). The upper images of the figure correspond to the normal engine cycle (H_2 + air mixture, equivalence ratio $\phi = 1.0$, initial pressure $P_0 = 60$ kPa, spark timing $\theta_{IT} = 360°$), whereas the lower images correspond to a knocking engine cycle ($H_2 + O_2 + Ar$ mixture, equivalence ratio $\phi = 1.0$, $P_0 = 40$ kPa, $\theta_{IT} = 360°$). From the appearance of the spark, a premixed flame propagated from the lower right side to the upper left side of the engine (images a–g) in the normal cycle. The flame front showed a wrinkled flame, and no auto-ignited kernels occurred in the end-gas region. For the knocking cycle, auto-ignition of the end gas appeared at the upper left corner of the engine in image D. This had grown in size in image E. Pressure wave movement from the upper left to the lower right can be visualized with the bluish-white high intensity in images D and E. A pressure wave passed through the visualization area in image F. The kernel at the upper left corner of the engine touched the propagating regular flame front in image E, indicating a higher luminous flame intensity. It seems strange to have a luminous flame in a hydrogen propagating flame due to the absent carbon content of the fuel. Strong auto-ignition and pressure waves caused the thermal boundary layer near the cylinder wall and the cylinder and piston head to erode. One should note that a contribution to luminescence may also be due to the combustion of lubricating oil grease.

After auto-ignition of the end-gas mixture, strong oscillations of the in-cylinder pressure occurred. Figure 13 shows the pressure wave movement, starting from the auto-ignition of the end-gas mixture ($H_2 + O_2 + Ar$ mixture, equivalence ratio

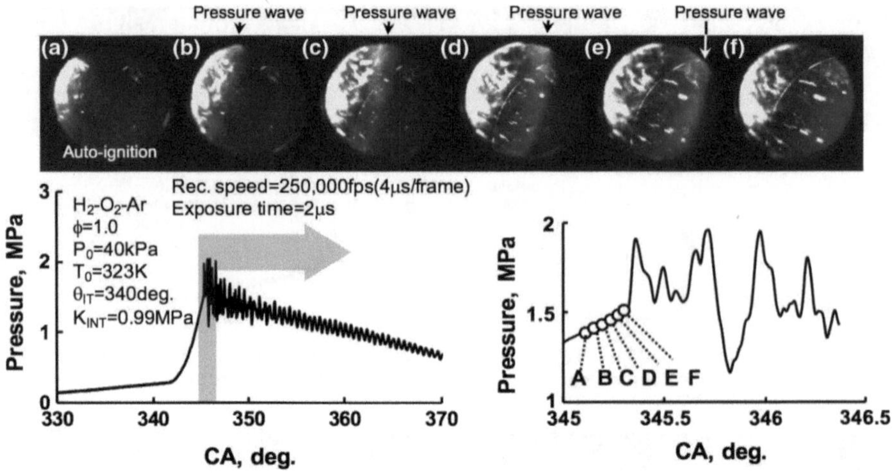

Fig. 13 Visualization of pressure wave induced by the initial auto-injection inside the end-gas [9], with the permission of Elsevier. The corresponding movie is available online

$\phi = 1.0$, $P_0 = 40$ kPa, $\theta_{IT} = 340°$), together with the in-cylinder pressure histories. Strong auto-ignition of the end-gas mixture occurred at the upper left side of the combustion chamber with timing A, due to the fast propagation of the regular flame front and rapid increase of the in-cylinder pressure. The large density gradient of the pressure wave then propagated to the lower right side of the engine and passed through the visualization area with timing F. The phase difference between the visualized images and the in-cylinder pressure shown in the figure is the result of different measurement locations and the sound velocity caused by the pressure wave.

Auto-ignition of the end gas, caused by flame propagation and intense oscillations due to pressure waves during knocking combustion in a hydrogen SI engine, was visualized using high-speed color and monochrome cameras. Auto-ignition in the end-gas region that was compressed by the propagating flame front was visualized using a high-speed video camera. Understanding auto-ignition inside the end-gas region during engine knocking is important in further developing new SI engines.

6 Conclusions

Very fast visualization of fuel injection and spray impingement on a wall, laser ignition processes, and auto-ignition of the end-gas region due to flame propagation during an engine knock cycle was demonstrated using a high-speed video camera and an ultra-high-speed camera. Ultra-high-speed visualization is a powerful tool for investigating atomization processes in liquid fuel injection, the ignition process in a laser-ignited spray flame, and auto-ignition processes in the end-gas region during engine knocking.

References

1. J.B. Heywood, *Internal Combustion Engine Fundamentals* (McGraw-Hill, New York, 1988)
2. F.F. Zhao, D.L. Harrington, M.-C. Lai, *Automotive Gasoline Direct-injection Engines* (Society of Automotive Engineers Inc., Warrendale, 2002)
3. C. Stan, *Direct Fuel Injection for Gasoline Engines* (Society of Automotive Engineers Inc, Warrendale, 1999)
4. H. Zhao, N. Ladommatos, *Engine Combustion Instrumentation and Diagnostics* (Society of Automotive Engineers Inc., Warrendale, 2001)
5. N. Kawahara, J.L. Beduneau, T. Nakayama, E. Tomita, Y. Ikeda, Spatially, temporally, and spectrally resolved measurement of laser-induced plasma in air. Appl. Phys. B **86**, 605–614 (2007). doi:10.1007/s00340-006-2531-4
6. J.L. Beduneau, N. Kawahara, T. Nakayama, E. Tomita, Y. Ikeda, Laser-induced radical generation and evolution to a self-sustaining flame. Combust. Flame **156**, 642–656 (2009). doi:10.1016/j.combustflame.2008.09.013
7. N. Kawahara, K. Tsuboi, E. Tomita, Laser-induced plasma generation and evolution in s transient spray. Opt. Express, **22**-S1, A44-A52 (2014). doi:10.1364/OE.22.000A44
8. N. Kawahara, E. Tomita, Y. Sakata, Auto-ignited kernels during knocking combustion in a spark-ignition engine. Proc. Combust. Instit. **31**, 2999–3006 (2007). doi:10.1016/j.proci.2006.07.210
9. N. Kawahara, E. Tomita, Visualization of auto-ignition and pressure wave during knocking in a hydrogen spark-ignition engine. Int. J. Hydrogen Energ. **34**, 3156–3163 (2009). doi:10.1016/j.ijhydene.2009.01.091

Part VII
Explosions and Safety

Protection of Buildings and Infrastructure Against Explosion

Alexander Stolz, Malte von Ramin and Daniel Schmitt

Abstract The protection of buildings and infrastructure against highly dynamic loadings caused by explosions requires a profound analysis of the dynamic structural and material behavior. High speed recording devices are a valuable instrumentation to support the respective studies. The present chapter gives different examples that emphasize the importance of high speed videos for the study of loadings, damage behavior of structures and the determination of dynamic material parameters. One of the main threats for humans and buildings are flying fragments for example vehicle components in case of a *Vehicle Born Improvised Explosive Device*. It is shown how the trajectory of such fragments can be calculated by tracing their launch conditions after an explosion. At the resistance side material parameters can differ significantly under highly dynamic conditions. Split-Hopkinson bar tests are often used to determine dynamic parameters of brittle materials. For the determination of the fracture energy the experimental analysis requires a high-speed video to evaluate the velocity of the specimen´s fragments. The respective coherences are explained. Furthermore, the design of members and the whole structure is based on the knowledge about their resistance behavior to extreme loadings like blast. Capturing the initiation and the propagation of a fracture process by high speed videos is therefore essential. It is also shown how in combination with other tools like the *Digital Image Correlation* system, it is possible to determine the dynamic displacement of a whole specimen's surface. The results are useful for the validation of numerical simulations.

A. Stolz (✉) · M. von Ramin · D. Schmitt
Fraunhofer Institute for High-Speed Dynamics, Ernst Mach Institute (EMI),
EMI Am Klingelberg 1, 79588 Efringen-Kirchen, Germany
e-mail: Alexander.Stolz@emi.fraunhofer.de

M. von Ramin
e-mail: Malte.von.Ramin@emi.fraunhofer.de

D. Schmitt
e-mail: Daniel.Schmitt@emi.fraunhofer.de

K. Tsuji (ed.), *The Micro-World Observed by Ultra High-Speed Cameras*,
DOI 10.1007/978-3-319-61491-5_13

1 Introduction

Accidental and attended malicious explosions represent an extraordinary loading case for buildings and infrastructures. The majority of these constructions was neither planned, designed nor built taking explosion as a loading case into account. Furthermore, explosions create physical stresses which differ significantly from ordinary loading cases in regard of the stress level and the time frame these stresses act on the components of a structure. These mentioned scenarios motivate researches in the development of highly resistant structures in order to reduce structural damage and to thereby increase the protection function of buildings.

Especially the younger past has shown that constructions and infrastructures summarized under the term critical infrastructure have a significant pronounced risk to become the target of a terroristic attack. Critical infrastructures are infrastructures with a high political, economic or societal relevance, which failure would lead to a break-down of the normal societal life. Hence the research for the protection of buildings and infrastructures concentrates also on these highly dynamic loading cases and their effects on structural elements and materials.

In order to protect these constructions a deep understanding to the significant physical phenomena and their effects on the material and construction is an essential part for an effective protective solution. This includes the understanding of the material behavior under the given special loading conditions as well as the knowledge of material and structural damage behavior and failure mechanisms of structural elements and even the whole construction.

The given chapter should highlight how modern high-speed video devices can support to understand the occurring phenomena and thereby optimize the structural resistance against these types of loading.

In order to protect effectively three main steps are required. At first the significant loading on the structure has to be determined quantitatively. Second the material should be able to compensate the loading and finally the structural component and the overall structure should be able to bear the resulting stresses without collapsing.

For all these steps high speed video coverage can provide a piece of the puzzle to improve the protection of buildings. Hence the following chapter will give an example for each of these steps.

2 Analysis of Fragment Projection Hazard

With the goals to protect from human injury after an explosion event and to implement efficient measures of protection for the built environment against explosive loading, it is essential to characterize the potential threat as accurate as possible. The potential threat of an explosion event is formed by several hazard types including the blast wave overpressure, fragment and debris projection, cratering, fireball, and ground shock. Specifically the blast wave, the projection of fragments, and debris pose a considerable threat at distances exceeding the other hazard sources by far [1].

The blast overpressure and its propagation are usually not assessed using optical recording equipment. Electrical high-pressure transducers capture overpressure-time-series in a very reliable fashion. Additionally, when it comes to describing the potential consequences of the blast pressure acting on an object, e.g., a structural component, a vehicle or a human body, the direction the pressure is acting in is essential. Commonly, the undisturbed free-flow "side-on" pressure and the face-on reflected pressure are distinguished. The use of optical equipment does not lend itself to determine a pressure value rather to quantify the velocity or acceleration of an object in motion. Nevertheless, high-speed video cameras can support the validation of recorded pressure histories by facilitating plausibility checks. For example, the sequence of explosion effects in time can be determined using high-speed video equipment. Figure 1 shows an exploding vehicle after the initiation. The pictures are still images of a high-speed video recording of an experimental test jointly conducted by the German Bundeswehr Technical Center for Protective and Special Technologies (WTD 52) and the Technical Center for Weapons and Ammunition (WTD 91) at the site of WTD 91. Clearly visible is the sequence of explosion effects: The blast front forming a ring around the explosive source, visible by the dust covering the ground (see yellow arrows in Fig. 1), with the debris still behind the blast front at 26 ms. At 252 ms the debris, progressing faster than the blast wave, passed the blast front, while the blast front starts to dilute with increasing distance from the explosive source. Also apparent is the development of the fireball, of which the spatial expansion is significantly smaller than the extent of the blast front and the debris throw.

As stated before, high-speed imaging and videos are more useful in capturing the motion of objects in a surrounding medium than in recording distinct pressure data. The potential injury from impacting fragments define the debris throw as one of the most hazardous effects of an explosion. A potentially fragmenting explosive source

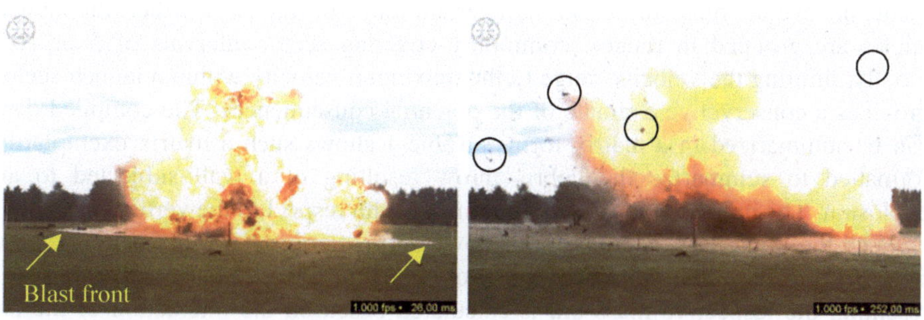

Fig. 1 Sequence taken from experimental explosion of a vehicle at 26 (*left*) and 252 ms (*right*). Blast front marked by *yellow arrows*, 4 fragments exemplarily marked by *circles*. Experiment conducted by WTD 52 and WTD 91, video stills used by kind permission of WTD 52

is characterized by an explosive substance, e.g., solid high-explosive chemicals or explosive gas mixtures, and a surrounding body. This potentially fragmenting surrounding body can be for example the casing shell of ammunition, the casing of an "improvised explosive device" (IED), a vehicle forming a "vehicle borne improvised explosive device" (VBIED), or a vessel containing explosive matter in an industrial processing plant. All of these enclosures encase the explosive substance more or less immediate and are thus subject to the sudden release of a very large amount of energy at the ignition of the explosion. Once the enclosure ruptures, fragments are projected outward from the explosive source, entering a flight phase, described as trajectories. At the end of the flight phase, the fragments impact the ground or an object in within the course of the trajectory, starting a new motion pattern, depending on the impact energy, the impact angle and the stiffness of the barrier. Further, during the flight phase, the fragments tend to describe a tumbling pattern and possibly break further apart as a result of damaged integrity at the onset of the flight phase. All these effects make it very hard to analytically describe a smooth and stable flight pattern. High-speed video footage enables a detailed investigation of the complex debris projection. The analysis of the video footage helps in understanding the different aspects influencing the flight behavior and the interaction of these aspects during the flight. Ideally, further analytical description and mathematical modelling afford a semi-empirical approach to describe the potential associated hazard in geospatial dimensions and content of kinetic energy.

2.1 Characterization of Fragment Projection

The course of a trajectory of a single fragment is determined by its initial launch conditions. These are the initial mass of the fragment, its initial launch velocity and the launch direction, i.e. the circumferential (azimuthal) and the vertical (meridian) launch angles. As the amount of fragments resulting from an explosion is—depending on the type of enclosure—most often rather large, for the purpose of describing the potential debris throw hazard it is sufficient to summarize the fragments in several "bins" representing a certain range of masses. Further, the launch angles are grouped in ranges, commonly covering sector intervals of 5 or 10°. Finally, limiting the velocity range to the maximum velocity within a launch sector provides a conservative estimate of the potential consequences. The compiled data can be summarized in a matrix format. Table 1 shows such a matrix exemplarily formatted to summarize the debris throw resulting of a wall subjected to an explosion. The initial launch angles are limited with respect to the vertical direction.

With the initial launch conditions known and assuming the flight path does not diverge from its initial azimuthal direction, the further course of a trajectory representing the respective mass bin is calculated based on the interaction of the air

Table 1 Excerpt of an example matrix compiling initial launch data from a masonry wall

Vert. launch angle [degrees]	Mass bin [kg]						Max. launch velocity [m/s]
	0.05–0.10	0.10–0.25	0.25–0.50	0.50–0.75	0.75–1.00	>1.00	
65							580
70							620
75							620
80							630
85							670
90							689
95							655
100							628
105							590
110							580
115							550

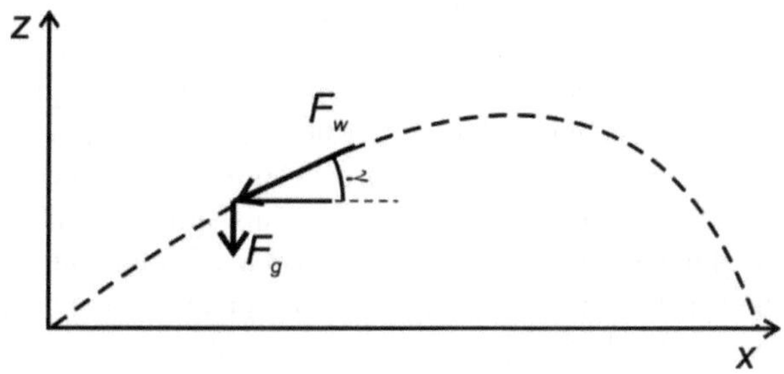

Fig. 2 Fragment trajectory determined from air drag force and gravity

drag force F_w, given by Eq. (1), acting against the direction of motion and the gravitational force F_g.

$$F_w = 0.5 \cdot c_w \cdot \rho \cdot v^2 \cdot A \quad (1)$$

With F_w = air drag force, c_w = air drag coefficient, v = velocity in direction of flight, A = exposed cross section, and ρ = density of air. The air drag coefficient depends on the projected shape of the fragment and on its velocity [2].

Figure 2 explains the interaction of forces shaping the course of a trajectory. The flight path is determined from the varying inclination, which can by deduced from the respective vertical and horizontal components of the velocity vector at the respective point in time by solving a coupled set of differential equations [1].

2.2 High-Speed Video to Capture Initial Launch Conditions

While the ensuing course of the flight path of fragments can be determined analytically, the key to an accurate calculation of the potential debris throw hazard lies in establishing the initial launch conditions. Especially for complex geometries of the explosive item, these are experimentally derived using software-based high-speed video analysis to determine the initial launch direction and velocity. For simple rotational-symmetric metal casings, semi-empirical analytical descriptions exist to define the initial launch conditions [3]. The use of high-speed video analysis and possible implications for the correct set-up are demonstrated on experimental investigations on "vehicle borne improvised explosive devices" (VBIEDs) . The experiments were jointly conducted by WTD 52 and WTD 91. The Fraunhofer Institute for High-Speed Dynamics, Ernst-Mach-Institut, EMI, supported the experimental investigations with respect to test planning and analytical modelling of the results with the goal to characterize the VBIED threat to the German armed forces [4].

Figure 3 shows the initial set-up of the experiment: The location of the vehicle and its direction are indicated by the central arrow. Four cameras are placed directed along the central axis of the vehicle and perpendicular to the object. A non-perpendicular set-up is possible, but it requires a correction of the velocity components with respect to the rotation of camera axis and central line of the exploding object. Furthermore, keeping the distance *d* from the center of the vehicle to the camera equal for all viewpoints simplifies adjusting the reference scale in the tracking software and later analysis of fragment motion, because the respective perspectives do not change. As the spatial velocity components are calculated based on the tracked objects within the video, it is essential the camera viewpoints are aligned properly. This is easily achieved under test lab conditions, but can become difficult in an outdoor environment where the cameras are up to several hundred meters away from the explosion. A GPS-supported aerial picture of the set-up can help aligning the cameras. Because of the size of the testing site, the aerial view needs to be "stitched" from several photographs. The picture in Fig. 4 was assembled by approximately 250 single aerial photographs taken from 30 m height with a 12 Megapixel resolution. A challenge in setting up the cameras is the correct distance to the explosive item. If the cameras are located too far from the explosive source, small fragments might not be in the resolution range of the picture and become invisible on the video (keeping in mind some fragments fly away from the camera into distances from several hundreds of meters to several kilometers). If the camera is too close, the blast wave reaches the camera, i.e. its protective enclosure, too early and moves the camera before a sufficient segment of the fragment propagation is captured. A camera in uncontrolled motion renders the determination of fragment velocities from correlated videos on a synced time track impossible.

To start the recording, the cameras are synced with the trigger of the explosion. That way it is ensured the time track is correct and the truthful velocities can be read from the video footage. As the explosion and the ensuing debris throw cover a time

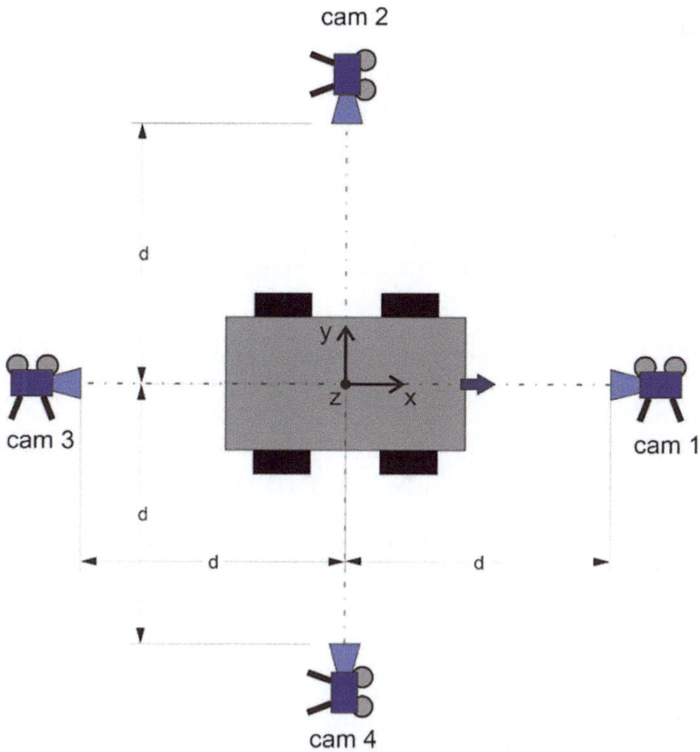

Fig. 3 Camera set-up for high-speed video analysis of explosion effects

range of milliseconds, the frame rate of the cameras needs to be sufficiently high. In the example, 1000 frames per second were used for all four full-HD video cameras (Camera Type: MegaVis; High Speed Vision GmbH). The videos are synced and correlated using a suitable tracking software. This requires proper definition of the reference scale in all video files prior to identifying the fragments and automatically tracking them. The tracking process involves the fragment recognition, the capturing of the fragment location in time and space and a forward-prognosis of the flight path, which enables tracking the fragment also through ranges where visibility is poor and/or the shape of the fragment changes due to tumbling or break-up during the flight. Figure 5 shows a screenshot of the tracking process taken from one camera perspective.

It is important to be aware of the goals of the high-speed video analysis. Most important is the qualitative impression of the experiment and the possibility to study the respective phenomena in slow motion. When it comes to quantitative assessment, the high-speed video footage is used to obtain the vertical launch angles and the corresponding maximum launch velocities as initial conditions for the trajectory calculations and therefore the analytical description of the kinetic energy content at impact. The fragment masses and the flight direction in azimuthal direction are

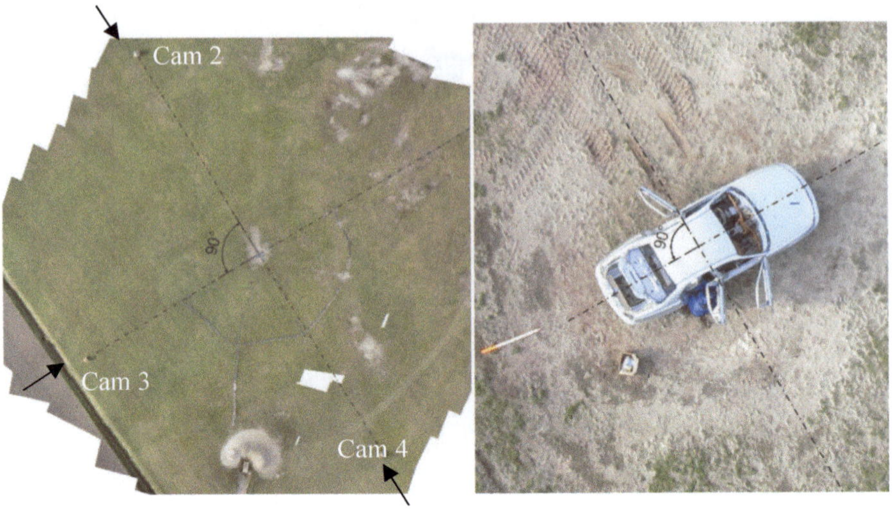

Fig. 4 *Left*, aerial view of experimental set-up. Explosion site in the *center*. *Right*, close-up of explosion site. Cameras perpendicular to each other, camera 1 beyond picture boundaries

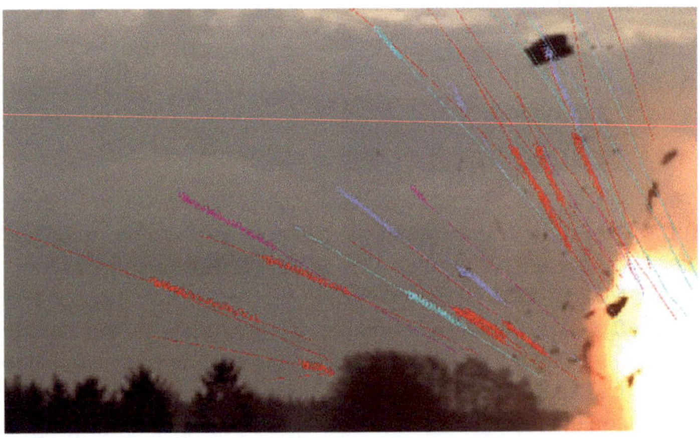

Fig. 5 Screen shot from tracking software. The fragments are distinguished by their masses, indicated by *line color* and *symbol* along the respective trajectory

assessed after the experiment by manually picking up the individual fragments from the ground in radial sectors around ground zero and sorting them into the respective mass bins and angular ranges. Hence, it is not necessary to track all fragments, but those with the maximum velocity within the respective angular range. The fragments are tracked using suitable video software. At the beginning of the explosion, the exact origin of the single fragments is not visible due to the fireball and dust development. However, with increasing flight distance and within the margins of inherent uncertainties, it is reasonable to assume all fragments originate at the center of gravity of the exploding object. Due to the fireball, the precise initial conditions can also not be determined—the initial launch angle and corresponding velocity are simply not visible. To overcome this problem, the initial conditions can be assumed by following a straight line from the origin to the first visible instance of the fragment. A more accurate method is to track the trajectory over the visible path and calculate backwards the invisible segment of the trajectory, which is slightly curved. The directional components of flight path and velocity need to be established by geometric correlations: In the different camera views, the same specific fragment needs to be located. With each trajectory only visible in the respective plane of view (x–z plane or y–z), the components v_x, v_y, v_z of the tracked velocity v_{xz}, resp. v_{yz} are calculated using the Pythagorean theorem (Eqs. 2 and 3) and solving for the unknown components.

$$v_{xz} = \sqrt{v_x^2 + v_z^2} \tag{2}$$

$$v_{yz} = \sqrt{v_y^2 + v_z^2} \tag{3}$$

The vertical velocity component v_z of the specific fragment is identical in all correlated videos. The fragment velocity in space is then finally given by the entire velocity vector $\vec{v} = [v_x, v_y, v_z]$.

The procedure bears a few challenges worth mentioning:

- Within the entire debris cloud and the fireball it is challenging to identify the respective identical fragments in at least two videos. Only very distinct fragments like the wheels of the cars in the VBIED example are suitable for this, but they do not need to be the ideal fragments to determine the maximum velocity for the corresponding launch angle. Usually, smaller body parts exhibit the highest velocities and flight distances, but they are hard to track.
- The fastest fragments within the projected plane of the video indicate a high velocity only within the plane in view. "Oncoming" fragments or debris flying away from the camera can appear rather slow in the respective view, with essentially only the transient vertical component visible. It is therefore essential to correlate the videos properly by linking the respective time stamps, setting the proper reference scale and identifying the correct objects.
- There are several software solutions available for simultaneously tracking multiple objects in videos that even support stereo vision to reconstruct two-dimensional images to three-dimensional environments. For the purpose of tracking fragments after an explosion in an outdoor environment and estimating their further motion, many software solutions quickly reach their limits. Especially the filter algorithms for object recognition are often insufficient to detect moving objects that

- tumble in a chaotic pattern (rapidly changing planar views of the same object in synced videos)
- change their color due to reflections from the fireball
- possibly break apart during the flight
- occasionally become invisible behind fireball, dust and debris clouds.

Furthermore, the changing background of the outdoor environment and the changing light conditions resulting from the fireball combined with thousands of small fragments drive the currently available software to its limits.

With all these challenges in mind, it is obvious that while high-speed video imaging has its central place in the assessment of fragment projection, currently still a considerable amount of effort and interpretation skills are necessary to produce meaningful and reliable results. Because the produced data is highly dependent on the specific characteristics of the respective test, statistical processing is necessary to prepare the data in more general form, for example as parameters in a fragment matrix such as Table 1. To conclude, Fig. 6 shows a sequence of video stills from a VBIED test conducted by WTD 52 and WTD 91 at the test site of WTD 91 in Meppen, Germany. Different from the images in Fig. 1, the stills are taken at later stages of the fragment projection, demonstrating visibly the challenges listed above.

3 Material Resistance of Concretes Under Spall Loading

Another interesting application is the determination of the tensile strength and fracture energy of concrete materials under dynamic loading cases which represent the transient stresses which occur in structural elements under explosive loadings. Especially the knowledge of the tensile properties of concrete materials under those loading conditions is on the one hand required for the evaluation of existing materials and structures but also on the other hand fundamental for the material development. The investigation of material properties at high strain rates (up to 180 s^{-1}) can be accomplished with high accuracy using split-Hopkinson-bar experiments.

The Hopkinson Bar in the so called spallation configuration is a laboratory test which creates comparable stress wave in a material like they can be observed for explosive loading cases. The spall tests enable to capture, the tensile strength, the Young's modulus and the fracture energy of concrete materials. Advantage of the laboratory test are the precise loading conditions and reproducible test series (Fig. 7).

The spallation configuration of the Hopkinson-bar consists of a loading facility and an incident bar of cylindrical shape. The test specimen is attached on the end of the incident bar. The load is generated by the impact of the striker projectile on the incident bar. The resulting pulse travels through the bar and loads the test specimen.

The uniaxial loading pulse consists of a compressive and decompressive wave portion. This stress wave is partly transferred and partly reflected at the interface

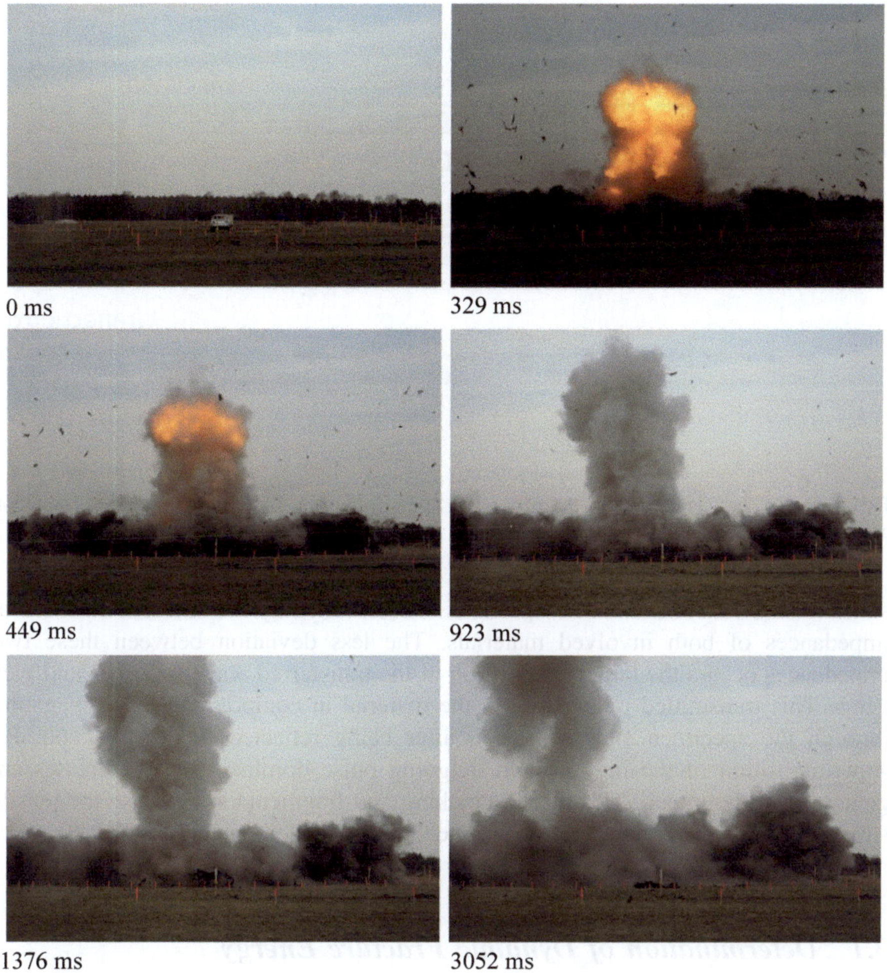

Fig. 6 Sequence of video stills from VBIED-test conducted by WTD 52 and WTD 91. Times below indicate the time span from initiation. Images used by kind permission of WTD 52

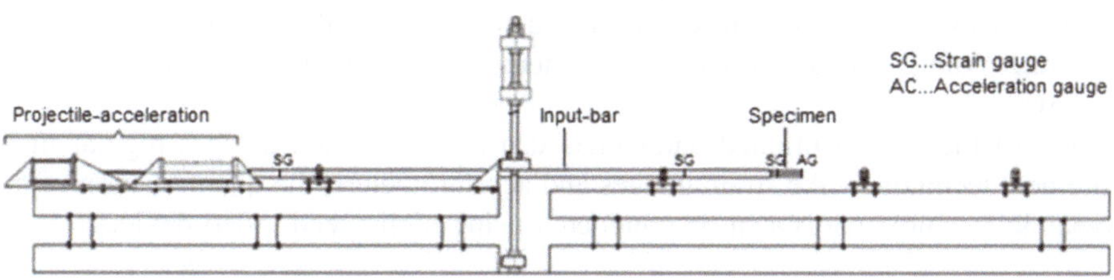

Fig. 7 Schematic setup of the EMI Hopkinson-bar in the spallation configuration

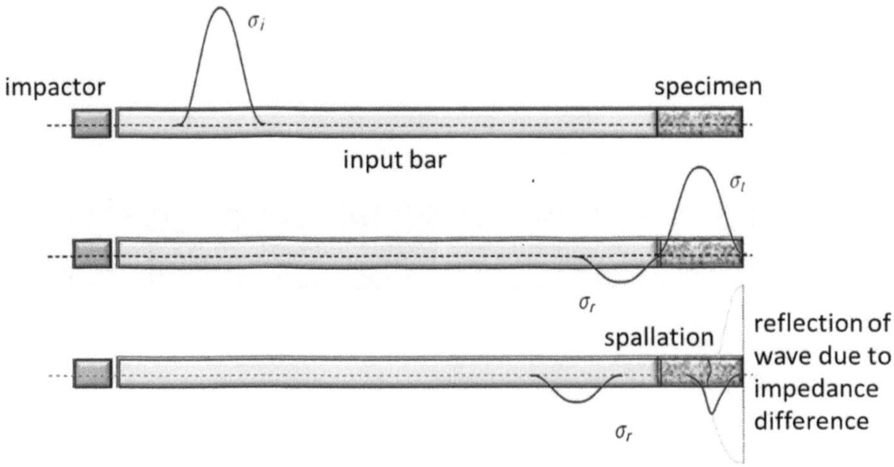

Fig. 8 Wave propagation and reflection in the Hopkinson-Bar device in the spallation configuration in accordance to [5]

surfaces. The relation of the two portions is determined by the different mechanical impedances of both involved materials. The less deviation between these two impedances occurs the larger the portion of the transferred wave becomes and vice versa. This transmitted pulse stresses the material in compression by propagating through the specimen and in tension after being reflected on its free end. By superimposition of the ingoing and outgoing pulse dominant tensile stresses are generated in the material leading to cracking and fragmentation in case the tensile strength is exceeded. This special fragmentation is called spallation (Fig. 8).

3.1 Determination of Dynamic Fracture Energy

The spallation configuration his type of experiment is restricted to brittle materials, which have a significant higher compressive strength than tensile strength since if not the specimen would be damaged by the compressive wave already [5].

In general, the experiment enable the determination of the wave velocity, the dynamic Young's modulus, the dynamic tensile strength and the dynamic fracture energy.

Whilst the first mentioned parameters can be determined using a regular measurement techniques like strain gauges and accelerometers, the quantification of the dynamic fracture energy can be conducted using high speed video devices.

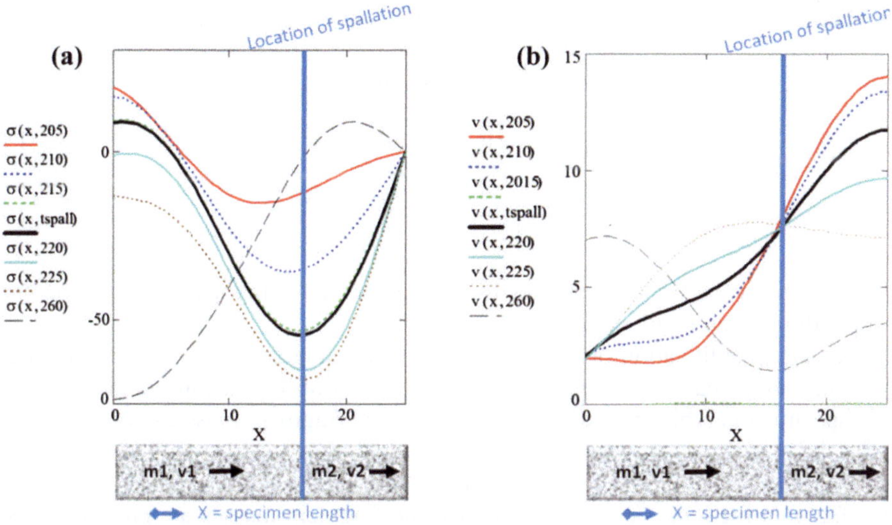

Fig. 9 Calculated stresses (**a**) and velocities (**b**) following approach in accordance to [5]

In contrast to static tension tests it is not possible to measure the stress over crack opening in these dynamic spall experiments directly, but the regarded energy can be calculated based on the time-history of the fragment velocities. This methodology was developed by Schuler [5]. Therefor two points in time have to be considered. The first point is defined by the start of the actual crack opening, the second defines the point when the opening process is completed and the crack is fully opened. The energy which is dissipated during this crack opening procedure defines the actual dynamic fracture energy of the brittle material. In order to determine the fracture energy Schuler [5] defined the following methodology:

First the stress and velocity distribution is calculated analytically following equations given in [5]. The equations describe the velocity and the stress within the test specimen as a function of the location x within the specimen and the time t. The results of these calculations for different time steps are shown in Fig. 9.

At a certain point in time the tensile stress reaches the dynamic tensile strength at a certain location within the test specimen and initiates there the fracture process (black solid line in Fig. 9a). For this moment also the velocities within the specimen can be calculated (Fig. 9b). The location in which the tensile strength is exceeded marks the interface between the two fragments which are separated by the spalling (blue line in Fig. 9). For both fragments the mean velocity can be calculated and represent the starting velocities for the fragmentation process. The actual velocities after the completion of the fracture process can then be captured using the high-speed video devices.

The fracture energy can then be calculated using Eqs. (4–6)

$$Gf = \Delta I \cdot \dot{\delta} \tag{4}$$

$$\Delta I = (v_{t1\_2} - v_{t2\_2})m_2 \tag{5}$$

$$\dot{\delta} = \frac{v_{t2\_2} - v_{t1\_2}}{2} - \frac{v_{t2\_1} + v_{t1\_1}}{2} \tag{6}$$

ΔI denotes the impulse transfer from fragment 1 to fragment 2 and $\dot{\delta}$ the corresponding mean crack opening velocity [5].

For the quantification of the fragment velocity single frames of the high-speed video coverage can used. Since the precise location of the fracture initiation within the specimen is arbitrary the determination of the fragment mass can only be done after the experiment by weighing the fragmented pieces. This requires a soft catch of the pieces. Hence notched specimen provide significant simplifications, due to the fact that the location of fracturing and the fragment masses are predetermined.

Figure 10 shows some exemplary still images taken from the high-speed video of a Split-Hopkinson bar experiment in spallation configuration on a normal strength concrete.

The high-speed videos have been conducted with 1000 fps with a resolution of 1024 by 512 pixels and a tone of 12 bit. The pixel size of the system in place is 20 μm and the shutter speed is 1 μs (Camera Type: FASTCAM APX 120KC; Photron).

With this procedure it is then possible to compute the dynamic fracture energy of the material and compare it to the static. A precise knowledge of this material behavior builds the basis for the assessment of the structural resistance to those kinds of loadings or an effective design approach to it.

4 Glass Facade Elements

In the field of high-speed dynamics often the location where a fracture process begins and the starting time of the fracture is important to study. Especially materials with very brittle behavior complicate the analysis of the dynamic fracture propagation. Since the facades of many buildings are often covered by glass curtain walls representing the first members of structures to withstand blast loadings due to explosive scenarios their resistance respectively fracture behavior is important to study for the further development of these building components regarding such extreme loading cases.

In comparison to other materials like concrete or steel, glass is very vulnerable and therefore is of special interest in the analysis of explosive scenarios. Depending on the structural set-up (e.g. single, safety or laminated glazing) and in particular on the bearing construction (line or punctual supported, with or without damper elements,

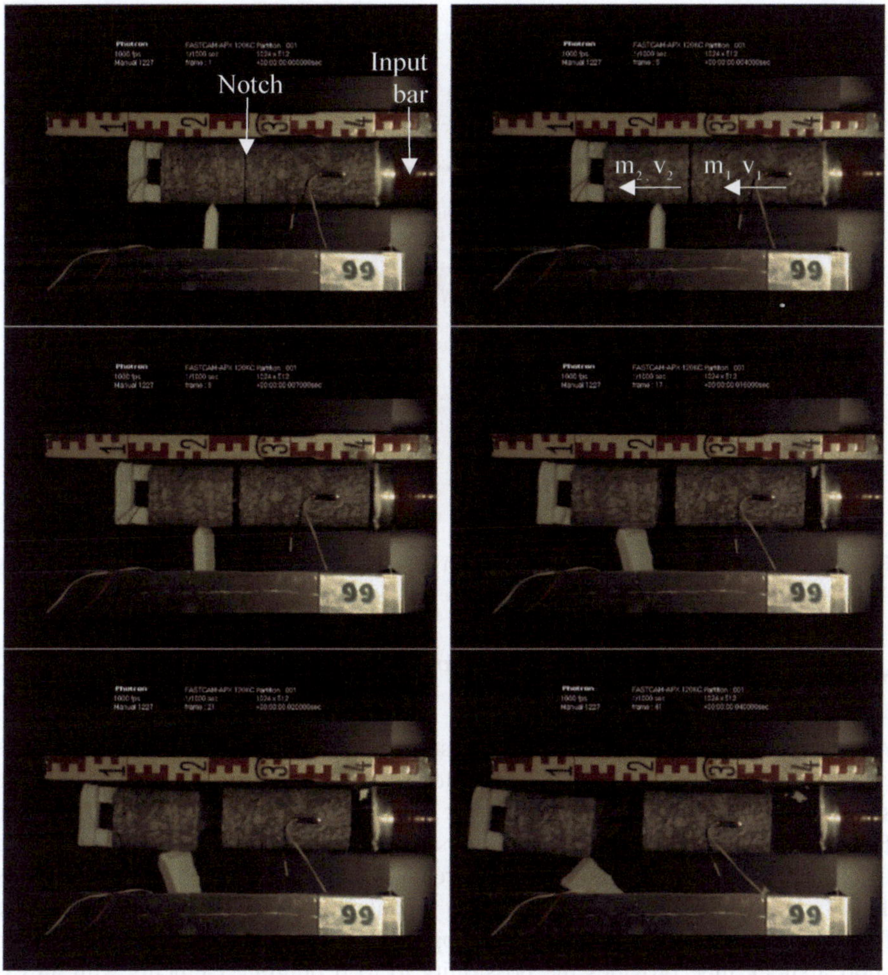

Fig. 10 Still images from a high-speed video coverage of a SHB experiment for the time t = 0, 4, 7, 16, 20, and 40 ms (from *left* to *right* and *top* to *bottom*). Notched specimen (normal concrete strength) tested in spallation configuration. Recording used for the determination of the fragment velocity

etc.) the fracture initiation of glass elements can be very different. Precisely because glass elements burst within only one or very few milliseconds the exact observation of the transient fracture process requires special instrumentation. By the use of high-speed cameras these highly dynamic mechanisms are possible to study.

In the following an example is given about how high-speed videos help to detect the local fracture initiation of glass curtain wall elements tested in shock tube experiments and how the starting time of the fracture process is determined and used for the further analysis.

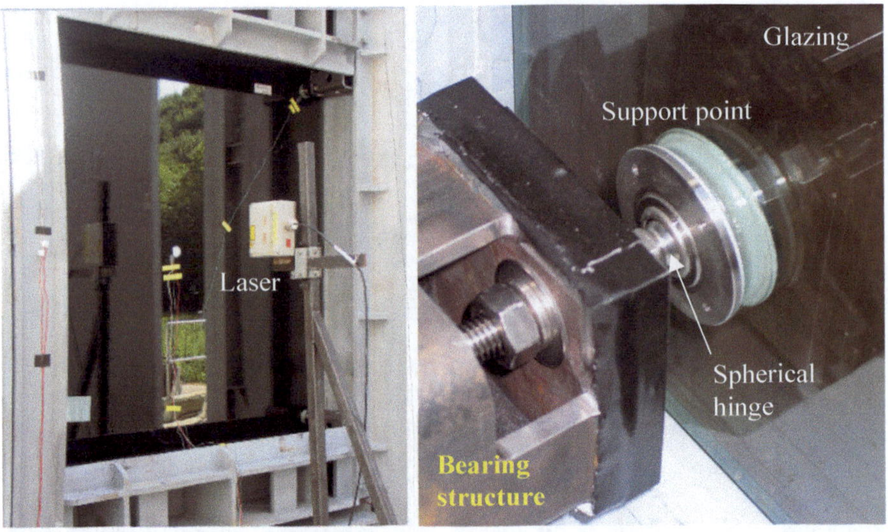

Fig. 11 Experimental set-up. Installed specimen in shock tube (*left*); Close-up image of a bearing point of the laminated glass element (*right*)

4.1 Punctual Supported Glass Panes

An established testing facility to investigate and quantify structural members regarding their resistance against blast loadings is the shock tube. One of these shock tubes is the BLAST-STAR (Blast Security Test And Research Facility) at Fraunhofer EMI where glass panes, doors or even shutters can be tested according to the national and international standards EN 13541, DIN EN 13123, DIN EN 12124, ISO 16934 and ASTM F1642. Detailed information about shock tube facilities, their working principles and application range can be found in [6, 7]. In short a shock tube provides the possibility to load building elements with an explosive loading which is created by using compressed air. A shock tube therefore needs no high explosives to create this kind of loading. The advantage of this test method is that the loading is completely plane (far field detonation scenarios) and highly reproducible.

Subsequently Fig. 12 shows four still images—taken from a high-speed recording—of a shock tube test on a point supported glazing element. The test was carried out with a multilayer laminated glass pane (Setup: tempered glass, PVB,[1] tempered glass) supported at four points. The support elements were installed into four circular openings of the pane with a vertical and horizontal distance of 100 mm to the edges of the specimen. A picture of the installed specimen and a close-up image of one of the aluminum-steel support points are given in Fig. 11.

[1]Polyvinylbutyral.

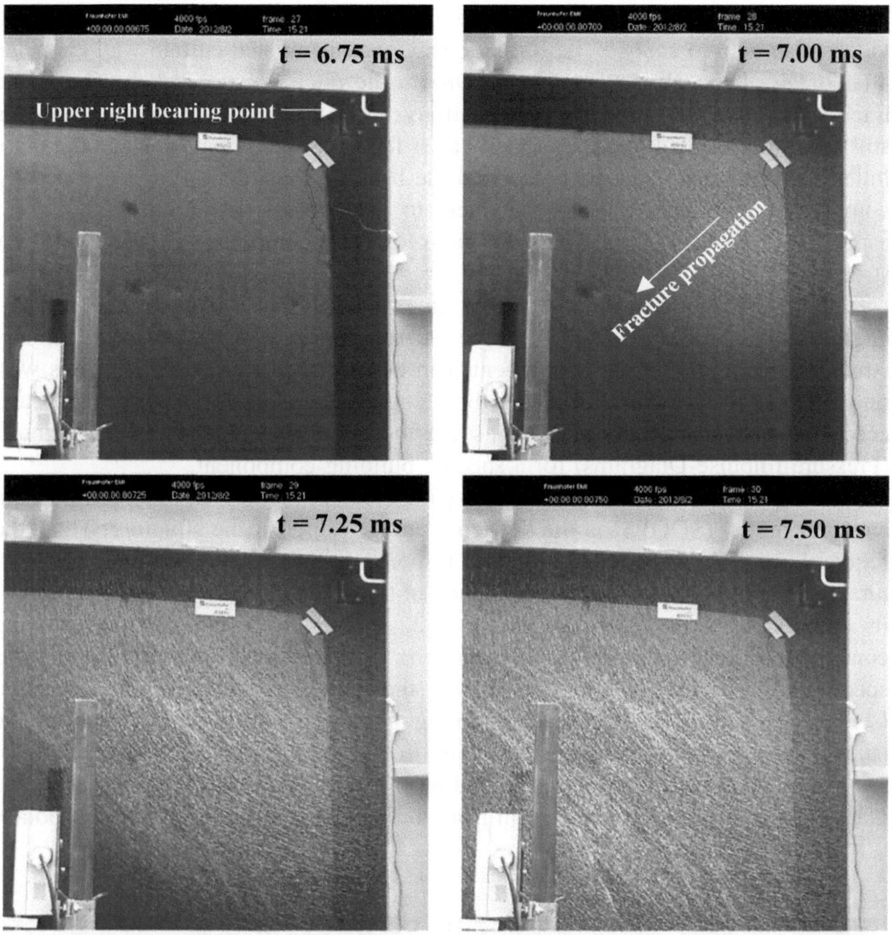

Fig. 12 Still images from a high-speed video recording the fracture process of a punctual supported laminated glass pane in a shock tube experiment for the time t = 6.75, 7.00, 7.25 and 7.50 ms (from *left* to *right* and *top* to *bottom*) from the beginning of the pressure build-up at the specimen

The still images in Fig. 12 show the upper right quarter of the glazing (from the protection side of view/camera position) including one of the upstands (bearing point) for the time t = 6.75, 7.00, 7.25 and 7.50 ms of the shock tube test. The trigger time (t = 0 ms) represents the beginning of the pressure build-up at the specimen (measured with gauges). For the high-speed video a configuration with 4000 fps and a resolution of 824 × 824 pixels was used (Camera Type: FASTCAM SA5; Photron). As the pictures show the fracture process starts clearly from the upper right bearing point and spreads out over the rest of the specimen. The combined tension and shear stresses at the bearing point led to the fracture of the curtain wall element.

Besides the localization of the fracture initiation of the glass elements high-speed videos also help to analyze the further behavior of laminated curtain wall elements after break of the glass components. Due to the ductile material behavior of the internal layers of the laminated constructions (PVB or others) curtain wall elements still present a residual loading capacity after fail of the glass panes. Once the glass failed the specimen can either stay into the bearing construction or get ripped off completely. High-speed videos can help to analyze this dynamic behavior and especially define the time when the break of each component of the multilayer structure occurs.

The latter information (definition of time) is important for the development of engineer models like a SDOFS (Single Degree of Freedom System). Since many structures like curtain wall components can be described by standby systems with only one or two degrees of freedom these engineer models are used for a fast and economic analysis of other blast load scenarios and the calculation of P-I-Diagrams (Pressure-Impulse-Diagrams) for a specific building component.

In the given example of a laminated glass element the degree of freedom of the engineer model (SDOFS) is the deflection at the center of the structure. Thus, the required input for developing a correct response function - representing the structural resistance behavior—is the deflection of the structure at the time when each of its components fails (1. break of glass; 2. internal layer rips off from bearing construction → complete fail of specimen). To determine the deflection of a specimen a laser measurement oriented at the center of the specimen is installed (see also Fig. 11). The time for determining the correct deflection is defined by the study of the high-speed video. Regarding this Fig. 13 shows another example of two still images taken from a high-speed recording for the time when the multilayer glass element starts to get ripped off from the bearing construction (support point).

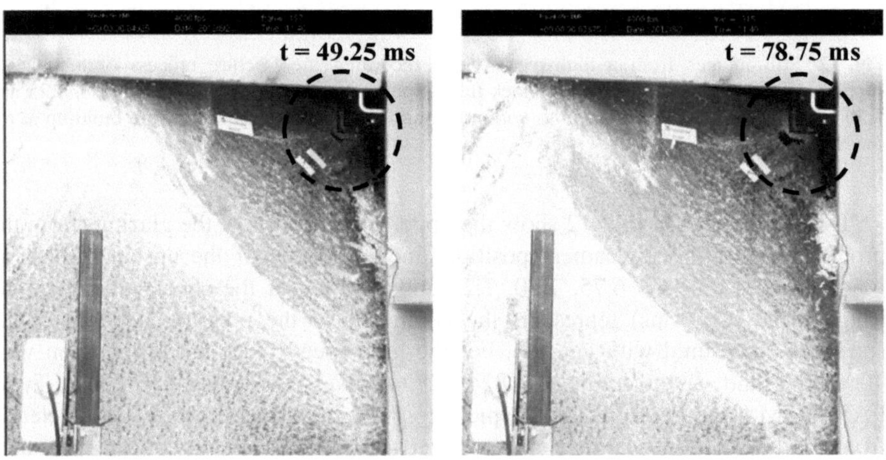

Fig. 13 Still images from a high-speed video recording the fracture process of a punctual supported laminated glass pane in a shock tube experiment for the time t = 49.25 and 78.75 ms (from *left* to *right*) from the beginning of the pressure build-up at the specimen

With the use of a finally defined SDOFS any number of pressure-impulse load combinations can be calculated to determine a surface of maximum deflection values in a P–I-diagram. Based on the experimental results then so called iso-damage curves (threshold curves regarding the fracture) can be determined and implemented into the diagram. Hence, the damage assessment of the structure is possible for any further load combination by just using the final P–I-diagram (for more information see [8]).

The described example about laminated glass elements illustrates how the fracture process of highly dynamic loaded structures can be investigated with the use of high-speed videos and finally helps for the development of engineer models. The explanations about the given curtain wall element is only an example. High-speed videos are also used in a similar way for the analysis of the fracture process of other building components and other scenarios as for instance impact. The upcoming chapter describes the combined application of two high-speed cameras with the DIC (Digital Image Correlation) method used for the analysis of the bending behavior and the study of buckling effects of profiled steel sheathing elements.

5 Validation for Numerical Simulation

Numerical simulations are nowadays also for blast loading scenarios effective and powerful tools to assess the resistance of structures to these loading cases. Nevertheless, numerical simulation always represent a simplification of the real life problem, although it is possible to simulate very complex system with modern technical equipment and computer power. These complex numerical simulations have to be conducted with great care and need a proof of reliability in terms of verification and validation of the numerically created results, see also [9–11].

Hence the validation of numerical simulations at experimental results is an essential requirement for the safe use of numerical simulations within the context of protective structure assessment and design.

For explosive loading cases normally two standard parameter are taken into account for the comparison between experiment and simulation: These parameters are the damage of the regarded component and the deformation in terms of displacements of a structure. The damage of the tested component can be tracked as shown in the paragraphs about the glass façade elements also with high speed camera devices.

The displacements are captured with conventional measurement devices like laser gauge devices. These devices can capture the transient time history for the displacement of one point. For bending problems the midpoint displacement is typically used as a significant value. Nevertheless, even if several laser gauges are used in one experiment those devices only capture the displacements for discrete points within the tested element.

For the majority of the applications these two characteristics deliver sufficient information for the validation of numerical simulations. In some special applications further information are needed to validate simulated the bearing behavior of a component to the experimentally overserved one.

5.1 Profiled Steel Sheeting Under Blast Loading

One example for such a system which needs further information for the validation of the numerical results is a profiled steel sheeting.

Figure 14 shows some still images from a profiled steel sheeting under a blast loading. The loading was created with a shock tube facility already mentioned in section before.

In addition, the measurement at the rear surface of the test element can be conducted without the disturbance of the explosion acting on the measurement techniques. The course of the frames taken from the high-speed video reveals that the profiled steel sheeting is not purely deformed in a bending mode. Instead in addition to the overall bending local buckling phenomena can be observed. The localized buckles are formed mainly in the dips of the steel plate and travel over the course of the time in axial direction from the middle of the plate to both ends of the plate. The local deformation of these buckles cannot be captured using the laser gauge equipment, but another measurement techniques in combination with a high-speed coverage can enable the quantification of the local deformations.

This measurement technique is the so called Digital Image Correlation (DIC) method. The DIC is an optical measurement technique which covers a full field of displacements without contact to the surface. The only requirement for this measurement techniques in the given application is a random speckle pattern on the steel shelter surface and a high-speed video device to capture this speckled surface over the time. Due to the fact that the DIC is not limited to an inherent length scale the method can also applied to the size of the profiled steel sheeting (2.3 by 1 m).

The DIC method bases on the comparison of digital images, by tracking blocks of pixels and their movement. An advantage of this method is also that it requires no specialized lighting as long as the pixel blocks provide a sufficient range of contrast and intensity levels. For more detail about the DIC see [12].

In order to profit from this technique the profiled steel sheets have been speckled with a black and white pattern as shown in Fig. 15. Since three dimensional deformations have to be captured two high-speed video devices have been installed. The two devices have been located in the same distance to the test object but in a certain angle to each other, so that movements in all three direction could be recorded. Using this experimental set-up the test have been conducted on the speckled sheeting.

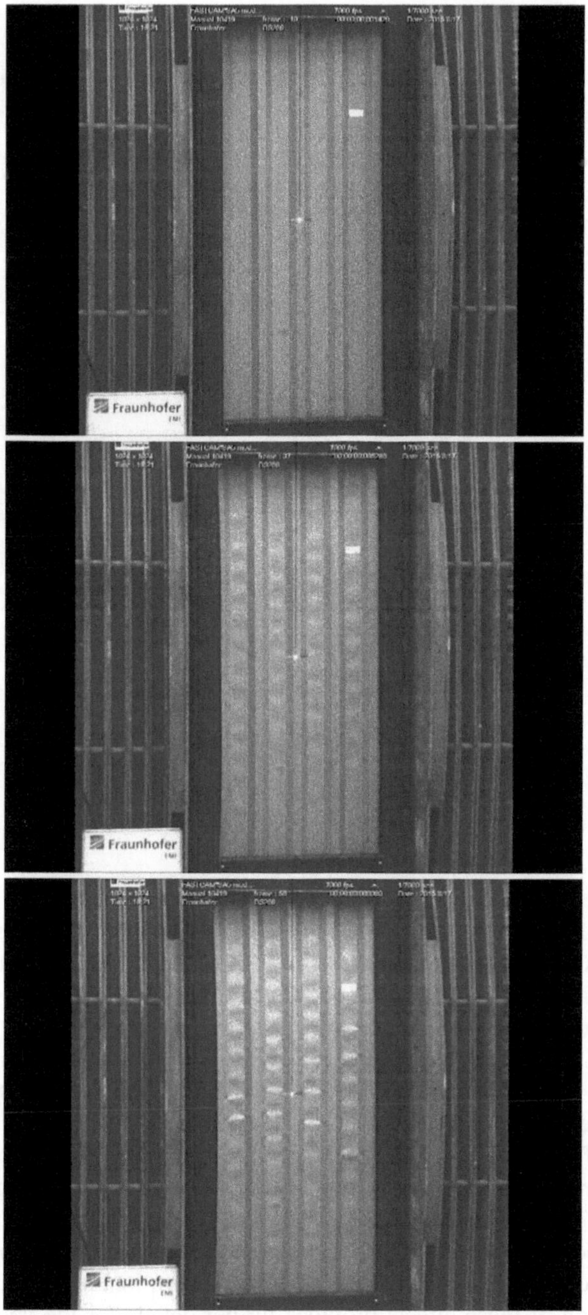

Fig. 14 Frames from the high-speed video capturing the buckling of the profiled steel sheeting for three different time steps

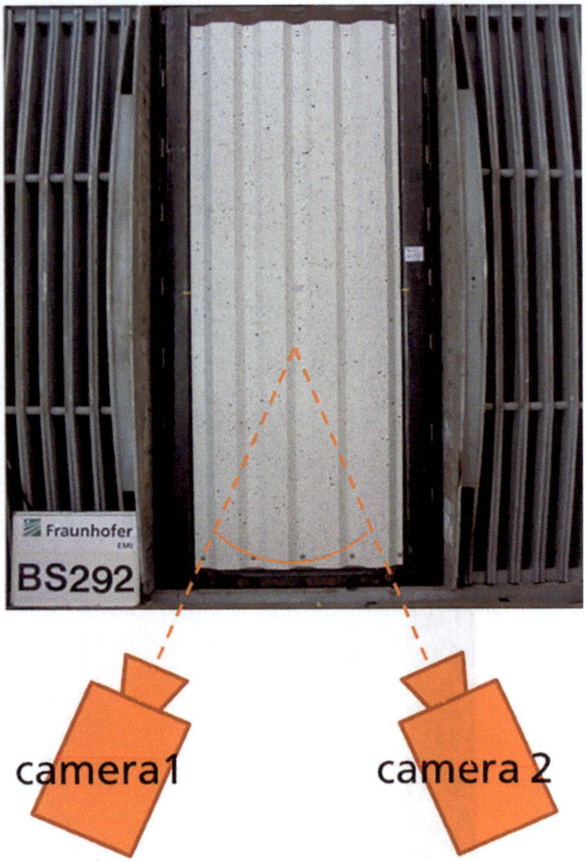

Fig. 15 Principal sketch of the experimental set-up for the DIC measurement

A first step in the analysis of the results is the determination of the level of precision of the DIC methodology applied to a 2.3 by 1 m large object. Therefor a standard laser displacement gauge was added to the set-up tracking the mid-field out of plane deformation of the profiled steel sheeting.

The comparison of the results in Fig. 16 reveals a very good precision of the DIC method. The values calculated by the DIC just differ about 2 percent from the laser gauge measurement. Hence the DIC provides reliable results also for the given large dimensions of the steel plate. The high-speed cameras used for the experiment including the DIC recorded the scene with a resolution of 1024 by 1024 pixels and 7000 fps (Camera Type: FASTCAM SA5 and SA1.1; Photron). Due to the combination of high-Speed camera and DIC technique it is now possible to quantify the local buckling effects of the profiled steel sheeting shown in Fig. 17. The newly created quantification of these effects provides further details for the validation of numerical simulations by comparing numerical results to experimental observations.

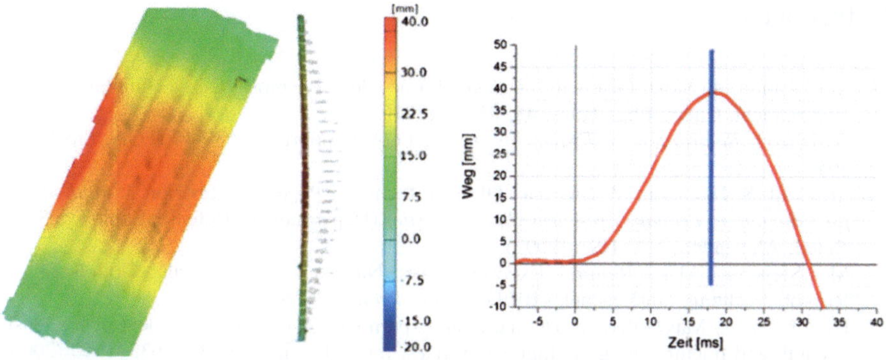

Fig. 16 Results of the DIC measurement in terms of out of plane displacement: distribution of displacement on the profiled steel sheeting in Front and side view (*left* and *middle*) at the time marked by the *blue cursor* in the laser gauge measurement (*right*)

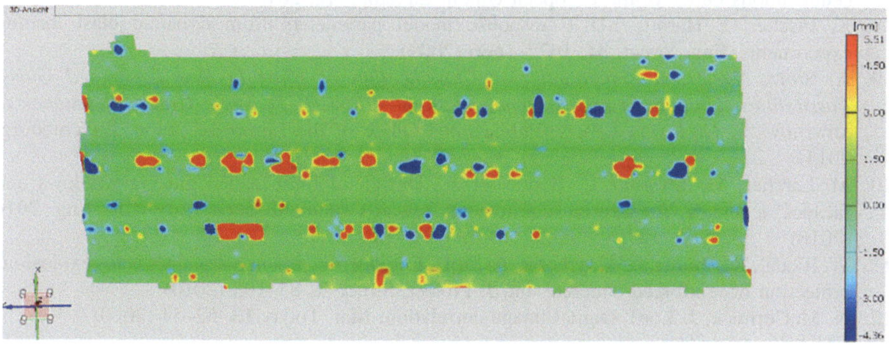

Fig. 17 Displacement plot including the local buckling effects

6 Summary

The given chapter underlines that moderns high speed video capturing devices can significantly support the protection of buildings and infrastructures to explosive loadings. On the one hand, they represent a powerful supporting tool to capture the actual loading which a building has to resist by tracing fragment initial launch conditions. On the other hand, the material resistance to such loadings can be observed with these cameras and damage initiation and propagation can be understood. Based on this knowledge effect design can be developed. For the design numerical simulations are state of the art. In combination with DIC techniques high speed video coverage offers a new quality for the validation of numerical results to experimental tests.

In summary, high speed video techniques are besides others one essential element to improve the protection of buildings for explosions.

References

1. M.v. Ramin, A. Stolz, Debris throw model for accidental explosions in a complex industrial environment. Chem. Eng. Trans. **48**, 85–90 (2016)
2. B. Janzon, *Retardation of Fragments in Air*, FOA Rapport A 2539-44 (FOA, Stockholm, 1971)
3. M. Crull, S. Hamilton, M. Swisdak, DDESB Technical Paper 16, Revision 3: *Methodology for Calculating Primary Fragment Characteristics* (Department of Defense, Explosive Safety Board, Alexandria, VA, USA, 2009)
4. M. Steyerer, M.v. Ramin, Experimentelle Nachweisverfahren als Grundlage der Risikobetrachtung am Beispiel VBIED, in *Bau-Protect 2016* (2016)
5. H. Schuler, C. Mayrhofer, K. Thoma, Spall experiments for the measurement of the tensile strength and fracture energy at high strain rates. Int. J. Impact Eng. **32**, 1635–1650 (2006)
6. A. Stolz, O. Millon, A. Klomfass, Analysis of the resistance of structural components to explosive loading by shock-tube tests and SDOF models. Chem. Eng. Trans. **48**, 151–156 (2016)
7. A. Stolz (ed.), *A Comparison of Existing Standards for Testing Blast Resistant Glazing and Windows—ERNCIP Thematic Group: Resistance of Structures to Explosion Effects* (Publications Office of the European Union, Luxembourg, 2015)
8. K. Fischer, I. Häring, SDOF response model parameters from dynamic blast loading experiments. Eng. Struct. **31**, 1677–1686 (2009)
9. A. Stolz, *Numerical Simulations for Classification of Blast Loaded Laminated Glass: Possibilities, Limitations and Recommendations—ERNCIP Thematic Group: Resistance of Structures to Explosion Effects* (Publications Office of the European Union, Luxembourg, 2015)
10. M. Larcher, M. Arrigoni, C. Bedon, et al., Design of blast-loaded glazing windows and facades: a review of essential requirements towards standardization. Adv. Civ. Eng. **2016** (2016)
11. W. Riedel, K. Thoma, C. Mayrhofer, A. Stolz, Engineering and numerical tools for explosions protection of reinforced concrete. Int. J. Protect. Struct. **1**, 85–103 (2010)
12. N. McCormick, J. Lord, Digital image correlation. Mat. Today **13**, 52–54 (2010)

Single Bubble Ignition After Shock Wave Impact

Hartmut Hieronymus

Abstract High-speed photography was applied to investigate the explosive behavior of bubble-containing systems of the type organic solvent—gaseous oxidizing agent. Knowledge about the explosive behavior of such systems is of great importance, for example, for the safe operation of chemical reactors in oxidation processes in the liquid phase. Examples of the complex dynamical reaction of bubbles are elucidated where bubble-containing liquids were subjected to the impact of shockwaves. The different stages of the shock-induced explosive behavior of oxygen and oxygen plus inert gas bubbles were studied experimentally and theoretically mainly in liquid cyclohexane. Other solvents, such as cumene and 2-ethylhexanal, have been found to show that shock-induced bubble explosions can occur even if the gas phase of the bubble is not in the explosive range before impact. The influence of different parameters on the bubble explosion process, such as the composition of the bubbles and the initial pressure, was investigated. Limiting conditions for bubble explosions were determined. In addition to the behavior of individual bubbles, the interaction of exploding bubbles with one another was observed, which provides information on the propagation mechanism of self-sustaining bubble-detonation waves. The results are important for the safety assessment of the explosion risks in corresponding two-phase systems.

Electronic supplementary material The online version of this chapter (doi:10.1007/978-3-319-61491-5_14) contains supplementary material, which is available to authorized users.

H. Hieronymus (✉)
Bundesanstalt für Materialforschung und -prüfung (BAM),
Unter den Eichen 87, 12205 Berlin, Germany
e-mail: hartmut.hieronymus@bam.de

K. Tsuji (ed.), *The Micro-World Observed by Ultra High-Speed Cameras*,
DOI 10.1007/978-3-319-61491-5_14

1 Introduction

Oxidation processes are of great importance in the chemical industry. They can be carried out in the gas phase as well as in heterogeneous systems with or without catalysts. Many of such heterogeneous systems consist of an organic solvent and oxygen where mostly the oxygen of the air is used. To increase the efficiency of an oxidation process, pure oxygen or air mixtures with an enlarged oxygen content under elevated pressure might be used instead of air. In systems containing an organic solvent—gaseous oxygen the presence of the liquid phase is an additional source of danger compared to pure gaseous systems. Apart from deflagrations and detonations in the gas phase as well as explosions of oxidatively formed by-products, various types of explosions are possible corresponding to the specific kind of the heterogeneous system. Explosions can occur on the smooth surface of the liquid, or in foams, aerosols or in bubbly liquids, each having specific behavior of occurrence and progression. Safety characteristics to assess the risks of pure gas-phase explosions had been well established whereas a need was reported in [1] for investigations of heterogeneous explosions.

The present contribution illustrates the value of high-speed photography for the investigation of highly dynamical explosion processes using the example of explosions in bubbly liquids. In [2], this type of explosive heterogeneous systems is subdivided into groups of different combinations of liquid and bubble gas where the liquid can be an inert, a fuel or an explosive and the bubble gas can be an inert, an oxidant or an explosive mixture. It was known, that under certain conditions bubbly liquids can support a shock induced formation of a strong self-sustaining pressure wave. This phenomenon is called a bubble detonation wave in [3]. Such bubble detonation waves were investigated in a system of the type organic liquid–oxygen bubbles, at an initial pressure of 100 kPa. Similar pressure amplitudes in other bubbly media were reported by other authors e.g. [4] or [5]. Little information was available on how the dynamics of the single exploding bubbles are related to the propagation of a bubble detonation wave. Therefore, highly resolved dynamic pressure recording and optical measurement with high time resolution up to one million frames per second was applied to concentrate on the observation of the explosion behavior of single or widely isolated bubbles and of small bubble clusters.

The investigation was focused on the system with cyclohexane as a liquid and with oxygen as bubble gas. In some cases, the oxygen was premixed with nitrogen. The investigated system was related e.g. to the oxidation of cyclohexane to cyclohexanone and cyclohexanol, which can be used as raw material for the production of nylon outlined in [6]. This process is used in the industry and is usually performed with air bubbles as oxidizer, but also a new trend towards the use of oxygen enriched mixtures or pure oxygen was noticeable when considering patents as e.g. [7]. The use of pure oxygen may enhance the efficiency of the oxidation process but, on the other hand, may increase the explosion risks and hence safety investigation became necessary.

2 Experimental Set-up for Bubble Generation and Observation

An equipment was used to observe the explosion behaviour of gas bubbles in a liquid after an impact of a shock wave onto the surface of the liquid. The bubbly system was prepared at initial pressures between 100 and 300 kPa in an autoclave schematically depicted in Fig. 1. The autoclave has the general shape of a vertical cylinder with an inner diameter of 100 mm and a length of 1100 mm.

In the bottom flange, a nozzle is mounted to inject the bubble gas into the liquid contained in the autoclave. Additionally, a pressure sensor P7 was mounted in the bottom flange. The autoclave has 4 bore holes perpendicular to the main cylinder. Two of them are closed by flanges carrying pressure sensors P3–P6. The other two holes are closed by an observation and illumination window on the front and rear side of the autoclave respectively.

The shock waves were generated by two different methods. In most of the experiments they were created by gas detonations of hydrogen-oxygen mixtures or of acetylene-oxygen-nitrogen mixtures in the gas phase above the bubbly liquid. Figure 1 shows the variant of the apparatus for this procedure. At the top flange a hot

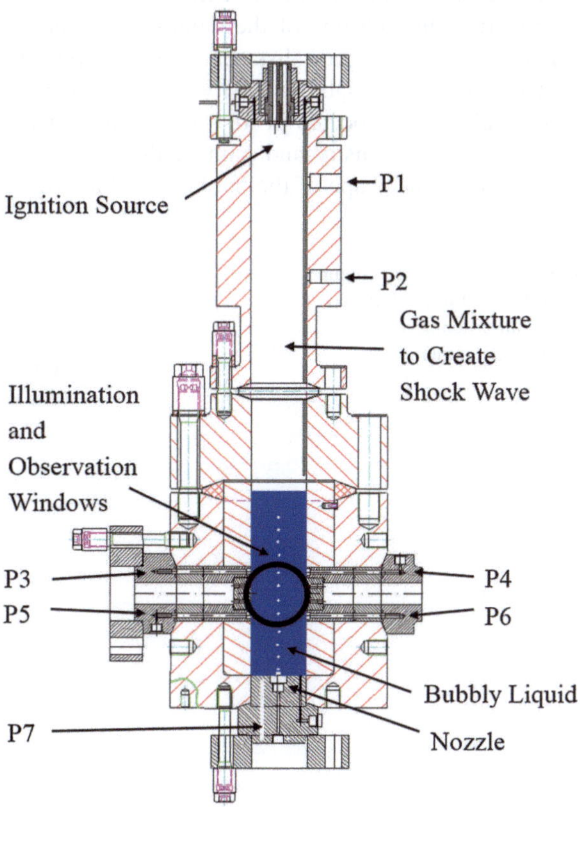

Fig. 1 Experimental set-up to generate shock induced bubble explosions. *P1–P7* denote positions of pressure sensors. Opposite windows serve to illuminate bubbly liquid and to observe it by a high-speed camera

wire igniter was installed to initiate the gas detonation which was monitored using two pressure sensors P1 and P2. Further tests were performed where the initial shock waves were generated mechanically by pushing a piston onto the liquid for a better control of the shock pressure independently of the system pressure.

The generated shock waves were detected by pressure measurements and especially by high-speed photography. Pressure measurements were performed using piezoelectric pressure transducers type 601H of the Kistler AG in combination with a multi-channel transient recorder operating at a sample rate of 1 MHz. The signal of the pressure transducers was fed to a Kistler 5001 SN amplifier. The amplified signal was recorded by a type TRA 800 transient recorder of the company W+W Instrumente AG Basel.

High-speed photography was applied to visualize the shock waves in the gas phase as well as in the liquid phase and further more to observe the dynamical behavior of the bubbles during their compression and explosion. For this purpose, a prototype of the digital high speed framing camera of the company Shimadzu, model ISIS V2 was used. It offers a maximum framing rate of 1,000,000 fps and a resolution of 312 × 260 pixels. Schlieren and shadow techniques are well known methods, see e.g. [8], to visualize gradients of the optical index in fluids as e.g. generated by shock waves. In the case under consideration, these techniques turned out to be even too efficient because the schlieren produced by the shock waves were so strong so that the shadows of the bubbles were perturbed. Therefore, a diffuse illumination through one of the windows was applied and the high-speed camera was mounted at a certain distance from the opposite window as shown in Fig. 2.

The light with variable intensity up to 2 kW was emitted by a quartz lamp onto an opal glass pane located in front of the illumination window. By the appropriate choice of light intensity and gain of the camera it was possible to visualize the shock waves, the shape of the bubbles, and the light emitted by exploding bubbles

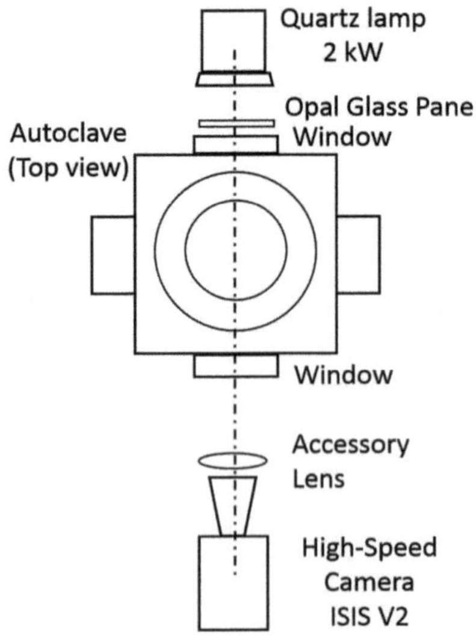

Fig. 2 Optical arrangement for diffuse illumination and observation using a high-speed camera

in each experiment. Triggering of the camera recording was derived from a signal of one of the pressure sensors. The passage of a shock wave across the sensor creates a sharp rise of the pressure signal that could be transformed electronically and fed into the trigger connectors of the transient recorder and of the high-speed camera. In combination with the pre-trigger feature of these devices an appropriate start of the relevant observation period was achieved. For illustration, Fig. 3 depicts a typical sequence of high-speed frames of shock induced bubble compression.

The observation window is seen to its full size at different points in time on each frame of Fig. 3. The dark horizontal line, slightly below the middle of the window, corresponds to the surface of the bubbly liquid. The charts at 24–32 μs show the gas detonation running towards the surface of the liquid. In the gas phase, the reflected shock wave can be seen running upwards in the sub-images at 36–50 μs. During this period the shock wave in the liquid, created by the impact of the gas detonation onto the surface, can be identified as a thin, bright horizontal line moving from the surface at 36 μs towards close to the lower part of the window at 50 μs. The bubbles, visible as dark spots in the liquid, shrink after passage of the shock wave in the liquid.

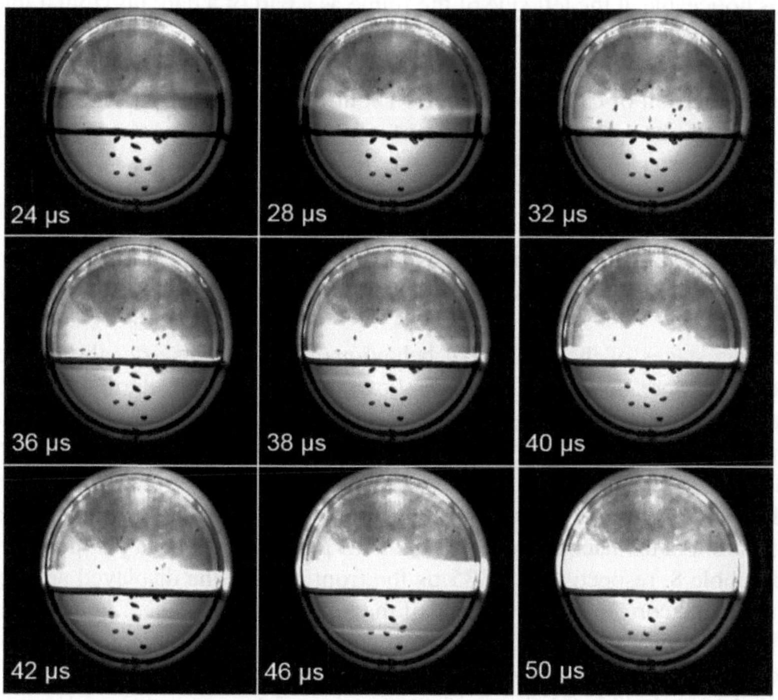

Fig. 3 Impact of shock wave on bubbly liquid and shock induced compression of bubbles. Framing rate 500,000 fps. The corresponding movie is available online

3 Results and Discussion

The behavior of bubble-containing fluids under the influence of a shock wave was investigated. As mentioned above, most of the studies were conducted with liquid cyclohexane. In addition, cumene, 2-ethylhexanal and methanol were used. Concerning the bubble gas, oxygen or oxygen-nitrogen mixtures were of particular interest. The mixture of the vapor of the liquid with these oxidizers inside the bubble are potentially explosive or could become explosive during the compression of the bubble. For comparative purposes experiments with nitrogen bubbles were carried out, from which conclusions were drawn about the dynamic behavior of the bubbles. The initial pressures of the bubbly systems were between 100 and 300 kPa.

For illustration, Fig. 4 shows the passage of the primary shock wave through a bubble-containing liquid and the subsequent bubble explosions. The liquid used here was cyclohexane. Pure oxygen was injected to create bubbles at an initial pressure of 100 kPa. The primary shock wave was generated by a detonation in the gas space above the bubble-containing cyclohexane. Due to the large scale of shades provided by the camera, the frame at 56 μs could be brightened for better visualization of the upper and lower rim of the observation window. The shock front is noticeable at the left side of the frame as a part of a dark horizontal line. The vertical position of the shock wave is indicated in each frame, except 120 and 122 μs where it is below the window already. The frames from 68 to 122 μs show the light emission of exploding bubbles about 25 μs after the passage of the shock front. The bright spot in the upper part of the frame at 56 μs originates from a reflection of external light and is not due to a bubble explosion in contrast to the other bright spots.

A more detailed recording is depicted in Fig. 5. The heterogeneous system under consideration is again cyclohexane-oxygen at an initial pressure of 100 kPa. The optical system in front of the camera was adjusted to image an area of about 30 mm by 30 mm.

Various stages of the dynamical behavior of the bubbles after the impact of a shock wave were observed. Figure 5 shows bubbles before, during and after the action of a primary shock. The captured period of time is so short that the general floating-up of the bubbles is not noticeable here.

At 11 μs, the primary shock wave just did not arrive at the visible area. The bubbles, numbered 1–8, all are in their initial state. They can be identified due to their position in the later frames. At 17 and 22 μs, the shock front just passes bubble 4 and bubble 8, respectively. At 35 μs the front has left the displayed area, but the pressure in the corresponding volume of the liquid is still elevated. In this frame, the individual bubbles were exposed to elevated pressure for different periods of time due to their position. Therefore, they are in different stages here. Bubble 1 is already exploded while bubble 8 is just in the beginning of the compression phase. Bubble 2 and 3 are compressed close to their minimum volume and start to explode. In the frame at 36 μs these two bubbles are close to the maximum of light emission

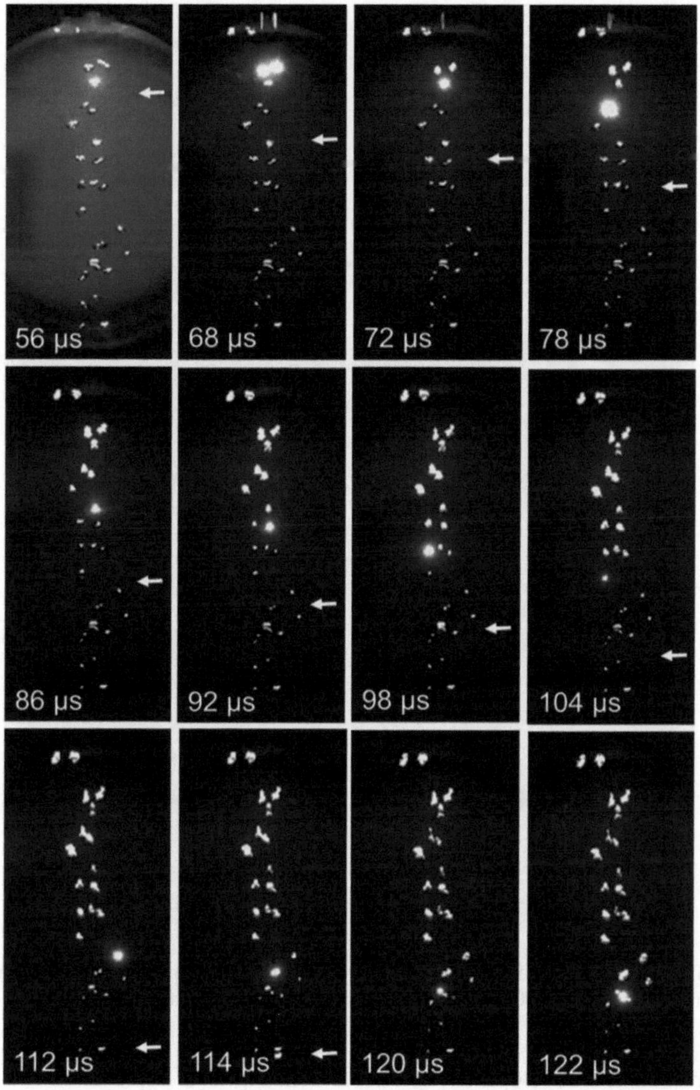

Fig. 4 Passage of pressure wave through bubbly liquid and subsequent bubble explosions. Framing rate 500,000 fps. The *arrows* mark the position of the primary shock-wave front. The corresponding movie is available online

generated by their explosion. At 37 μs the light emission has already weakened. During the explosion, the bubbles emit spherical shock waves recognizable e.g. as dark rings around bubble 2 and 3. Bubble 5 exhibits a so called jet formation during the compression phase, seen in frames at 35–37 μs of Fig. 5. This phenomenon will be discussed in more detail further down.

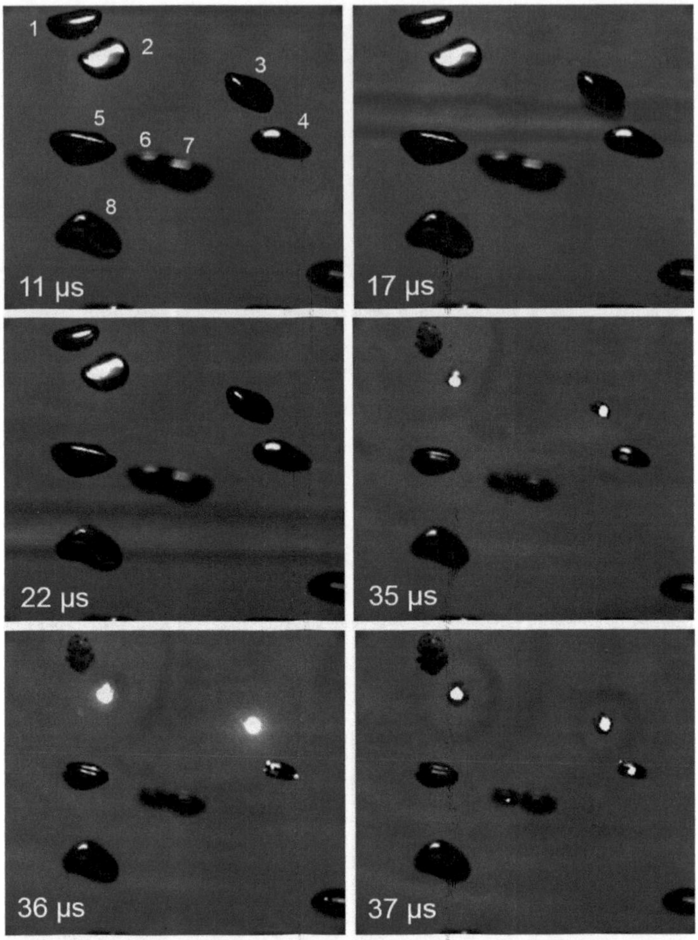

Fig. 5 Different stages of bubble explosion following passage of pressure wave through bubbly liquid. Framing rate 1,000,000 fps. The corresponding movie is available online

A more detailed development of bubble 5 in Fig. 5 is depicted in the frames of Fig. 6. The frame at 16 μs shows the incident front of the primary shock front slightly above the unperturbed bubble. The bubble here possesses a horizontal longitudinal extension of about 6 mm. After the passage of the shock front, the compression phase of the bubble begins. In the first phase, the bubble shrinks relatively slowly. At 29 μs the corresponding length dimension has decreased to about 5 mm while the relative geometrical shape of the bubble is essentially the same. Generally, during bubble compression, a jet, which penetrates the bubble, is often, but not always, observed. In the frame at 32 μs a jet has become obvious by the horizontal constriction which is not seen in the frame at 29 μs. This phenomenon can be seen clearly in the frames 37–39 μs of Fig. 6 where the shrinking

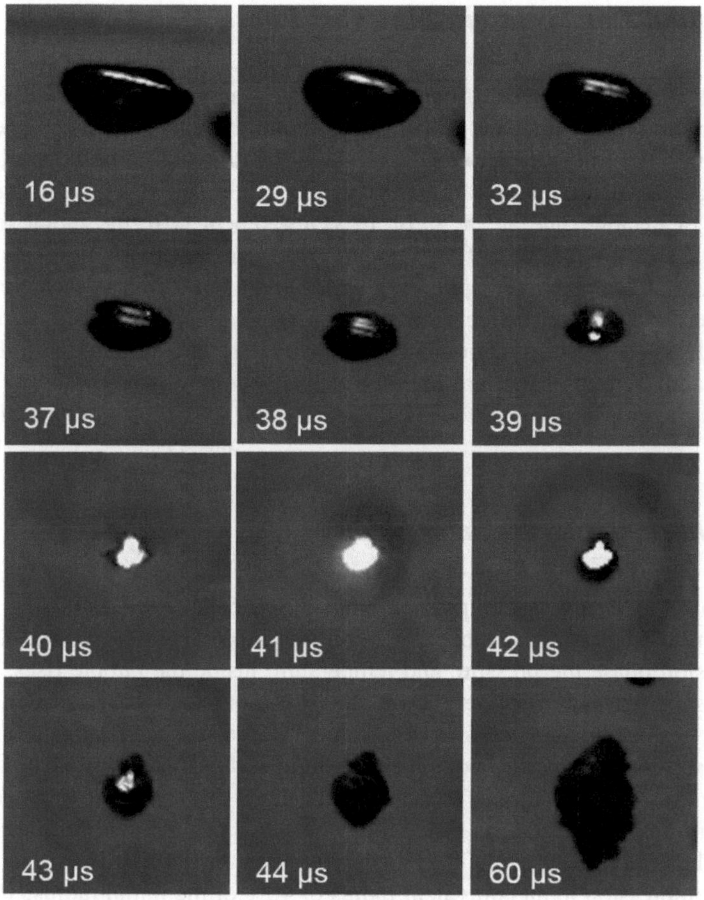

Fig. 6 Different stages of bubble 5 of the previous figure. Framing rate 1,000,000 fps

of the bubble continues at higher rate. At 40 μs, the beginning of the bubble's explosion is in evidence due to its light emission. The following frame, at 41 μs, shows the maximum light emission and the emission of a shock wave that appears as a dark shadow around the bright exploding bubble. At 42 μs the progress of the emitted shock can be identified as a ring around the weakening light. Frames at 43–60 μs show the end of the explosion and the expansion of reaction products.

Due to the non-spherical shape of the bubbles, it is useful to define the equivalent bubble radius r. The shape of the bubbles is approximately an ellipsoid with rotational symmetry around the vertical axis with a vertical principal axis. The equivalent bubble radius r is defined by the radius r of a sphere containing the same volume as the ellipsoid. The initial equivalent radius before the impact of the primary shock is denoted by r_0 in the following. The initial equivalent radius of the bubble in Fig. 6 is 2.5 mm. The initial equivalent radii of the cyclohexane-oxygen system under investigation were between 1.2 and 3.6 mm.

3.1 Initial Composition of the Gas Phase

The bubbles move upwards through the liquid for about half a second before they are met by the incident shock. In [9] it was found that this is enough time to reach more than 97% of the saturation value of vapor by diffusion from the liquid into the bubble gas at an initial pressure of 100 kPa of the bubble gas, using a vapor pressure of cyclohexane at 25 °C of 13 kPa given in [10], the molar fraction of cyclohexane in the bubble before the shock wave impact is 13 mol%. In this case the mixture of cyclohexane vapor and oxygen in the bubble is nearly stoichiometric and therefore it is well inside the explosion range. The situation is similar for a methanol-oxygen system. On contrast to that system, the systems with the liquids cumene and 2-ethylhexanal are initially well below the lower explosion limit due their low vapor pressure even when they reach the saturation limit by diffusion.

3.2 Shock Induced Compression

The bubble in Fig. 6 is compressed from $r_0 = 2.5$ mm (see Fig. 6 at 16 μs) to $r = 1$ mm just before the bubble explosion starts (see Fig. 6 at 39 μs). During the compression phase the pressure and temperature inside the bubble rises. Since the compression phase is relatively short, there is not much heat transfer from the bubbles to the liquid. Therefore, the values of pressure and temperature after a compression to a particular equivalent radius can be approximated iteratively by applying Poisson's equations for isentropic changes of state taking into account that the isentropic index of the gas inside the bubble changes with temperature as shown in [9]. Thus, a pressure of 2900 kPa and a temperature of 550 K can be estimated after a compression to an equivalent radius of 40% of the initial radius. This would lead to a new saturation value of the vapor, but this value cannot be reached by diffusion during the relatively short compression phase. It can be estimated, that concentration of the cyclohexane in the cyclohexane-oxygen system in the compressed bubble remains below 18 mol%. This means this system remains well inside the explosion range reported on in [11] during the complete compression phase at least as long as it is governed by diffusion processes. Similarly, the methanol-oxygen systems remain well inside the explosion limits. Due to the high temperature and pressure an explosion in the bubble can be induced. As mentioned above, the bubble in the liquid cumene and 2-ethylhexanal are initially below the lower explosion limit. They cannot reach the explosion range via diffusion alone during the compression phase. Nevertheless shock induced bubble explosions were observed in these systems, too. Details to this are reported in [12, 13]. Turbulences in the bubble and especially a jet formation during the compression phase can be responsible for this phenomenon.

3.3 *Jet Formation*

The main reason for the formation of a jet are inhomogeneities in the pressure field around the bubble, surface instabilities and the non-symmetrical shape of the bubbles, as suggested for the first time in [14]. It seems as if the jet was formed earlier in the case of closely adjacent bubbles than in others. In the example shown in Fig. 6, the equivalent radius r equals about 0.8 r_0, when the jet becomes noticeable at 32 µs. For the investigated cyclohexane-oxygen systems, jet formation was observed during compression when the equivalent bubble radius r has come within the range of 0.5 r_0–0.9 r_0. An additional mass and heat transport is associated with the penetration of the jet into the bubble, which has an influence on the explosive behavior of the bubble, further details are described in [9].

After the jet formation, the compression phase continues. Either the bubble is split into two parts or reaches a minimum equivalent radius without being split. In the case of oxygen-nitrogen bubbles, which have not been split, there is the possibility of ignition of an explosion as can be seen in the frame at 40 µs in Fig. 6. In this example, bubble ignition occurred when the equivalent radius r had decreased to about 0.4 r_0. Bubble ignition was observed only when the bubble was not split. For the cyclohexane-oxygen systems, the equivalent radius of the bubble during ignition was in the range of 0.2 r_0–0.6 r_0.

As mentioned above, vapor-oxidizer mixtures in the bubbles in the liquids cumene and 2-ethyhexanal are initially below the lower explosion limit. Nevertheless, shock induced explosions in these systems were observed. This can be explained by jet-formation and related mass transport from the liquid to the bubble gas, so that it reaches the explosion range and can be ignited by the heat generated in the compression.

3.4 *Influence of Various Parameters on Characteristics of Bubble Explosion*

The explosive behavior of individual bubbles under the impact of a shock wave is influenced by different parameters. Due to the dynamics of ascending bubbles, not all parameters, such as the bubble size and the distance between adjacent bubbles, could be set with high accuracy in the experiment. For this reason, statements about the systematic behavior when varying a single parameter must always be made taking into account the other parameters. Important parameters are initial pressure, shock pressure, gas composition of bubbles, bubble diameter, properties of the liquid, and distance between the bubbles. Parallel to the experimental studies, theoretical treatment was performed and discussed in detail in [15]. In Fig. 7, calculated limiting curves are depicted for bubble explosions in the cyclohexane-oxygen system.

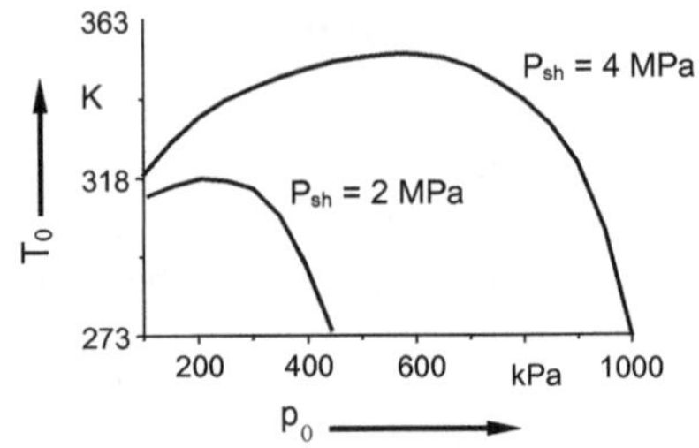

Fig. 7 Limiting conditions for possible bubble explosion in the plane of initial temperature T_0 and initial pressure p_0 of the system for pressures P_{sh} of 2 and 4 MPa of incident shock waves, respectively

For fixed shock wave pressures of 2 and 4 MPa, respectively, the corresponding region of initial temperature and pressure where no bubble explosion can occur is situated above the corresponding curve. The values are lower than the experimental ones. It must be borne in mind that the theoretical determination of the limits should be conservative in the sense of safety. In addition, idealized conditions are assumed for the theoretical determination, such as an idealized pressure jump to a constant level, spherical shape of the bubbles, and no jet formation during the collapse phase.

For air as bubble gas injected into cyclohexane, a bubble explosion was only found at an averaged shock wave pressure of at least 7.5 MPa. For an oxygen content of 10 and 90% of nitrogen injected into cyclohexane, no bubble explosion was observed even when exposed to shock waves of 10 MPa.

A few experiments have been performed using argon instead of nitrogen addition to the oxygen. Theoretically, nitrogen addition should have a stronger inerting effect than argon, since argon has a larger isentropic exponent and therefore tends to higher temperatures during compression enhancing bubble explosions. This trend was observed in the experiments with the corresponding gas bubbles in methanol as liquid.

3.5 Triggering Bubble Explosions by Nearby Exploding Bubbles

Apart from light, exploding bubbles emit spherical symmetrical shock waves into the surrounding liquid, as shown in Figs. 5 and 6. The shock pressure of such waves decreases inversely proportional to the distance from the bubble as outlined in [13]. In Fig. 8, a situation is depicted where such spherical shock waves trigger the ignition of another pre-compressed bubble.

The initial incident shock wave had passed the depicted area from 15 to 25 μs. At 45 μs two bubbles exploded, recognizable by the light emission. The following frame shows the emission of the spherical shock waves at 46 μs by these two bubbles. At 47 μs these waves have expanded so that their front is close to the lower left bubble. Both shock waves impact onto that bubble and trigger its ignition at 48 μs.

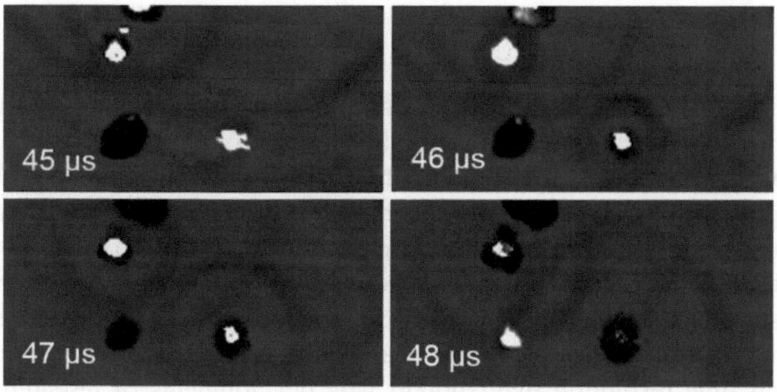

Fig. 8 Triggering of bubble explosion by explosion of neighboring bubbles. Framing rate 1,000,000 fps

The mechanism of the bubble ignition under consideration is based on the penetration of the spherical shock waves into the already pre-compressed bubble. The ignition is caused by the increase in temperature and pressure behind the refracted shock wave. The compressed state of the gas bubble caused by the first incident shock wave before the impact of the secondary shock waves plays an essential role for this type of ignition process. In this case, only due to the pre-compression, the difference in the acoustic impedance between the bubble and the surrounding liquid is sufficiently small, so that the attenuation of the shock wave when penetrating the bubble is relatively weak. The fractured shock wave within the bubble in this case was found to be strong enough to initiate an ignition in [16, 17].

The parameters of the gas behind the fractured shock wave were calculated. Thereafter, the shock wave penetrating into the bubble can initiate the bubble explosion when the involved bubbles are close enough to each other. In particular, the temperature and pressure behind the fractured shock wave may be so high that the ignition can start promptly as described in [16].

It follows that a synchronization of bubble explosions is possible. This result implies a better understanding of the safety-relevant phenomenon of the so-called bubble-detonation waves, i.e. self-sustaining shock waves in bubble-containing liquids, which were reported on in [3]. Such waves imply high risks in corresponding industrial processes, especially those using pure oxygen instead of air as oxidizing gas. Due to the high spatial and especially time resolution offered by the ISIS V2 high-speed camera a contribution to explain parts of such self-sustaining bubble detonation waves could be achieved. Further applications of this high-speed camera in the field of safety techniques are reported, e.g. to clarify the mechanism of explosions near the surface of combustible liquids reported in [18] and to investigate the origin and development of detonations in micro reactors presented in [19].

4 Conclusions

The elucidation of risks in the process industry can take advantages of the development of new optical methods. Using the example of explosion risks in oxidizer containing bubbles in organic liquids, high-speed photography has proven to be very helpful for the investigation of complex systems with the potential of fatal explosions. High spatial resolution and especially high time resolution up to one million frames per second of video photography has contributed to clarify the mechanism of explosions in heterogeneous systems and to find safer regimes of operation in systems relevant to the chemical industry. Further developments in spatial and time resolution may support new developments of research in safety technology.

References

1. H. Hieronymus, K. Mitropetros, J. Bender, H. Seeger, S. Seifert, R. Wendler, B. Plewinsky, Heterogene Explosionen. Technische Überwachung **43**, 39–45 (2002)
2. A.I. Sychev, A.V. Pinaev, Self-sustaining detonation in liquids with bubbles of explosive gas. J. Appl. Mech. Tech. Phys. **27**, 119–123 (1986)
3. A.V. Pinaev, A.I. Sychev, Structure and properties of detonation in a liquid–gas bubble system combustion. Explosion Shock Waves **22**, 360–368 (1986)
4. A.E. Beylich, A. Gülhan, Waves in reactive bubbly liquids, in *Proceedings of IUTAM Symposium on Adiabatic Waves in Liquid-Vapour Systems*, ed. by G. Meier, P. Thompson (Göttingen, 1989), pp. 39–48
5. K.H. Overhoff, H.G. Schecker, J. Fellensiek, U. Onken, Explosion behavior of bubbles ignited under water. DECHEMA-Monogr. **111** (Fortschr. Sicherheitstech. 2), 221–235; ISSN: 0070-315X (1988)
6. *Ullmann's Encyclopedia of Industrial Chemistry*. 5th ed., vol. A8 (VCH Verlagsgesellschaft mbH, Weinhaim, 1987), p. 219
7. M.I. Greene, C. Summer, R.J. Gartside, Cyclohexane oxidation, US Patent Nr. 6,008,415 (1999)
8. G.S. Settles, *Schlieren and Shadowgraph Techniques, Visualizing Phenomena in Transparent Media* (Springer, Heidelberg, 2001)
9. K. Mitropetros, H. Hieronymus, J. Steinbach, Single bubble ignition after shock wave impact. Chem. Eng. Sci. **61**, 397–416 (2006)
10. D.R. Lide (ed.), *CRC Handbook of Chemistry and Physics*, 81st edn. (CRC Press, Boca Raton, 2000)
11. W. Apel, K. Mitropetros, V. Schröder, and H. Hieronymus, Experimentelle Bestimmung von Explosionsgrenzen von Cyclohexan in reinem Sauerstoff. Chem. Ing. Tech. **77**, 133–137 (in German) (2005)
12. W. Apel, *Explosionen von Sauerstoffhaltigen Blasen in organischen Lösemitteln* (Technische Universität Berlin, Fakultät III – Prozesswissenschaften, 2004)
13. K. Mitropetros, *The Explosion Behavior of Oxygen Containing Bubbles In Cyclohexane, BAM-Dissertationsreihe*, vol. 11 (Technische Universität Berlin, Fakultät III - Prozesswissenschaften, 2005)
14. M. Kornfeld, L. Suvorov, On the destructive action of a cavitation. J. Appl. Phys. **15**, 495 (1944)

15. P.A. Fomin, K. Mitropetros, H. Hieronymus, Explosion limits of a single oxygen containing bubble in organic solvent with high saturated vapor pressure, in *Proceedings of 11th International Symposium Loss Prevention and Safety Promotion in the Process Industries (Loss Prevention 2004)* Prag, 2004, pp. 2179–2185
16. K. Mitropetros, P.A. Fomin, J. Steinbach, H. Hieronymus, Bubble ignition induced by nearby bubble explosion inside a multi-dispersed bubbly medium, in *Proceedings of the European Combustion Meeting* 2005, Louvain-la-Neuve, 3–6 April 2005 (CD-ROM)
17. K. Mitropetros, P.A. Fomin, H. Hieronymus, Safety aspects of a bubbly medium inside a chemical reactor. Chem. Eng. J. **107**, 27–32 (2005)
18. J. Dengel, R. Wendler, W. Denzer, J. Buchholz, M. Beckmann-Kluge, J. Steinbach, B. Plewinsky, H. Hieronymus, Explosions near the surface of organic liquids. Process Saf. Environ. **83**, 452–458 (2005)
19. E. Balcazar, F. Neher, C. Liebner, H. Hieronymus, E. Klemm, Determination of ignition temperature in micro reactors. Chem. Eng. Trans. **48**, 547–552 (2016)

Part VIII
Droplets

High-Speed Interferometry Under Impacting Drops

Kenneth R. Langley, Er Q. Li and Sigurdur T. Thoroddsen

Abstract Over the last decade the rapid advances in high-speed video technology, have opened up to study many multi-phase fluid phenomena, which tend to occur most rapidly on the smallest length-scales. One of these is the entrapment of a small bubble under a drop impacting onto a solid surface. Here we have gone from simply observing the presence of the bubble to detailed imaging of the formation of a lubricating air-disc under the drop center and its subsequent contraction into the bubble. Imaging the full shape-evolution of the air-disc has required μm and sub-μs space and time resolutions. Time-resolved 200 ns interferometry with monochromatic light, has allowed us to follow individual fringes to obtain absolute air-layer thicknesses, based on the eventual contact with the solid. We can follow the evolution of the dimple shape as well as the compression of the gas. The improved imaging has also revealed new levels of detail, like the nature of the first contact which produces a ring of micro-bubbles, highlighting the influence of nanometric surface roughness. Finally, for impacts of ultra-viscous drops we see gliding on ~ 100 nm thick rarified gas layers, followed by extreme wetting at numerous random spots.

1 Introduction

The iridescence caused by light interference in thin films is prevalent in everyday life. Walking through a parking lot after a rainstorm will reveal rainbow-like patterns of the oil slicks on top of puddles of water. These same patterns can be seen in soap films generated by children blowing bubbles in the summer or in the foam

K.R. Langley · E.Q. Li · S.T. Thoroddsen (✉)
King Abdullah University of Science and Technology (KAUST),
Thuwal, Saudi Arabia
e-mail: sigurdur.thoroddsen@kaust.edu.sa

K.R. Langley
e-mail: kenneth.langley@kaust.edu.sa

E.Q. Li
e-mail: eqli@ustc.edu.cn

K. Tsuji (ed.), *The Micro-World Observed by Ultra High-Speed Cameras*,
DOI 10.1007/978-3-319-61491-5_15

when hand washing dishes after a meal. To the uneducated observer these might appear as a simple trick of the light; however, this simple effect can convey precise information about the thickness and profiles of these films.

These colorful patterns have been observed and studied by scientists for centuries. Robert Hooke in his 1665 *Micrographia* noted the iridescence of peacock feathers and attributed the effect to alternating layers of plate and air. Sir Isaac Newton further studied this effect on peacock feathers but also investigated other thin films. In his 1707 book *Optiks*, he wrote:

> It has been observed by others, that transparent Substances, as Glass, Water, Air, &c. when made very thin by being blown into Bubbles, or otherwise formed into Plates, do exhibit various Colours according to their various thinness, altho' at a greater thickness they appear very clear and colourless… I took two Object-glasses, the one a Plano-convex for a fourteen Foot Telescope, and the other a large double Convex for one of about fifty Foot; and upon this, laying the other with its plane side downwards, I pressed them slowly together, to make the Colours successively emerge in the middle of the Circles, and then slowly lifted the upper Glass from the lower to make them successively vanish again in the same place.

Figure 1 shows Newton's sketch of the interference rings he observed. Although Newton had a primitive, corpuscular understanding of the nature of light, he still recognized that these bands represented the thickness of the air layer between the planar side of the plano-convex lens and the adjacent double convex lens. This phenomenon is now known as Newton's rings.

In 1801, Thomas Young proposed the wave theory of light and demonstrated the wave properties of light in his famous double slit experiment that revealed constructive and destructive optical interference. Young continued to mature the wave theory and was able to measure with reasonable accuracy the wavelengths of several colors, which are fundamental to the quantitative nature of interferometry.

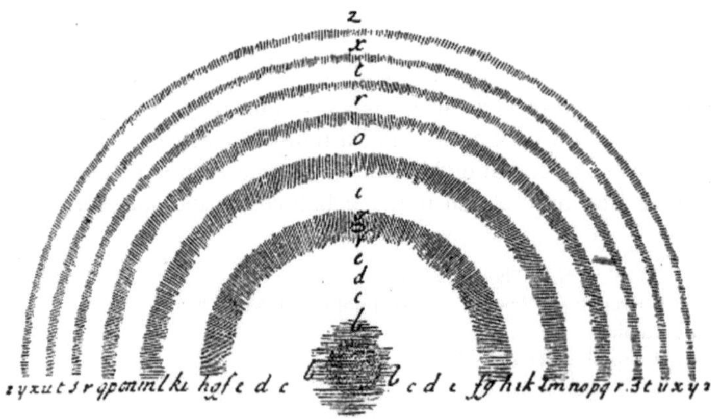

Fig. 1 Sketch from Isaac Newton's *Optiks* showing the interference patterns between a flat plate and a convex lens. The distance between the fringes becomes progressively shorter, away from the centerline, as the slope of the lens increases and the separation of the two surfaces increases more rapidly

The first application of light interference to quantitative measurements was developed by American physicist Albert A. Michelson in the 1880s and used in the famous Michelson-Morley experiment, which was meant to determine the relative motion of the Earth through the luminous aether by detecting changes in the speed of light in perpendicular directions by shifts in interference patterns. Although this experiment yielded negative results, this had immense implications for the development of relativity. Michelson later used his interferometer to measure other quantities eventually leading to his being awarded the Nobel Prize in Physics in 1907. In the more than one hundred years since Michelson's first interferometer, scientists have applied the basic principles of wave interference to a wide variety of disciplines and measurements. This includes measurement of strain-fields and vibrations within solids [1]; to measure rotation in ring laser gyroscopes in inertial navigation systems [2]; for testing the quality of optical surfaces [3]; for improving

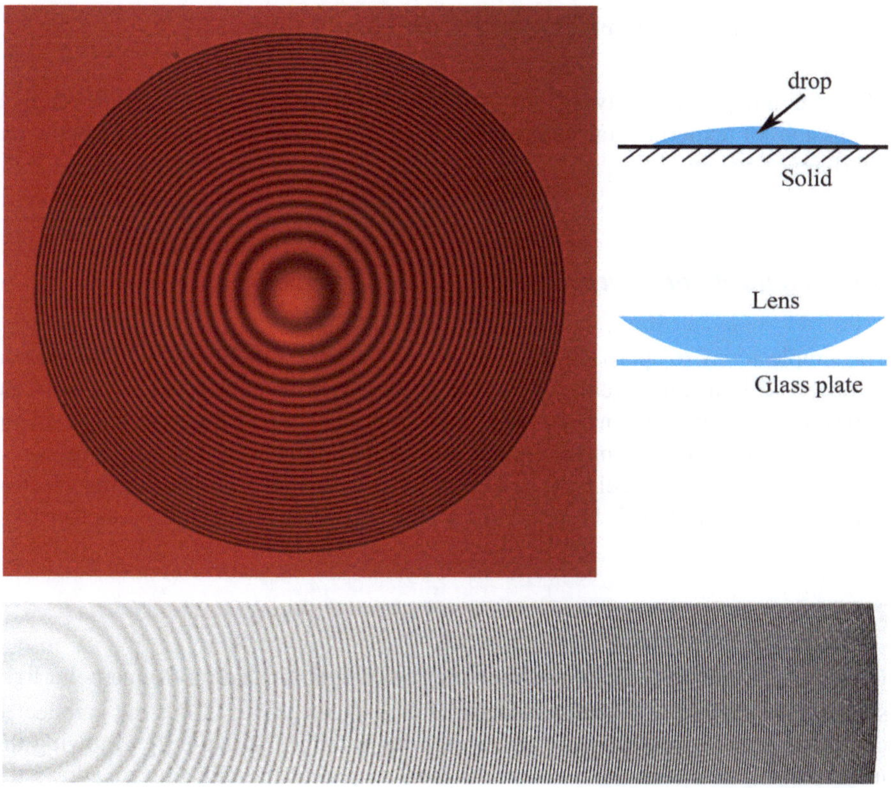

Fig. 2 Interferometry of an evaporating drop. *Top* Transmission interferometry of a water drop sitting on a hydrophilic silicone wafer. *Bottom* Larger drop with 183 fringes between its center to the contact line, on the right edge. Images taken with a 6 k video camera with ~ 0.8 μm/px spatial resolution and the contrast has been greatly enhanced. Sketches show the evaporating drop and the setup to obtain Newton rings

resolution of astronomical telescopes [4–6] as well as in biology and medicine for measuring biomolecules, sub-cell structures and tissue [7, 8], to name a few. One can argue the application of interferometry has culminated in last year's detection of gravitational waves generated by the merger of two black holes, using the two LIGO-devices, with their kilometers-long L-shaped interferometric arms [9, 10].

Figure 2 shows a typical application of interferometry in fluid mechanics, to measure the shape of a drop evaporating on a smooth solid surface. The drop is here strongly wetting and the contact angle is very shallow, exhibiting separate fringes over the entire surface. Despite showing 183 fringes the central thickness of the drop in the bottom part of Fig. 2 is only 44 μm. A similar approach has been used to study the shape of small inkjet droplets as they hit a solid [11].

Herein, we will show how high-speed interferometry is used to unravel the dynamics of the thin layer of air caught under a drop impacting on a solid surface.

2 Principles of Interferometry of Thin Films

The section begins by deriving a model for interference in a thin film, followed by a discussion of experimental setups and other considerations used to obtain the interference images.

2.1 *Theoretical Foundations*

The simplest way to approach the derivation of relative film thickness as a function of observed light interference is to look at a geometric model and determine the optical path-length difference between a ray of light reflected from the first interface and the first reflection from the second interface of the thin film [12]. Using results from the geometric model, a generalized model can be developed using electromagnetic theory [12, 13]. Figure 3 shows the important geometric parameters.

2.1.1 Geometric Model

First, consider a coherent, monochromatic ray of light emanating from a point light source traveling along $\overline{AB}$ with an angle of incidence θ_0. Part of this ray is reflected at surface **I** and part is transmitted through the interface at angle θ_1. Knowing the indices of refraction of the ambient medium n_o and of the thin film n_1 allows the determination of the relationship between θ_0 and θ_1 using Snell's law,

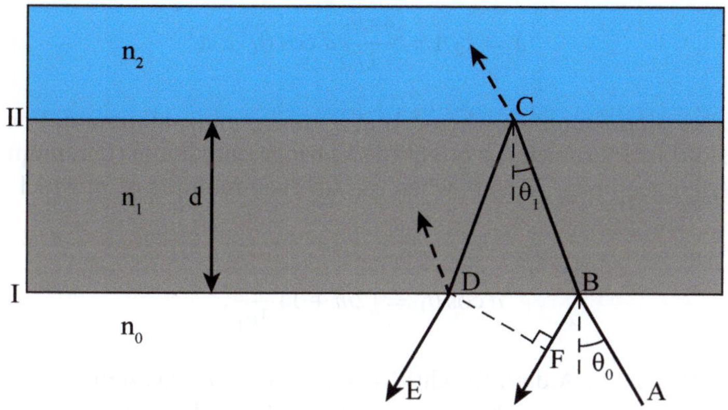

Fig. 3 Line drawing of light rays shining on and being reflected from a thin film

$$n_0 \sin \theta_0 = n_1 \sin \theta_1. \tag{1}$$

The light then passes through the thin film (gray area) where once again part is reflected and part is transmitted at interface **II**. The reflected light traveling along $\overline{CD}$ will again be reflected or transmitted when it reaches interface **I**. Note that only the first reflection from interface **II** is considered here. The difference in the optical path length, between the first reflection $\overline{BF}$ and second transmission $\overline{DE}$, can be calculated by

$$\Lambda = n_1 \left[|\overline{BC}| + |\overline{CD}| \right] - n_0 |\overline{BF}| \tag{2}$$

The lengths of segments $\overline{BC}$ and $\overline{CD}$ are equal and can be determined using the cosine of the transmission angle and the thickness of the film, d, as $|\overline{BC}| = |\overline{CD}| = d/\cos \theta_1$. Likewise, the length of segment $\overline{BD}$ can be determined using the tangent of the transmission angle. This combined with the angle of incidence and Snell's law (1) yields the length of segment $\overline{BF}$,

$$|\overline{BF}| = |\overline{BD}| \sin \theta_0 = 2d \tan \theta_1 \frac{n_1}{n_0} \sin \theta_1 \tag{3}$$

Combining the above segment lengths into (2) yields

$$\Lambda = \frac{2n_1 d}{\cos \theta_1} - n_0 2d \tan \theta_1 \frac{n_1}{n_0} \sin \theta_1 = 2n_1 d \cos \theta_1. \tag{4}$$

The optical path length difference can be related to the phase difference of the light, δ, by multiplying by the free-space propagation number. For angles of incidence up to $\sim 30°$, there is also an additional relative phase shift of π radians. This results in

$$\delta = k_0 \Lambda = \frac{4\pi n_1}{\lambda_1} d \cos\theta_1 \pm \pi \tag{5}$$

This expression simplifies when δ is at a maximum or minimum. For reflection (transmission) interferometry, a bright (dark) fringe, maximum (minimum) intensity in the interference image, occurs when δ is an even multiple of π, which simplifies (5) to

$$d \cos\theta_1 = (2m+1)\frac{\lambda_1}{4n_1}, \tag{6}$$

where $m = 0, 1, 2, \ldots$ A dark (bright) fringe, minimum (maximum) intensity in the interference image, occurs when δ is an odd multiple of π, which simplifies (5) to

$$d \cos\theta_1 = (2m)\frac{\lambda_1}{4n_1}. \tag{7}$$

By using light perpendicular to the film, $\theta_1 = 0$ and measuring only the maxima and minima of the fringe-intensity pattern, we can therefore obtain a thickness resolution of $\lambda/4$ between an adjacent maximum and minimum. Keep in mind that obtaining the absolute thickness corresponding to a particular fringe requires a known reference. In Sect. 3.7 we explain how this can be obtained for the problem at hand.

2.1.2 Generalized Model

Using electromagnetic theory, the interferometric model can be generalized by considering the complete electric and magnetic fields at each interface [12, 13] of the thin film. This generalization allows for better tailoring of the model to specific experimental conditions and can easily be adapted to multiple layers of films.

The boundary conditions at each interface require that the tangential components of both the electric, $\vec{E}$, and magnetic, $\vec{H}$, fields are continuous across the interface. This coupled with the fact that the electric and magnetic fields can be related through the index of refraction and unit propagation vector for nonmagnetic materials (i.e., $\vec{H} = \sqrt{(\varepsilon_0/\mu_0)} n\vec{k} \times \vec{E}$, where ε_0 and μ_0 are the electric permittivity and magnetic permeability in free space), yields the following relations at interface **I**,

$$E_I = E_{Ii} + E_{Ir} = E_{It} + E'_{IIr} \tag{8}$$

$$H_I = \sqrt{\frac{\varepsilon_0}{\mu_0}} (E_{It} - E_{Ir}) n_0 \cos\theta_0 \tag{9}$$

$$H_I = \sqrt{\frac{\varepsilon_0}{\mu_0}}(E_{It} - E'_{IIr})n_1 \cos\theta_1, \tag{10}$$

and at interface **II**,

$$E_{II} = E_{IIi} + E_{IIr} = E_{IIt} \tag{11}$$

$$H_{II} = \sqrt{\frac{\varepsilon_0}{\mu_0}}(E_{IIt} - E_{IIr})n_1 \cos\theta_1 \tag{12}$$

$$H_{II} = \sqrt{\frac{\varepsilon_0}{\mu_0}}E_{IIt}n_2 \cos\theta_{IIt}, \tag{13}$$

where E'_{IIr} in (10) is the light reflected from interface **II** when it has traversed the film back to interface **I** and the subscripts i, r, and t represent the incident, reflected and transmitted components at each interface. From the geometric model, a wave will have a phase shift of $k_0\Lambda/2$ for each time it traverses the thin film. Applying this to (11) and (12) yields

$$E_{II} = E_{It}e^{-ik_0\Lambda/2} + E'_{IIr}e^{ik_0\Lambda/2} \tag{14}$$

$$H_{II} = \sqrt{\frac{\varepsilon_0}{\mu_0}}(E_{It}e^{-ik_0\Lambda/2} - E'_{IIr}e^{ik_0\Lambda/2})n_1 \cos\theta_1. \tag{15}$$

Equations (14) and (15) can be solved for E_{It} and E'_{IIr} and then substituted into (8) and (10) forming the following linear system:

$$\begin{bmatrix} E_I \\ H_I \end{bmatrix} = M \begin{bmatrix} E_{II} \\ H_{II} \end{bmatrix}, \tag{16}$$

where the characteristic matrix, M, is

$$M = \begin{bmatrix} \cos(k_0\Lambda/2) & i\sin(k_0\Lambda/2)/Y_1 \\ Y_1 i \sin(k_0\Lambda/2) & \cos(k_0\Lambda/2) \end{bmatrix} \tag{17}$$

where for s polarization

$$Y_1 = \sqrt{\frac{\varepsilon_0}{\mu_0}}n_1 \cos\theta_1 \tag{18}$$

and for p polarization

$$Y_1 = \sqrt{\frac{\varepsilon_0}{\mu_0}} n_1 \frac{1}{\cos\theta_1}. \tag{19}$$

The characteristic matrix relates the fields at adjacent interfaces and can thus be defined for any pair of interfaces. In a system with p layers, a characteristic matrix can be defined relating the first interface to the pth interface by multiplying the characteristic matrices for each pair of adjacent interfaces (i.e., $M_{sys} = M_1 M_2 M_3 \ldots M_p$).

Reflection and transmission coefficients can be defined such that $r = E_{Ir}/E_{Ii}$ and $t = E_{It}/E_{Ii}$ and by expanding the system in (16) can be written in terms of the relevant physical parameters (i.e., indices of refraction, film thickness, angle of incidence and light wavelength) yielding

$$r = \frac{Y_0 m_{11} + Y_0 Y_2 m_{12} - m_{21} - Y_2 m_{22}}{Y_0 m_{11} + Y_0 Y_2 m_{12} + m_{21} + Y_2 m_{22}} \tag{20}$$

and

$$t = \frac{2Y_0}{Y_0 m_{11} + Y_0 Y_2 m_{12} + m_{21} + Y_2 m_{22}}, \tag{21}$$

where m_{ij} are entries in the characteristic matrix, M, and Y_k are defined as in (18) or (19) with the index of refraction and angle from the corresponding layer properties. If there is no interference from s/p polarization from an incoherent light, then the image intensity of the reflected light can be related to the square of the reflection coefficient (i.e., $I \sim |r_s|^2 + |r_p|^2$) [14, 15].

The power of this generalized model lies in its ability to be adapted to a diverse set of experimental conditions. De Ruiter et al. [13] use the reflection coefficient to adapt their model for differences in angles of incidence of the originating light beam, as experienced in microscope illumination, and to account for wavelength distributions in a finite bandwidth of light, as might be experienced from using LED illumination or an interference optical filter with a polychromatic light.

For idealized situations of monochromatic light incident perpendicular to the interfaces, the simplified geometric relations from the previous section apply. The intensity of the interference signal between the extrema then changes simply as:

$$I(r) \propto \sin^2\left(\frac{\delta(r)\pi}{\lambda/2}\right). \tag{22}$$

3 Uses of Interferometry to Measure Thin Fluid Films

There are limited ways to precisely measure the thickness of thin fluid films, that range in thickness from 100s of nanometers to a few micrometers. Interferometry has been an instrumental technique in obtaining both relative and absolute thicknesses of these films. When interferometry is coupled with high-speed imaging, the time evolution of the entire thickness profiles can be elucidated. For the air-film under an impacting drop, this requires sub-microsecond temporal resolution, as will be shown below.

3.1 Early Work

For more than half a century, researchers have been studying the critical thickness at which liquid films rupture. This has primarily been motivated by emulsion dynamics, where droplets of one liquid inside another immiscible liquid come in contact and can coalesce if the intervening film is broken. The stability of these films determines whether the emulsion coarsens with time or stays intact. For very thin films, surfactants and exotic non-hydrodynamic molecular forces, such as van der Waals, start playing a key role. It is therefore important to measure how thick these films are and how they drain with time [16], as well as measure these forces directly versus distance [17].

The earliest versions of these experiments [18–20] used photomultipliers to detect the intensity of the light reflected from the film. The thickness of the film, h, was determined using

$$h = \left(\frac{\lambda}{2\pi n}\right) \arcsin\left\{\frac{\Delta}{1+[1+4Q/(1-Q^2)](1-\Delta)}\right\}, \tag{23}$$

where $\Delta = (I - I_{min})/(I_{max} - I_{min})$ and $Q = (n-1)^2/(n+1)^2$ and I_{max} and I_{min} correspond to the last maximum and minimum interference values recorded during the dynamics [21]. Here n is the index of refraction of the film. These experiments were able to determine the absolute thickness of the film by tracing the fringe evolution from the time of rupture backward in time. With these photomultipliers it is only possible to measure the thickness at a single location. While some more recent studies have continued the use of photomultipliers [21], some have begun using high-speed cameras to track the temporal and spatial evolution of the thin film profile [22], while initially being limited to slowly evolving films.

Such interfacial air-films can become very extended, as shown in Fig. 4 for a highly viscous drop impacting onto a pool of low-viscosity silicone oil. Here the air-film covers the entire interface, even up to the pool surface along the thin cylindrical thread trailing behind the drop. Here the interferometry can only be applied along the vertical axis of symmetry, but must take into account the

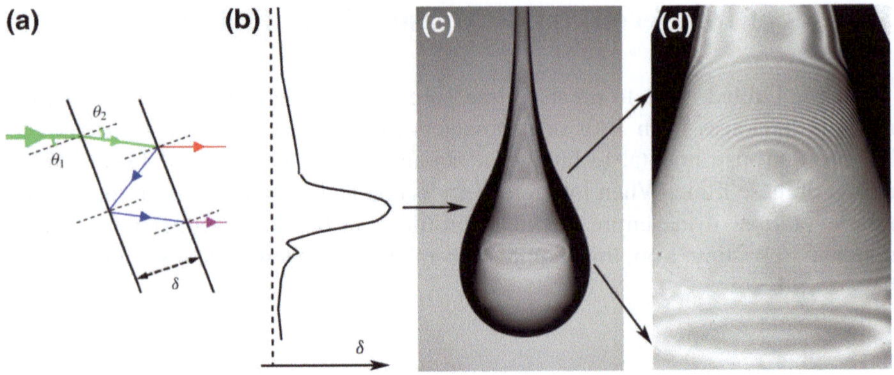

Fig. 4 Interferometry of a thin air-film wrapped around a viscous drop penetrating into a low-viscosity pool. The maximum in the air-film thickness in (**b**) is ~6 μm. The drop in (**c**) is about 2.3 mm across. Modified from Beilharz et al. [23]

changing angle of the incident light, as indicated by the sketch. The localized thickening of the air-layer on the top-side of the drop, requires separate experiments at larger magnification, to resolve the multitude of fringes.

3.2 *The Air Film Under an Impacting Drop*

When a drop impacts perpendicularly onto a flat, solid surface, a small axisymmetric disc of air is entrapped under its center. This entrapment occurs much faster than for the emulsions described above. On the other hand, the impact timing is repeatable and the drop can cut a laser-beam to precisely control the imaging [24]. The air-disc contracts, by surface tension, ultimately resulting in an air bubble at the solid-liquid interface. As the droplet approaches the surface, there exists a stagnation point beneath its center in the surrounding gas. Thus the droplet is unable to push aside all of the gas and the lubrication pressure within this gas-flow becomes sufficiently strong to create a dimple in the bottom of the drop. This phenomenon has real-world implications in fields such as spray coating, particularly with conductive materials, or for inkjet printing, where electrical or thermal contact can be detrimentally affected by the bubble.

Chandra and Avidisian [25] first photographed the entrapped bubble from above using short duration flash photography. Thoroddsen and Sakakibara [26] were able to image the air disc from below through a glass substrate, while studying the expanding lamella of an impacting drop, using multiple strobe flashes. Van Dam and Le Clerc [27] noticed the same entrapment phenomenon under much smaller droplets from inkjet printing. However, none of these early studies presented a clear explanation for the formation of this bubble. Cavitation was even suggested as a possible cause.

The advent of digital high-speed video cameras significantly advanced the study of this and related phenomenon [28]. Thoroddsen et al. [24] studied the initial size of the air disc and its subsequent contraction using high speed imaging at frame rates ranging up to 100 kfps. Although they were not able to measure the shape of the air disc, by measuring its radial size and the volume of the resulting central bubble, they could estimate the average initial thickness of the disc, as being between 1 and 4 μm. They also successfully modeled the contraction rate of the disc into the central bubble, based on capillary-inertial dynamics. Using X-ray imaging Lee et al. [29] observed that the central air disc was not of uniform thickness during the contraction, but grew thicker at the edges. They also verified a peculiar mechanism, shown in Fig. 5, where a small droplet is pinched off inside the central bubble [24].

Concurrently with the above experiments, many researchers were working on theories and numerical simulations to describe the air-disc dynamics. This phenomenon is essentially a balance between the inertia of the drop and the viscous lubrication of the gas layer. As such, the most relevant non-dimensional parameter becomes the inverse Stokes number [30, 31]:

$$St = \frac{\mu_g}{\rho_\ell RV}, \tag{24}$$

where μ_g and ρ_ℓ are respectively the gas dynamic viscosity and liquid density, R is the drop radius and V is its impact velocity.

The problem was first simplified and analyzed in only 2D with the viscous gas deforming the drop inviscidly [30, 31] and was later expanded into a 3D axisymmetric model [32]. These studies all neglected any compression of the air, which is

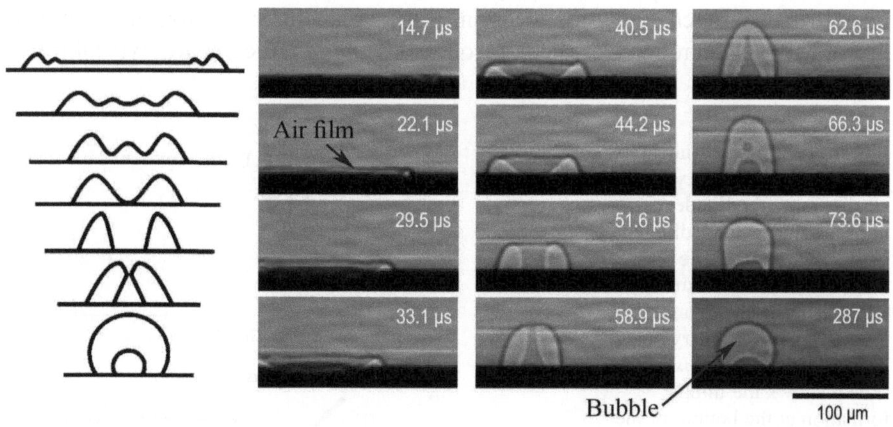

Fig. 5 Contraction of the air-disc into a bubble. Capillary waves travel from the edge towards the center where they amplify and touch the substrate, thereby pinching off a micro-drop inside the bubble inside the drop. Sketch modified from [24] and X-ray images from [29]

a reasonable assumption for impact velocities $\lesssim 1$ m/s. Mandre et al. [33] formulated a compressibility parameter

$$\varepsilon = \frac{P_{atm}}{\left(RV^7 \rho_\ell^4 / \mu_g\right)^{1/3}},$$

that predicts when the gas will be significantly compressed, if $\varepsilon < 3$.

The characteristic distance between the drop and the solid surface, H, when the lubrication pressure in the gas $\sim \mu_g VR/H^2$ can counteract the inertia of the drop $\sim \rho_\ell V^2/H$, scales as $H \sim RSt^{2/3}$, with the use of Eq. (24). For a millimetric water drop impacting at 1 m/s, this gives a distance ~ 1 μm. The corresponding time-scale is then $\tau \sim H/V \sim 1$ μs. Furthermore, H reduces at higher impact velocities and τ becomes even smaller, on account of both the decreasing H and increasing V.

It is only recently that high-speed video sensors have achieved imaging at these time-scales to allow interferometry for direct comparison between experiments and theoretical models [34, 35].

3.3 Experimental Setup

There are many experimental configurations for interferometers, from the rather simple ones to the quite complex, such as Mach-Zehnder setups [36, 37]. Figure 6 shows a typical interferometry setup for measuring the air disc that is entrapped under an impacting drop as used by [38]. This basic setup is quite simple consisting of a coherent, monochromatic light source, a high-speed video camera and a dichroic mirror. This particular setup also allows the use of two different modes of interferometry by changing the location of the light source, i.e. transmissive when lit from above and reflective when lit from below. Figure 7 compares typical fringes

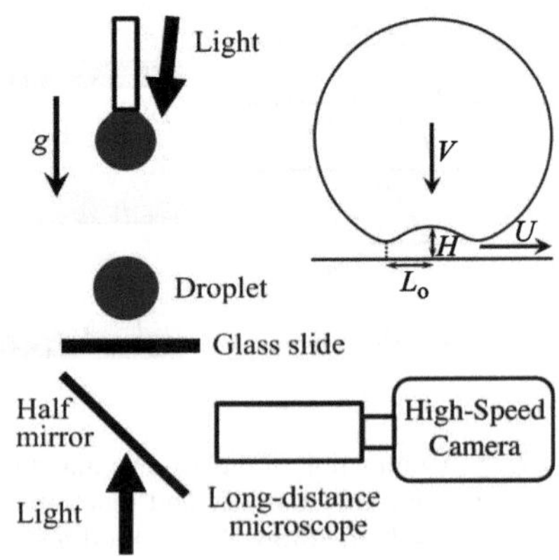

Fig. 6 Typical experimental setup for interferometry of the air disc entrapped under an impacting drop. Two different modes can be used, i.e. transmissive interferometry when lit from above and reflective interferometry when lit from below. The *inset* sketch shows the dimple formation at the bottom of the drop. Modified from [38]

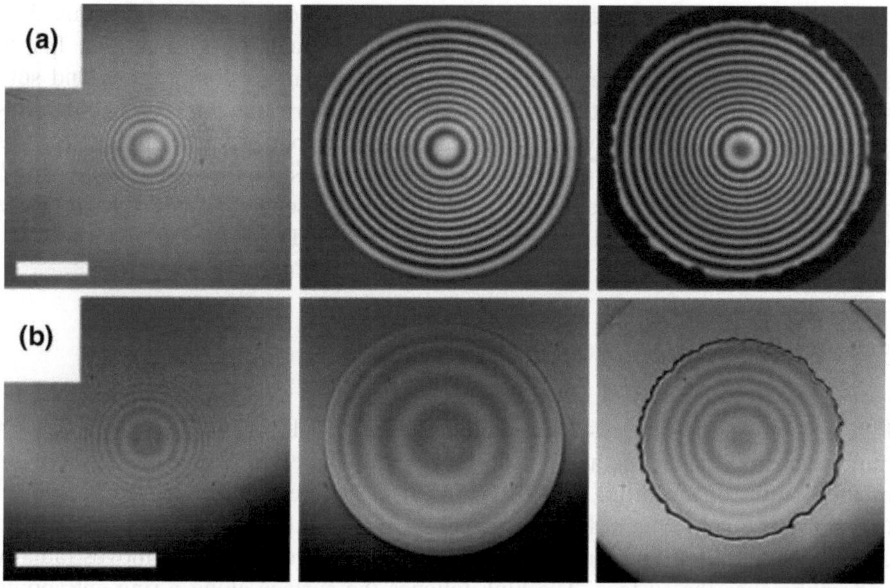

Fig. 7 The air-disc under an impacting drop. Comparison of *reflective* (**a**) and *transmissive* (**b**) interferometry, for two different impact conditions. From *left* to *right*, images show fringes before, at first contact and after formation of a wetting ring. For reflective interferometry, the contact between drop and glass surface appears as a black ring. The *scale bars* are 200 μm long. Modified from [38]

obtained with the two different illumination modes. The contrast is higher for reflective interferometry, but the transmitted one shows more details near the contact line.

At intermediate frame-rates the absolute thickness of a particular fringe cannot be determined from monochromatic images, unless there is a known reference, like a contact with the plate. This can be achieved with more sophisticated setups, such as by using more than one wavelength of the light. Van der Veen et al. [39] used frame rates of 10 kfps to extract snapshots of the evolving film profile, for $V = 0.22$ m/s. They used white-light interferometry, which requires a delicate calibration using colored fringes for a known air-gap between a lens and a plane, i.e. Newton rings (see Fig. 1). Butler et al. [40] have used a similar white-light interferometry setup to study slow drainage flows. Subsequently, de Ruiter et al. [13] used two-color interferometry to pin down the absolute film thicknesses, obtaining sub-fringe thickness accuracy of ~ 20 nm. The two wavelengths can be obtained using a polychromatic light source and use optical filters or a monochromator to obtain a single wavelength output [13].

The sophisticated color methods described above, can be sidestepped by "brute force" when using sufficiently high video frame-rates that individual monochromatic fringes can be followed from one frame to the next. In this way fringes can be followed back in time from the first contact to determine the absolute thicknesses for each fringe. This was demonstrated by Li and Thoroddsen [38] who used an ultra-fast high-speed video camera [35] with frame rates up to 5 million fps, to obtain time-resolved profiles of the air layer formation, its compression and subsequent rapid expansion. At the very short 200 ns interframe duration one requires in situ image storage [34], which limits the total number of frames to 180 for this camera [35]. The limited exposure of each frame makes the illumination intensity a challenge. The current system solves this by using a separate diode laser for each video frame. Each light pulse duration can be adjusted down to 50 ns.

3.4 Shape-Evolution of the Air Film

Figure 8 shows the time-resolved evolution of the air-layer profile under a $R_b = 4$ mm water drop, impacting at $V = 1.06$ m/s [38]. In panel (a) we see the deceleration of the bottom of the drop, from ~20 to 10 μs before contact (red curve). Panel (b) shows the subsequent dimple formation, with the centerline surface reversing course, while a kink forms away from the center, becomes sharper, moves radially and approaches the surface. The kink touches the surface along a ring entrapping the air-disc.

At a higher impact velocity of ~6 m/s, the deceleration at the bottom of the drop, can reach an astonishing 300,000 g's. The radial speed U of the kink can also be measured from the video frames and can become much larger than the impact velocity. It is in good agreement with theoretical predictions [33, 41], i.e. $U/V \sim St^{-1/3}$, with the radial velocity reaching values as high as 55 times the impact velocity [38]. For experiments at atmospheric pressure, the initial radius of

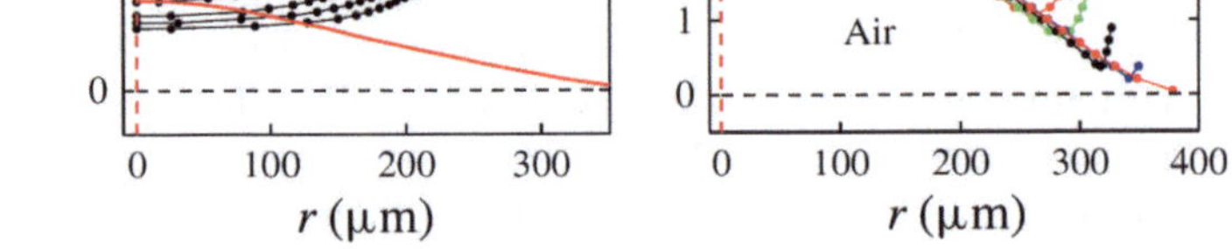

Fig. 8 Drop deceleration and dimple formation under a water drop impacting at 1.06 m/s, $R_b = 4.1$ mm ($St = 4.3 \times 10^{-6}$, $\varepsilon^{-1} = 0.69$). **a** Flattening of the bottom of the drop over a total duration of ~10 μs. *Red curve* shows shape at first contact. **b** Dimple and kink formation. Times shown at $t = -10.4, -9.8, -9.2, -8.6, -8.0, -7.4, -5.6, -4.6, -3.8, -2.8, -1.6$ and 0 μs relative to first contact (*red curve*). The vertical dashed red line marks the axis of symmetry. Modified from [38]

the air disc L_o is also in perfect agreement with the axisymmetric theory of Hicks and Purvis [32], without any adjustable constants, i.e.

$$L_o = 3.8\left(\frac{4\mu_g}{\rho_\ell V}\right)^{1/3} R^{2/3},$$

where R must be the bottom radius of curvature of the drop at impact R_b. This bottom shape was measured with a concurrent side-view camera.

3.5 Compression of the Gas

For large impact velocities the lubrication pressure in the air-layer, becomes strong enough to compress the gas significantly. The amount of compression can be measured directly by integrating the volume in the disc. Theoretically, the amount of compression depends on whether it progresses adiabatically, or isothermally [33].

Liu et al. [42] used monochromatic interferometry to first measure the profile of the air layer during compressible impacts, with time-resolution of 7 μs. They made an argument for isothermal compression, while the time-resolved profiles from Li and Thoroddsen [38] were more consistent with the adiabatic theory presented by Mandre et al. [33]. For $V = 4$ m/s the very rapid expansion following the first contact only lasts ~2 μs, which highlights the need for very rapid imaging. The actual dynamics are probably a combination of the two idealized situations and full thermodynamic treatment will be needed, including the heat conduction into the drop and the solid [43].

3.6 Effect of Nanometric Surface Roughness

Thoroddsen et al. [24] noted that a ring of microbubbles is often left behind on the solid following the drop impact. This ring marks the initial size of the air disc, which might be related to the first contact dynamics. Li et al. [44] investigated this with the high-speed interferometry. They found that even a few nanometer surface roughness influences the formation of these micro-bubbles, which are observed for 10 nm roughness of a microscope glass slide, while they are absent for impacts onto molecularly smooth, freshly cleaved mica sheets. This is shown in Fig. 9. The proposed mechanism is that when the surface kink travels radially away from the center, it makes localized contacts with substrate asperities, slightly before the fully wetting contact, as sketched in panel (a). When the different wetting spots touch they entrap the microbubbles, which thereby mark the original touchdown of the drop. At higher impact velocities the initial film thickness becomes progressively thinner and one will expect surface roughness to become even more important.

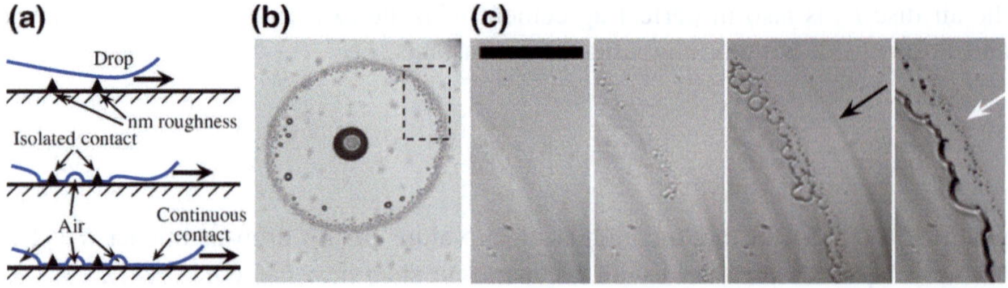

Fig. 9 Transmission interferometry under a water drop impacting a glass slide. **a** Conceptual model, where the radially moving kink in the free surface makes the first localized contacts with asperities on the solid surface. **b** Ring of micro-bubbles mark the radius of the first contact of the drop. **c** Close-up frames showing the first contact. Times are t = −0.2, 0, 0.4 and 1.8 μs with respect to first contact. *Scale bar* is 100 μm long. Modified from [44]

3.7 Skating or Gliding on Top of the Air Film

Mandre et al. [33] made the intriguing suggestion that an impacting drop does not touch the solid substrate, but rather skates on a thin layer of air while it jets out radially. They suggested that surface tension prevents the kink in the drop surface from touching down on the solid. This can indeed occur for very low impact Weber numbers, where the drop can rebound without ever touching the surface. For water drops, de Ruiter et al. [45] have shown such rebounding for $We = \rho RV^2/\sigma < 4$. For a millimetric water drop this corresponds to an impact velocity of 0.48 m/s. However, at higher impact velocities, the kink must contact the solid as the air-disc contracts into the central bubble [24]. The skating, therefore, is not relevant to the splashing of the drop, as discussed in [46].

Driscoll and Nagel [47] used sub-fringe interferometry to show that, for moderate impact velocities, if there were any air-layer present, its thickness would have to be less than 3.5 nm. This would certainly make the air-layer highly unstable to van der Waals forces.

In contrast, when the viscosity of the drop is much larger than that of water, the surface kink is suppressed and gliding can occur [48]. For the largest viscosities, this motion is more akin to draping onto the air cushion. Prior theoretical models assumed the liquid to be inviscid and most experiments were done with low viscosity liquids. Langley et al. [48] systematically investigated the effect of the droplet viscosity on the formation of the air disc. They used silicone oils to vary the liquid viscosity across seven orders of magnitude, from 10 to 20 million times that of water. As the drop viscosity increases the center of the air-disc becomes progressively thinner as internal viscous stresses limit the drop deformation. In addition to obtaining the profile of the air layer as the central dimple is formed, they also discovered, as the viscosity increased, that the drop forms an extended gliding layer of air near the outer edge, instead of making contact around a ring as with the lower viscosity liquids. Now contact between the viscous drop and the substrate is made when the thinnest air layer ruptures in numerous random locations, as shown in

Fig. 10. Two-color high-speed interferometry was here needed to determine the thickness of the layer of air on which the droplet was gliding when the ruptures first occur. Typical video images taken with the two colors, are shown in Fig. 10. This layer of air was in all cases less than 160 nm thick when it ruptures. Figure 10a compares the dimple shape for the two colors, which are essentially independent measurements. The symbols mark the maxima and minima in the fringe patterns, while the continuous curves are evaluated from Eq. (22). The two deviate by less than 10 nm. For thicknesses less than $\lambda/4$ the accuracy of the measurement is reduced and perhaps other techniques may work better, as discussed in the next sub-section.

Once the thickness of the extended air layer was determined with the two colors, monochromatic interferometry, which could be performed at much higher frame rates, was used to determine the radial profile of the air layer and its shape evolution. Once the droplet started gliding, the central dimple remained approximately frozen in shape, even before full contact around a ring was achieved. This shows that minimal gas flow is driven away from the center out through the thin film. The film is simply too thin for this flow to be significant. The regions between the wetting contacts (dark spots in Fig. 10a) entrap a myriad of micro-bubbles when they meet. Keep in mind that this entrapment mechanism occurs before a full circular contact line is established around the whole center. This mechanism is therefore somewhat different from the local contacts in the ejected lamella, levitating ahead of the spreading contact line, which was discovered in [49] and further studied by [50, 51].

Surprisingly, the spreading velocity of these localized wetting contacts, appear to be independent of the drop viscosity, which is inconsistent with earlier theories [52]. However, the air-film has thinned below 100 nm and rarified gas effects may play a role here, as the mean-free-path is of similar size $\sim$70 nm. This extreme wetting

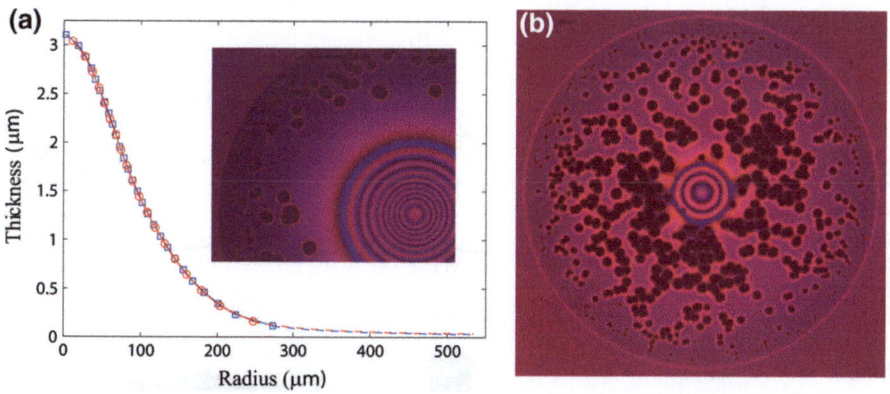

Fig. 10 Two-color transmission interferometry under a very viscous drop. **a** Shape of the bottom dimple measured with interferometry for both *red* and *blue light*. For drop viscosity of 100 cSt and $V \simeq 1.2$ m/s. **b** Numerous contacts between the bottom of the drop and the glass plate. For drop viscosity of 10,000 cSt and $V = 2.8$ m/s. Modified from [48]

may be assisted by enhanced slip, or the liquid draping onto the surface ahead of the contact line. We suggest that drop impacts may be the best way to study wetting of such high-viscosity liquids, as other methods, like plunging plates, will not work at very high viscosities owing to the ellipticity of the governing equations.

3.8 Frustrated Total Internal Reflection

Finally, we mention a related optical method that is also used to measure the thickness of the thinnest films during droplet impact. This is Frustrated Total Internal Reflection (FTIR) [53–56]. As was shown in Fig. 3, when a typical light beam encounters an interface of two medium, part of the light is reflected from the interface and part is transmitted through it. If the incident light approaches the interface at an angle greater than some critical angle θ_c (i.e., $\theta_0 > \theta_c$), all of the light will be reflected from the interface resulting in total internal reflection. When this occurs, there is an evanescent wave of light that penetrates to some minuscule depth into the second medium. If a third medium, such as an impacting drop, approaches the interface of the first two mediums (interface **I**) the light from the evanescent wave will tunnel into the third medium, frustrating the total internal reflection [57]. This method is a perfect way of detecting contact of a drop with the substrate, shown by dark regions in Fig. 11c. However, the transmission coefficient of light tunneling into the third medium must be known as a function of the distance between the interface and the third medium, to allow the image intensities to be used to determine the thickness of the intervening air films. This is particularly important for the dynamics of the first contact, where the separation becomes much smaller than $\lambda/4$. Also for cases where no contact occurs, like for impacts on very hot surfaces, i.e. the so-called Leidenfrost effect, where film boiling lubricates with

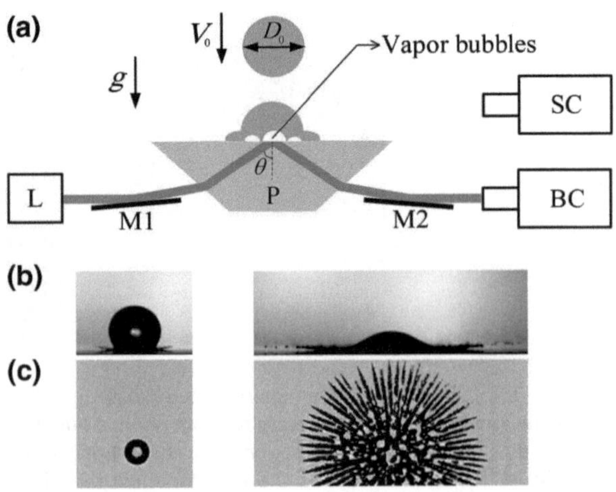

Fig. 11 **a** Setup for Total Internal Reflection (TIR) of a drop impact on a super-heated surface, using laser illumination (L) with a side-view (SC) and bottom-view (BC) cameras. **b** High-speed side-view frames and **c** bottom-view frames showing radial vapor streaks during. Modified from Khavari et al. [58]

a vapor layer. Figure 11 shows FTIR through a bottom prism as well as side-view images taken by Khavari et al. [58]. When a water drop impacts on a hot glass substrate, at the boundary between nucleate and film boiling, the vapor forms radial channels interspersed with contacting fingers.

3.9 Summary and Outlook

The rapid advancement of high-speed video technology has now reached large-pixel-area imaging at sub-μs time-resolution, with frame rates up to 10 million fps. This has enabled detailed study of many small-scale free-surface phenomena. One of these is the entrapment of a bubble under a drop impacting on a solid surface. Exploiting 200 ns time-resolution interferometry, we can resolve the time-evolution of the air-thickness profile. This has revealed new details during the first contact of the drop with the solid. It has also highlighted the importance of even nanometric surface roughness. Impacts onto more complex and compliant surfaces still await.

Finally, experiments at higher impact velocity or under reduced surrounding air pressure, will result in higher values of the compressibility factor, which should achieve larger compression of the gas and even thinner air films. Their rupture and dynamics are bound to bring new surprises.

References

1. G. Pedrini, W. Osten, M.E. Gusev, High-speed digital holographic interferometry for vibration measurement. Appl. Opt. **45**, 3456–3462 (2006)
2. R. Anderson, H.R. Bilger, G.E. Stedman, Sagnac effect: A century of earth-rotated interferometers. Am. J. Phys. **62**, 975–985 (1994)
3. M.V. Mantravadi, D. Malacara, Newton, Fizeau, and Haidinger interferometers. Opt. Shop Test **59**, 1–45 (2006)
4. A.A. Michelson, F.G. Pease, Measurement of the diameter of Alpha-Orionis by the interferometer. Proc. Natl. Acad. Sci. **7**, 143–146 (1921)
5. M.A. Johnson, A.L. Betz, C.H. Townes, 10-μm heterodyne stellar interferometer. Phys. Rev. Lett. **33**, 1617 (1974)
6. A. Labeyrie, Interference fringes obtained on VEGA with two optical telescopes. Astrophys. J. **196**, L71–L75 (1975)
7. D.D. Nolte, *Optical Interferometry for Biology and Medicine*, vol. 1 (Springer, Heidelberg, 2011)
8. D. Huang, E.A. Swanson, C.P. Lin, J.S. Schuman, W.G. Stinson, W. Chang, M.R. Hee, T. Flotte, K. Gregory, C.A. Puliafito, J.D. Fujimoto, Optical coherence tomography. Science **254**, 117881 (1991)
9. W.E. Althouse, M.E. Zucker, LIGO: the laser interferometer gravitational-wave observatory. Science **256**, 325 (1992)
10. B.P. Abbott, R. Abbott, LIGO scientific consortium: observation of gravitational waves from a binary black hole merger. Phys. Rev. Lett. **116**, 061102 (2016)

11. C.W. Visser, P.E. Frommhold, S. Wildeman, R. Mettin, D. Lohse, C. Sun, Dynamics of high-speed micro-drop impact: numerical simulations and experiments at frame-to-frame times below 100 ns. Soft Matter **11**, 1708 (2015)
12. E. Hecht, *Optics* (Addsion Wesley, Boston, 2002)
13. J. de Ruiter, F. Mugele, D. van den Ende, Air cushioning in droplet impact. I. Dynamics of thin films studied by dual wavelength reflection interference microscopy. Phys. Fluids (1994-present) **27**, 012104 (2015)
14. G. Wiegand, K.R. Neumaier, E. Sackmann, Microinterferometry: three-dimensional reconstruction of surface microtopography for thin-film and wetting studies by reflection interference contrast microscopy (RICM). Appl. Opt. **37**, 6892–6905 (1998)
15. O. Theodoly, Z.H. Huang, M.P. Valignan, New modeling of reflection interference contrast microscopy including polarization and numerical aperture effects: Application to nanometric distance measurements and object profile reconstruction. Langmuir **26**, 1940–1948 (2010)
16. D.Y.C. Chan, E. Klaseboer, R. Manica, Film drainage and coalescence between deformable drops and bubbles. Soft Matter **7**, 2235–2264 (2011)
17. J.N. Isrellachvili, G.E. Adams, Measurement of forces between two mica surfaces in aqueous electrolyte solutions in the range 0–100 nm. J. Chem. Soc. Faraday Trans. **74**, 975–1001 (1978)
18. A. Scheludko, D. Platikanowa, Untersuchung Dunner Flussiger Schichten Auf Quecksilber. Kolloid-Z. Z. Polym. **175**, 150 (1961)
19. A. Scheludko, D. Platikanov, E. Manev, Disjoining pressure in thin liquid films and the electro-magnetic retardation effect of the molecule dispersion interactions. Discuss. Faraday Soc. **40**, 253–265 (1965)
20. I. Ivanov, B. Radoev, E. Manev, A. Scheludko, Theory of the critical thickness of rupture of thin liquid films. Trans. Faraday Soc. **66**, 1262–1273 (1970)
21. J. Angarska, B. Dimitrova, K. Danov, P. Kralchevsky, K. Ananthapadmanabhan, A. Lips, Detection of the hydrophobic surface force in foam films by measurements of the critical thickness of the film rupture. Langmuir **20**, 1799–1806 (2004)
22. S.I. Karakashev, A.V. Nguyen, E.D. Manev, A novel technique for improving interferometric determination of emulsion film thickness by digital filtration. J. Colloid Interf. Sci. **306**, 449–453 (2007)
23. D. Beilharz, A. Guyon, E.Q. Li, M.-J. Thoraval, S.T. Thoroddsen, Antibubbles and fine cylindrical sheets of air. J. Fluid Mech. **779**, 87–115 (2015)
24. S.T. Thoroddsen, T.G. Etoh, K. Takehara, N. Ootsuka, Y. Hatsuki, The air bubble entrapped under a drop impacting on a solid surface. J. Fluid Mech. **545**, 203–212 (2005)
25. S. Chandra, C. Avedisian, On the collision of a droplet with a solid surface. Proc. Roy. Soc. Lond. A Math. Phys. Eng. Sci. **432**, 13–41 (1991)
26. S.T. Thoroddsen, J. Sakakibara, Evolution of the fingering pattern of an impacting drop. Phys. Fluids (1994-present) **10**, 1359–1374 (1998)
27. D.B. van Dam, C. Le Clerc, Experimental study of the impact of an ink-jet printed droplet on a solid substrate. Phys. Fluids (1994-present) **16**, 3403–3414 (2004)
28. S.T. Thoroddsen, T.G. Etoh, K. Takehara, High-speed imaging of drops and bubbles. Ann. Rev. Fluid Mech. **40**, 257–285 (2008)
29. J.S. Lee, B.M. Weon, J.H. Je, K. Fezzaa, How does an air film evolve into a bubble during drop impact? Phys. Rev. Lett. **109**, 204501 (2012)
30. F. Smith, L. Li, G. Wu, Air cushioning with a lubrication/inviscid balance. J. Fluid Mech. **482**, 291–318 (2003)
31. A. Korobkin, A. Ellis, F. Smith, Trapping of air in impact between a body and shallow water. J. Fluid Mech. **611**, 365–394 (2008)
32. P.D. Hicks, R. Purvis, Air cushioning and bubble entrapment in three-dimensional droplet impacts. J. Fluid Mech. **649**, 135–163 (2010)
33. S. Mandre, M. Mani, M.P. Brenner, Precursors to splashing of liquid droplets on a solid surface. Phys. Rev. Lett. **102**, 134502 (2009)

34. T.G. Etoh, D. Poggemann, G. Kreider, H. Mutoh, A.J.P. Theuwissen, A. Ruckelshausen, Y. Kondo, H. Maruno, K. Takubo, H. Soya, K. Takehara, T. Okinaka, Y. Takano, An image sensor which captures 100 consecutive frames at 1000000 frames/s. IEEE Trans. Electron Devices **50**, 144–151 (2003)
35. J. Crooks, B. Marsh, R. Turchetta, K. Taylor, W. Chan, A. Lahav, A. Fenigstein, Kirana: a solid-state megapixel uCMOS image sensor for ultrahigh speed imaging. Proc. SPIE **8659**, 865903 (2013)
36. E. Mach, Ueber einen interferenzrefraktor. Z. Instrumentenkd. **12**, 89–93 (1892)
37. L. Zehnder, Ein neuer interferenzrefracktor. Z. Instrumentenkd. **11**, 275–285 (1891)
38. E.Q. Li, S.T. Thoroddsen, Time-resolved imaging of a compressible air disc under a drop impacting on a solid surface. J. Fluid Mech. **780**, 636–648 (2015)
39. R.C.A. van der Veen, T. Tran, D. Lohse, C. Sun, Direct measurements of air layer profiles under impacting droplets using high-speed color interferometry. Phys. Rev. E **85**, 026315 (2012)
40. C.S. Butler, Z.L. Seeger, T.D. Bell, A.I. Bishop, R.F. Tabor, Local determination of thin liquid film profiles using colour interferometry. Eur. Phys. J. E **39**, 1–7 (2016)
41. L. Duchemin, C. Josserand, Curvature singularity and film-skating during drop impact. Phys. Fluids **23**, 091701 (2011)
42. Y. Liu, P. Tan, L. Xu, Compressible air entrapment in high-speed drop impacts on solid surfaces. J. Fluid Mech. **716**, R9 (2013)
43. P.D. Hicks, R. Purvis, Liquid-solid impacts with compressible gas cushioning. J. Fluid Mech. **735**, 120–149 (2013)
44. E.Q. Li, I.U. Vakarelski, S.T. Thoroddsen, Probing the nanoscale: the first contact of an impacting drop. J. Fluid Mech. **785**, R2 (2015)
45. J. de Ruiter, R. Lagraauw, D. van den Ende, F. Mugele, Wettability-independent bouncing on flat surfaces mediated by thin air films. Nat. Phys. **11**, 48–53 (2015)
46. C. Josserand, S.T. Thoroddsen, Drop impact on a solid surface. Ann. Rev. Fluid Mech. **48**, 365–391 (2016)
47. M.M. Driscoll, S.R. Nagel, Ultrafast interference imaging of air in splashing dynamics. Phys. Rev. Lett. **107**, 154502 (2011)
48. K.R. Langley, E.Q. Li, S.T. Thoroddsen, Impact of ultra-viscous drops: air-film gliding and extreme wetting. J. Fluid Mech. **813**, 647–666 (2017)
49. S.T. Thoroddsen, K. Takehara, T.G. Etoh, Bubble entrapment through topological change. Phys. Fluids **22**, 051701 (2010)
50. M.M. Driscoll, C.S. Stevens, S.R. Nagel, Thin film formation during splashing of viscous liquids. Phys. Rev. E **82**, 036302 (2010)
51. J. Palacios, J. Hernandez, P. Gomez, C. Zanzi, J. Lopez, On the impact of viscous drops onto dry smooth surfaces. Exp. Fluid **52**, 1449–1463 (2012)
52. A. Marchand, T.S. Chan, J.H. Snoeijer, B. Andreotti, Air entrainment by contact lines of a solid plate plunged into a viscous fluid. Phys. Rev. Lett. **108**, 204501 (2012)
53. J.M. Kolinski, S.M. Rubinstein, S. Mandre, M.P. Brenner, D.A. Weitz, L. Mahadevan, Skating on a film of air: drops impacting on a surface. Phys. Rev. Lett. **108**, 074503 (2012)
54. J.M. Kolinski, L. Mahadevan, S.M. Rubinstein, Lift-off instability during the impact of a drop on a solid surface. Phys. Rev. Lett. **112**, 134501 (2014)
55. M. Shirota, M.A. van Limbeek, C. Sun, A. Prosperetti, D. Lohse, Dynamic leidenfrost effect: relevant time and length scales. Phys. Rev. Lett. **116**, 064501 (2016)
56. M.A. van Limbeek, M. Shirota, P. Sleutel, C. Sun, A. Prosperetti, D. Lohse, Vapour cooling of poorly conducting hot substrates increases the dynamic leidenfrost temperature. Int. J. Heat Mass Transf. **97**, 101–109 (2016)
57. S. Zhu, A. Yu, D. Hawley, R. Roy, Frustrated total internal reflection: a demonstration and review. Am. J. Phys. **54**, 601–607 (1986)
58. M. Khavari, C. Sun, D. Lohse, T. Tran, Fingering patterns during droplet impact on heated surfaces. Soft Matter **11**, 3298–3303 (2015)

Observation of Chemical Reactions Induced by Impact of a Droplet

Stefan C. Müller

Abstract In order to investigate, how a chemical reaction starts and develops, we select the moment of the impact of a droplet falling into a reactive solution and observe the initial stage of the reaction with a high-speed camera. As examples, we use the pH indicator system with bromothymol blue (BTB) reaction, as well as the Belousov-Zhabotinsky (BZ) reaction forming an excitable medium, in which superthreshold disturbance propagates as an excitation wave and thus gives rise to a reaction-diffusion structure. Focusing on the BTB solution, we observe the color change caused by the droplet containing a pH indicator when impinging on the surface of alkaline solution. Contrary to our expectation, this reaction starts at the equatorial line, and not at the protruding edge of the droplet, where it first gets into touch with its reaction partner. Small vertical fingers emerge from the front line within 1.5 ms. Some arguments make it is likely that heat diffusion is responsible for the finger formation. For the BZ reaction, where due to a redox reaction the color changes from red to blue and vice versa, we do not observe this color change in our experiment. However, the effects on the drop shape (from spherical to ellipsoidal) is the same as observed in the BTB solution. Independently of the chemical systems, a thin needle-like tip developed from the protruding edge within about 300 μs after the drop has touched the solution surface. The emergence of this thin pin cannot be due to chemical processes, but to the physical impact of the droplet.

1 Introduction

Slow or very fast processes are common in chemical reactions. The slow ones may be readily observed with the eyes, the fast ones need more sophisticated measuring techniques. Let us focus on reactions occurring in the liquid or solid phase and, to be even more restrictive: to the liquid phase with water as the solvent. Here we find

S.C. Müller (✉)
Institut für Experimentelle Physik, Otto-von-Guericke-Universität Magdeburg,
Universitätsplatz 2, 39106 Magdeburg, Germany
e-mail: stefan.mueller@ovgu.de

K. Tsuji (ed.), *The Micro-World Observed by Ultra High-Speed Cameras*,
DOI 10.1007/978-3-319-61491-5_16

an extremely wide scale of kinetic constants which determine relevant reactions in the realm of physico-chemical processes as well as so many of biochemical or biological nature. Of great interest is to study the moment when a reaction really starts and what its course is like right after the first moments when the participating reactions get “into touch with each other”. This is the question we wish to ask here with appropriate reaction partners and observation techniques.

Measuring fast kinetics of a chemical reaction has been a major goal in many scientific fields, and impressive techniques have been developed to extend our knowledge of chemical processes occurring on very short time scales, for instance rapid flow [1] or relaxation [2] methods. More recently, laser applications make it possible to investigate reactions in the femtosecond range or even shorter [3]. Mostly, these reactions take place in a volume under spatially homogeneous conditions.

In this contribution we will focus on reactions that develop in spatially extended systems, such as in a reactive layer in a petri dish or in a cylindrical container without any stirring. Inhomogeneities may readily arise and reactions may propagate there in form of wave fronts or occur at the boundary of different phases [4, 5]. Of course, the dynamics of the travelling waves as well as the evolution of reactive states at a phase boundary are dependent on the time scales of the participating reactions and on the transport processes involved.

A high-speed video camera can be used to analyze the particular behavior of such inhomogeneous reaction systems. We will use image sequences up to 1,000,000 frames per second for our purposes.

As an experimental approach to grasp the early moments of a reaction we observe the time course of a falling drop. The drop is loaded with one of the reaction partners and impinges on the surface of its counterpart, whereby a color change takes place indicating the formation of reaction product. This process is observed with our fast camera, thus allowing for the evaluation of reaction velocity. Many parameters can be tuned appropriately in this procedure, such as concentrations, drop volume and size, falling distance, and others.

While performing these investigations we noticed that shape and speed of the falling drop exhibit properties which apparently develop independently of the chemical processes taking place in the drop volume. Beyond some drop distortions known from hydrodynamic studies, we observed the formation of a small needle-like sprout at the bottom of the drop which will be described later in some detail.

In the theoretical domain there have not been satisfactory explanations yet for most of the described phenomena.

2 Experimental

2.1 Materials

Two reactions with a color indicator were investigated.

(A) A pH-indicator system: a solution for the droplets is prepared from 0.1 g bromothymol blue (BTB) (Sigma-Aldrich, Germany) dissolved in 20 ml ethanol, followed by adding distilled water, with the whole volume amounting to 100 ml. The bromothymol blue concentration is 1.6×10^{-3} M. The color of this solution is yellow. 0.1 M NaOH solution was purchased from Bernd Kraft GmbH (Germany).

(B) Belousov-Zhabotinsky (BZ) system: A solution of 75 mM ferroin in 25 mM H_2SO_4 is prepared for the droplet. The counter solution (BZ solution) in the cuvette is a 1:1 mixture of the solutions I and II. Solution I consists of $NaBrO_3$ and NaBr in water with concentrations of 0.673 M and 0.061 M respectively. Solution II contains 0.235 M malonic acid in 0.51 M H_2SO_4. These two solutions are mixed with volume ratio 1:1 and it is waited until the brown color of initially produced bromine has disappeared.

2.2 Experimental Set-up

The experimental set-up is shown in Fig. 1. A glass cuvette of 1 cm optical path length (1 cm $\times$ 1 cm $\times$ 4 cm), or alternatively 2.5 cm optical path length (2.5 $\times$ 2.5 $\times$ 6 cm), was filled with NaOH or the BZ solution. The height of the pipette was variable from 4 to 17 cm above the surface of the liquid in the cuvette. The radius of the droplet was 1.6 mm. The cuvette was illuminated with a 150 W halogen lamp both from the back and the front through optical fibers. A sheet of white paper was attached to its back side in order to get close to homogeneous illumination. An ultra high-speed camera HPV-1 (Shimadzu Corp. Japan) for the BTB system and HPV-X (Shimadzu Corp. Japan) was placed in front of the cuvette. The spatial resolution of the camera per image is 312 $\times$ 256 pixels for the HPV-1 and 400 $\times$ 256 pixels for the HPV-X. The dynamic range is 10 bit (monochrom). The recording speed is varied up to 1×10^6 frames/s and 100 frames are taken successively for the HPV-1. Corresponding values are 1×10^7 frames/s and 256 frames for the HPV-X. The camera was triggered by a TTL signal, which was caused by an optical sensor at the moment when the drop passed across a red laser diode beam ($\approx$638 nm).

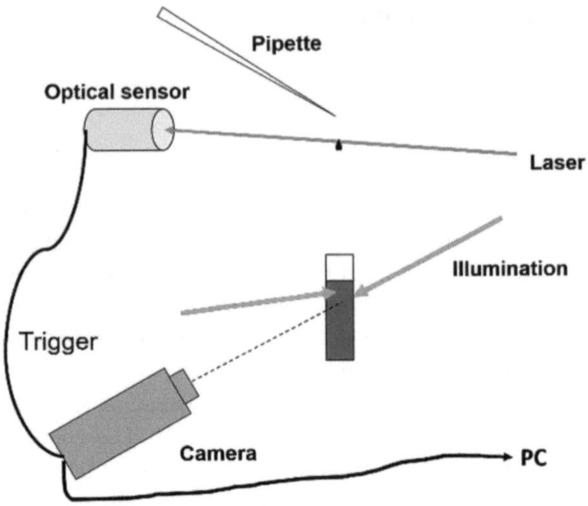

Fig. 1 Experimental set-up

3 Results

3.1 Bromothymol Blue pH-Reaction

Our first color reaction involves the pH indicator bromothymol blue, colored yellow in acid solution and turning to dark blue in alkaline environment: the pK value of bromothymol blue is 7.0. In the experiment a small drop (diameter 3.2 mm) of BTB falls from a height of 7 cm on the surface of a 0.1 M NaOH solution in a square cuvette of 1 cm optical path length. At the moment of drop impact, the color change starts and can be observed with a camera. We took a series of images with various camera speeds: with 1000 frames/s we observed the behavior of the entire drop, but a more interesting scenario appears when the frame rate is higher than 8000 frames/s.

The behavior of the droplet containing BTB in the initial phase is presented in Fig. 2 showing its color change when touching the surface of a solution containing NaOH. Note that the used camera is monochrome, and therefore, the color change results in the images as difference in gray level. In the left image (at 0.5 ms) the color change cannot be seen yet probably because of interference of the meniscus. In the subsequent image (at 1.5 ms) the droplet has changed its shape from a sphere to an ellipsoid. The thin dark band has developed in the center image to an equatorial band of a certain width. Small fingers appear at the lower edge of the blue band. The band moves downward and becomes broader in time. Later on (right image and following, not shown) the volume below this dark region remains always yellow. A needle-like sprout is observed as early as at 200 μm (figure not shown) and it is developing time by time (from left to right in Fig. 2).

Different from the early evolution, the later phase (after 3 ms) is affected strongly by the size of the cuvette, as well as the height of the pipette, suggesting

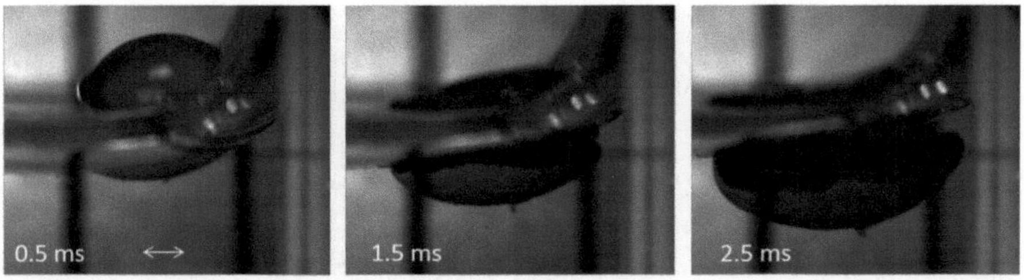

Fig. 2 Images of the bromothymol blue droplet entering a 0.1 M NaOH solution. Times after impact on the surface 0.5 ms (*left*), 1.5 ms (*middle*), and 2.5 ms (*right*). The thick curved zone extending from *left to the upper right* shows the meniscus between the reactive liquids. Camera speed: 16000 frames/s, scale bar 1 mm

that the later development of the chemical reaction is affected by the drop dynamics [5]. Since the droplet behavior in this later phase is similar to that in the BZ system, we will skip its explanation here and return to the issue later.

3.2 BZ Reactive System

Our second example for a reaction with a color indicator is the autocatalytic BZ reaction [6–8]. Here the redox catalyst ferroin plays a major role as a catalyzer of the reaction and at the same time a redox indicator with color changes between blue (oxidized form Fe^{3+}, ferriin) and red (reduced form Fe^{2+}, ferroin). Beyond performing autonomous oscillations, an important aspect of the BZ reaction is that under appropriate conditions the reaction starts locally and develops long ranging spatial inhomogeneity in form of traveling excitation waves.

The idea is to introduce a drop containing the catalyst ferroin into the bulk of the BZ reaction, thus initiating the full reaction scenario which consists of several steps: starting from a reduced (red), excitable state, a fast activation of an autocatalytic step leads to an oxidized (blue), excited state, followed by a much slower process of refractoriness during which the initial excitable state is recovered. This closes one oscillatory cycle with a change from red to blue and back to red, representing the characteristic dynamics of this reaction.

Figure 3 shows four images selected during the initial phase of a droplet impinging on the surface of the BZ-reactive solution, up to 800 μs. Here, a square cuvette with 1 cm optical path length was used and the height of the tip of the pipette to the liquid surface was 8 cm. The time $t = 0$ is defined as the moment when the droplet touches for the first time the solution of the cuvette. At $t = 50$ μs the protruding edge becomes well visible under the meniscus, and over time the portion of the droplet in the reactive medium gets larger. At 300 μs a thin needle-like tip is observed (see the arrow in Fig. 3). Already at 200 μs it is predictable from the shape of the protruding edge, which forms a kind of cone. Similar

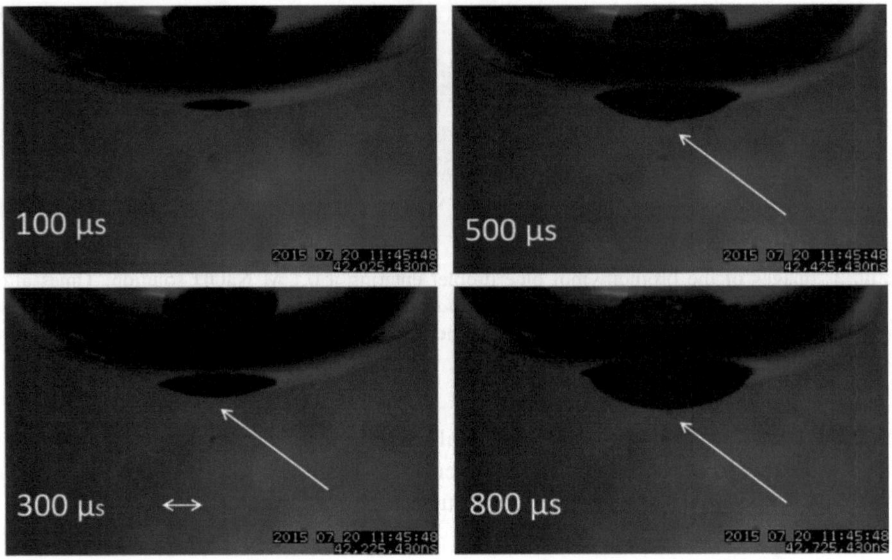

Fig. 3 Initial phase of a ferroin droplet, up to 1 ms, impinging on the surface of the BZ reactive solution. Cuvette size: 1 cm × 1 cm × 4 cm, height of the pipette: 8 cm, scale bar: 1 mm, camera speed: 20,000 frames/s. The dark arc is the meniscus of the liquid in the cuvette

to our former experiments with the pH indicator bromothymol blue, the droplet changes its shape from a quasi-sphere to a flat spheroid (t = 300 to 800 μs).

The needle-like tip becomes clearer as time proceeds. We remark that a quite similar tiny tip had also appeared in the BTB experiment, detectable in Fig. 2. We will return to this interesting feature in the following Section.

Thus, on the whole, the scenario of the falling BZ drop resembles that of the BTB solution except for the pronounced color change.

What happens with the falling drop at later observation times, until about 100 ms? As illustrated in Fig. 4, the smoothly curved, cap-like interface between drop and liquid medium is lost at about 20 ms and a sequence of rather irregular shapes and forms increasing in complexity follows.

A larger cuvette of 2.5 cm × 2.5 cm basis, filled to 6 cm height, was used, and the height of the tip of the pipette was 8 cm. Up to about 10 ms the cup shape of the droplet entering into the BZ solution is preserved. We notice that between 10 and 20 ms the small needle-like pin at the bottom of the cup has become more pronounced (compare with the similar remarkable protrusion in Fig. 2). At this the whole volume of the falling droplet appears to be contained in the cuvette solution, now starting to change its shape. The needle-like pin becomes more prominent. At t = 30 ms the original regular "cup" changes to a conical form in that the protrusion is now the tip of a cone. Later, at 40 ms an overshoot of the droplet appears above the surface of the liquid in the cuvette, and correspondingly the cone is transformed to be almost a cylinder. Subsequently, from 60 to 100 ms, the major portion of the

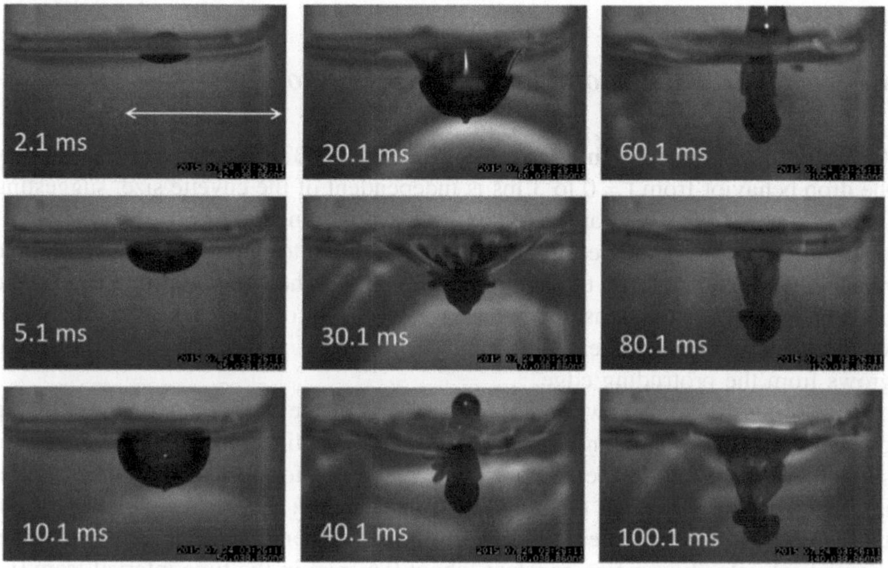

Fig. 4 Later phase of a ferroin droplet impinging on the surface of the BZ reactive solution. Size of the (large) cuvette: 2.5 cm × 2.5 cm × 6 cm, height of the pipette: 8 cm, scale bar: 1 cm, camera speed: 1000 frames/s. (The bright arcs appearing after 10 ms are due to light reflected from the surface of the droplet.)

original droplet forms a cap of mushroom-like form before being caught in turbulence (100–150 ms, not shown).

Similar to observations in the BTB reaction [5], the drop behavior from t = 0 to 3 ms was independent of the cuvette size. On the other hand, after 3 ms the droplet dynamics in the solution is affected by the surrounding walls, so that the behavior of the droplet is slightly different from that in the large cuvette, forming a sprout of conic form around 30 ms, as well as a cylindric shape after 60 ms. We note that in this case there occurs no color change.

When the droplet falls from 17 cm height, a clear sprout at the protruding edge is observed after only 5 ms. The overshoot appears earlier (around 30 ms) and reaches a higher position. During the later phase, both above and below the solution surface, more turbulent images appear: from the part of the overshoot some drops fall, from the tip a mushroom grows and its cap is separated. The influence of boundary conditions is obvious. It requires more experimental efforts to learn about the multitude of behaviors that a simple drop containing chemicals may develop in time.

4 Discussion

4.1 Chemical Reaction: Early Phases up to 3 ms

As has been mentioned in the Results, for both cases (BTB system and BZ system), the drop behavior from t = 0 to 3 ms is independent of the cuvette size, suggesting that no effects of the wall are involved in this time range.

In the BTB system we observe (see Fig. 2 center image at 1.5 ms) three remarkable phenomena: 1. the color change starts at the equatorial line (not at the protruding edge), 2. patterns appear at the lower front of the reaction: a number of vertically oriented fingers emerge from the front line, and 3. a needle-like sprout grows from the protruding edge.

Regarding the point 1, we return to the shape of the drop, which is the moving object in our study. Differently from the splash of a solid sphere impacting a liquid surface [9], the shape of the liquid sphere changes to form a flat spheroid, when it invades the liquid medium (see Fig. 2). Consequently, the vertical velocity of the falling drop is at that moment different from the horizontal spreading speed. While the vertical velocity remains constant (about 0.6 m/s) in the time interval from 0.3 to 1.5 ms, the horizontal one is, at the beginning, about 15 times higher. It then decreases rapidly and, after 1 ms, becomes quasi-constant (about 1.5 m/s). Thus, during the initial phase, the indicator solution in the droplet is displaced mainly in the horizontal direction. And therefore, the chemical reaction is expected to occur most often in this region, giving rise to the observed equatorial band.

Except for the area of the equatorial line, we can assume that the reaction takes place on the basis of diffusion. The diffusion constant of the used species in aqueous solution is on the order of 10^{-9} m^2/s. Thus, for the time interval of 1 ms we get a diffusion length of about 1 μm, meaning that during this short time the reaction takes place at the interface in a sheet of only 1 μm thickness. Then, the color change would remain below detectability, even with a very sensitive spectrophotometer, and therefore we do not see any color change at the protruding edge of the droplet.

There are reports that a thin air film is formed between a droplet and its counter liquid, when the droplet falls into the liquid [10, 11]. Our result that no color change is observed at the protruding edge could be explained by formation of a thin air film. However, this alone cannot explain why the reaction starts at the equatorial line.

As to the finger formation (the point 2), these fine structures suggest that some type of instability of the chemical front exists on the drop surface: we argue below that an instability due to heat conduction is likely.

For comparison, finger formation caused by concentration-dependent density gradients were found in other reactive systems [12–15]. For example, Almarcha et al. [12, 13] shows finger formation along a chemical front occurring in an acid-base reaction:

$$HCl + NaOH \rightarrow NaCl + H_2O.$$

Fingers appeared in the time interval of 40–200 s: about 10^5 times slower than in our case of BTB [4].

Such finger formation has been analyzed on the basis of a reaction-(material) diffusion-convection model [12, 15]. We recall that most neutralization reactions are exothermic [16]. While the effect of heat diffusion is negligible in a long time range like in the work of Ref. [9], it may play a significant role in a short time range of a few milliseconds. Since the equation of heat conduction is formally analogous to the diffusion equation, heat can cause a fingering instability similar to that due to material diffusion, just on quite different time scales. The thermal diffusion coefficient of water is calculated from the thermal conductivity of water [17] (0.6 W/mK) divided by its volume specific heat capacity [17], to be 1.4×10^{-4} m^2/s. This value is about 10^5 times larger than the material diffusion coefficient of water (10^{-9} m^2/s). It is, therefore, likely that an instability due to heat conduction occurs much faster than one due to material diffusion. This argument is supported by the fact that the factor 10^5 between the considered time ranges (milliseconds vs. 100 s) coincides with that between the thermal and the material diffusion coefficient.

Point 3 will be discussed in 4.2 together with the results for the BZ system.

For the BZ system no clear color change was observed through the whole observation time (0–100 ms). However, this does not mean that there is no chemical reaction during this time interval. Reactions start at the moment when the liquid of droplet meet the counter liquid. We could not observe any color effect, which does not mean that there is no chemical reaction during our observation. The red color remains red, because reactions taking place in this time range do not show any color change, or because the spectra of ferroin and ferriin is overlapped such that a tiny spectrum change could not be detected. An acceleration of reaction initiation would be desirable as a necessary task for future investigations. We need also further investigation by using, for example, appropriate filters to separate the absorption peaks of ferroin and ferrin.

4.2 A Needle-Like Sprout

For discussing what we have seen in terms of dynamical behavior of a reactive droplet in a solution of its reactive counterpart, we wish to emphasize that the falling process involves both chemical and physical aspects. As described in 4.1, during a relatively early phase the chosen chemistry appears to dominate the scene. On the other hand, for the later stage we realize that one finds phenomena which are partially known from hydrodynamics, such as the overshoots or the sprout or mushroom formation. Factually, this work is closely related to a plethora of investigations on nonreactive falling droplets, which offer a number of well-established scenarios in hydrodynamic behavior [18–20].

This does not include the thin needle-like tip which appears around 200–300 μs, without a significant difference between the experiments with the pH indicator BTB [4] (see Fig. 2), and the BZ reactive solution and non-reactive systems. Thus, the appearance of this needle-like tip does not seem to depend on the type of chemical system. These results suggest that the needle is formed at the point where the droplet first touches the solution in the cuvette. It seems that this particular point of the surface of the droplet is disrupted and forms a tiny hole, and that from this hole the liquid of the droplet escapes. The difference in density, viscosity as well as surface tension between BZ solution and water against ferroin solution, as well as that in BTB system does not affect the formation of the thin tip.

When the initial height of the falling drop is 17 cm, a thin needle-like tip is not observed in the early phase and a thicker sprout appears after 5–6 ms. With this larger height the impact may be so strong that not only the point where the droplet first touches the solution but a larger area of the surface of the droplet is disturbed. This could be an explanation why a single thin needle-like tip was not observed.

Yet we cannot really exclude the existence of the thin needle at this early phase: it just could be hidden due to splash. The question is, whether the thin tip becomes thicker and rounder to be the sprout, or the thin tip and the thick sprout are not of the same origin but are two different phenomena.

Figure 5 shows a droplet flying askew by chance and falling on the liquid at an oblique angle. Here a thin needle-like tip is observed at 2 ms along the direction of the falling line, and it develops further in the same direction, as seen in the images at 15 and 30 ms. At 15 ms one additional, thicker (round) tip appears at the protruding edge and this develops further, too. This suggests that the thin needle-like tip which appears in the early phase is different from the thicker sprout observed later.

The formation of the thicker sprout occurs at the protruding edge, independent of the falling angle. This could be due to the mostly hydrodynamic origin: the liquid of the droplet which is pushed by air and forms a cup goes down toward the protruding edge. For the later phase (5–150 ms) small differences in physical parameters (density, viscosity and surface tension) play important roles, because the

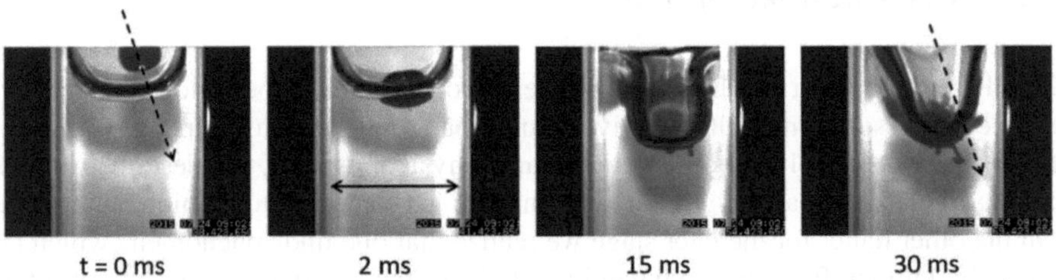

Fig. 5 A ferroin droplet impinging on the surface of the BZ solution at an oblique angle (see the *arrow with broken line*). Size of the cuvette: 1 cm × 1 cm × 4 cm, height of the pipette: 8 cm, scale bar: 1 cm, camera speed 1000 frames/s

droplet behavior depends on the size of the cuvette and the height of the falling droplet.

Our findings unravel chemical and physical effects to be responsible for the observed behavior. More experiments would be helpful to understand the role of additional parameters for the problem, in particular surface tension, viscosity, density, drop curvature, and various other geometrical quantities.

5 Concluding Remark

In this chapter applications of high-speed cameras to chemical reactions are shown. This is a challenge for chemistry as well as for high-speed visualization. In an overall assessment, our experimental system, as simple as it may appear, closely combines problems of chemistry and physics. Especially using color reactions in a small droplet volume does not only offer a most useful method to extract information about fast chemical processes, but our examples also show how color indicators serve, in this context, to elucidate important hydrodynamical processes. A satisfactory explanation of the observed phenomena needs both an improvement of speed and flexibility of observation techniques and progress in a theoretical description of the reported effects, questions arising for future research.

Acknowledgements The author thanks Kinko Tsuji for helpful discussions.

References

1. B. Chance, The accelerated and stopped-flow apparatus II. J. Franklin Inst. **229**, 613–640 (1940)
2. M. Eigen, L. De Maeyer, Relaxation methods, in *Technique of Organic Chemistry*, vol. 8, ed. by A. Weissberger (John Wiley and Sons, New York, 1963)
3. J.P. Connelly, M.G. Müller, R. Bassi, R. Groce, A.R. Holzwarth, Femtosecond transient absorption study of carotenoid to chlorophyll energy transfer in the light harvesting complex II of photosystem II. Biochemistry **36**, 281–287 (1997)
4. K. Tsuji, S.C. Müller, Chemical reaction evolving on a droplet. J. Phys. Chem. Lett. **3**, 977–989 (2012)
5. K. Tsuji, S.C. Müller, Observation of a falling droplet with pH indicator. Interfacial Phenom. Heat Transf. **1**, 289–299 (2013)
6. A.N. Zaikin, A.M. Zhabotinsky, Concentration wave propagation in two-dimensional liquid-phase self-oscillating system. Nature **225**, 535–537 (1970)
7. J. Ross, S.C. Müller, C. Vidal, Chemical waves. Science **240**, 460–465 (1988)
8. R. Kapral, K. Showalter (eds.), *Chemical Waves and Patterns* (Kluwer Academic Press, Dordrecht, 1995)
9. M. Do-Quang, G. Amberg, The splash of a solid sphere impacting on a liquid surface: numerical simulation of the influence of wetting. Phys. Fluids **21**, 022102 (2009)
10. Y. Couder, E. Fort, C.-H. Gautier, A. Boudaoud, From bouncing to floating: noncoalescence of drops on a fluid bath. Phys. Rev. Lett. **94**, 177801 (2005)

11. D. Beilharz, A. Guyon, E.Q. Li, M.-J. Thoraval, S.T. Thoroddsen, Antibubbles and fine cylindrical sheets of air. J. Fluid Mech. **779**, 87–115 (2015)
12. C. Almarcha, P.M.J. Trevelyan, P. Grosfils, A. De Wit, Chemically driven hydrodynamic instabilities. Phys. Rev. Lett. **104**, 044501 (2010)
13. C. Almarcha, P.M.J. Trevelyan, L.A. Riolfo, A. Zalts, C. El Hasi, A. D'Onofrio, A. De Wit, Active role of a color indicator in Buoyancy-driven instabilities of chemical fronts. J. Phys. Chem. Lett. **1**, 752–757 (2010)
14. M. Böckmann, S.C. Müller, Growth rates of the Buoyancy-driven instability of an autocatalytic reaction front. Phys. Rev. Lett. **85**, 2506–2509 (2000)
15. K. Matthiessen, H. Wilke, S.C. Müller, Influence of surface tension changes on hydrodynamic flow induced by traveling chemical waves. Phys. Rev. E **53**, 6056–6060 (1996)
16. R.H. Lambert, L.J. Gillespie, Heat of neutralization at constant concentration and the heat of ionization of water. J. Am. Chem. Soc. **53**, 2632–2639 (1931)
17. R.C. Weast, M.J. Astle (eds.), *CRC Handbook of Chemistry and Physics*, 60th edn. (CRC Press, Boca Raton, 1980)
18. P.V. Hobbs, A.J. Kezweeny, Splashing of a water drop. Science **155**, 1112–1114 (1967)
19. K. Cai, Phenomena of a liquid drop falling to a liquid surface. Exp. Fluids **7**, 388–394 (1989)
20. A. Prosperetti, H.N. Oguz, The impact of drops on liquid surfaces and the underwater noise of rain. Ann. Rev. Fluid Mech. **25**, 577–602 (1993)

Part IX
Microbubbles and Medical Applications

Dynamics of Coated Microbubbles in Ultrasound

Valeria Garbin

Abstract The stability and dynamics of microbubbles coated with an interfacial layer of adsorbed material, ranging from phospholipids, to proteins and nanoparticles, are central to food products, biomedical imaging applications, and controlled release. The dynamics of coated microbubbles in ultrasound fields are of particular relevance to food production and biomedical imaging. High-speed imaging has proven to be an invaluable tool to reveal micromechanical phenomena of the coating during ultrasound-driven microbubble dynamics, so as to gain a fundamental understanding of the factors affecting microbubble durability and performance. This Chapter includes an introduction to the basic concepts of microbubble stability (Sect. 1), and to the dynamics of coated microbubbles in ultrasound (Sect. 2). An overview of recent research advances is then provided, focusing on the following topics: Dynamics of biomedical microbubbles in ultrasound studied by combined optical trapping and ultra-high speed imaging (Sect. 3); Buckling and expulsion of coating material from ultrasound-driven microbubbles (Sect. 4); Shape oscillations of coated bubbles (Sect. 5).

1 Introduction: Coated Microbubbles

Formulations based on microscopic gas bubbles are central in the consumer product and pharmaceutical industries [1]. Many food and personal care products are formulated as foams, and therefore owe their texture to a dense packing of microscopic gas bubbles. In the pharmaceutical industry, microbubble dispersions are produced for use as contrast agents in ultrasound imaging. In both examples, the stability of microbubbles in the surrounding liquid is central to the shelf-life and performance of the formulation.

V. Garbin (✉)
Department of Chemical Engineering, Imperial College London,
London SW7 2AZ, UK
e-mail: v.garbin@imperial.ac.uk

K. Tsuji (ed.), *The Micro-World Observed by Ultra High-Speed Cameras*,
DOI 10.1007/978-3-319-61491-5_17

1.1 Mechanisms of Microbubble Dissolution

Ensuring stability of microscopic gas bubbles in a liquid presents a significant challenge, since they tend to dissolve very rapidly. Gas diffuses from the bubble into the surrounding liquid if the concentration of dissolved gas in the liquid, c, is below the saturation concentration, c_s. The saturation concentration is proportional to the partial pressure of gas acting on the liquid interface, p_g, as stated by Henry's law [2]:

$$c_s = k_H M p_g, \tag{1}$$

where k_H is the Henry's constant, which measures the solubility of the gas in the liquid, and M the gas molar mass. Gas diffuses out of the bubble until the saturation concentration in the liquid is reached. It is important to note that the pressure of the gas inside a bubble depends on the size of the bubble:

$$p_g(R) = p_0 + \frac{2\gamma}{R}, \tag{2}$$

where p_0 is the ambient pressure far from the bubble and γ the surface tension of the gas-liquid interface. $\Delta p = 2\gamma/R$ is the Laplace pressure, caused by the curvature of the interface of a spherical bubble with radius R [3]. As a consequence of Eq. (2), the saturation concentration in the liquid surrounding a bubble depends on the radius of the bubble, and bubbles can dissolve even if $c > c_{s,0}$, where we have denoted as $c_{s,0}$ the saturation concentration for a planar interface, $c_{s,0} = k_H M p_0$ [4]. The driving force for dissolution due to the curvature of the interface is particularly pronounced for very small bubbles. For an air bubble with radius $R_0 = 1$ μm in water, the Laplace pressure is $\Delta p \approx 1.44 \times 10^5$ Pa, in excess of the atmospheric pressure, $p_0 = 10^5$ Pa. The time to complete dissolution for a bubble of initial radius R_0 can be estimated from [4]:

$$t_d = \frac{R_0^2}{3Dk_H}\left(\frac{R_0\rho_g}{2M\gamma} + \frac{1}{R_g T}\right), \tag{3}$$

where D is the diffusivity of the gas in the liquid, ρ_g the density of gas in the bubble, R_g the universal gas constant, and T the absolute temperature. For a 1 μm air bubble in water, the time to dissolution is $t_d \approx 30 \times 10^{-3}$ s. Clearly, for practical applications, it is necessary to stabilize microbubbles against dissolution.

1.2 Microbubbles Stabilized by Surfactant Coatings

Microbubbles in food, personal care, and pharmaceutical applications are typically stabilized by coating their interface with surfactants. A surfactant coating helps stabilize the bubbles against dissolution by three main mechanisms. Firstly, surfactants decrease the surface tension of the interface. A decrease in the surface tension γ helps stabilize microbubbles since the driving force for dissolution due to the curvature of the interface, i.e. the Laplace pressure $\Delta p = 2\gamma/R$, decreases [5, 6]. Secondly, the surfactant molecules adsorbed on the interface provide a steric barrier to gas transport, therefore creating a resistance to gas permeation which also contributes to stability [7, 8]. Finally, the rheological properties of an interface coated with surfactants contribute to the long-term stability of coated bubbles. Interfaces with sufficiently large elasticity can completely halt bubble dissolution [9]. The Gibbs criterion states that the elasticity of the interface can stop bubble dissolution, even if the surface tension is non-zero, if the elastic modulus of the interface, ε, satisfies $\varepsilon > \gamma/2$ [10, 11]. Soluble surfactants only have a weak influence on the dissolution of gas bubbles, whereas insoluble surfactants such as lipids and proteins can considerably reduce the dissolution rate [7, 12–14]. Proteins are extensively used in foam stabilization in the food industry [15]. In medical ultrasound, dispersions of micron-sized gas bubbles are used as imaging contrast agents. Commercial microbubble formulations are stabilized by either protein coatings, such as albumin (Optison, GE Healthcare Life Sciences) or by phospholipids (SonoVue, Bracco Imaging). For two-component mixtures of phospholipids and poly-ethylene glycol, which is added to prevent microbubble aggregation, phase separation of the monolayer is observed [16], as shown in Fig. 1a.

1.3 Microbubbles Stabilized by Solid Particles

Solid particles can be used in place of molecular surfactants to stabilize bubbles for several days [17]. Nanoparticles and microparticles can adsorb at fluid-fluid interfaces, much like molecular surfactants, but with a much larger energy barrier to removal [18, 19]. Foams stabilized by solid particles instead of molecular surfactants are exploited in food products and biomedical applications [15, 20, 21] and in advanced materials [22, 23]. The stability imparted by solid particles can be more effective than for the case of surfactants, a striking example being the dissolution arrest of particle-coated bubbles [24]. Using microfluidic methods, it is possible to generate monodisperse, stable nanoparticle-coated bubbles, with functional coatings composed of mixtures of hydrophobic materials and nanoparticles with unique properties [25]. Figure 1b shows an example of monodisperse microbubbles with a silica nanoparticle shell.

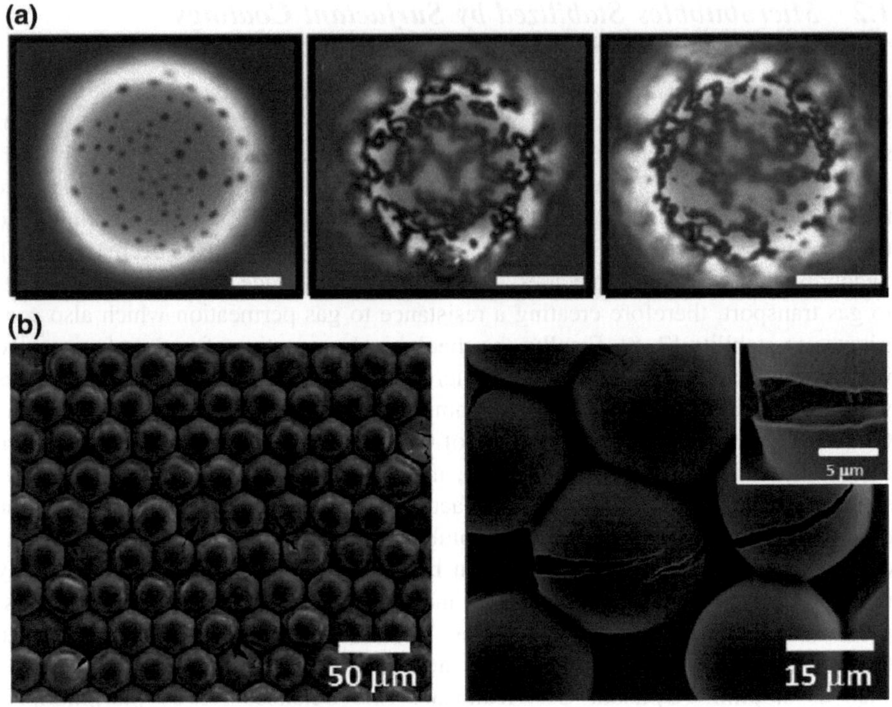

Fig. 1 **a** Microbubbles coated with a two-component phospholipid monolayer comprising polyethylene glycol as emulsifier. The coating undergoes phase separation, with the bright and dark domains indicating the coexistence of condensed and expanded phases. Adapted with permission from Ref. [16]. Copyright (2004) Elsevier. **b** Monodisperse microbubbles coated with a layer of silica nanoparticles. Adapted with permission from Ref. [25]. Copyright (2010) American Chemical Society

2 Theory: Dynamics of Coated Microbubbles in Ultrasound

Bubble dynamics in sound fields are central in underwater acoustics, surface cleaning, and medical ultrasound [26, 27]. Bubbles in a liquid respond to the periodic changes in pressure associated with an acoustic wave, $p_a(t) = \Delta p \sin(\omega t)$, where Δp is the pressure amplitude, by undergoing periodic compression and expansion at the driving frequency $\omega = 2\pi f$, where f is the frequency in s^{-1}. The phase shift between the driving ultrasound wave and the volumetric oscillations of the bubble depends on whether the driving frequency is above or below the resonance frequency of the bubble (see below). Figure 2 shows the temporal evolution of the acoustic pressure $p_a(t)$, the bubble radius $R(t)$, and an image sequence of bubble oscillations.

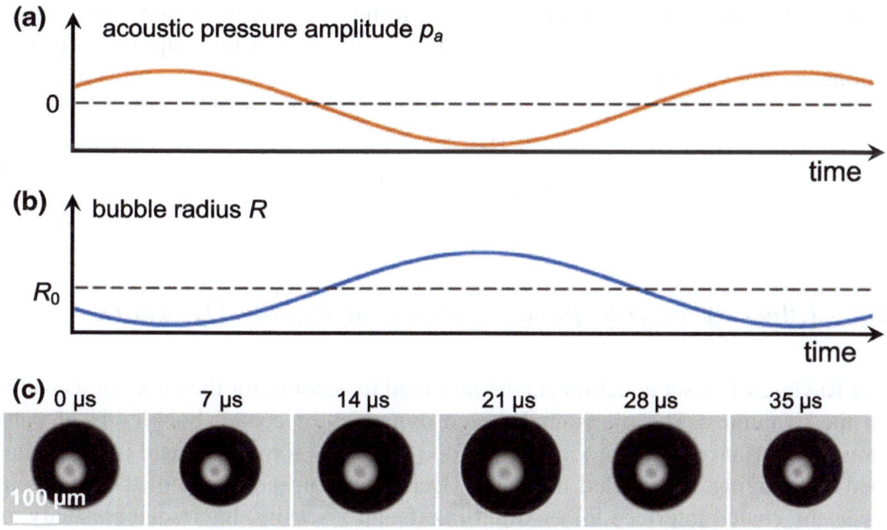

Fig. 2 **a** The ultrasound wave transmitted to the fluid produces a sinusoidal variation of the pressure relative to ambient pressure. **b** The bubble undergoes radial oscillations at the frequency of the ultrasound wave. Below resonance, the bubble undergoes compression ($R<R_0$) when the pressure increases, and expansion ($R > R_0$) when the pressure decreases. **c** High-speed image sequence at 350,000 fps of a bubble driven by ultrasound at 37 kHz (V. Garbin, unpublished)

2.1 *Bubble Dynamics in Ultrasound*

The dynamics are governed by the Rayleigh-Plesset equation [28], which describes the temporal evolution of the bubble radius, $R(t)$:

$$\rho\left(R\ddot{R}+\frac{3}{2}\dot{R}^2\right)=\left(p_0+\frac{2\sigma}{R_0}\right)\left(\frac{R_0}{R}\right)^{3\kappa}-p_0-p_a(t)-\frac{2\sigma}{R}-\frac{4\mu\dot{R}}{R}. \tag{4}$$

The dots denote derivatives with respect to time, ρ is the density of the liquid, p_0 is the ambient pressure, σ the surface tension of the gas-liquid interface, R_0 the equilibrium radius of the bubble, κ the polytropic exponent, and μ the viscosity of the liquid. The terms on the left hand side account for the inertia of the liquid during bubble motion. The first term on the right hand side describes the gas pressure inside the bubble as a function of its radius R through a polytropic relationship [26]. The fourth term on the right hand side accounts for the Laplace pressure, which is important for micron-sized bubbles. The last term on the right hand side accounts for viscous dissipation in the fluid. For small driving amplitude, $\Delta p \ll p_0$, the bubble behaves as a forced harmonic oscillator [29], where the mass of the system is due to the inertia of the fluid, and the restoring force to the compression of the gas, with an associated resonance frequency ω_0 for which the amplitude of oscillations is a maximum:

$$\omega_0 = \frac{1}{R_0}\sqrt{\frac{3\kappa}{\rho}\left(p_0 + \frac{2\sigma}{R_0}\right) - \frac{2\sigma}{\rho R_0}}. \tag{5}$$

2.2 *Effect of an Interfacial Coating on Bubble Dynamics*

The Rayleigh-Plesset equation can be modified to account for the effect of a coating on the dynamics. The ultrasound-driven dynamics of coated bubbles is of paramount importance in the production of food foams by sonication, and in ultrasound medical imaging, as detailed in Sect. 3. Depending on the nature of the interfacial layer, the coated interface may exhibit interfacial viscosity, interfacial elasticity, or both. Early models describe the shell as a continuous viscoelastic solid of finite thickness [30–32], which is appropriate for capsules. More recently, a model accounting for the molecular or particulate nature of the interfacial coating has been put forward, which accounts for the changes in surface tension during compression-expansion of the interface, as well as for viscosity of the interfacial layer [33]. Equation (4) is modified as follows:

$$\rho\left(R\ddot{R} + \frac{3}{2}\dot{R}^2\right) = \left(p_0 + \frac{2\sigma(R_0)}{R_0}\right)\left(\frac{R_0}{R}\right)^{3\kappa} - p_0 - p_a(t) - \frac{2\sigma(R)}{R} - \frac{4\mu\dot{R}}{R} - \frac{4\kappa_s\dot{R}}{R^2}, \tag{6}$$

with κ_s the surface dilatational viscosity, and the constitutive equation for the dependence of the surface tension on area of the bubble given, in the linear regime, by:

$$\sigma(R) = \sigma(R_0) + \chi\left(\frac{R^2}{R_0^2} - 1\right), \tag{7}$$

with χ the surface dilatational elasticity. Outside of the linear regime of deformation of the coating, this model accounts for non-linear phenomena empirically [33]: the surface tension is set to approach the value of the uncoated interface upon large expansion, and to approach zero upon strong compression when the interfacial layer buckles [34].

3 Dynamics of Biomedical Microbubbles in Ultrasound Studied by Combined Optical Trapping and Ultra-high Speed Imaging

Micron-sized gas bubbles coated with a stabilizing surfactant monolayer of either phospholipids or proteins are commonly used as contrast agents in ultrasound medical imaging [27]. Biomedical microbubbles have average diameters between 1 and 10 μm and, correspondingly, resonance frequencies in the MHz range, see Eq. (5). When they are subjected to a diagnostic ultrasound field, at typical medical imaging frequencies between 1 and 10 MHz, the bubbles undergo radial oscillations (see Sect. 2). The resulting acoustical response allows discriminating the blood pool from the surrounding tissues [27]. In the past decades, fundamental studies of the response of biomedical microbubbles in ultrasound have revealed complex non-linear phenomena [35]. The bubble dynamics have been recorded acoustically using ultrasound transducers, and, more recently, bubble dynamics at MHz frequencies were recorded optically using high-speed cameras [36–38], with the advantage of providing direct evidence of phenomena involving non-linear oscillations [33], bubble rupture, surface modes and interactions with neighboring objects [39, 40]. A problem common to all experiments on microbubbles is the lack of control on the position and distance from interfaces and neighboring bubbles, due to buoyancy and flow. To enable more controlled studies, optical tweezers have been developed to trap and position biomedical microbubbles [41–43]. Optical trapping to isolate a single microbubble and study its dynamics in an ultrasound field has been successfully combined with high-speed imaging [44, 45].

3.1 *Optical Trapping of Biomedical Microbubbles*

Since the early demonstrations of optical trapping of dielectric particles [46] and living cells [47], "optical tweezers" have become a well-established technique for the contactless manipulation of microscopic particles.

3.1.1 Optical Trapping of Dielectric Particles

In the standard mode of operation, a laser beam, which has a Gaussian intensity profile, is brought to a tight focus by a high-numerical-aperture microscope objective. Particles with a higher refractive index than the surrounding medium, such as solid particles or cells, are subjected to optical gradient and scattering forces that confine them to the focus of the laser [48]. Optical trapping of gas bubbles presents the challenge that these particles have a lower refractive index than the surrounding medium. As a consequence, the gradient optical force has the opposite sign than for high-index particles, and therefore bubbles are repelled by a focused,

Gaussian laser. The method that has been developed to trap low-index microparticles consists in modifying the intensity profile of the laser from Gaussian, with a maximum on the optical axis, to a Laguerre-Gaussian profile, with a minimum of intensity on the optical axis. Low-index particles are repelled by the annular high-intensity region surrounding the optical axis of a Laguerre-Gaussian beam, and are effectively trapped in the low-intensity region on the beam axis [49].

3.1.2 Optical Trapping of Microbubbles

Optical trapping using Laguerre-Gaussian beams has been successfully applied to biomedical microbubbles, which behave as low-index particles even if they have a high-index coating because the surfactant monolayer is only a few nanometers thick, and therefore the interaction with the laser beam is dominated by the gas core. Optison microbubbles have been trapped in three-dimensions using a static diffractive optical element to modify the laser intensity profile to a Laguerre-Gaussian profile [41]. In the same work, trapping of multiple microbubbles in a 4×4 array was also achieved, by adding a second diffractive optical element to split the incident beam into multiple beams. The diffractive optical elements can be combined and implemented dynamically on a liquid crystal display [42], thus enabling real-time adjustments to the position and number of traps, and therefore the configuration of trapped microbubbles. Another method to trap multiple microbubbles in a dynamic configuration is to use a circularly scanning optical tweezers: a time-averaged potential well is created, and stable trapping is achieved because of the slow diffusion of the microbubbles in the liquid [43].

3.2 *High-Speed Imaging Studies of Biomedical Microbubble Dynamics in Controlled Confinement*

Optical trapping has been used in combination with high-speed imaging at up to 15,000,000 frames per second (fps) to provide direct visualizations and reveal new phenomena of biomedical microbubble dynamics in controlled confinement.

3.2.1 Interaction of Microbubbles with Cells

Prentice et al. combined optical trapping with high-speed imaging at 500,000 fps (Imacon 468 camera) or 1,000,000 fps (Cordin 550 camera) to study mechanisms of cell membrane disruption by biomedical microbubbles in ultrasound. Single Optison bubbles were positioned in the vicinity of, but non in contact with, a cell layer. The bubbles were activated by ultrasound at 1 MHz at a peak negative pressure of more than 1 MPa. The high-speed image sequences revealed cavitation, jetting, and local deformation of the cell layer (see Fig. 3a) [44].

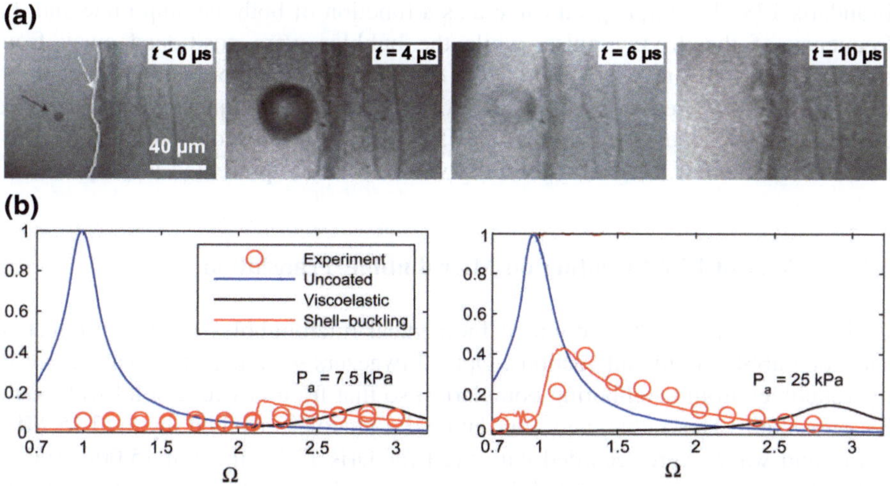

Fig. 3 **a** Cell damage from microbubble cavitation. Initially ($t < 0$ μs) a microbubble (*black arrow*) having diameter 4.5-μm is optically trapped at a distance of 26.5 μm from the cell layer (*outlined in white*). Upon driving with ultrasound at 1 MPa, the 10-fold increase in bubble radius is followed by bubble collapse and deformation of the cell layer. Adapted with permission from Ref. [44]. Copyright (2005) Nature Publishing Group. **b** Resonant behaviour of lipid-coated microbubbles undergoing small-amplitude oscillations upon driving with ultrasound at 7.5–25 kPa. The non-linearity introduced by the coating on the bubble interface causes the pronounced skewing of the resonance curve observed in experiment (*symbols*) and predicted by the model in Ref. [33]. Adapted with permission from Ref. [50]. Copyright (2010) Elsevier

3.2.2 Effect of Boundaries on Bubble Dynamics

Garbin et al. combined optical trapping with the ultra-high speed camera Brandaris 128, described in Chapter "Brandaris Ultra High-speed Imaging Facility"—Lajoinie et al., to study the effect of neighboring boundaries on bubble dynamics. The Brandaris 128 camera was operated at 15,000,000 fps, to resolve the radial dynamics of BR-14 microbubbles (Bracco S.A., Geneva, Switzerland) excited by a 2.25-MHz ultrasound burst at peak negative pressures between a few tens and a few hundreds of kPa. The authors reported optical observations of the change in the dynamics of the very same microbubble due to the influence of interfaces and neighboring bubbles [45]. Using the setup combining optical trapping and the Brandaris 128 camera, Overvelde et al. studied the influence of the stabilizing phospholipid coating on the nonlinear dynamics of BR-14 microbubbles, while ensuring the bubbles were positioned sufficiently far from solid boundaries to avoid confinement effects. The viscoelasticity of the interfacial monolayer was found to enhance the nonlinear bubble response at acoustic pressures as low as 10 kPa [50], in keeping with the predictions of the model presented in [33] (see Fig. 3b). The predictions of this model were further explored by Sijl et al. using the same setup. The radial subharmonic response of the microbubbles was recorded with the

Brandaris 128 ultra high-speed camera as a function of both the amplitude and the frequency of the driving pulse, while the bubbles were positioned away from boundaries to prevent unwanted effects of confinement. Subharmonic generation was observed for pressures as low as 5 kPa for a driving frequency equal to twice the resonance frequency of the bubble [51]. The low threshold pressure for subharmonic behaviour is again in keeping with the model presented in [33].

3.2.3 Effect of Lipid Coating on Microbubble Translation

Garbin et al. studied the motion of biomedical microbubbles under the effect of acoustic bubble-bubble interactions. Optical tweezers were used to isolate a pair of microbubbles from neighboring boundaries, so that friction with a solid wall could be prevented. The radial and translational dynamics, excited by a 2.25 MHz ultrasound wave, were recorded with the Brandaris 128 camera at 15,000,000 fps [52]. The measurements revealed the importance of unsteady viscous effect, due to the coating on the microbubbles. The setup combining optical tweezers and the Brandaris 128 camera was also used to study shape oscillations of biomedical microbubbles. This topic is covered in Sect. 5.

4 Buckling and Expulsion of Coating Material from Ultrasound-Driven Microbubbles

High-speed video microscopy of coated microbubbles has revealed that, under ultrasonic driving, the coating material can be expelled in the surrounding medium [21, 53, 54]. The relevance of this finding is two-fold: on the one hand, loss of material from the coating can negatively affect bubble stability and performance; on the other hand, this phenomenon can be exploited for novel controlled release applications in drug delivery, lab-on-a-chip, and catalysis.

4.1 Lipid-Coated Microbubbles

4.1.1 Buckling of Lipid Monolayer

During microbubble oscillations in ultrasound, the surfactant coating undergoes periodic compression and expansion. For planar lipid monolayers, it is well known that there exists a critical area per lipid molecule below which the monolayer cannot further compress, and it deforms out of plane, exhibiting buckling, wrinkling and folding [34]. This phenomenon has been observed for lipid-coated microbubbles undergoing dissolution, and therefore a significant decrease in area available per lipid molecule [5]. Bubbles coated with lipids having different chain lengths were all seen to undergo buckling upon dissolution, over a timescale of tens of seconds. Interestingly, upon further bubble dissolution the interface was seen to return to a spherical shape, indicating that lipids must have been expelled from the monolayer. A challenging question to address was whether these phenomena would be observed also during the, much faster, compression of lipid monolayers on microbubbles oscillating in ultrasound. In this case, the compression of the lipid monolayer is periodic, and it occurs over a timescale of microseconds. Sijl et al. used optical tweezers combined with the Brandaris 128 ultra high-speed camera to study the effect of buckling on the nonlinear bubble response [55]. High-speed recordings at 15,000,000 fps revealed for the first time the occurrence of reversible buckling of the monolayer on the microsecond timescale (see Fig. 4a), a phenomenon that would have remained elusive with acoustic measurements only.

4.1.2 Lipid Shedding

With regards to expulsion of lipids from the coating, high-speed fluorescence imaging has provided time-resolved, direct evidence of this phenomenon. High-speed fluorescence imaging was performed at 150,000 fps (Fastcam SA1.1, Photron) to capture the instantaneous dynamics of lipid shedding. Lipid detachment was observed within the first few cycles of ultrasound, and the detached lipids were subsequently transported by the surrounding flow field (see Fig. 4b). The threshold for lipid expulsion was determined using the ultra high-speed camera Brandaris 128, running at 10,000,000 fps. The threshold in relative amplitude of bubble oscillations for lipid shedding was found to be 30%, pointing to a decrease in area for expulsion of lipids of approximately 10%. Expulsion of lipids was found to be reproducible, indicating that the phenomenon can be controlled [53].

4.2 Particle-Coated Microbubbles

Microbubbles coated with a monolayer of nanoparticles hold promise for combined ultrasound diagnostic imaging and drug delivery, hence their dynamics in ultrasound has received significant attention. Through acoustic measurements, Stride et al. found that an interfacial layer of gold nanoparticles on the surface of a microbubble increases the nonlinear response of the bubbles at low ultrasound pressure amplitudes. This finding was ascribed to the fact that close packing of the nanoparticles restricts bubble compression [56], leading to a significant asymmetry of the compression and expansion phases. Magnetic microbubbles (see Chapter "Characterisation of Functionalised Microbubbles for Ultrasound Imaging and Therapy"—Stride et al.), that is microbubbles stabilized by an interfacial layer of magnetic nanoparticles, have been shown to have the additional advantage that they can be manipulated both with acoustic and magnetic fields [57]. High-speed video

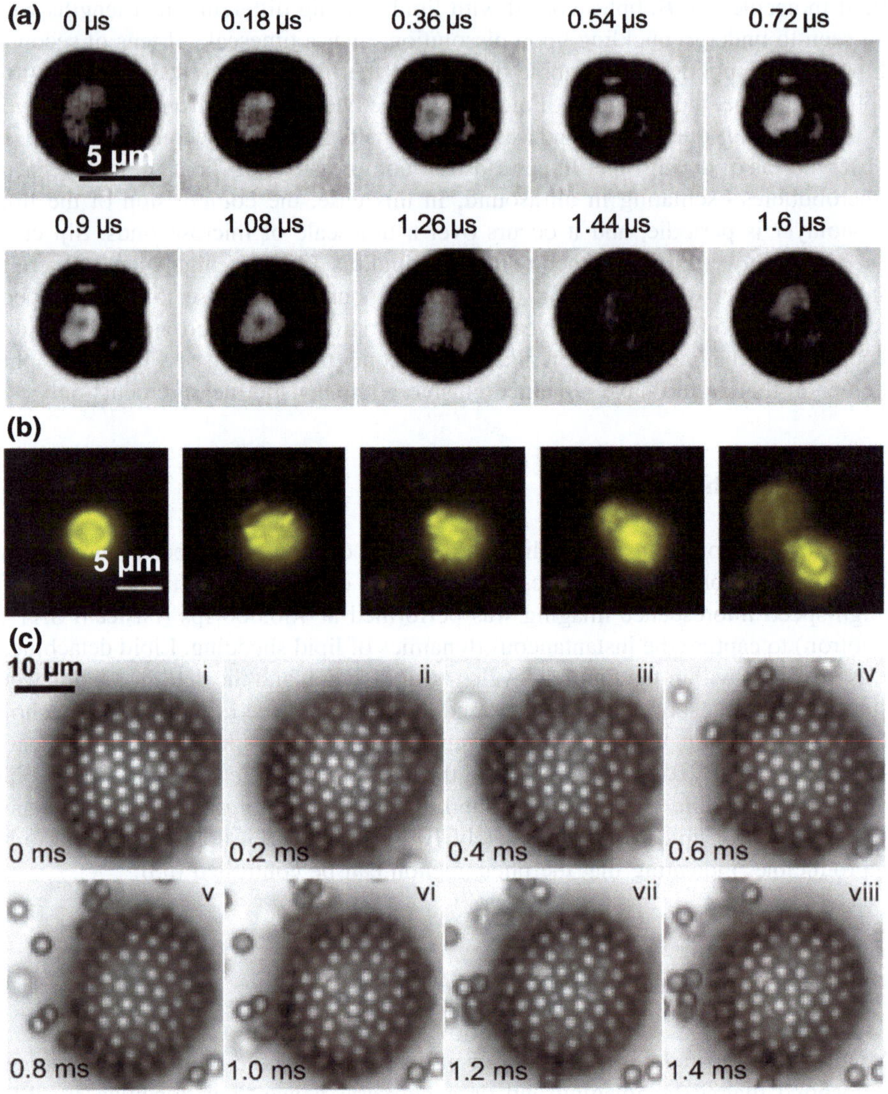

Fig. 4 **a** Buckling of a lipid-coated microbubble driven by ultrasound at 1.5 MHz. One period of oscillations is shown in the image sequence. Adapted with permission from Ref. [55]. Copyright (2011) American Institute of Physics. **b** Lipid shedding from ultrasound-driven microbubble revealed by high-speed fluorescence microscopy. Adapted with permission from Ref. [53]. Copyright (2014) Elsevier. **c** Volumetric oscillations of microbubbles coated with a monolayer of 3-µm colloids. During the compression phase, the particle monolayer undergoes buckling, and colloids are expelled in the surrounding liquid. From Ref. [54]

microscopy of magnetic microbubbles at up to 2,500,000 fps (Cordin 550) revealed that the amplitude of oscillation is very similar for magnetic and nonmagnetic microbubbles of the same size in the same conditions [58].

Upon area compression, monolayers of solid particles can buckle like solid-like films if attractive interparticle interactions are dominant [59, 60], or particles can be expelled in the surrounding liquid if they are well-dispersed [61]. Direct visualization of particle expulsion from ultrasound-driven microbubbles was performed by Poulichet and Garbin [54]. Bubbles coated with micron-sized particles were driven at 40–50 kHz and the dynamics recorded at 300,000 fps (Fastcam SA5, Photron). The monolayer undergoes buckling, and subsequently the particles are expelled (see Fig. 4c). This newly observed phenomenon is due to the fact that the ultrafast deformation of the monolayer can re-disperse the particles, even if attractive interparticle interactions are dominant. Magnetic microbubbles were also found to shed nanoparticles upon ultrasonic driving and the nanoparticles were found to be projected at distances of the order of 100 μm, with potential applications to drug delivery into tissues that are not easily penetrated [21].

5 Shape Oscillations of Coated Bubbles

Spherical oscillations of bubbles become unstable above a threshold in acoustic forcing, above which the bubbles exhibit shape oscillations [62]. An initially small perturbation of the spherical shape grows in amplitude through a parametric instability for a driving frequency $\omega = 2\omega_n$, with ω_n the resonance frequency of a spherical harmonic distortion of order n ($n > 1$), given by Lamb [63]:

$$\omega_n^2 = \frac{(n-1)(n+1)(n+2)\sigma}{\rho R_0^3}. \quad (8)$$

Using high-speed imaging at 2000 fps, Trinh et al. have studied large-amplitude non-spherical oscillations of millimeter-sized bubbles, simultaneously levitated and driven by two acoustic fields at different frequencies, and reported non-linear phenomena such as coupling between modes of different order [64]. Versluis et al. have studied micron-sized bubbles in the linear regime, using high-speed imaging at 1 Mfps (Brandaris 128 ultra-high speed camera, Chapter "Brandaris Ultra High-speed Imaging Facility" Lajoinie et al.). The bubbles were injected from the bottom of the observation chamber, and their dynamics was recorded while they were rising by buoyancy through the focal plane of the imaging system. The experiments showed mode selectivity depending on the bubble radius R_0, consistent with Eq. (8), and subharmonic behaviour with $\omega_n = \omega/2$.

High-speed imaging has also provided the first evidence that coated microbubbles can exhibit nonspherical shapes [65, 66], but in both studies the bubbles were in contact with a wall because of buoyancy (see Fig. 5a). Dollet et al.

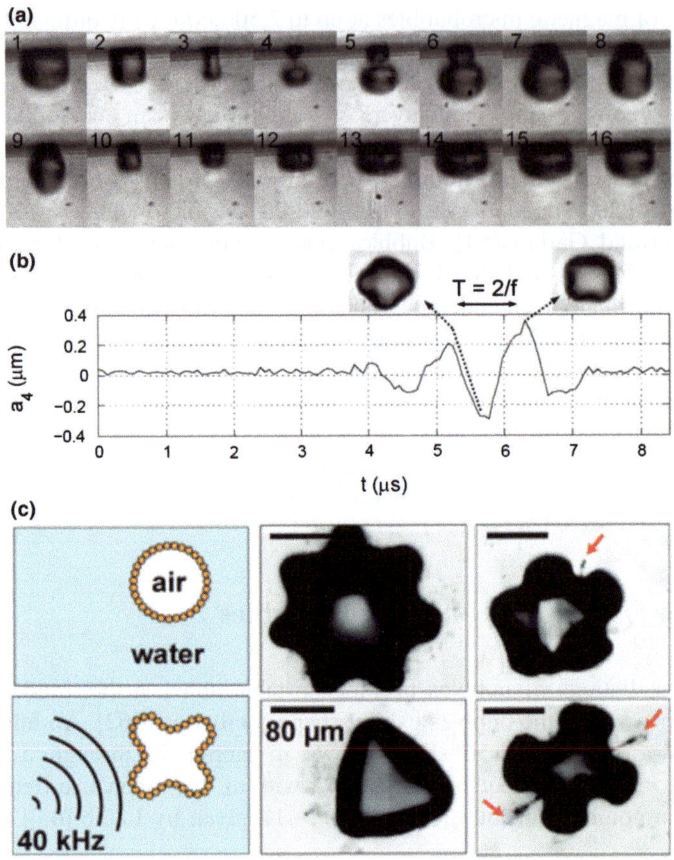

Fig. 5 **a** Side-view high-speed image sequence of a lipid-coated microbubble oscillating in contact with a solid boundary during two cycles of ultrasonic driving. Non-spherical oscillations are induced by the confinement due to the boundary. Adapted with permission from Ref. [65]. Copyright (2008) Elsevier. **b** Lipid-coated microbubbles isolated from boundaries using optical tweezers exhibit shape oscillations with characteristic subharmonic behavior. Adapted with permission from Ref. [67]. Copyright (2008) Elsevier. **c** Nanoparticle-coated bubbles undergoing shape oscillations. Expulsion of plumes of nanoparticles (marked by *red arrows*) is observed preferentially from the antinodes of the shape oscillations. From Ref. [68]

used optical tweezers combined with ultra-high speed imaging to overcome this difficulty [67], and studied the occurrence of nonspherical oscillations of lipid-coated microbubbles. Shape oscillations of lipid-coated bubbles were found to also exhibit subharmonic behavior (see Fig. 5b), but in contrast to uncoated bubbles, mode selectivity was not observed [67].

Particle-coated bubbles undergoing shape oscillations were found to exhibit directional particle desorption from the antinodes of the shape oscillations [21, 54]. High-speed microscopy at 300,000 fps was used to record the dynamics of non-spherical bubble deformation and of particle expulsion [68]. It was found that particle-coated bubbles do not exhibit mode selectivity, in contrast with uncoated bubbles, but in analogy with lipid-coated bubbles. Desorption of nanoparticles from the antinodes of the shape oscillations was ascribed to the fact that the antinodes are the locations where the acceleration of the interface, the interface curvature, and the rate of change of area are a maximum. Decomposition of the bubble shape into spatial Fourier modes revealed that the interplay of modes of different order in the non-linear regime results in preferential desorption from the antinodes where the modes are in phase (see Fig. 5c).

6 Summary

Coated microbubbles driven by acoustic fields exhibit a rich variety of phenomena that have been revealed by high-speed imaging. A large body of work on coated bubble dynamics has been contributed by the biomedical ultrasound community, who have studied extensively lipid- and protein-coated microbubbles used as contrast agents for ultrasound medical imaging. In this field, the use of direct, high-speed optical visualization, as opposed to acoustic measurements only, has enabled researchers to explain certain signatures in the acoustic response in terms of micromechanical phenomena of the coating, such as buckling. In addition, high-speed fluorescence microscopy has revealed the occurrence of shedding of coating material, with important consequences for microbubble stability. An emerging area of application is in nanoparticle-coated microbubbles, which offer the opportunity to combine imaging with drug delivery if the nanoparticles consist of an active pharmaceutical ingredient. The modes of expulsion of nanoparticles from the coating of ultrasound-driven microbubbles have been characterised by high-speed microscopy. These studied have provided the first design rules for controlled release of nanoparticles from ultrasound-driven bubbles, which may find applications beyond drug delivery, for instance in catalysis and sonochemistry.

Acknowledgements The writing of this book chapter, and part of the work by the Author described therein, was supported by the European Research Council, Starting Grant No. 639221.

References

1. M. Lee, E.Y. Lee, D. Lee, B.J. Park, Stabilization and fabrication of microbubbles: applications for medical purposes and functional materials. Soft Matter **11**, 2067–2079 (2015)
2. P.S. Epstein, M.S. Plesset, On the stability of gas bubbles in liquid-gas solutions. J. Chem. Phys. **18**, 1505–1509 (1950)
3. P. de Gennes, F. Brochard-Wyart, D. Quere, *Capillarity and Wetting Phenomena: Drops, Bubbles, Pearls, Waves* (Springer, New York, 2003)
4. P.B. Duncan, D. Needham, Test of the Epstein-Plesset model for gas microparticle dissolution in aqueous media: effect of surface tension and gas undersaturation in solution. Langmuir **20**, 2567–2578 (2004)

5. M.A. Borden, M.L. Longo, Dissolution behavior of lipid monolayer-coated, air-filled microbubbles: effect of lipid hydrophobic chain length. Langmuir **18**, 9225–9233 (2002)
6. G. Pu, M.A. Borden, M.L. Longo, Collapse and shedding transitions in binary lipid monolayers coating microbubbles. Langmuir **22**, 2993–2999 (2006)
7. J. Hanwright, J. Zhou, G.M. Evans, K.P. Galvin, Influence of surfactant on gas bubble stability. Langmuir **21**, 4912–4920 (2005)
8. A. Katiyar, K. Sarkar, P. Jain, Effects of encapsulation elasticity on the stability of an encapsulated microbubble. J. Colloid Interf. Sci. **336**, 519–525 (2009)
9. W. Kloek, T. van Vliet, M. Meinders, Effect of bulk and interfacial rheological properties on bubble dissolution. J. Colloid Interf. Sci. **237**, 158–166 (2001)
10. B.S. Murray, R. Ettelaie, Foam stability: proteins and nanoparticles. Curr. Opin. Colloid Interface Sci. **9**, 314–320 (2004)
11. A. Stocco, E. Rio, B.P. Binks, D. Langevin, Aqueous foams stabilized solely by particles. Soft Matter **7**, 1260–1267 (2011)
12. K. Soetanto, M. Chan, M. Okujima, Change in size and number of sodium laurate microbubbles with time in saline at different air concentrations. Jpn. J. Appl. Phys. **36**, 3238 (1997)
13. E. Dickinson, R. Ettelaie, B.S. Murray, Z. Du, Kinetics of disproportionation of air bubbles beneath a planar air–water interface stabilized by food proteins. J. Colloid Interface Sci. **252**, 202–213 (2002)
14. A.R. Cox, F. Cagnol, A.B. Russell, M.J. Izzard, Surface properties of class II hydrophobins from Trichoderma reesei and influence on bubble stability. Langmuir **23**, 7995–8002 (2007)
15. B.S. Murray, K. Durga, A. Yusoff, S.D. Stoyanov, Stabilization of foams and emulsions by mixtures of surface active food-grade particles and proteins. Food Hydrocolloids **25**, 627–638 (2011)
16. M.A. Borden, G. Pu, G.J. Runner, M.L. Longo, Surface phase behavior and microstructure of lipid/PEG-emulsifier monolayer-coated microbubbles. Colloids Surf. B Biointerfaces **35**, 209–223 (2004)
17. Z. Du, M.P. Bilbao-Montoya, B.P. Binks, E. Dickinson, R. Ettelaie, B.S. Murray, Outstanding stability of particle-stabilized bubbles. Langmuir **19**, 3106–3108 (2003)
18. B.P. Binks, Particles as surfactants—similarities and differences. Curr. Opin. Colloid Interface Sci. **7**, 21–41 (2002)
19. V. Garbin, Colloidal particles: surfactants with a difference. Phys. Today **66**, 68 (2013)
20. S. Lam, K.P. Velikov, O.D. Velev, Pickering stabilization of foams and emulsions with particles of biological origin. Curr. Opin. Colloid Interface Sci. **19**(5), 490–500 (2014)
21. Y. Gao, C.U. Chan, Q. Gu, X. Lin, W. Zhang, D.C.L. Yeo, A.M. Alsema, M. Arora, M.S.K. Chong, P. Shi, C.D. Ohl, C. Xu, Controlled nanoparticle release from stable magnetic microbubble oscillations. NPG Asia Mater **8**, e260 (2016)
22. Y. Zhang, M.C. Allen, R. Zhao, D.D. Deheyn, S.H. Behrens, J.C. Meredith, Capillary foams: stabilization and functionalization of porous liquids and solids. Langmuir **31**, 2669–2676 (2015)
23. Y. Zhang, J. Wu, H. Wang, J.C. Meredith, S.H. Behrens, Stabilization of liquid foams through the synergistic action of particles and an immiscible liquid. Angew. Chem. **126**, 13603–13607 (2014)
24. M. Abkarian, A.B. Subramaniam, S.H. Kim, R.J. Larsen, S.M. Yang, H.A. Stone, Dissolution arrest and stability of particle-covered bubbles. Phys. Rev. Lett. **99**, 188301 (2007)
25. M.H. Lee, V. Prasad, D. Lee, Microfluidic fabrication of stable nanoparticle-shelled bubbles. Langmuir **26**, 2227–2230 (2010)
26. A. Prosperetti, Thermal effects and damping mechanisms in the forced radial oscillations of gas bubbles in liquids. J. Acoust. Soc. Am. **61**, 17 (1977)
27. J.R.B.B. Goldberg, F. Forsberg, *Ultrasound Contrast Agents: Basic Principles and Clinical Applications*, 2nd edn. (Dunitz, London, 2001)
28. M.S. Plesset, A. Prosperetti, Bubble dynamics and cavitation. Annu. Rev. Fluid Mech. **9**, 145–185 (1977)

29. T.G. Leighton, *The Acoustic Bubble* (Academic Press, USA, 1994)
30. C.C. Church, The effects of an elastic solid surface layer on the radial pulsations of gas bubbles. J. Acoust. Soc. Am. **97**, 1510–1521 (1995)
31. P.J.A. Frinking, N. de Jong, Acoustic modeling of shell-encapsulated gas bubbles. Ultrasound Med. Biol. **24**, 523–533 (1998)
32. D.B. Khismatullin, A. Nadim, Radial oscillations of encapsulated microbubbles in viscoelastic liquids. Phys. Fluids **14**, 3534–3557 (2002)
33. P. Marmottant, S. van der Meer, M. Emmer, M. Versluis, N. de Jong, S. Hilgenfeldt, D. Lohse, A model for large amplitude oscillations of coated bubbles accounting for buckling and rupture. J. Acoust. Soc. Am. **118**, 3499–3505 (2005)
34. K.Y.C. Lee, Collapse mechanisms of langmuir monolayers. Annu. Rev. Phys. Chem. **59**(1), 771–791 (2008)
35. N. de Jong, A. Bouakaz, P. Frinking, Basic acoustic properties of microbubbles. Echocardiography **19**, 229–240 (2002)
36. J. Chomas, P. Dayton, D. May, J. Allen, A. Klibanov, K. Ferrara, Optical observation of contrast agent destruction. Appl. Phys. Lett. **77**, 1056–1058 (2000)
37. N. de Jong, P. Frinking, A. Bouakaz, M. Goorden, T. Schourmans, X. Jingping, F. Mastik, Optical imaging of contrast agent microbubbles in an ultrasound field with a 100-MHz camera. Ultrasound Med. Biol. **26**, 487–492 (2000)
38. N. Kudo, T. Miyaoka, K. Kuribayashi, K. Yamamoto, Study of the mechanism of fragmentation of a microbubble exposed to ultrasound using a high-speed observation system. J. Acoust. Soc. Am. **108**, 2547 (2000)
39. K. Tachibana, T. Uchida, K. Ogawa, N. Yamashita, K. Tamura, Induction of cell-membrane porosity by ultrasound. Lancet **353**, 1409 (1999)
40. P. Marmottant, S. Hilgenfeldt, Controlled vesicle deformation and lysis by single oscillating bubble. Nature **243**, 153–156 (2003)
41. P.A. Prentice, M.P. MacDonald, T.G. Frank, A. Cuschieri, G.C. Spalding, W. Sibbett, P.A. Campbell, K. Dholakia, Manipulation and filtration of low index particles with holographic Laguerre-Gaussian optical trap arrays. Opt. Express **12**, 593–600 (2004)
42. V. Garbin, D. Cojoc, E. Ferrari, R.Z. Proietti, S. Cabrini, E. Di Fabrizio, Optical micromanipulation using Laguerre-Gaussian beams. Jpn. J. Appl. Phys. **44**, 5772–5775 (2005)
43. P.H. Jones, E. Stride, N. Saffari, Trapping and manipulation of microscopic bubbles with a scanning optical tweezer. Appl. Phys. Lett. **89**, 081113 (2006)
44. P. Prentice, A. Cuschieri, K. Dholakia, M. Prausnitz, P. Campbell, Membrane disruption by optically controlled microbubble cavitation. Nat. Phys. **1**, 107–110 (2005)
45. V. Garbin, D. Cojoc, E. Ferrari, E. Di Fabrizio, M.L.J. Overvelde, S.M. van der Meer, N. de Jong, D. Lohse, M. Versluis, Changes in microbubble dynamics near a boundary revealed by combined optical micromanipulation and high-speed imaging. Appl. Phys. Lett. **90**, 114103 (2007)
46. A. Ashkin, J. Dziedzic, J. Bjorkholm, S. Chu, Observation of a single-beam gradient force optical trap for dielectric particles. Opt. Lett. **11**, 288 (1986)
47. A. Ashkin, J. Dziedzic, T. Yamane, Optical trapping and manipulation of single cells using infrared laser beams. Nature **330**, 769–771 (1987)
48. P. Jones, O. Maragó, G. Volpe, *Optical Tweezers: Principles and Applications* (Cambridge University Press, Cambridge, 2015)
49. K. Gahagan, G. Swartzlander, Optical vortex trapping of particles. Opt. Lett. **21**, 827 (1996)
50. M. Overvelde, V. Garbin, J. Sijl, B. Dollet, N. de Jong, D. Lohse, M. Versluis, Nonlinear shell behavior of phospholipid-coated microbubbles. Ultrasound Med. Biol. **36**, 2080–2092 (2010)
51. J. Sijl, B. Dollet, M. Overvelde, V. Garbin, T. Rozendal, N. de Jong, D. Lohse, M. Versluis, Subharmonic behavior of phospholipid-coated ultrasound contrast agent microbubbles. J. Acoust. Soc. Am. **128**, 3239–3252 (2010)

52. V. Garbin, B. Dollet, M. Overvelde, D. Cojoc, E. Di Fabrizio, L. van Wijngaarden, A. Prosperetti, N. de Jong, D. Lohse, M. Versluis, History force on coated microbubbles propelled by ultrasound. Phys. Fluids **21**, 092003 (2009)
53. Y. Luan, G. Lajoinie, E. Gelderblom, I. Skachkov, A.F. van der Steen, H.J. Vos, M. Versluis, N.D. Jong, Lipid shedding from single oscillating microbubbles. Ultrasound Med. Biol. **40**, 1834–1846 (2014)
54. V. Poulichet, V. Garbin, Ultrafast desorption of colloidal particles from fluid interfaces. Proc. Nat. Acad. Sci. **112**, 5932–5937 (2015)
55. J. Sijl, M. Overvelde, B. Dollet, V. Garbin, N. de Jong, D. Lohse, M. Versluis, Compression-only, behavior: a second-order nonlinear response of ultrasound contrast agent microbubbles. J. Acoust. Soc. Am. **129**, 1729–1739 (2011)
56. E. Stride, K. Pancholi, M. Edirisinghe, S. Samarasinghe, Increasing the nonlinear character of microbubble oscillations at low acoustic pressures. J. Royal Soc. Interface **5**, 807–811 (2008)
57. X. Zhao, P.A. Quinto-Su, C.D. Ohl, Dynamics of magnetic bubbles in acoustic and magnetic fields. Phys. Rev. Lett. **102**, 024501 (2009)
58. H. Mulvana, R.J. Eckersley, M.X. Tang, Q. Pankhurst, E. Stride, Theoretical and experimental characterisation of magnetic microbubbles. Ultrasound Med. Biol. **38**, 864–875 (2012)
59. S.S. Datta, H.C. Shum, D.A. Weitz, Controlled buckling and crumpling of nanoparticle-coated droplets. Langmuir **26**, 18612–18616 (2010)
60. S. Razavi, K.D. Cao, B. Lin, K.Y.C. Lee, R.S. Tu, I. Kretzschmar, Collapse of particle-laden interfaces under compression: buckling vs particle expulsion. Langmuir **31**, 7764–7775 (2015)
61. V. Garbin, J.C. Crocker, K.J. Stebe, Forced desorption of nanoparticles from an oil-water interface. Langmuir **28**(3), 1663–1667 (2012)
62. A. Francescutto, R. Nabergoj, Pulsation amplitude threshold for surface waves on oscillating bubbles. Acustica **41**(3), 215–220 (1978)
63. H. Lamb, *Hydrodynamics*, 6th edn. (Cambridge University Press, Cambridge, 1932)
64. E.H. Trinh, D.B. Thiessen, R.G. Holt, Driven and freely decaying nonlinear shape oscillations of drops and bubbles immersed in a liquid: experimental results. J. Fluid Mech. **364**, 253–272 (1998)
65. H. Vos, B. Dollet, J. Bosch, M. Versluis, N. de Jong, nonspherical vibrations of microbubbles in contact with a wall—a pilot study at low mechanical index. Ultrasound Med. Biol. **34**, 685–688 (2008)
66. S. Zhao, K.W. Ferrara, P.A. Dayton, Asymmetric oscillation of adherent targeted ultrasound contrast agents. Appl. Phys. Lett. **87**, 134103 (2005)
67. B. Dollet, S.M. van der Meer, V. Garbin, N. de Jong, D. Lohse, M. Versluis, Nonspherical oscillations of ultrasound contrast agent microbubbles. Ultrasound Med. Biol. **34**, 1465–1473 (2008)
68. V. Poulichet, A. Huerre, V. Garbin, Shape oscillations of particle-coated bubbles and directional particle expulsion. Soft Matter **13**, 125–133 (2016)

Characterisation of Functionalised Microbubbles for Ultrasound Imaging and Therapy

Eleanor Stride, Helen Mulvana, Paul Rademeyer, Dario Carugo, Joshua Owen, Richard Browning, Mengxing Tang and Robert Eckersley

Abstract Functionalised microbubbles have shown considerable potential both as contrast agents for ultrasound imaging and as a means of enhancing ultrasound mediated therapy. With the development of advanced techniques such as quantitative ultrasound imaging and targeted drug delivery, the accurate prediction of their response to ultrasound excitation is becoming increasingly important. Characterising microbubble behavior represents a considerable technical challenge on account of their small size (<10 μm diameter) and the ultrasound frequencies used to drive them in clinical applications (typically between 0.5 and 20 MHz). This chapter examines the three main techniques used for the characterization of microbubble dynamics: ultra-high speed video microscopy, laser scattering and acoustic attenuation and back scattering measurements. The principles of the techniques are introduced with examples of their applications and their relative advantages and disadvantages are then discussed. In the second half of the chapter magnetically functionalized microbubbles are used as a case study and results obtained using each of the three techniques are presented and compared. The chapter concludes with recommendations for combining different methods for microbubble characterization.

E. Stride (✉) · P. Rademeyer · D. Carugo · J. Owen · R. Browning
Institute of Biomedical Engineering, University of Oxford, Oxford OX3 7DQ, UK
e-mail: eleanor.stride@eng.ox.ac.uk

H. Mulvana
School of Engineering, University of Glasgow, Glasgow G12 8QQ, UK

M. Tang
Department of Bioengineering, Imperial College London, London SW7 2AZ, UK

R. Eckersley
Biomedical Engineering Department, Kings College London, London SE1 7EH, UK

K. Tsuji (ed.), *The Micro-World Observed by Ultra High-Speed Cameras*,
DOI 10.1007/978-3-319-61491-5_18

1 Introduction

Ultrasound contrast agents consisting of gas microbubbles stabilized by a surfactant or polymer coating have now been in clinical use for several decades. Despite being of similar volume to red blood cells, they are able to scatter ultrasound far more effectively on account of their high compressibility. In addition, when driven even at moderate ultrasound pressure amplitudes their response is highly nonlinear. This leads to an imaging signal that contains significant harmonic components which in turn enables microbubbles to be readily distinguished from the surrounding tissue. Consequently, microbubbles are extensively used for imaging tissue perfusion particularly in cardiovascular applications and cancer diagnosis [1].

Microbubble agents have also been widely investigated for therapeutic applications, including as nucleation agents to promote thermal ablation in high intensity focused ultrasound (HIFU) surgery [2], for opening of the blood brain barrier (BBB) [3] and as vehicles for targeted drug delivery and gene therapy [4]. For this last application, material can be attached to or incorporated within the microbubble coating (Fig. 1) to temporarily deactivate it. The passage of the microbubbles can be traced through the body using diagnostic ultrasound imaging and the therapeutic material then released by destroying the microbubbles with higher intensity ultrasound at a target site. Localising the treatment in this manner reduces the risk of harmful side effects. Moreover, it has been shown that the motion of the microbubble in the ultrasound field can promote both distribution of drugs throughout a tissue volume [5] and uptake at the individual cellular level [6], so called "sonoporation".

Treatment localisation can be enhanced by functionalising the microbubble surface to promote binding to specific types of cell, e.g. endothelial cells expressing biomarkers such as vascular endothelial growth factor (VEGF) that may be present at a disease site [7]. The same approach can be used to facilitate molecular imaging [8]. In addition, physical manipulation of the microbubbles can be exploited to achieve target accumulation, for example using acoustic radiation force [9] or an

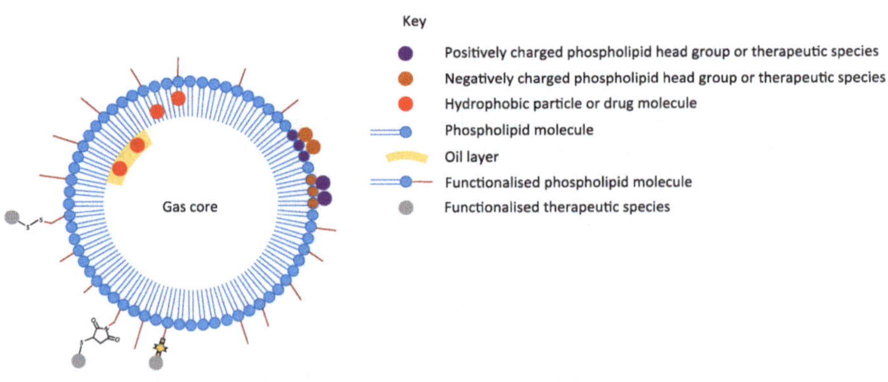

Fig. 1 Schematic of a functionalised microbubble

externally applied magnetic field. In the case of the latter it is necessary to modify the microbubble composition to make them magnetically responsive. This can be done through incorporation of magnetic nanoparticles such as superparamagnetic iron oxide [10]. Nanoparticles may also be conjugated to microbubbles for therapeutic applications. For example, drug loaded liposomes or solid polymeric particles may be attached to the microbubble surface to increase the effective "payload" of individual bubbles [11].

Modifying microbubbles in this way will influence their response to ultrasound. Microbubble dynamics are a primarily a function of their size and the ultrasound exposure parameters (predominantly the frequency and pressure amplitude) but they are also affected by the density, elasticity and viscosity of the bubble coating and to a lesser extent the immediate tissue environment [12]. Any change in the microbubble response will also affect their visibility under ultrasound imaging and potentially their ability to deliver therapeutic effects. It is therefore important to understand how microbubble response and surface composition are related. In this chapter the use of high speed imaging for microbubble characterisation is described and compared with two complementary techniques, laser scattering and measurement of acoustic attenuation and backscatter.

2 Characterisation Techniques

2.1 Ultra High-Speed Imaging

Direct observation of microbubble dynamics under ultrasound excitation at MHz frequencies is challenging on account of the short time and length scales involved which are on the order of microseconds and micrometres respectively. In order to achieve the required temporal resolution, frame rates of several million frames per second are needed. The minimum frame rate must be at least twice and preferably several times the driving frequency in order to capture nonlinear behaviour. Similarly, large magnifications are needed in order to resolve microbubble oscillations and hence high illumination intensities are required.

The spatial resolution and illumination requirements can be met with conventional microscope objectives and high power light sources. The temporal resolution requirements are more challenging, particularly since it is desirable to image microbubble dynamics over the full extent of an ultrasound pulse or indeed over several pulses. This means not only are high imaging frame rates needed but also relatively large numbers of frames. There are several different types of camera available that partially meet these requirements: streak cameras, shutter controlled cameras and rotating mirror cameras. Until very recently only rotating mirror cameras could offer an adequate combination of frame rate and number of frames for microbubble characterisation and so will be the main focus of discussion for this section.

There are several different types of rotating mirror camera in use in research laboratories around the world offering frame rates up to 10 million frames per second and up to 128 frames. The Brandaris 128 ultra-high speed imaging camera was developed specifically for microbubble research by Chin et al. [13] in collaboration with the Cordin Co. and can capture 6 blocks of 128 frames (or 12 blocks of 64 frames) at up to 25 million frames per second. A similar design has recently been used for the (UPMC Cam) [14] which offers improved sensitivity, spatial resolution and also fluorescence imaging capability.

The camera can be used to capture images through a standard microscope objective through appropriate optical coupling. For imaging microbubble dynamics, specially designed apparatus is required that enables simultaneous optical imaging and ultrasound exposure. An example of a typical experimental set up is shown in Fig. 2. One disadvantage of rotating mirror cameras is that the turbine requires a finite time to reach its operating speed. This dictates the time at which the ultrasound pulse can be transmitted and consequently it is necessary to confine microbubbles within the focal volume of the microscope objective and ultrasound transducer to ensure they are in position at the appropriate time. Various methods can be used, including optical tweezers [15], adhesion to a surface or fibre or trapping the microbubbles within an optically and acoustically transparent capillary tube immersed in a water bath.

The output from each experiment consists of a series of images of the microbubble at different times. The change in the microbubble diameter as a function of time can be extracted from these images via edge detection. The data can then either be compared directly for different microbubble formulations and/or fit to a theoretical model of microbubble dynamics. The image sequences also enable other phenomena such as non-spherical behaviour, coating shedding or fragmentation to be observed [16–18].

2.2 *Laser Scattering*

Ultra high-speed imaging has provided invaluable insights into the dynamics of microbubbles. It does have certain drawbacks, however. The first is the need to confine the microbubble. The majority of theoretical models describe a spherically symmetric, unconfined microbubbles and proximity to a surface will alter a microbubble's response considerably [19]. Optical tweezers offer a solution to this problem but implementing them significantly complicates the experimental set up. The second drawback is the difficulty involved in measuring the sound field to which the microbubble is exposed. It is extremely challenging to insert a suitable instrument (hydrophone) into the very small volume in which the microbubble is confined. Hence there will be some uncertainty as to the exact pressure to which the microbubble is exposed. Third, due to the need to confine and position individual microbubbles for each experiment, the rate at which measurements can be made is

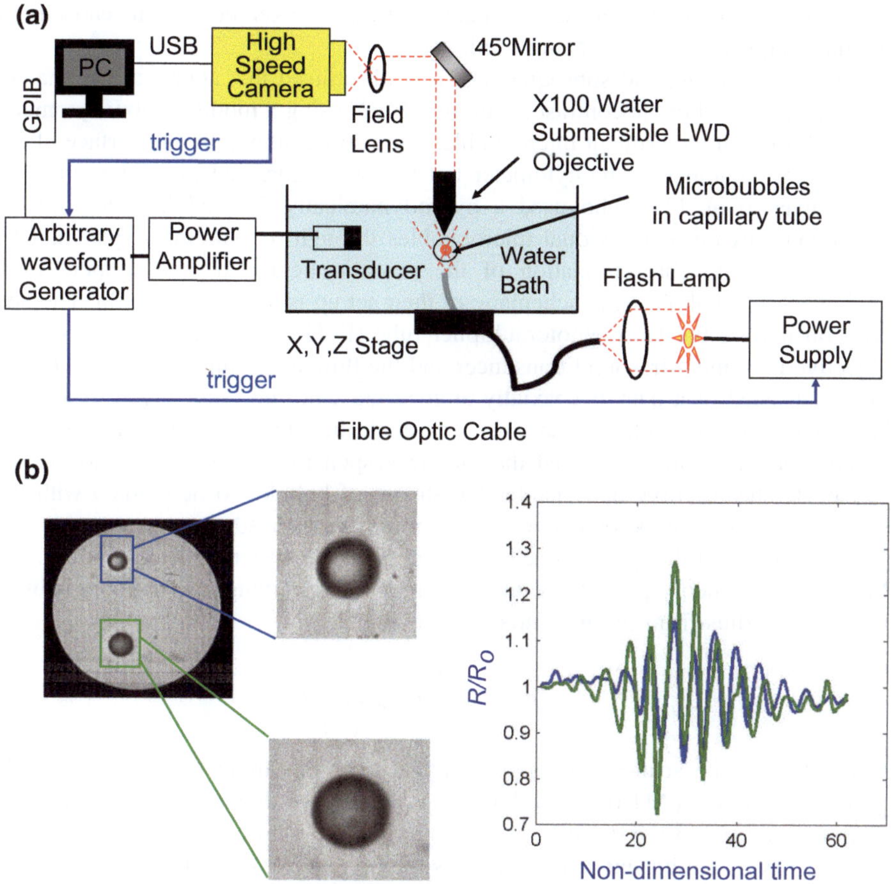

Fig. 2 **a** Schematic of experimental apparatus for ultra high-speed imaging of microbubbles **b** Example of microbubble image obtained (single frame) and corresponding radius-time curve extracted by image analysis (reproduced from [23] with permission of Elsevier)

quite low. This limits the size of date sets and hence effective statistical analysis of microbubble population characteristics.

An alternative means of capturing microbubble dynamics is to use the fact that a microbubble oscillating in an ultrasound field will strongly scatter any incident light and the amplitude of that scattered signal will be proportional to the microbubble volume. Thus, the change in the microbubble diameter can be inferred without the need to form an image. The advantage of this approach is that the experimental system can be considerably simpler, essentially consisting of a light source and detector. Importantly, optical sensors such as photomultiplier tubes offer much higher sensitivity and faster response times than rotating mirror cameras. This allows data acquisition to be triggered much more flexibly and there is also no restriction on the number of measurements that can be taken beyond the capacity of the data acquisition system being used.

Guan and Matula and subsequently Tu et al. utilised this approach for characterising the microbubble contrast agent SonoVue® using a modified flow cytometer [10, 20]. In their experiment microbubbles were still confined by the surface of the observation chamber in the cytometer but this was addressed in a later study by Rademeyer et al. [21] who used a hydrodynamically focused flow to direct a continuous stream of individual microbubbles through measurement system. This enabled accurate characterization of the acoustic field and measurement rates equivalent to 6 bubbles/s. A schematic of their set up is shown in Fig. 3a. The key components are the laser, photomultiplier tube (PMT), co-focally aligned optical objective lens and ultrasound transducer and the flow focussing device. The last of these consisted of a pair of coaxially aligned glass nozzles submerged in a water bath. The flow through each nozzle was controlled by an independent syringe pump. The inner nozzle carried the bubble suspension and the outer nozzle the so-called "sheath" flow that enables the stream of bubbles to be aligned with the focus of the microscope objective and the ultrasound transducer.

An example of the type of trace obtained using the system is shown in Fig. 3b. Mie scattering theory provides a direct means of determining microbubble volume and hence radius from the measured optical signal.

$$I_{scat}(\theta) = I_0 \frac{1}{R^2} \sigma'_{scat}(\lambda, r, \theta) \tag{1}$$

where I_{Scat} is the scattered light intensity at over the angle θ which is directly proportional to the PMT output voltage, I_O is the incident intensity, R is the distance from the particle to the detector, σ'_{Scat} is the particle's differential scattering cross section, r is its equivalent radius and λ is the optical wavelength.

2.3 Acoustic Scattering and Attenuation

Both high speed imaging and laser scattering are limited by the maximum signal to noise ratio that can be achieved and hence the minimum bubble size and/or amplitude of oscillation that can be measured. In addition, neither technique provides direct information as to how the acoustic signal from a microbubble changes with composition. For comparison this can be determined by performing acoustic scattering and attenuation measurements.

The transmission of ultrasound through a microbubble suspension can be characterised by measuring the attenuation and scattering of the signal. The speed of sound may also be determined. A typical experimental set up for performing these measurements is shown in Fig. 4. An acoustically transparent measurement chamber containing the microbubble suspension is submerged within a larger vessel

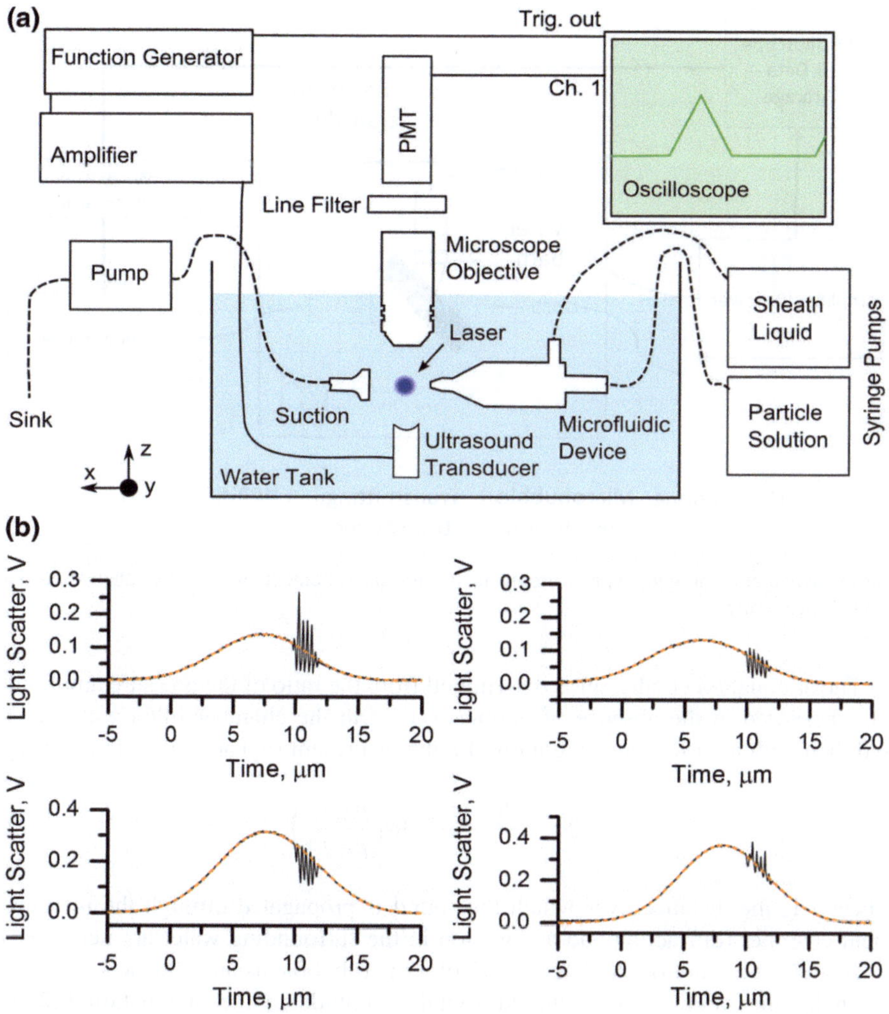

Fig. 3 **a** Schematic of experimental apparatus for laser scattering measurements from microbubbles **b** Examples of photomultiplier tube signals obtained from individual microbubbles. Reproduced from [21] with permission from the Royal Society of Chemistry

of water. A pair of ultrasound transducers[1] are positioned on either side of the chamber. For transmission measurements the transducers are aligned coaxially. For scattering measurements they are aligned with their axes perpendicular.

[1]Single element transducers may be used as shown in Fig. 4 or alternatively the transmitting transducer may be a clinical ultrasound probe and the receiving transducer a hydrophone.

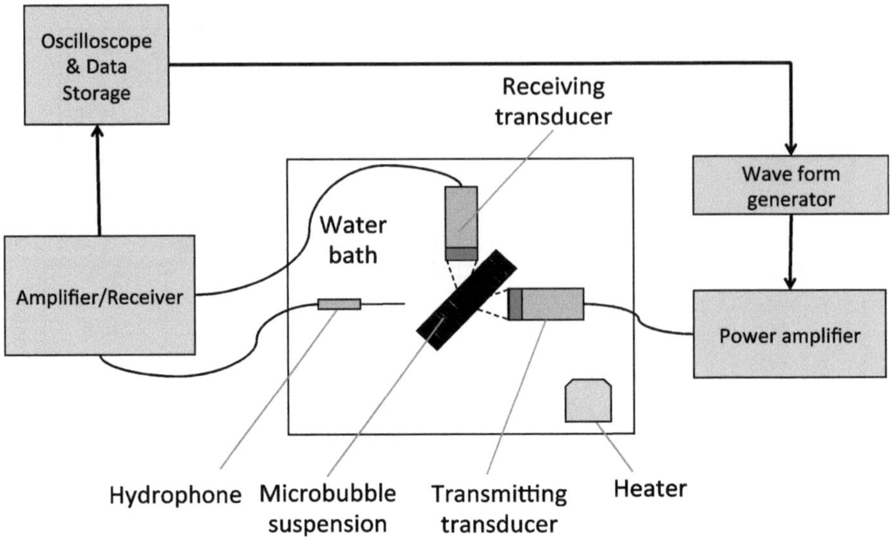

Fig. 4 Schematic showing typical arrangements for measurement of acoustic attenuation and acoustic scattering

The attenuation coefficient is determined from the ratio of the pressure amplitude (p_1) measured in the absence of bubbles (i.e. with the chamber filled with water) and then with a known concentration of bubbles present (p_2) at a given frequency f.

$$\alpha(f) = \frac{20 \log_{10} e}{x} \ln\left(\frac{p_2(f)}{p_1(f)}\right) \tag{2}$$

where x is the distance over which the sound is propagated through the measurement chamber (diffraction and attenuation in the surrounding water are neglected).

A useful measure of the "efficiency" of a microbubble as an acoustic scatterer is given by the ratio of the scattered signal amplitude to the attenuation [22] or "STAR." This may be expressed in terms of the amplitude at the central frequency of the driving pulse or the harmonic content of the signal, i.e. the nonlinear scattering to attenuation ration (nSTAR).

3 Case Study—Magnetic Microbubbles

3.1 Ultra High-Speed Imaging Observations

As mentioned above, microbubble targeting may be desirable for both diagnostic and therapeutic applications. Magnetic targeting has been used demonstrated both in vitro and in vivo for microbubble mediated delivery of a range of different molecules, including DNA [23, 24]. Magnetic nanoparticles are of a much higher density and elastic modulus than typical microbubble coating components. They are moreover often suspended in a relatively viscous oil. Their incorporation into a microbubble coating would therefore be expected to influence microbubble dynamics.

Mulvana et al. [25] undertook a high speed camera study of magnetic microbubble dynamics in order to investigate this. Microbubbles were prepared according to the protocol described in Stride et al. [23]. 1,2-Distearoyl-sn-glycero-3-phosphocholine (DSPC) was purchased from Avanti Polar Lipids Inc. (Alabaster, AL, USA). A ferrofluid suspension of 50 nm (hydrodynamic diameter) spherical magnetite nanoparticles in isoparaffin (10% volume fraction) was purchased from Liquids Research Ltd. (Bangor, UK). DSPC (15 mg) was weighed into a vial previously rinsed with surgical spirit (BP Unichem, Surrey, UK). Filtered deionised water (15 ml) was then added to the vial and the mixture sonicated using an ultrasonic cell disruptor (XL2000, probe diameter 3 mm; Misonix Inc., Farmingdale, NY, USA) at power setting 4 (15 s) followed by sonication at the air water interface (15 s). 15 μL of the 10 nm magnetite nanoparticle suspension was added followed by sonication (15 s in the liquid and 15 s at the air water interface) at power setting 4. The solution was then manually shaken for 30 s to produce magnetic microbubbles.

The high speed imaging apparatus used was as shown in Fig. 2. Dilute microbubble suspensions were drawn by capillary action into thin-walled cellulose tubes of internal diameter 200 μm (Membrana GmbH, Wuppertal-Oberbarmen, Germany) and immersed in a saline bath. The capillary was mounted on a position system so that single bubbles to be located at the joint focus of an ultrasound transducer (Videoscan, Panametrics-NDT, Waltham, MA, USA) and a microscope objective (Olympus LUMPLFL × 100 1.00NA). The oscillations of the microbubble were then observed using a rotating mirror high speed camera (Cordin 550, Salt Lake City, UT, USA) operating between 2 and 3 million frames/s with illumination provided by a 5 kV flash lamp (Cordin 659, Salt Lake City, UT, USA) was coupled to a fibre optic cable via a condenser lens (Comar Scientific, Reading, UK).

The transducer was driven by an arbitrary waveform generator (AWG2021 Sony-Tektronix, Tokyo, Japan) and a power amplifier (ENI 240L) to produce single 12-cycle Gaussian enveloped pulses with a centre frequency of 0.5 MHz and peak negative pressure of approximately 100 kPa. The sound field was characterised using a needle hydrophone (Precision Acoustics Ltd, Dorchester, U.K.) positioned at the transducer and objective focal point. The selection of the driving frequency was dictated by the maximum camera frame rate available ($\sim$2.5 Mfps). This also determined the average size of the microbubbles selected for the measurements as it was desirable to match the driving frequency to the bubble resonance frequency.

Data were acquired for the following: magnetic bubbles, nonmagnetic bubbles and also liquid micelles containing magnetic particles but no gas. In addition, the response of the magnetic microbubbles was recorded in the presence of a permanent magnet (rectangular block 10 mm × 10 mm × 25 mm N52 grade NdFeB, transversal magnetisation 1.5 T; NeoTexx, Berlin, Germany). For each microbubble studied, a series of 58 images were recorded at 2.5 million frames per second (Mfps) with a final image resolution of 20 pixels per micrometre. The minimum time between experiments was ~6 min due to charging of the flash lamp. Images were downloaded to a PC for analysis using software developed in house using MATLAB®, to determine the microbubble diameter for each frame in the recording and hence the microbubble radius as a function of time [26].

Figure 5 shows the key results from the experiments in the form of a resonance curve for the different types of microbubble, i.e. how the amplitude of oscillation varied with microbubble radius. As expected the magnetic micelles showed no response to ultrasound excitation. Unexpectedly, however, there was no significant difference between the response of the magnetic and non-magnetic microbubbles under these exposure conditions. Application of the magnetic field also did not appear to have any measurable effect on their behaviour.

3.2 Laser Scattering Measurements

In order to investigate the response of magnetic microbubbles at different frequencies and pressures, laser scattering measurements were performed using the set up shown in Fig. 3. Magnetic and non-magnetic microbubbles of the same formulation and also microbubbles coated with gold nanoparticles with hydrodynamic diameters of either 2 or 50 nm were excited with a 5 cycle pulse having a centre frequency of 3.5 MHz at different peak negative pressures.

As shown in Fig. 6a, with increasing nanoparticle size the median amplitude of oscillation over the measured population of bubbles decreased at a given driving pressure as might be expected due to the increased density, stiffness and viscosity of the coating. In Fig. 6a however, the influence of the microbubble size is not shown as all of the data are included (each point represents several thousand bubbles). Figure 6b shows the size distributions for the microbubbles and Fig. 6c how the

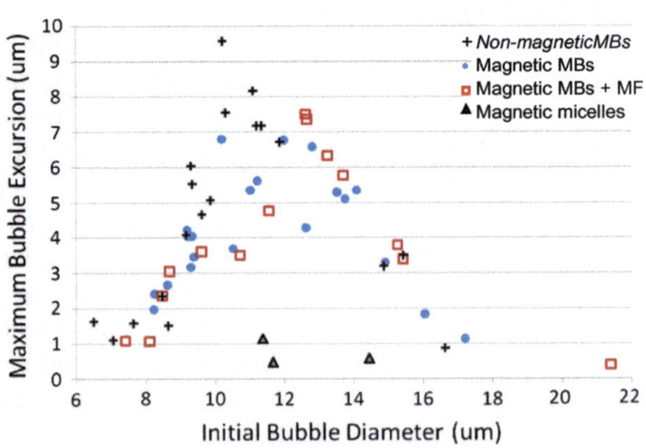

Fig. 5 Microbubble response to ultrasound excitation (12-cycle, 0.5 MHz, Gaussian windowed pulse 100 kPa focal peak negative pressure). Variation in the amplitude of radial expansion with initial bubble diameter as measured from the high speed camera footage for different types of microbubble and with or without an imposed magnetic field (MF)

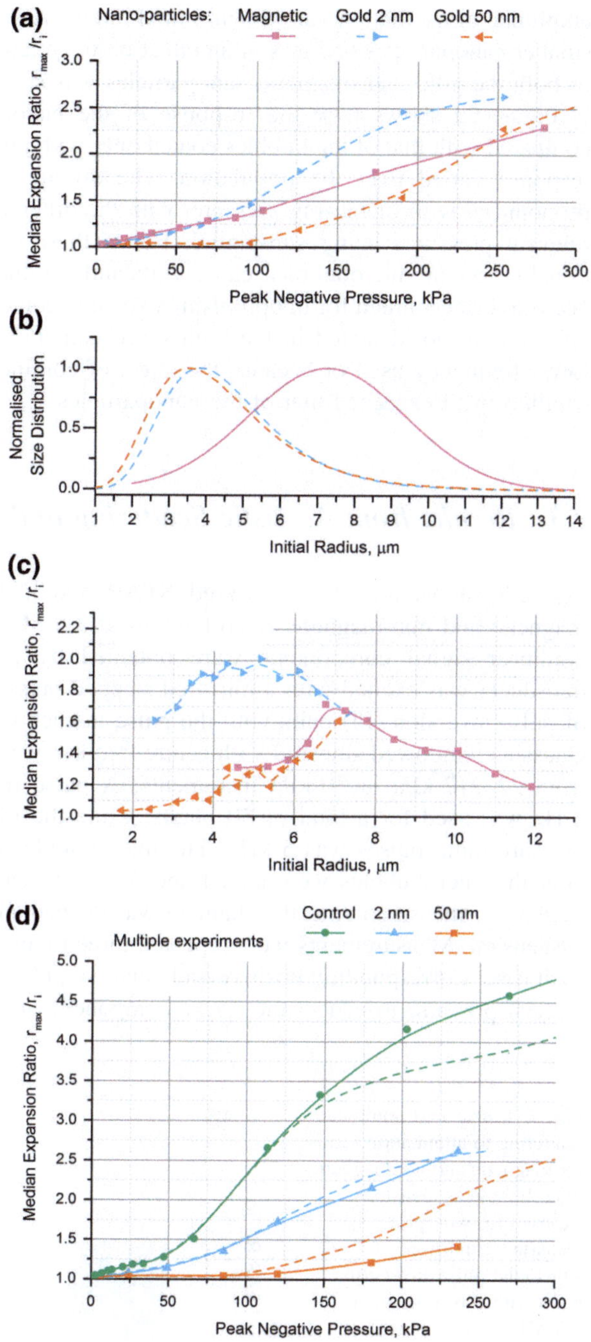

Fig. 6 **a** Normalised median amplitude of radial response of microbubbles coated with different types of nanoparticle as measured by laser scattering at 3.5 MHz and different peak negative pressures. **b** Size distributions of the microbubbles from part (**a**). **c** Normalised median amplitude of radial response of microbubbles as a function of initial bubble size. **d** Normalised median amplitude of radial response of microbubbles with and without a nanoparticle coating (*dotted lines* indicate experimental variability)

amplitude of oscillation varied with initial radius. Again these data indicate that the smaller nanoparticles had less of an effect on the microbubble dynamics. The effect of both the gold and magnetic nanoparticles was similar.

Figure 6d shows how the response of the nanoparticle coated microbubbles compared with that of the bubbles coated only with phospholipids and also how the response varied over the population. The difference in response became more pronounced with increasing pressure, with the difference in amplitude at clinically relevant pressures being substantial. Hence these data imply that surface functionalisation of microbubbles can significantly influence their response and this needs to be accounted for in optimising exposure conditions. It is possible that these effects were not detected in the high-speed camera experiments due to the much lower frequency used or because the effect of confinement near the surface of the capillary wall exceeded that of the nanoparticles.

3.3 Results from Acoustic Scattering and Attenuation

Figure 7 shows how the measured STAR and nSTAR values varied between magnetic and non-magnetic microbubbles at 3.5 MHz. For these experiments the nonlinear signal components were obtained by pulse inversion. The driving transducer was excited with a mirrored stepped pressure ramp made up of a series of pulse-inversion (PI) pairs with the pulse in each case consisting of a two-cycle Gaussian windowed sinusoid with centre frequency 3.5 MHz over a pressure range from 22–217 kPa in 30 kPa increments. A pulse repetition frequency (PRF) of 2 Hz was used for a total of 50 bursts, while the PRF between the incremented pressure ramp pulses was 5 kHz. The microbubble suspension was briefly stirred after the microbubbles were added and following each data acquisition. Between each set of measurements, the chamber was flushed and refilled with a fresh bubble suspension. Measurements were repeated three times to obtain an average value in each case. Corresponding positive and negative pulse inversion pairs were summed at each pressure to cancel the linear component of the pulse, leaving a mirrored

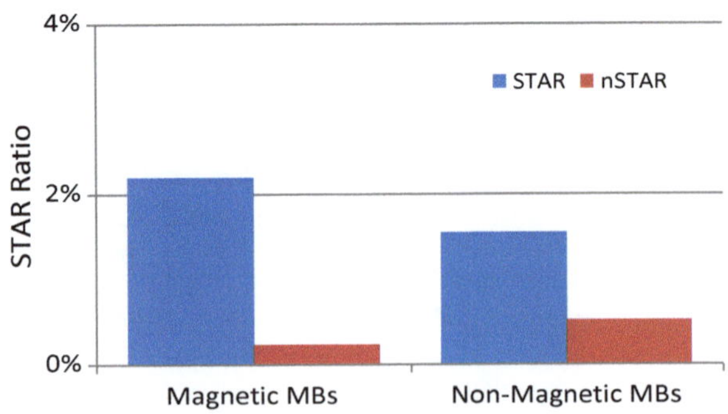

Fig. 7 Linear and nonlinear scattering to attenuation ratios for magnetic and nonmagnetic microbubble suspensions exposed to two-cycle Gaussian windowed sinusoidal pulse inversion pairs with centre frequency 3.5 MHz over a pressure range of 22–217 kPa

pressure ramp of residual pulses for further processing. The mean pulse-inverted residual signal was calculated over 50 repeats at each pressure using a fast Fourier transform. In each case, the amplitude spectra were calculated with reference to the background noise, so that measurements were in decibels greater than the background signal for a given acoustic excitation pressure and then used to calculate the scattering to attenuation (STAR) and nonlinear STAR (nSTAR) ratios. These data further indicate that the presence of the magnetic particles does influence the acoustic response of the microbubble. In particular it reduces the non-linear scattering response which will potentially limit microbubble detectability and would need to be compensated for in designing imaging/treatment monitoring protocols.

4 Concluding Remarks

In this chapter three complementary techniques for studying the behaviour of functionalised microbubbles were discussed and data from studies of microbubbles coated with solid nanoparticles were compared. The results indicate that each of the different techniques has advantages and disadvantages and currently that no one technique provides a comprehensive means of characterising microbubble response.

The main advantage of the ultra high-speed imaging method is that provides an image of the microbubble and thus phenomena such as non-spherical behaviour and importantly interactions with other structures (other bubbles, biological cells) can be captured. As above, it is possible to limit the influence of microbubble confinement by using optical tweezers to position microbubbles at the start of an experiment. Similarly it is also possible to simultaneously image microbubbles in two perpendicular planes (at the cost of either spatial resolution or number of frames) [27]. With the current generation of rotating mirror cameras experiments are relatively labour intensive which limits the size of data sets that can be obtained and/or the range of parameters that can be investigated in a single study.

The laser scattering approach has the advantage that very large numbers of bubbles can be measured over the course of an experiment enabling sufficiently large data sets for statistical analysis to be obtained. The maximum excitation frequency that can be used is much higher enabling data to be gathered over a wider range of clinically relevant exposure conditions. As above the microbubbles are not physically constrained by a boundary and therefore their measured responses may be more relevant to clinical applications. It is not however possible, at least with the set up shown in Fig. 3 to distinguish between spherical and non-spherical bubbles. The minimum bubble size that can be measured is also limited (although this can be overcome with relatively simple modifications to the optical set up and laser wavelength) and it is not possible to directly visualise phenomena such as lipid shedding which may be very relevant in therapeutic applications.

Measurements of acoustic scattering and attenuation have the advantage of being relatively simple and inexpensive to implement; and providing information that is directly related to the signals produced by microbubbles in diagnostic and therapeutic applications. The main drawback is that they do not provide any information regarding the variation in acoustic response across a microbubble population. Both the high-speed camera and laser scattering data indicate that this can be substantial.

Thus, it must be concluded that with the currently available technology it is desirable to combine multiple techniques to investigate microbubble behaviour. Recent work has demonstrated that it is possible to measure the acoustic emissions from single bubbles in both high speed camera [28] and laser scattering experiments and it is possible that with improvements in camera technology it may soon be possible to perform all three simultaneously.

References

1. D. Cosgrove, Ultrasound contrast agents: an overview. Eur. J. Radiol. **60**, 324–330 (2006). doi:10.1016/j.ejrad.2006.06.022
2. W. Luo, X.D. Zhou, X.L. Ren, M.J. Zheng, J. Zhang, G.B. He, Enhancing effects of SonoVue, a microbubble sonographic contrast agent, on high-intensity focused ultrasound ablation in rabbit livers in vivo. J. Ultras. Med. **26**, 469–476 (2007)
3. S. Meairs, A. Alonso, Ultrasound, microbubbles and the blood-brain barrier. Prog. Biophys. Mol. Bio. **93**, 354–362 (2007). doi:10.1016/J.Pbiomolbio.2006.07.019
4. J.L. Bull, The application of microbubbles for targeted drug delivery. Expert Opin. Drug Del. **4**, 475–493 (2007). doi:10.1517/17425247.4.5.475
5. M. Bazan-Peregrino, C.D. Arvanitis, B. Rifai, L.W. Seymour, C.C. Coussios, Ultrasound-induced cavitation enhances the delivery and therapeutic efficacy of an oncolytic virus in an in vitro model. J. Control. Release **157**, 235–242 (2012). doi:10.1016/j.jconrel.2011.09.086
6. I. Lentacker, I. De Cock, R. Deckers, S.C. De Smedt, C.T. Moonen, Understanding ultrasound induced sonoporation: definitions and underlying mechanisms. Adv. Drug Deliv. Rev. **72**, 49–64 (2013). doi:10.1016/j.addr.2013.11.008
7. H. Leong-Poi, J. Christiansen, A.L. Klibanov, S. Kaul, J.R. Lindner, Noninvasive assessment of angiogenesis by ultrasound and microbubbles targeted to alpha(v)-integrins. Circulation **107**, 455–460 (2003). doi:10.1161/01.CIR.0000044916.05919.8B
8. P.A. Dayton, D. Pearson, J. Clark, S. Simon, P.A. Schumann, R. Zutshi et al., Ultrasonic analysis of peptide-and antibody-targeted microbubble contrast agents for molecular imaging of alpha(V)beta(3)-expressing cells. Mol. Imaging **3**, 125–134 (2004). doi:10.1162/1535350041464883
9. M.J. Shortencarier, P.A. Dayton, S.H. Bloch, P.A. Schumann, T.O. Matsunaga, K.W. Ferrara, A method for radiation-force localized drug delivery using gas-filled lipospheres. IEEE T. Ultrason. Ferr. **51**, 822–831 (2004). doi:10.1109/TUFFC.2004.1320741
10. J. Tu, J. Guan, Y. Qiu, T.J. Matula, Estimating the shell parameters of SonoVue microbubbles using light scattering. J. Acoust. Soc. Am. **126**, 2954–2962 (2009). doi:10.1121/1.3242346
11. B. Geers, I. Lentacker, S. Cool, J. Demeester, S.C. De Smedt, N.N. Sanders, Ultrasound responsive doxorubicin-loaded microbubbles; towards an easy applicable drug delivery platform. J. Control. Release **148**, e59–e60 (2010). doi:10.1016/j.jconrel.2010.07.023
12. E. Stride, The influence of surface adsorption on microbubble dynamics. Philos. T. Roy. Soc. A **366**, 2103–2115 (2008). doi:10.1098/rsta.2008.0001
13. C.T. Chin, C. Lancee, J. Borsboom, F. Mastik, M.E. Frijlink, N. de Jong et al., Brandaris 128: a digital 25 million frames per second camera with 128 highly sensitive frames. Rev. Sci. Instrum. **74**, 5026–5034 (2003). doi:10.1063/1.1626013

14. E.C. Gelderblom, H.J. Vos, F. Mastik, T. Faez, Y. Luan, T.J.A. Kokhuis et al., Brandaris 128 ultra-high-speed imaging facility: 10 years of operation, updates, and enhanced features. Rev. Sci. Instrum. **83**, 103706 (2012). doi:10.1063/1.4758783
15. V. Garbin, D. Cojoc, E. Ferrari, E. Di Fabrizio, M.L.J. Overvelde, S.M. van der Meer et al., Changes in microbubble dynamics near a boundary revealed by combined optical micromanipulation and high-speed imaging. Appl. Phys. Lett. **90**, 114103 (2007). doi:10.1063/1.2713164
16. B. Dollet, S.M. van der Meer, V. Garbin, N. de Jong, D. Lohse, M. Versluis, Nonspherical oscillations of ultrasound contrast agent microbubbles. Ultrasound Med. Biol. **34**, 1465–1473 (2008). doi:10.1016/j.ultrasmedbio.2008.01.020
17. M.A. Borden, D.E. Kruse, C.F. Caskey, S. Zhao, P.A. Dayton, K.W. Ferrara, Influence of lipid shell physicochemical properties on ultrasound-induced microbubble destruction. IEEE T. Ultrason. Ferr. **52**, 1992–2002 (2005). doi:10.1109/TUFFC.2005.1561668
18. J. Viti, R. Mori, F. Guidi, M. Versluis, N. de Jong, P. Tortoli, Nonlinear oscillations of deflating bubbles. IEEE T. Ultrason. Ferr. **59**, 2818–2824 (2012). doi:10.1109/Tuffc.2012.2524
19. M. Overvelde, V. Garbin, B. Dollet, N. de Jong, D. Lohse, M. Versluis, Dynamics of coated microbubbles adherent to a wall. Ultrasound Med. Biol. **37**, 1500–1508 (2011). doi:10.1016/j.ultrasmedbio.2011.05.025
20. J. Guan, T.J. Matula, Using light scattering to measure the response of individual ultrasound contrast microbubbles subjected to pulsed ultrasound in vitro. J. Acoust. Soc. Am. **116**, 2832–2842 (2004). doi:10.1121/1.1795334
21. P. Rademeyer, D. Carugo, J.Y. Lee, E. Stride, Microfluidic system for high throughput characterisation of echogenic particles. Lab Chip **15**, 417–428 (2015). doi:10.1039/c4lc01206b
22. A. Bouakaz, N. De Jong, C. Cachard, Standard properties of ultrasound contrast agents. Ultrasound Med. Biol. **24**, 469–472 (1998). doi:10.1016/S0301-5629(97)00290-1
23. E. Stride, C. Porter, A.G. Prieto, Q. Pankhurst, Enhancement of microbubble mediated gene delivery by simultaneous exposure to ultrasonic and magnetic fields. Ultrasound Med. Biol. **35**, 861–868 (2009). doi:10.1016/j.ultrasmedbio.2008.11.010
24. D. Vlaskou, O. Mykhaylyk, P. Pradhan, C. Bergemann, A.L. Klibanov, K. Hensel et al., Magnetic microbubbles as mediators of gene delivery. Human Gene Ther. **21**, 1429–1430 (2010). doi:10.1063/1.3530059
25. H. Mulvana, R.J. Eckersley, M.-X. Tang, Q. Pankhurst, E. Stride, Theoretical and experimental characterisation of magnetic microbubbles. Ultrasound Med. Biol. **38**, 864–875 (2012). doi:10.1016/j.ultrasmedbio.2012.01.027
26. K. Chetty, E. Stride, C.A. Sennoga, J.V. Hajnal, R.J. Eckersley, High-speed optical observations and simulation results of SonoVue microbubbles at low-pressure insonation. IEEE T. Ultrason. Ferr. **55**, 1333–1342 (2008). doi:10.1109/tuffc.2008.796
27. H.J. Vos, M. Versluis, N. de Jong, Orthogonal observations of vibrating microbubbles. Ultrasonics **1–6**, 765–768 (2007). doi:10.1109/Ultsym.2007.196
28. J. Sijl, H.J. Vos, T. Rozendal, N. de Jong, D. Lohse, M. Versluis, Combined optical and acoustical detection of single microbubble dynamics. J. Acoust. Soc. Am. **130**, 3271–3281 (2011). doi:10.1121/1.3626155

Therapeutic Applications of Cavitation Phenomena in the Medical Fields

Nobuki Kudo

Abstract Cavitation is an important phenomenon that has been intensively studied in the field of fluid dynamics. The formation of cavitation bubbles is induced under the condition of pressure fluctuations in a fluid, and the collapse of bubbles elicits strong effects in their surroundings. Normally, cavitation does not occur inside a biological body; however, the use of an artificial organ and irradiation of acoustic energy may induce the formation of cavitation bubbles inside the body. In this chapter, medical equipment is introduced, that induces or utilizes cavitation phenomena. A mechanical heart valve is known to generate bubbles that may cause thrombosis and valve failure, and various types of ultrasound exposure are used for medical purposes such as surgical equipment, non-invasive therapy, and drug delivery. High-speed observation plays an important role in visualizing cavitation activities generated in various applications. Results of observations provide valuable insights into the mechanisms by which cavitation bubbles elicit effects and also insights into how the effects can be minimized or maximized.

1 Introduction

In the field of medicine, ultrasound is utilized both in diagnosis and therapy. In diagnostic application, the patient's body is irradiated with short-pulsed ultrasound, and echo signals from tissue boundaries are used to make a tomographic image of the body. The technique is widely used in clinical fields as a noninvasive technique for diagnostic imaging. In therapeutic application, continuous or long-burst ultrasound is utilized for heating and eroding biological tissues and also for disintegrating calculi.

Biological effects of ultrasound are categorized into thermal and non-thermal effects. A thermal effect is caused by absorption of ultrasound energy by tissue.

N. Kudo (✉)
Graduate School of Information Science and Technology,
Hokkaido University, Sapporo, Hokkaido 060-0814, Japan
e-mail: kudo@ist.hokudai.ac.jp

K. Tsuji (ed.), *The Micro-World Observed by Ultra High-Speed Cameras*,
DOI 10.1007/978-3-319-61491-5_19

Temperature elevation above 43 °C can cause lethal damage to biological cells. Non-thermal effects include a wide range of mechanisms such as mechanical and chemical effects. Cavitation, a phenomenon in which bubbles generated in water undergo radial oscillation and collapse, plays important roles as mechanisms to produce non-thermal effects.

Rarefactional pressures of ultrasound and lateral pressures generated beside a high-speed fluid flow can produce cavitation bubbles. Oscillation of pre-existing bubbles is also a cavitation phenomenon. Bubble oscillation can cause various types of adverse effects on surrounding biological tissues. Direct observation of bubble dynamics is helpful for understanding the mechanisms by which adverse effects occur. Since ultrasound therapy utilizes ultrasound waves in a frequency range of several tens of kHz to several MHz, high-speed observation is essential to elucidate the rapid activities of cavitation bubbles.

In this section, medical equipment that induces or utilizes cavitation phenomena is introduced and high-speed images taken to elucidate activities of cavitation bubbles are shown.

2 Artificial Heart Valve

The blood flow in heart chamber is a representative example of violent high-speed mechanical actions inside an animal body. In the human heart, the right ventricle contracts to pump blood throughout the body. In this phase, the aortic valve opens for blood to flow into the vessel system and the mitral valve closes to stop backflow. Insufficient closure of the mitral valve causes a decrease in stroke volume, and a chronic decrease in stroke volume results in cardiac insufficiency. Cardiac ultrasonography is the first choice for diagnosis of cardiac insufficiency. A jet generated in the backflow of a mitral valve with incomplete closure can be visualized using color Doppler mode images of diagnostic ultrasound equipment.

In cases in which cardiac insufficiency becomes serious, surgery to replace the diseased mitral valve by an artificial heart valve is performed. Arterial valves are categorized into two types: mechanical and bioprosthesis valves. Mechanical valves have higher risks of valve thrombosis but have longer lifetime and do not require periodic replacement. Bioprosthesis valves require periodic replacement but have a low risk of thrombosis. Figure 1 shows a bileaflet mechanical valve used for replacement of a mitral valve [1]. In the diastolic phase, blood flows from the right-atrial side (upper surface) to the right-ventricle side (undersurface). In systolic phase, two valve leaflets are closed to stop the backflow. Water hammer phenomenon and squeeze flow that flow through a narrow channel between the leaflet and valve sheet are generated in this phase, resulting in occurrence of cavitation at the right-atrial side of the valve.

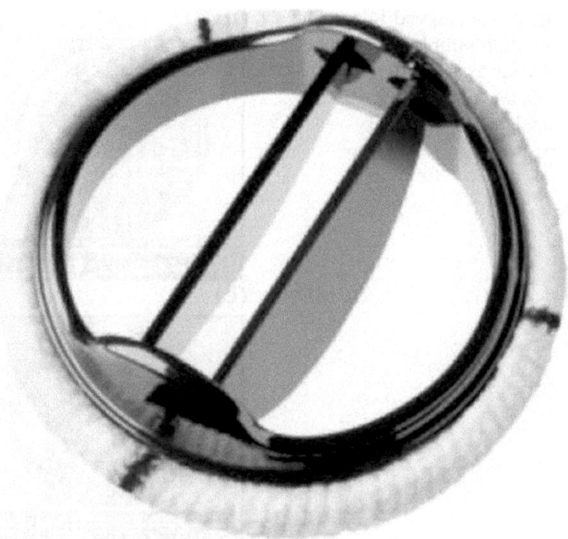

Fig. 1 Bileaflet mechanical valve for the mitral valve [1]. © 2017 HeartValveSurgery.com

2.1 *Bubble in Heart*

When patients with a mechanical heart valve are examined by echocardiography, high-intensity spotty echoes are frequently observed inside the heart chamber. This observation indicated that operation of the mechanical valve induces cavitation bubbles, which may lead to valve failure and valve thrombosis. To decrease the occurrence of such events, dynamics of blood flow in a heart chamber have been intensively studied by visualizing cavitation dynamics.

Figure 2 shows high-speed images of cavitation bubbles generated adjacent to a bileaflet mechanical valve [2]. An EktaPro Motion Analyzer (HRC1000, Kodak, Rochester, NY) was used to take images of 512 × 384 pixels at a maximum framing rate of 1000 fps. Five sequential frames were captured from just before valve closure to 8 ms after closure. (a) Endo-diastolic phase (<1 ms before complete closure of the valve). No bubble was observed. (b) Complete closure phase (t_0). Cavitation clouds were observed at several locations indicated by white arrows: a gap between the two leaflets and a gap between the leaflet and its surroundings. Water hammer phenomenon and squeeze flows produce the cavitation clouds. (c) Dissipation of clouds at about t_0+5 ms. (d), (e) Pressure recovery phase. The cavitation clouds become transparent because the cavitation bubbles filled with water vapor quickly disappear in this pressure condition. Growth of persisting gas-filled bubbles is observed at approximately t_0+8 ms. The gas-filled bubbles are built up during repeating heartbeats, and they can be visualized as bright spots in diagnostic ultrasound images.

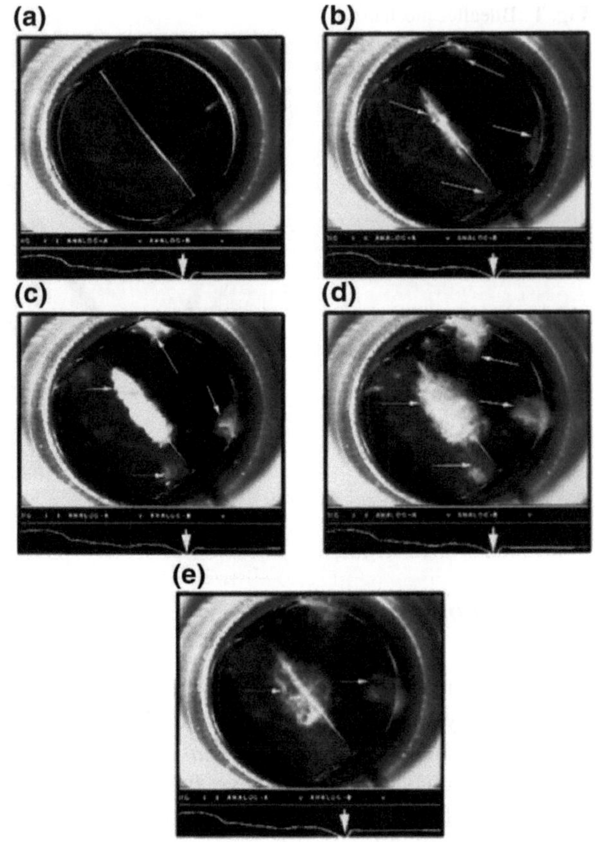

Fig. 2 High-speed images of cavitation bubbles generated adjacent to a bileaflet mechanical valve [2]. © 1999 Annals of Biomedical Engineering

3 Cavitation Ultrasonic Surgical Aspirator

Ultrasound is also utilized for surgical equipment. An ultrasonic scalpel uses the thermal effect of ultrasound, which is achieved by absorption of frictional heat generated by ultrasonic vibration. A blade vibrating at an ultrasound frequency of several tens of kHz is pressed against a biological tissue, enabling tissue coagulation, ablation, and stopping bleeding.

A Cavitation ultrasonic surgical aspirator (CUSA) is a device that uses ultrasound for erosion of soft tissues. Erosion is obtained not by the thermal effect but by the non-thermal effect of cavitation. Cavitation bubbles generated by a vibrating hollow needle tip cause a mechanical effect on biological cells, and eroded tissue is aspirated via a hole in the needle. A CUSA is unique in its selectivity for target tissue. Since the tissue is mainly destroyed by the mechanical effect of cavitation bubble activities, erosion mainly occurs in tissues with a high water content that have a lower cavitation threshold. This is the reason why a CUSA is used for brain

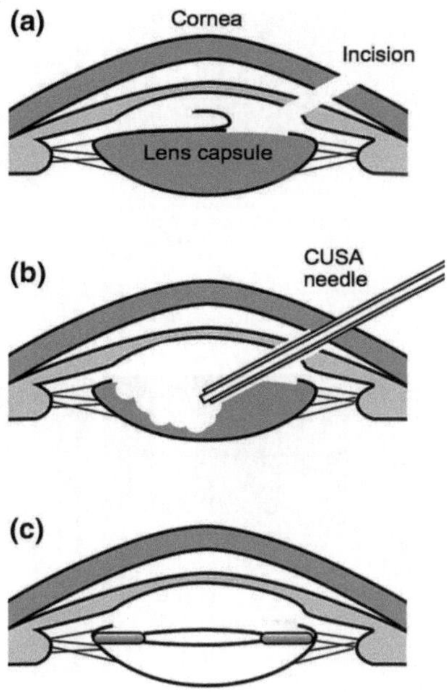

Fig. 3 Procedure for ultrasonic phacoemulsification surgery using a cavitation ultrasonic surgical aspirator (CUSA)

and liver surgery to remove parenchyma and neoplastic lesions with preservation of collagen-rich structures such as blood vessels and nerves.

A CUSA is also used for ultrasonic phacoemulsification [3], which is surgery to replace a clouded eye lens with an intraocular lens to restore clear vision. Figure 3 shows the procedure for the surgery. (a) A small incision is made at the cornea, and a CUSA needle is inserted from the incision. (b) The needle is then introduced inside an anterior lens capsule, and a clouded lens is destroyed by cavitation generated at the needle tip. The eroded tissue emulsion is aspirated to the outside via a hole in the needle. (c) A folded intraocular lens is inserted into the lens capsule, and then the lens is spread. The incision is naturally closed within 2–3 days after surgery.

Figure 4 shows high-speed images of pressure fields and cavitation bubbles generated around the tip of a hollow titanium needle [4]. The needle tip submerged in water is illuminated using light flashes of 10 ns in duration and 5 kHz in repetitive frequency from a copper vapor laser, and high-speed schlieren images were captured by a video camera. The six photos were taken in the duty cycle of 40 μs: the upper three panels show cavitation bubbles generated on the surface of the tip, and the lower three panels show subsequent generation of a circular shock front during implosion of the bubbles. High-speed observation of bubble dynamics and the resulting shock fronts is important because it provide keys for understanding the mechanisms of tissue erosion and for finding a way to facilitate the effect.

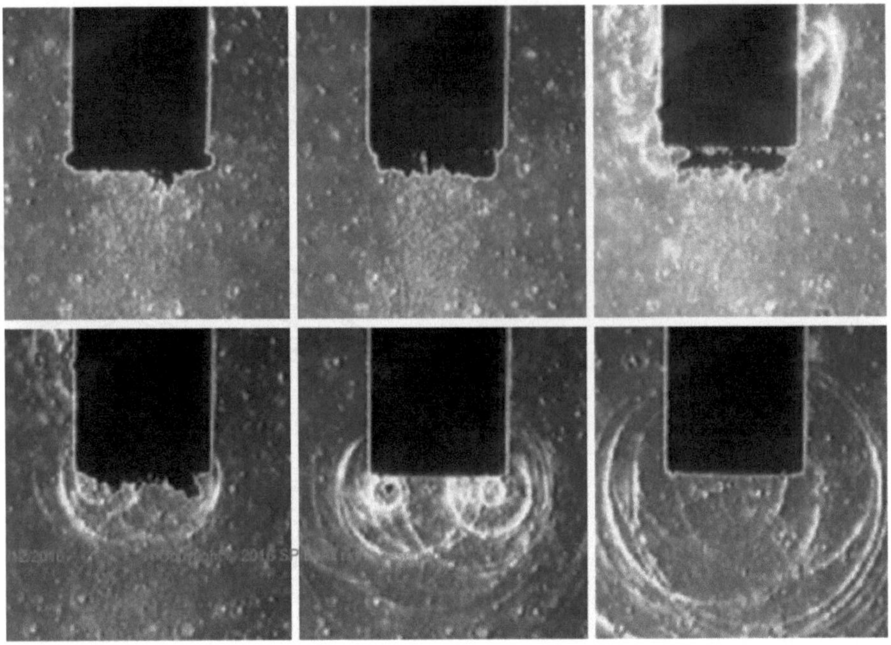

Fig. 4 Sequence through the duty cycle (40 μs) of one vibration of the CUSA tips captured using ultra high speed Schlieren imaging showing collapsing cavitation bubbles and shock waves [4]. © 2006 Journal of Biomedical Optics

4 Extracorporeal Shock Wave Lithotripsy

Calculi produced in the kidney, urinary duct and urinary bladder are called urinary tract stones. The main components of the stones are uric acid and calcium carbonate. Other types of calculi are produced in the biliary system including the bile duct and gallbladder. The main components of biliary calculi are cholesterol and bilirubin.

Extracorporeal shock wave lithotripsy (ESWL) is a noninvasive technique for utilizing a shockwave field to fragment calculi [5]. A shockwave or a short ultrasound pulse is generated outside the human body and focused on the location of a calculus. The shockwave generated at the focus has a steep pressure increase up to 20–100 MPa in peak positive pressure followed by sustained negative pressure. Hundreds to thousands of shockwaves are irradiated to disintegrate the calculus into fragments that can be excreted from the tract.

Renal calculi are rigid but fragile. Cholesterol gallstones are also rigid and easily disintegrated by ESWL. Pigment gallstones containing bilirubin are soft but difficult to fragment by ESWL. For effective disintegration of a calculus, optimization of the exposure conditions is important considering their difference in acoustic nature. For this reason, high-speed observation is widely utilized to examine the disintegration process.

Figure 5 shows mechanisms of stone disintegration. The first mechanism is compressive fracture observed at the frontal surface of the stone (Fig. 5a) [6]. The fracture occurs when the peak positive pressure of a shockwave exceeds the compressive breaking strength of the stone. The second mechanism is fracture caused by the shockwave reflection at the posterior interface between the stone and water (Fig. 5b). Since the phase of a pressure waveform is inverted at the interface, high compressive pressure of the incident shockwave is converted into high rarefactional pressure propagating in the opposite direction, which produces microcracks in the stone near the exiting end. Repetitive application of shockwaves results in splitting off of the stone fragments from the posterior surface, which is called the Hopkinson effect. The third mechanism is cleavage of a stone by squeezing. In this mechanism, the shockwave propagating outside of the stone surface produces compressive pressure inside the stone in directions both perpendicular and parallel to the direction of shockwave propagation (Fig. 5c) [6].

Figure 6 shows a high-speed photograph of the destruction process of a model stone mimicking a renal stone [6]. The model stone has a cylindrical shape of 10 mm in height and 10 mm in diameter. A focused shockwave of 31 MPa in peak positive pressure is insonified to the stone. Cleavage of the stone was observed in the directions perpendicular and horizontal to the direction of shockwave propagation, indicating stone destruction by the squeezing mechanism.

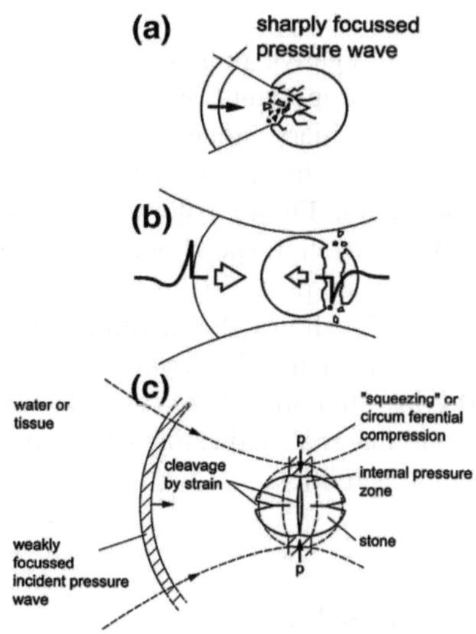

Fig. 5 Mechanisms of stone disintegration in extracorporeal shockwave lithotripsy. **a** Compressive fracture at the frontal stone surface [6], **b** fracture caused by shockwave reflection at the posterior stone surface, and **c** cleavage of a stone by squeezing [6]. © 2001 Ultrasound in Medicine and Biology

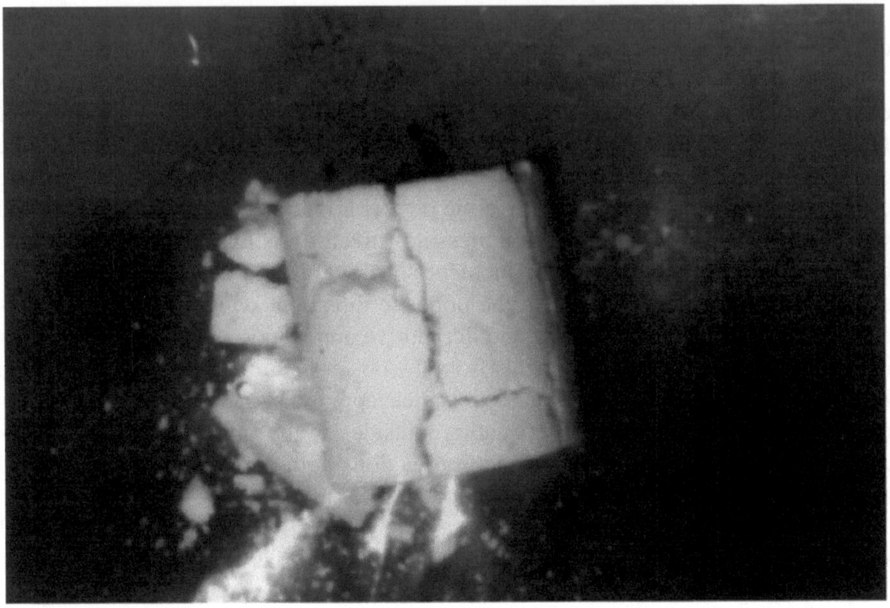

Fig. 6 First cleavage of a cylindrical artificial stone (HMT) into several fissures parallel and perpendicular to the wave propagation direction. The stone dimensions were 10 × 10 mm. The positive pulse pressure was 31 MPa. The cylinder axis was perpendicular to the wave propagation direction [6]. © 2001 Ultrasound in Medicine and Biology

The negative pressure region of a shockwave induces cavitation, which also plays an important role in stone disintegration. Cavitation mechanisms are important for biliary stones because the mechanisms depending on pressure waves propagating inside the stone are less effective for the calculi with larger ultrasound attenuation. Figure 7 shows high-speed images of a human gallstone taken using an NAC-E-10 high-speed prism camera that can take 10,000 fps [7]. A shockwave of 64 MPa in peak positive pressure is insonated from the bottom of the stone. Figure 7a shows the cavitation bubbles observed in front of the stone surface 0.5 ms after the shockwave hits the stone. Implosion of the bubbles was observed 0.2 ms after that (Fig. 7c). The following frames (Figs. 7d–h) show outburst of stone material, indicating the significance of the mechanisms depending on cavitation activities.

5 Histotripsy

High-intensity focused ultrasound can generate cavitation bubbles in a deep region of the body. Violent oscillation and the resulting collapse of bubbles under intense ultrasound exposure cause serious mechanical damage to surrounding biological

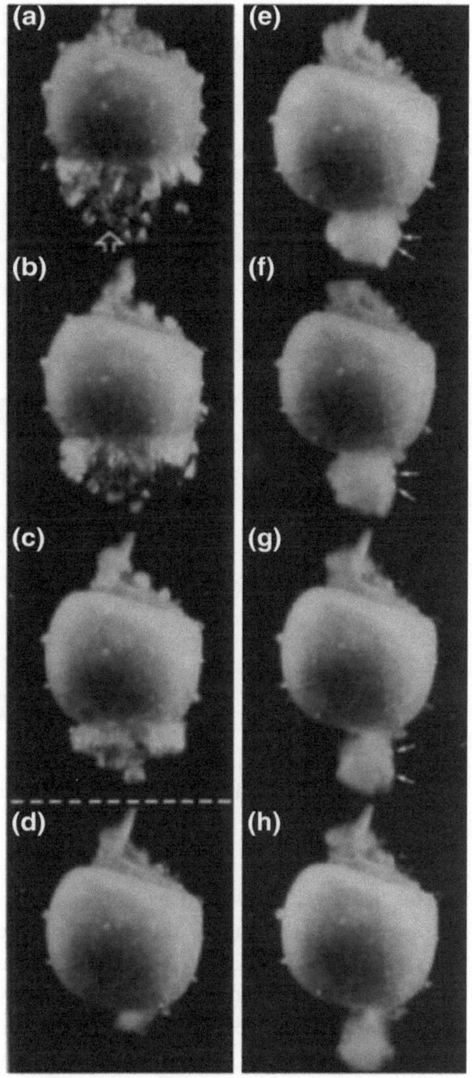

Fig. 7 High-speed sequence of a human gallstone ($0.7 \times 0.9 \times 0.9$ cm^3) being hit by a shock wave from below (*arrow*). Each frame = 0.1 ms. (**a–c**) represent the end of the first phase of cavitational activity; **a** 0.5 ms after the shock wave has hit the stone; **c** imploding circle of cavitational bubbles 0.7 ms after shock wave exposure simultaneously indicating the end of the first phase. **d**–**h** represent the second phase: material outburst. Note the cavitation bubbles again inside the material jet (*small arrows*) [7]. © 1991 Ultrasound in Medicine and Biology

tissues. The technique that utilizes the phenomena is called histotripsy [8, 9]. Figure 8a shows a through hole generated at a porcine atrial wall. Figure 8b shows a histology image of tissue erosion made in a bulk ventricle tissue, indicating that histotripsy can destroy biological tissue at the cellular level. Figure 8c shows "M" shape lesion generated by histotripsy shown in ultrasound imaging.

Whereas ESWL is a technique for disintegrating a stone with minimal damage to surrounding tissue, histotripsy is a technique that positively utilizes maximized tissue damage. The difference in tissue damage is realized by changing the ultrasound exposure conditions. Both techniques use repetitive exposure of ultrasound

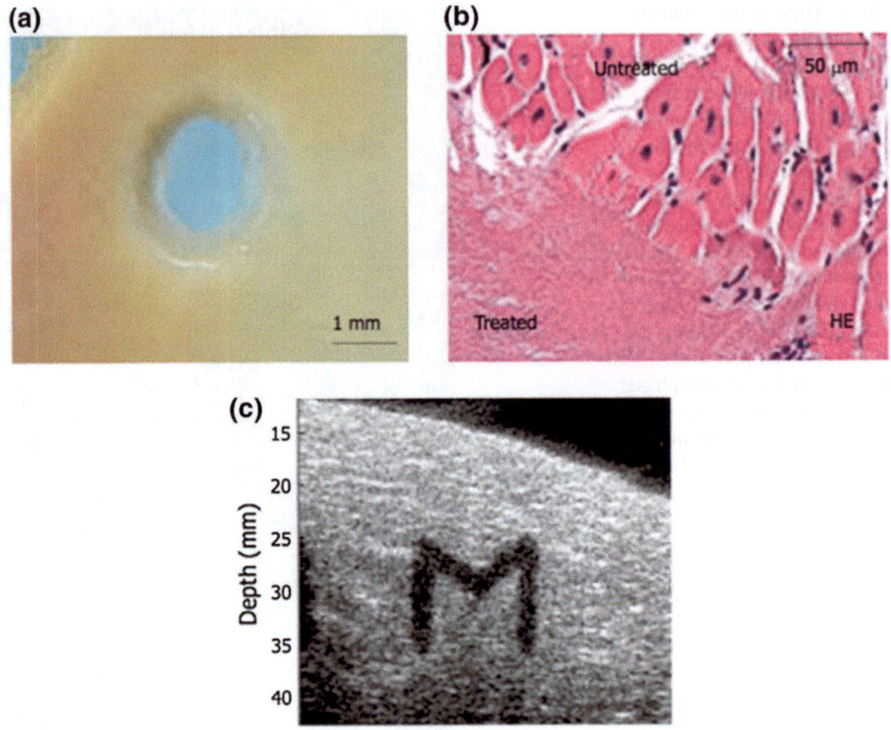

Fig. 8 Representative **a** image and **b** histology of tissue erosion after histotripsy, and **c** "M" shape lesion generated by histotripsy shown in ultrasound imaging [9]. © 2011 World Journal of Clinical Oncology

pulse; however, ESWL uses single-cycle pulses with low duty ratios, while histotripsy uses burst pulses that contain several to several tens of cycles with higher duty ratios. In ESWL, cavitation bubbles are generated at the stone surface, and the bubbles play a role in erosion of a calculus into small fragments, but they may also cause tissue damage. On the other hand, histotripsy induces a larger amount of cavitation bubbles, i.e., cloud cavitation, to homogenize tissue effectively. Cavitation phenomena can promote both thermal and non-thermal effects of ultrasound exposure. Thermal mechanisms are utilized in high-intensity focusing ultrasound therapy for tissue heating, and non-thermal mechanisms are utilized in histotripsy for tissue erosion.

Figure 9 shows shadowgrams of a cavitation cloud generated inside a transparent tissue-mimicking phantom [10]. The images were taken using a high-speed camera (SIM 02, Specialized Imaging, UK) that can take16 frames at rates up to 2×10^8 fps. In the shadowgrams, wavefronts of focused ultrasound are visualized as a set of dark oblique lines in parallel, and cavitation bubbles are visualized as an aggregate of dark points.

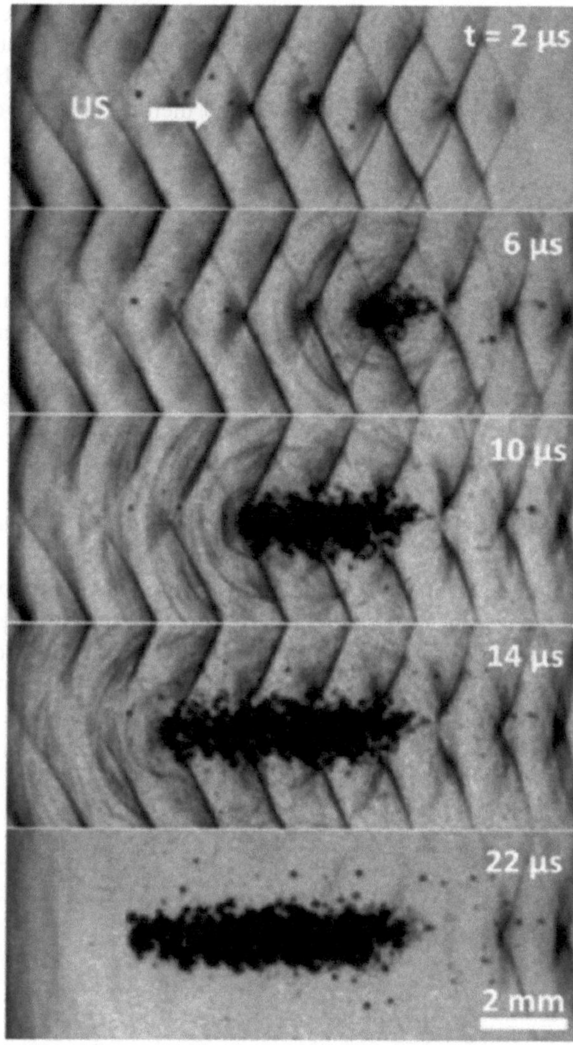

Fig. 9 Growth of a bubble cloud at the focus during application of a 20-cycle pulse. Ultrasound propagation was from *left* to *right*. The cloud started from a distal location within the focal zone and grew toward the transducer along the acoustic axis. The *dark lines* in each frame are a shadowgraph pattern created by the shock fronts of each cycle of the wave [10]. © 2011 The Journal of the Acoustical Society of America

In the first image ($t = 2$ μs), the two early cycles in the burst pulse have already arrived at the focus, but no cavitation bubble is visualized. In the second image ($t = 6$ μs), a cavitation cloud is generated at the focus area, and concentric circular wavefronts are observed around this area, indicating that the cloud causes ultrasound reflection and cavitation bubble implosion. In the following frames, the area of the cloud has expanded in the direction toward the ultrasound transducer. This means that a large rarefactional pressure generated by reflection of a shockwave at the water-bubble interface promotes growth of the cloud.

High-speed observations of a cavitation cloud have led to a better understanding of histotripsy mechanisms. Observation have suggested the importance of creating initial bubbles in a tissue because it triggers the growth of clouds. The importance is now widely accepted in ultrasound therapy utilizing cavitation phenomena.

6 Sonoporation

Sonoporation is a technique for temporally increasing the permeability of a biological cell membrane by ultrasound exposure for transduction of a foreign gene or drug for which normally there is no permeability [11]. The technique typically uses continuous or long-burst ultrasound of several MHz in frequency. It is well recognized that sonoporation efficiency is significantly improved in the presence of bubbles of several microns in size adjacent to cells; however, the mechanisms of sonoporation have not been fully elucidated yet.

Many studies have been carried out to try to elucidate the mechanisms. Both high-speed and microscopic observation techniques are needed to observe interaction between biological cells and micron-sized bubbles oscillating in ultrasound frequency [12, 13]. Figure 10 shows a typical high-speed microscopic system developed for observation of cell-bubble interaction. The observation system consists of three parts: a microscope, an ultrasound exposure system with an observation chamber, and a high-speed camera with a high power illuminating light source.

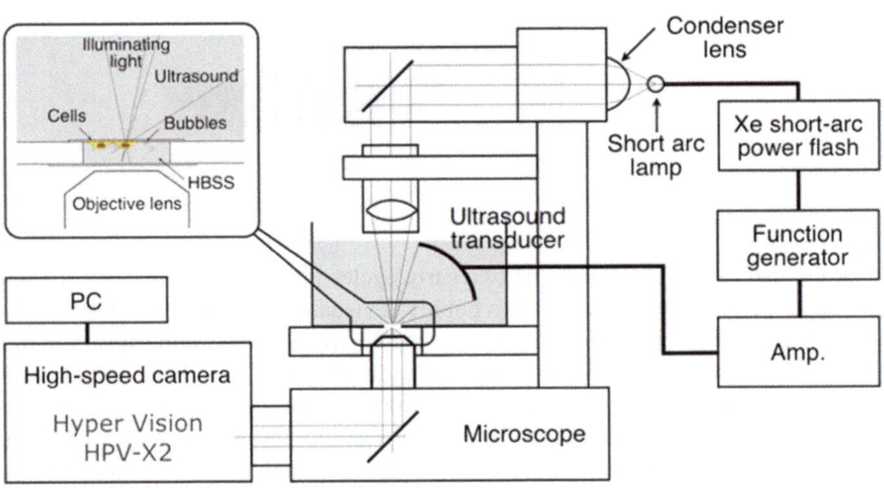

Fig. 10 Experimental setup for high-speed observation of bubble-cell interaction during sonoporation

An inverted-type microscope (IX70, Olympus, Tokyo, Japan) with a 40× objective lens (SLCPlanFI 40XPH; NA = 0.55, WD = 6.4–8.3 mm; Olympus) was used. A water bath filled with water was placed on a stage of the microscope, and a hole of 10 mm in diameter was made at its bottom. The upper side of the hole was covered with a coverslip seeded with monolayer cells facing down and the lower side was covered with a plain coverslip to make an observation chamber. The chamber was filled with Hank's balanced salt solution (HBSS) in which lipid-coated microbubbles of several microns in diameter were suspended. In this arrangement, bubbles that flow up by buoyancy can make contact with the cells. A laboratory-assembled focused transducer was used to generate pulsed ultrasound of 1 MHz in center frequency.

High-speed observation of cell-bubble interaction is difficult because a bright field image is used to ensure sufficient illuminating light intensity, but the monolayer cells are almost transparent in the illumination mode. To visualize cells with better contrast, a high-speed video camera (Hyper Vision HPV-X2, Shimadzu Corporation) equipped with a COMS burst image sensor that has a wide dynamic range in image contrast was used in the system [14]. The camera was connected to the front port of the microscope to take 256 frames at a framing rage of 10 million fps. A short-arc power flash (SA-200F, Nissin electronic) that generates a light pulse of about 200 μs in duration was used for an illuminating light source.

Figure 11 shows frames selected from a high-speed movie taken during sonication of the cells and a bubble. The transducer was driven by a 10-cycle sinusoidal-wave burst pulse, and the peak positive and negative pressures at the focus were +0.7 and −0.5 MPa, respectively. Each frame shows the oscillating bubble in maximum expansion or minimum contraction phase in each pressure cycle. Only one bubble exits in the initial frame; however, repetitive oscillation of the bubble produced a large number of daughter bubbles, and the shapes of the daughter bubbles were always changing due to their fragmentation and coalescence in each cycle.

Various types of cell deformation caused by bubble activities can be observed in the frames. After five cycles of bubble oscillation (#11), Cells 2−4 showed significant changes from their original shapes. The lower part of Cell 2 showed a bent shape, which started at frame #5. A depression was produced on the right side of Cell 3, which was first observed at frame #9. The bubble activities shown in frames #2−#10 indicate that cells are partially peeled off from the substrate because the expanding bubble pushed the cells aside. The deformation of Cells 1 and 3 in frame #11 became more obvious in frames #13 and #17. Deformation of Cell 5 in a stretching shape was also confirmed in frame #11. The deformation occurred in frames #8 to #10, indicating that a liquid flow generated by the non-uniform contraction of bubbles can cause stretching deformation of a cell.

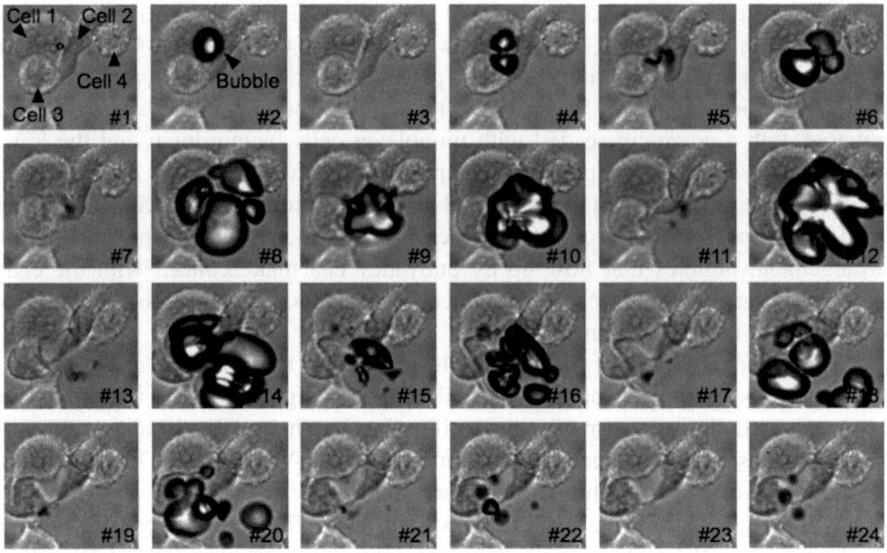

Fig. 11 High-speed frames selected from a high-speed movie captured during sonication. The movie was taken using half pixel mode at a framing rate of 10 million fps. Intervals between adjacent frames shown in this figure are not equal

7 Summary

The importance of high-speed observation in the field of medicine was reviewed with focus on cavitation activities utilized in various applications. Since the effects of cavitation involve a wide range of mechanisms, it is important to find which mechanism is dominant in practical conditions and also how the mechanism will change depending on various conditions such as sonication and gaseous conditions. Further efforts are needed to observe cavitation activities in a more physiological condition mimicking an in vivo situation.

References

1. http://www.heart-valve-surgery.com/mechanical-prosthetic-heart-valve.php
2. E. Rambod, M. Beizaie, M. Shusser, S. Milo, M. Gharib, A physical model describing the mechanism for formation of gas microbubbles in patients with mitral mechanical heart valves. Ann. Biomed. Eng. **27**, 774–792 (1999)
3. H. Tsuneoka, T. Shiba, Y. Takahashi, Feasibility of ultrasound cataract surgery with a 1.4 mm incision. J. Cataract Refract. Surg. **27**, 934–940 (2001)
4. R.M. Verdaasdonk, C.F.P. van Swol, M.C.M. Grimbergen, A.I. Rem, Imaging techniques for research and education of thermal and mechanical interactions of lasers with biological and model tissues. J. Biomed. Opt. **11**, 41110-1–41110-15 (2006)

5. C. Chaussy, E. Schmiedt, Extracorporeal shock wave lithotripsy (ESWL) for kidney stones. An alternative to surgery? Urol. Radiol. **6**, 80–87 (1984)
6. W. Eisenmenger, The mechanisms of stone fragmentation in ESWL. Ultrasound Med. Biol. **27**, 683–693 (2001)
7. W. Sass, M. Bräunlich, H.-P. Dreyer, E. Matura, W. Folberth, H.-G. Priesmeyer, J. Seifert, The mechanisms of stone disintegration by shock waves. Ultrasound Med. Biol. **17**, 239–243 (1991)
8. J.E. Parsons, C.A. Cain, G.D. Abrams, J.B. Fowlkes, Pulsed cavitational ultrasound therapy for controlled tissue homogenization. Ultrasound Med. Biol. **32**, 115–129 (2006)
9. Y.-F. Zhou, High intensity focused ultrasound in clinical tumor ablation. World J. Clin. Oncol. **2**, 8–27 (2011)
10. A.D. Maxwell, T.-Y. Wang, C.A. Cain, J.B. Fowlkes, O.A. Sapozhnikov, M.R. Bailey, Z. Xu, Cavitation clouds created by shock scattering from bubbles during histotripsy. J. Acoust. Soc. Am. **130**, 1888–1898 (2011)
11. D.L. Miller, S.V. Pislaru, J.E. Greenleaf, Sonoporation: mechanical DNA delivery by ultrasonic cavitation. Somat. Cell Mol. Genet. **27**, 115–134 (2002)
12. N. Kudo, K. Okada, K. Yamamoto, Sonoporation by single-shot pulsed ultrasound with microbubbles adjacent to cells. Biophys. J. **96**, 4866–4876 (2009)
13. N. Kudo, High-speed in situ observation system for sonoporation of cells with size- and position-controlled microbubbles. IEEE Trans. Ultrason. Ferroelectr. Freq. Control **64**, 273–280 (2016)
14. R. Kuroda, Y. Tochigi, K. Miyauchi, T. Takeda, H. Sugo, F. Shao, S. Sugawa, A 20 Mfps global shutter CMOS image sensor with improved light sensitivity and power consumption performances. ITE Trans. Media Technol. Appl. **4**, 149–154 (2015)

List of Movies

Here is a list of movies, which are electronic supporting materials for the following articles of this book. They are available to authorized users.

Chapter 1 *History of Curiosity* (doi: 10.1007/978-3-319-61491-5_1)
1.1: Fig. 3 Beer bubble collapse observed at 16000 fps. White dots are random reflections of illumination light. A black semicircle at the right side is the tip of an optical fiber used for illumination.
1.2: Fig. 4 Beer bubble collapse observed at 32000 fps. White dots are random reflections of illumination light.

Chapter 2 *The Bubble Challenge for High-Speed Photography* (doi: 10.1007/978-3-319-61491-5_2)
2.1: Fig. 16 (1) Stereoscopic view of bubble motion in an ultrasonic field. Viewing angle between the two cameras = 35.4°, ultrasonic frequency = 22.8 kHz, pressure amplitude = 132 kPa, frame size: 1 cm × 1 cm per camera, frame rate: 2250 fps, exposure time: 440 μs.
2.2: Fig. 16 (2) Tracking of individual bubbles in Video 2.1. Different bubbles are marked by different colors. The data serve for determining their 3D positions. The data also serve for determining velocities of the bubbles.
2.3: Fig. 16 (3) A 3D plot of the bubble positions from Video 2.1 according to the identification in Video 2.2 and their pathways. The rotation is for better perception of the overall 3D bubble arrangement. It is inhomogeneous and forms a network of branches (called streamers).

Chapter 3 *Brandaris Ultra High-Speed Imaging Facility* (doi: 10.1007/978-3-319-61491-5_3)
3.1: Fig. 10 Bubble oscillation: Polydisperse bubbles viewed under the microscope of the Brandaris 128 camera, frame rate: 10.31 Mpfs.
3.2: Fig. 13 Surface modes: Microbubble shape oscillations excited through ultrasonic parametric driving taken with the Brandaris 128 camera, frame rate: 1.13 Mfps.

K. Tsuji (ed.), *The Micro-World Observed by Ultra High-Speed Cameras*,
DOI 10.1007/978-3-319-61491-5

3.3: Fig. 14 Acoustic droplet vaporization. Brandaris 128 recording of the ultrafast vaporization of a 5-μm perfluoropentane microdroplet, frame rate: 12.73 Mfps.

Chapter 4 *Evolution of High-Speed Image Sensors* (doi: 10.1007/978-3-319-61491-5_4)
4.1: Fig. 4 Experimental thunderbolt. Frame rate: 1 Mfps, taken with NHK's high-speed color video camera (Brush discharges propagate downward seeking for a landing site; Images of every 20 frames are shown; Figure 4 in the text, by Arai et al. [9]).
4.2: Fig. 5 A shock wave around a cylinder. Frame rate: 330 kfps, taken with color Mach-Zehnder interferometry with the NHK's ultra-high-speed color video camera (the diameter of the cylinder: 20 mm, the Mach number: 1.31; images of every 25 frames are shown) (Figure 5 in the text; by Kleine [10]).

Chapter 5 *Cameras with On-chip Memory CMOS Image Sensors* (doi: 10.1007/978-3-319-61491-5_5)
5.1: Fig. 15 (a) Glass crack generation by bullet shooting captured by 5 Mfps full resolution.
5.2: Fig. 15 (b) Glass crack generation by bullet shooting captured by 20 Mfps half resolution.
5.3: Fig. 16 (b) Water mist injected from an air brush captured by 10 Mfps half resolution with performance improved chip.

Chapter 6 *High-Speed Imaging of Shock Waves and their Flow Fields* (doi: 10.1007/978-3-319-61491-5_6)
6.1: Fig. 10 (a) Laboratory explosions visualised by omnidirectional schlieren: reflection of the blast wave generated by a 10 mg charge of silver azide for a height of burst of 35 mm; frame rate: 500,000 fps.
6.2: Fig. 11 Omnidirectional schlieren visualisation with front lighting of a rifle bullet flying at $M_\infty = 1{:}07$; frame rate: 500,000 fps.
6.3: Fig. 14 Shearing interferometry visualisation of a shock wave (MS = 1:33) interacting with a diamond cylinder; frame rate: 460,000 fps.

Chapter 8 *On the Use of Digital Image Correlation for the Analysis of the Dynamic Behavior of Materials* (doi: 10.1007/978-3-319-61491-5_8)
8.1: Fig. 9 Gray level pictures in the deformed configuration of a tensile test on a T700 carbon fiber/M21 epoxy matrix composite (top left). The acquisition rate is 65,000 fps. Gray level correlation residuals (bottom left). The bright areas correspond to cracks developing in the composite. Vertical (top right) and horizontal (bottom right) displacement maps (expressed in pixels) measured via global spatio-temporal digital image correlation.

8.2: Fig. 11 Cylinder expansion test of copper. Images acquired with a single high-speed rotating mirror framing camera (left). 3D shapes reconstructed via stereocorrelation (right). The blue mesh corresponds to the 3D reconstructed positions, and the red dots to the interpolated 3D surface.

Chapter 10 *Real-time Hard X-ray Imaging* (doi: 10.1007/978-3-319-61491-5_10)
10.1: Fig. 3 Dynamics in an aqueous foam obtained by means of high-speed phase contrast radioscopy, acquired every 25 μs. The collapse of two cell walls can be followed as well as the rearrangement of the pores in the immediate neighborhood.
10.2: Fig. 4 Laser processing of a polystyrene foam: Interaction of an isolated laser irradiation (800 mJ, 20 ns pulse) with the aluminium-coated surface of a polystyrene foam is seen. The frame acquisition rate: 1.4 MHz, the integration time of the camera: 200 ns. (Contrast in the movie is dominated by X-ray phase contrast.)

Chapter 11 *Development and Application of High-Speed Laser Visualization Techniques in Combustion Research* (doi: 10.1007/978-3-319-61491-5_11)
11.1: Fig. 11 Example of temporally resolved 3D iso-concentration surfaces of OH radicals in a laboratory flame.

Chapter 12 *Visualization of Combustion Processes of Internal Combustion Engines* (doi: 10.1007/978-3-319-61491-5_12)
12.1: Fig. 1 Time series of high-speed direct images and related in-cylinder pressure histories for a normal engine cycle, a camera speed 60,000 fps.
12.2: Fig. 2 Time series of high-speed direct images and related in-cylinder pressure histories for a knocking engine cycle, a camera speed 60,000 fps.

Chapter 14 *Single Bubble Ignition after Shock Wave Impact* (doi: 10.1007/978-3-319-61491-5_14)
14.1: Fig. 3 Shock wave impact onto bubbly liquid. Capture speed: 500,000 fps. A shock wave generated by a detonation in a gaseous mixture of acetylene and oxygen impacts onto the surface of liquid cyclohexane. The shock wave is reflected to the gas phase and generates a shock wave in the liquid. The bubbles in the liquid are compressed and emit spherical shock waves. Shock waves were visualized by diffuse light illuminating a window opposite to the observation window.
14.2: Fig. 4 Shock induced explosions of bubbles in a column. Capture speed: 500,000 fps An initial shock wave propagates through a column of bubbles in liquid cyclohexane. The bubbles are composed of oxygen and vapor of cyclohexane. They are compressed and explode after the passage of the shock wave. Exploding bubbles emit flashes. The position of the bubbles is visualized by diffuse light illuminating a window opposite to the observation window.

14.3: Figs. 5 and 6 Dynamics of shock induced bubble explosions. Capture speed: 1 000,000 fps An initial shock wave propagates through liquid cyclohexane containing bubbles that are composed of oxygen and vapor of cyclohexane. The bubbles pass through different stages such as compression, jet formation, explosion with emission of flashes and spherical shock waves, and expansion. Shock waves are visualized by diffuse light illuminating a window opposite to the observation window.

Index

K. Tsuji (ed.), *The Micro-World Observed by Ultra High-Speed Cameras*,
DOI 10.1007/978-3-319-61491-5

MIX
Papier aus verantwortungsvollen Quellen
Paper from responsible sources
FSC® C105338

If you have any concerns about our products,
you can contact us on
ProductSafety@springernature.com

In case Publisher is established outside the EU,
the EU authorized representative is:
Springer Nature Customer Service Center GmbH
Europaplatz 3, 69115 Heidelberg, Germany

Printed by Libri Plureos GmbH
in Hamburg, Germany